W9-AEW-228
DISCARDED
JENKS LRC
GORDON COLLEGE

Art in Organic Synthesis

Art in Organic Synthesis

Second Edition

Nitya Anand
Central Drug Research Institute
Lucknow, India

Jasjit S. Bindra
Pfizer Central Research
Groton, Connecticut

Subramania Ranganathan
Indian Institute of Technology
Kanpur, India

JENKS L.R.C.
GORDON COLLEGE
255 GRAPEVINE RD.
WENHAM, MA 01984-1895

A WILEY-INTERSCIENCE PUBLICATION

JOHN WILEY & SONS

New York Chichester Brisbane Toronto Singapore

QD
262
.A528
1988

Copyright © 1970, 1988 by John Wiley & Sons, Inc.

All rights reserved. Published simultaneously in Canada.

Reproduction or translation of any part of this work beyond that permitted by Section 107 or 108 of the 1976 United States Copyright Act without the permission of the copyright owner is unlawful. Requests for permission or further information should be addressed to the Permissions Department, John Wiley & Sons, Inc.

Library of Congress Cataloging-in-Publication Data:

Anand, Nitya.
Art in organic synthesis.

"A Wiley-Interscience publication."
Includes bibliographical references and indexes.
1. Chemistry, Organic—Synthesis. I. Bindra, Jasjit S. II. Ranganathan, Subramania. III. Title.

QD262.A528 1987 547′.2 87-14762
ISBN 0-471-88738-2

Printed in the United States of America

10 9 8 7 6 5 4 3 2 1

To

Soniya, Naveen, Neeraj, and Swarn
Ratna, Ranjit, and Ranjna
Anand and Darshan

Preface

Art in Organic Synthesis, originally published in 1970, was conceived as a novel method of presenting the great achievements of synthetic organic chemistry, mainly in the form of flow charts. It was hoped that this type of uninterrupted graphic presentation—new at the time—would have a better impact on the reader and would facilitate the understanding of strategies for synthesis of complex structures. The flow chart presentation is particularly suited to illustrating the dissection of complex structures into smaller building blocks or synthons, which can then be joined together by appropriate reactions, as was first verbalised by E. J. Corey in his formal retrosynthetic analysis approach. We are happy that our expectations for *Art in Organic Synthesis* were fulfilled and that the volume was welcomed enthusiastically by practitioners of the art and, much to our delight, served for many years as an important teaching tool.

The art of organic synthesis resides in the optimum and coordinated utilisation of each element of the complete ensemble of conception of strategy, and in the selection of appropriate reactions and reagents. The potential and capabilities of each element are best illustrated by a total synthesis, which provides a clear comprehension of the integrated construction and rupture of bonds. A total synthesis is like a symphony orchestra, where each instrument plays its own and important role but the combined effect transcends the sum of the contributions of all.

In the 17 years since the publication of the first edition, major developments have taken place in the art of organic synthesis which underline the shifting focus of attention on different facets of the art. Some of the new facets relate to the use of chiral synthons and chiral reagents, strategies for greater steric control, selective functional group transforms, and yield optimisation. Another important facet of the art is the delineation and use of synthetic pathways that simulate those encountered in nature. The challenges posed by total synthesis have led to the development of many new reagents and the discovery of new reactions. These developments have generated a new confidence in organic chemists in their ability to design and construct complex molecular structures at will.

The need for revision of the volume to include these newer developments is, therefore, obvious. In the planning of this second edition, our uppermost concern has been to make it as illustrative as the earlier volume. This attempt is manifested in the choice and analysis of syntheses from the numerous excellent ones available in the literature. Some of the classic examples in the earlier volume have been retained with little change, several syntheses have been completely revised or updated, and a large number of new syntheses have been added to reflect the changing emphasis in the area. In every case the art work and the text have been completely redone.

NITYA ANAND
JASJIT S. BINDRA
SUBRAMANIA RANGANATHAN

Lucknow, India
Groton, Connecticut
Kanpur, India
December 1987

Preface to the First Edition

Art is associated with the creation of new concepts and ideas and their expression in different perceptible forms, and this facet of human endeavour has contributed much to our civilization. In the area of organic synthesis, art has found expression in the synthesis of a wide variety of molecular structures. The earliest striking example of this art was Robinson's elegant synthesis of tropine using simple bond-forming reactions. The middle of this century saw the art of organic synthesis in full bloom, and many outstanding syntheses have been achieved of which special mention may be made of Woodward's chlorophyll synthesis. More than any other branch of organic chemistry, synthesis has led to a better understanding of the principles of bond-making and bond-breaking, the consequences of dissymetry, the preferred conformation of molecules and the stereoelectronic features that govern their transition state in reactions. An appreciation of these factors and the availability of sophisticated tools coupled with human ingenuity has elevated organic synthesis to a highly challenging intellectual adventure where imagination, inspiration, design, and experimental skill all converge for the achievement of the objective. "Art in Organic Synthesis" attempts to illustrate these various aspects of organic synthesis. In a hundred or so examples we have tried to highlight synthetic achievements in terms of ingenuity in design, extent of stereochemical control, new reactions and new reagents. The examples chosen describe the synthesis of a natural product or of an unusual or strained molecule prepared to check certain predicted properties of the structures or simply because it presented a unique challenge to the skill and ingenuity of the organic chemist. Due to limitation on space we have had to be rather selective, but this does not in any way imply that syntheses not included in this book are any less important. In the presentation the emphasis has been on economy of words and liberal use of flow sheets and perspective structural formulae to illustrate the force of arguments predicting the stereochemical outcome of important steps. Obscure or unusual reactions have been discussed at greater length in the footnotes. In addition to other indices, a type-transformation index is included to highlight some of the less common or new reactions. The literature coverage is up to the early part of 1969.

The authors hope that this small volume, which illustrates the creativity of organic synthesis, the tools available to it, its potentiality and predictive capacity, will provide an insight into the basic philosophy and strategy employed in the design and execution of an organic synthesis.

We wish to acknowledge the many helpful comments by Dr. P. C. Dutta, who read the entire manuscript, and by Drs. M. L. Dhar and M. M. Dhar, who read parts of it. We are also grateful to Dr. B. Singh and Mr. H. Raman for some reference work, and to Dr. (Miss) Ranjna Saxena who shared the burden of checking all references and greatly assisted in proof reading and in preparation of the indices.

All the perspective formulae have been drawn by one of us (J.S.B.). The formidable task of lettering in the formulae and typing the entire text suitable for direct reproduction of the manuscript was skillfully executed by Mr. H. C. Chhabra, who kept his patience and composure under very trying conditions: the authors would like to express their deep sense of gratitude to him.

One of us (N.A.) would like to record his profound appreciation to Swarn, his wife, for her understanding and for sparing him the time, which truly belonged to her, that went into the completion of this work.

NITYA ANAND
JASJIT S. BINDRA
SUBRAMANIA RANGANATHAN

Central Drug Research Institute,
Lucknow (India)
Indian Institute of Technology,
Kanpur (India)
December, 1969

Acknowledgments

First and foremost our thanks are due to Mr. Ali Kausar who drew all the formulae and perspective diagrams, which are a distinctive feature of this book. The formidable task of accommodating and adjusting the text to the diagrams/graphics has been admirably carried out by Mr. M. K. Thapar, for which we thank him sincerely. The computerised preparation of Author and Reagent Indices has been done very efficiently by Mr. Ahsan Jamal and Dr. R. K. Sharma, and we are most grateful to them.

Dr. D. K. Dikshit went through the entire manuscript meticulously during the various stages of its preparation. His suggestions helped to improve the quality of the presentation, and for this we are grateful to him. We would also like to thank Dr. Naveen N. Anand for his help in structuring the chapter on Gene Synthesis.

Nitya Anand is grateful to the Indian National Science Academy for the award of a Senior Scientist position during the tenure of which this manuscript was prepared. He also thanks the Director, Central Drug Research Institute, Dr. M. M. Dhar for helpful discussions, Dr. S. S. Iyer for use of Library facilities, and the Library staff for their most forthcoming assistance. He expresses his gratitude to Ranbaxy Laboratories for their invaluable support. Above all he acknowledges his profound appreciation to Swarn, his wife, for her understanding, unstinted cooperation, and encouragement for the completion of this work.

NITYA ANAND
J. S. BINDRA
S. RANGANATHAN

Contents

Abbreviations xvii

Adrenosterone 1
Aflatoxins 8
Ajmaline 13
Aldosterone 18
Amyrin 27
Androsterone 35
Annulenes 37
Antheridiogen 44
Aromatic Anions 48
Aspidospermine, Aspidospermidine 50
Asteranes 57
Atisine 58
Avermectin 64
Benzene Dimer 70
Benzene Oxides 71
Benzocyclopropene 73
Betweenanenes 74
E-Bicyclo[4,1,0]heptane 77
Bongkrekic Acid 78
10,9-Borazaronaphthalene 81
Bullvalene 82
Cantharidin 83
Capped Porphyrins 84
Carpanone 86
Carpetimycin A 87
Catenanes 90
Catharanthine 93
Cavitands 95
Cephalosporin-C 97
Chelidonine 100
Chlorophyll 103
Cholesterol 109
Clavulones 114
Coenzyme A 116

Conessine. 118
Coriolic Acid, Dimorphecolic Acid 123
Corrin Template 126
Cortisone 127
Cyclobutadiene, Cubane 134
Cyclosporine 136
Cytochalasine B 140
Dewar Benzene 146
E,E-1,4-Diacetoxy-1,3-butadiene 147
"Diamond" Structures 148
Dodecahedrane 150
Endriadric Acids 153
Erythromycin 158
Estrone 165
Gene Synthesis. 179
Gibberellic Acid 189
Helicenes 193
Histidine 196
Iceane 197
Kekulene 199
Kopsanone,10,22-Dioxokopsane 201
Longifolene 204
Luciduline 210
Lycopodine 214
Lysergic Acid, Ergotamine 224
O-Methylorantine 231
Monensin 233
Octalene 244
Out-Out, In-In Bicyclic Systems 245
Ovalicin 247
Pagodane 249
Pentalenene 252
Pentaprismane 254
Perannulanes 256
Peristylanes 258
Picrotoxinin, Coriamyrtin 262
Prismane 269
Progesterone 270
[1.1.1]Propellane 277
Prostaglandins 278
Qinghaosu 287
Quassin 291

Quninine 293
Reserpine. 303
Resistomycin 312
Rifamycin S 314
Sexipyridine 325
Sigma Directed pi-Systems 327
Sporidesmin-A 330
Strychnine 333
Superphane 337
Tetracyclines 338
Tetrahedrane, Tetra-tert-Butyl 345
Tetrodotoxin 346
Thienamycin. 351
Tigogenin, Diosgenin 354
Tropavalene 360
Tropinone 361
Verrucarin A 362
Vinblastine, Vincristine. 370
Vindoline. 372
Vitamin B_{12} 375

Subject Index 387
Author Index 390
Reagent Index 407
Reaction Type Index 423

Abbreviations

Ac	acetyl
Am	amyl
Ar	aryl, substituted phenyl
Aq	aqueous
BBN	9-borabicyclo[3.3.1]nonane
BOC	*t*-butoxycarbonyl
BPCOCl	4-phenylbenzoyl chloride
Bs	benzenesulfonyl
Bu	butyl
Bz	benzoyl
Bzl	benzyl
ca	about
CE	-cyanoethyl
Chf	chloroform
CSA	camphorsulfonic acid
CSC	camphorsulfonyl chloride
DBN	1,5-diazabicyclo[4.3.0]non-5-ene
DBU	1,5-diazabicyclo[5.4.0]undec-5-ene
DCB	*o*-dichlorobenzene
DCC	dicyclohexylcarbodiimide
DDQ	2,3-dichloro-5,6-dicyano-1,4-benzoquinone
DEA	N,N-diethylaniline
$CpCo(CO)_2$	Cyclopentadienyl cobalt carbonyl
DEG	diethylene glycol
DHF	dihydrofuran
DHP	dihydropyran
DIBAL	diisobutylaluminumhydride
DIC	diisopropylcarbodimide
Diox	dioxane
DMA	N,N-dimethylaniline
DMAP	4-dimethylaminopyridine
DME	dimethoxyethane, glyme
DMF	N,N-dimethylformamide
DMSO	dimethylsulfoxide
DMT	4,4′-dimethoxytriphenyl-methyl, 4,4′-dimethoxytrityl
EAA	ethyl acetoacetate
Et	ethyl
FPh	9-phenylfluorenyl
HMDS	hexamethyldisilazane
HMPT	hexamethylphosphoric triamide, hexamethylphosphoramide
H_3O^+	aqueous acid
Imid	imidazole
Imid-co-imid	1,1′-carbonyldiimidazole
INOC	isonitrileoxide cycloaddition
LAH	lithium aluminum hydride
LDA	lithium diisopropylamide

Liq	liquid	PDC	pyridinium dichromate
LTA	lead tetraacetate	PDS	dipyridyl disulfide
Lindlar catalyst	deactivated palladium on carbon/ $BaSO_4$	Ph	phenyl
		Phth	phthaloyl
Lut	lutidine	PNB	*p*-nitrobenzyl
Me	methyl	PP	4-pyrrolidylpyridine
MCPBA	*m*-chloroperbenzoic acid	PPA	polyphosphoric acid
		Pr	propyl
MeBmt	(4R)-4[(E)-2-butoxyl]-4-N-dimethyl-1-threonine	Py	pyridine
		RT	room temperature
		SEM	-trimethylsilylethoxymethyl
MEM	β-methoxyethoxymethyl chloride	s	secondary
MIRC	Michael initated ring closure	t	tertiary
		TBDMS	*t*-butyldimethylsilyl
MOM	methoxymethyl	TBDPS	*t*-butyldiphenylsilyl
MSNT	1-(mesitylenesulfonyl)-3-nitro-1,2,4-triazole	TBS	*t*-butyldimethylsilyl
		TBSOTf	*t*-butyldimethylsilyloxyl triflate
Ms	mesyl, methanesulfonyl	TCAA	trichloroacetic anhydride
MsOH	methanesulfonic acid	Tce	1,1,1-trichloroethyl
MTM	methylthiomethyl	TEA	triethylamine
MVK	methyl vinyl ketone	TfA	trifluoromethanesulfonic acid
n	normal		
NBA	N-bromacetamide	TFA	trifluoroacetic acid
NBS	N-bromosuccinimide	TFAA	trifluoroacetic anhydride
NCS	N-chlorosuccinimide		
NHBT	N-hydroxybenzotriazole	THF	tetrahydrofuran, tetrahydrofuranyl
NHS	N-hydroxysuccinimide	THP	tetrahydropyranyl
		TIPS	triisopropylsilyl
NMMNO	N-methylmorpholine-N-oxide	TMB	1,3,5-trimethoxybenzoyl
NPS	*o*-nitrophenylsulfenyl	TMEDA	N,N,N′,N′-tetramethylethylenediamine
ODCB	*o*-dichlorobenzene		
ONP	*p*-nitrophenyl ester		
PBPCO	4-phenylbenzoyl	TMS	trimethylsilyl
PCC	pyridinium chlorochromate	Tol	toluene
		TPST	1-(1,3,5-triisopropyl)-phenylsulfonyl-1,2,3,4-tetrazole
PCP	pentachlorophenol		
pCP	*p*-chlorophenol		

Triflate	trifluoromethanesulfonate	TsOH	*p*-toluenesulfonic acid
tRNA	transfer ribonucleic acid	Xy	xylene
		Z	benzyloxycarbonyl
Ts	*p*-toluenesulfonyl	Δ	reflux or heat

Although only one enantiomer is depicted in structural formulae, all compounds are racemates, unless otherwise stated. The following conventions have been followed:

1. A short line without a letter at the end denotes a CH_3 group.
2. A solid thick line indicates β-configuration.
3. A broken line indicates α-configuration.
4. A wavy or straight line indicates either unknown or unspecified configuration.
5. Commonly accepted amino acid abbreviations have been used.

1

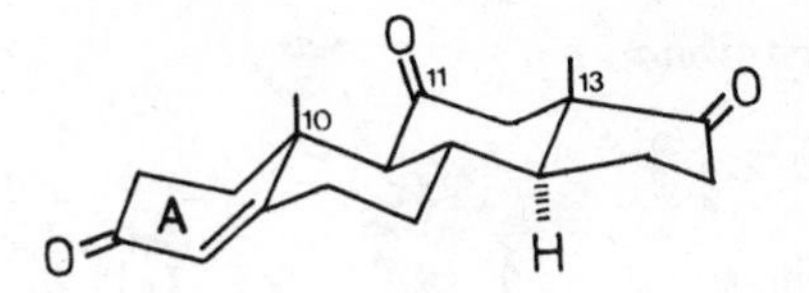

ADRENOSTERONE

The first synthesis of adrenosterone illustrates the application of a versatile steroid synthesis developed by Leon Velluz and his colleagues at Roussel-Uclaf, Paris, for the direct preparation of optically active steroids on an industrial scale(1-3).

The linear version of this Scheme(4), outlined below, is based on construction of a B-C-D tricyclic compound already possessing the five-membered D-ring and the methyl group at C-13. Introduction of the asymmetric centre at C-10 has been accomplished by reaction with 1,3-dichlorobutene-2 followed by stereoselective introduction of the angular 19-methyl group [transformation B→C].

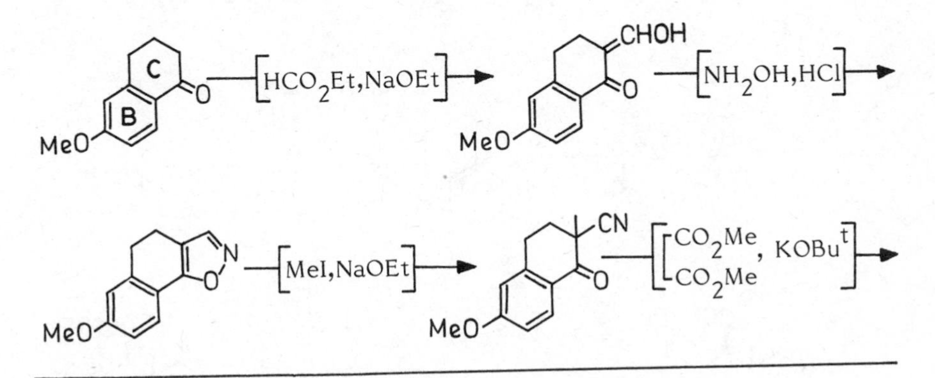

1. Velluz, L., Nominé, G., Mathieu, J. Angew. Chem., 1960, 72, 725.

2. This route is quite general for the synthesis of both aromatic and non-aromatic steroids. Suitable modifications of the reaction scheme from the key tricyclic ketone(A) permit the preparation of either estradiol, 19-norsteroids or adrenosterone and cortisone.

3. Unlike other syntheses which are generally carried out with racemates and in which optical resolution is attempted only at the end of the synthesis, this approach involves resolution of the intermediates at the earliest possible opportunity. Unwanted optical isomers are thus excluded quite early from the reaction mixture, avoiding the expense of processing material which has anyhow to be discarded at the end of the sythesis. Such a strategy was adopted earlier in the synthesis of cortisone by Barkley, L.B., Farrar, M.W., Knowles, W.S., Raffelson, H., Thompson, Q.E. J. Am. Chem. Soc., 1954, 76, 5014.

4. Velluz, L., Nominé, G., Mathieu, G., Toromanoff, E., Bertin, D., Tessier,J., Pierdet,A. Compt. Rend. 1960, 250, 1084; Velluz, L., Nominé, G., Mathieu, J., Toromanoff, E., Bertin, D., Bucourt, R., Tessier, J. Compt. Rend. 1960, 250, 1293.

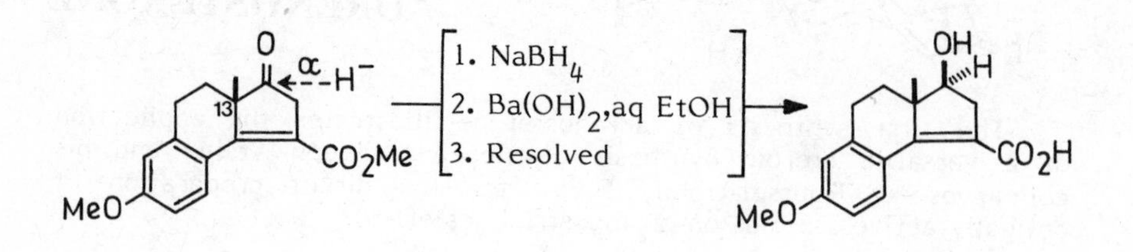
O
α
H⁻
13
CO2Me
MeO
1. NaBH4
2. Ba(OH)2,aq EtOH
3. Resolved
OH
H
CO2H
MeO

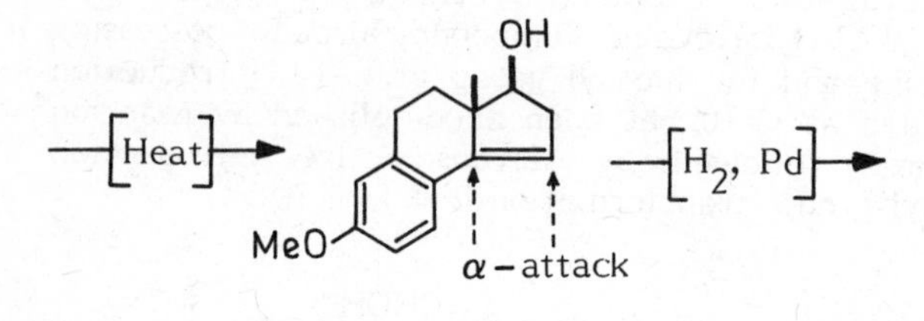
Heat
OH
MeO
α – attack
H2, Pd

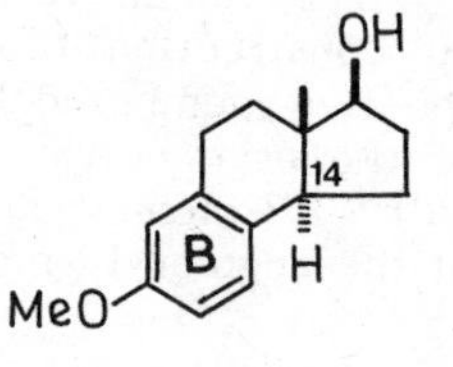
OH
14
B
H
MeO

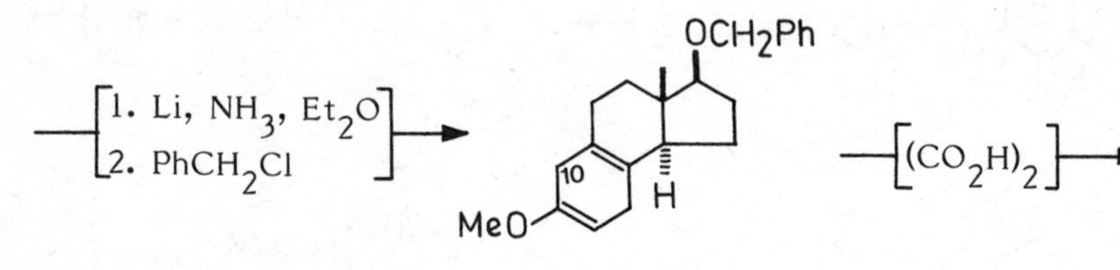
1. Li, NH3, Et2O
2. PhCH2Cl
OCH2Ph
10
H
MeO
(CO2H)2

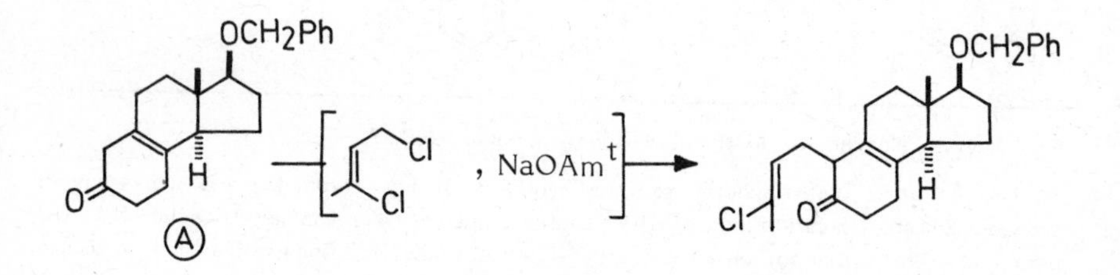
OCH2Ph
H
O
A
Cl
Cl
, NaOAmt
OCH2Ph
H
Cl
O

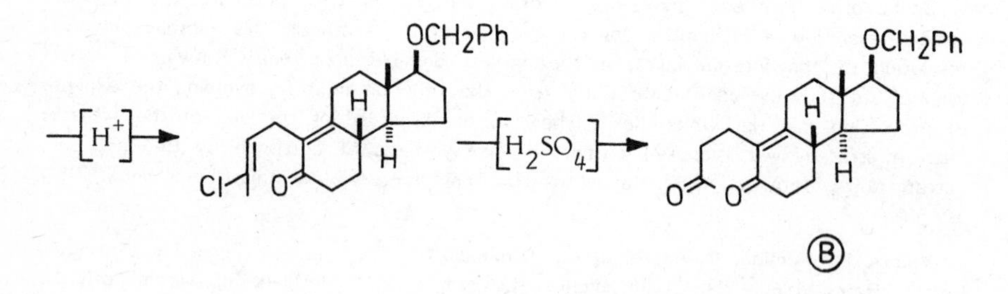
H+
OCH2Ph
H
H
Cl
O
H2SO4
OCH2Ph
H
H
O
O
B

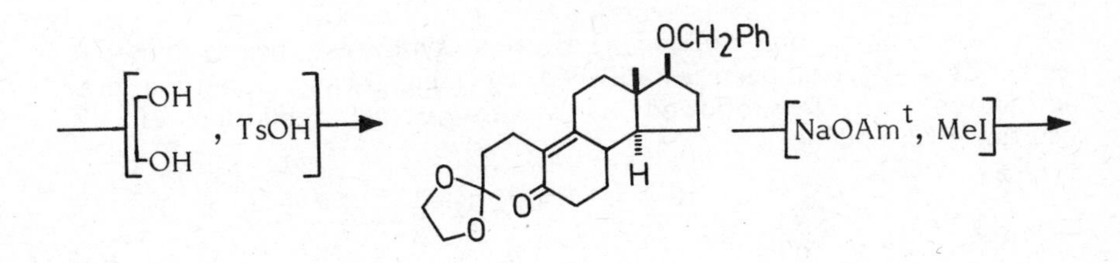

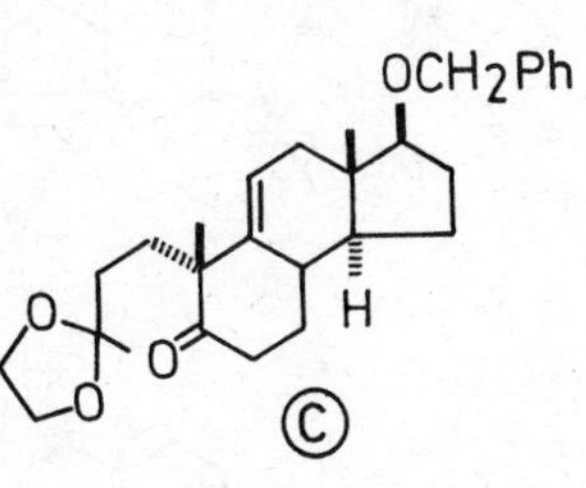

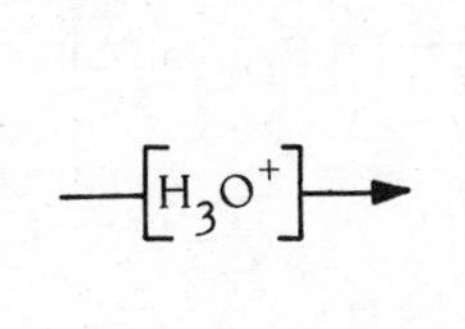

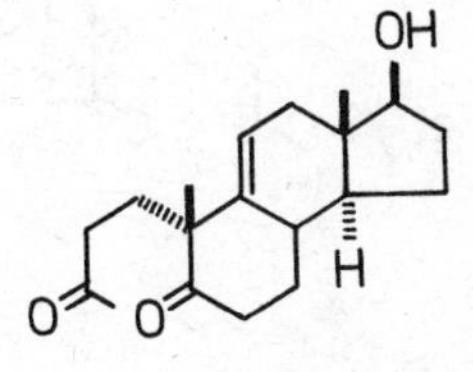

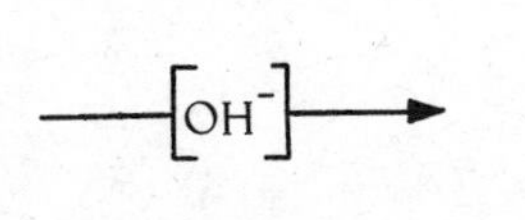

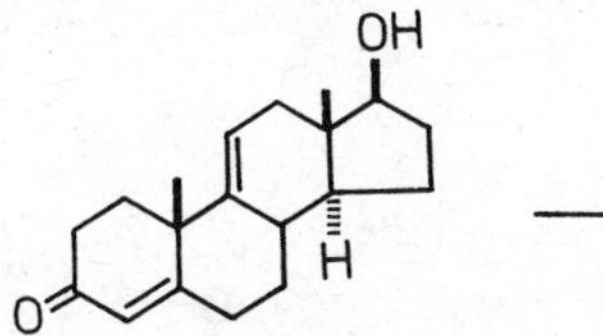

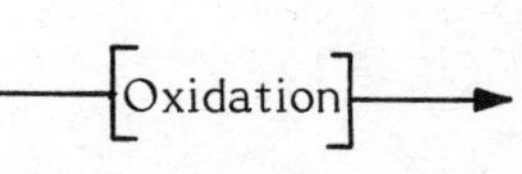

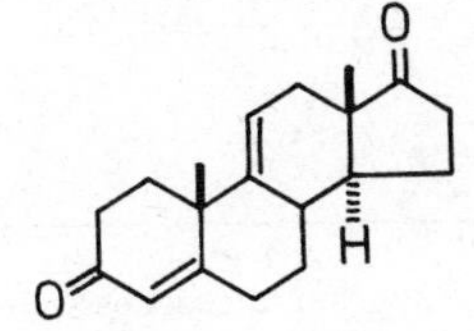

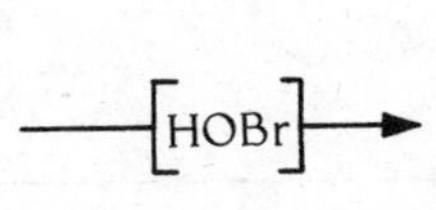

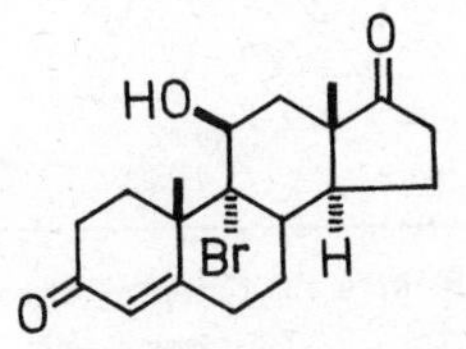

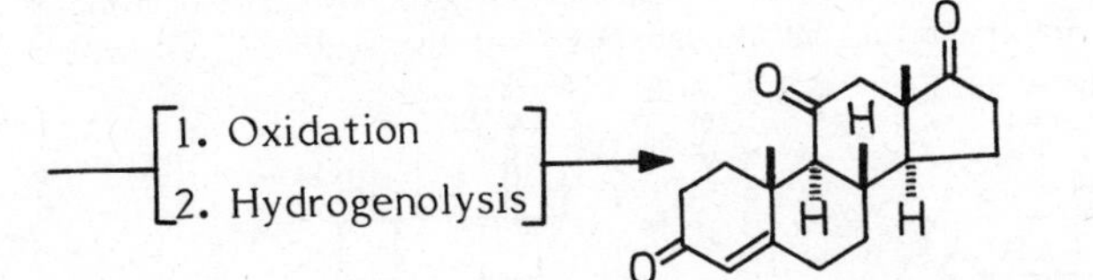

Adrenosterone

In the converging version(5) of this synthesis the asymmetric centre at C-13 has been established by joining ring C onto a pre-methylated ring D, followed by stereoselective reduction of the resulting hydroindene(6).

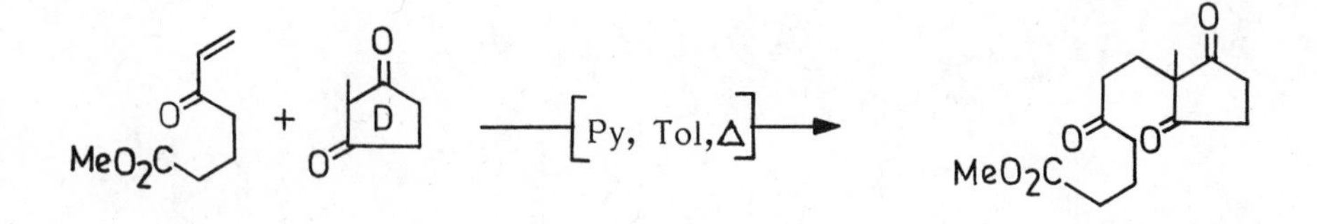

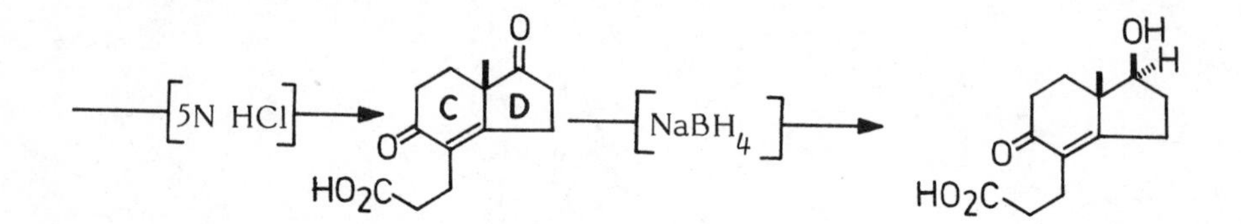

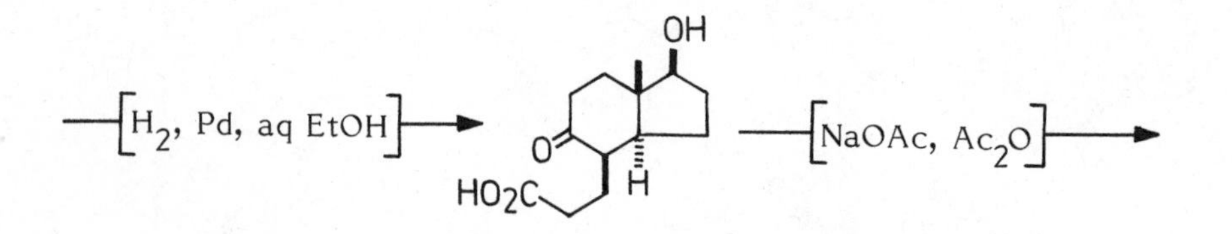

5. In a multistep synthesis the overall yield drops rapidly with the number of successive steps. This results in loss of increasingly expensive intermediates (i). A converging synthesis, on the other hand, is more economical of material in the sense that each product of the reaction sequence is obtained by combination of two precursors delaying the formation of expensive intermediates until a late stage of the synthesis, Velluz,L. Valls,J.; Mathieu,J. Angew. Chem. Int. Ed., 1967, 6, 778.

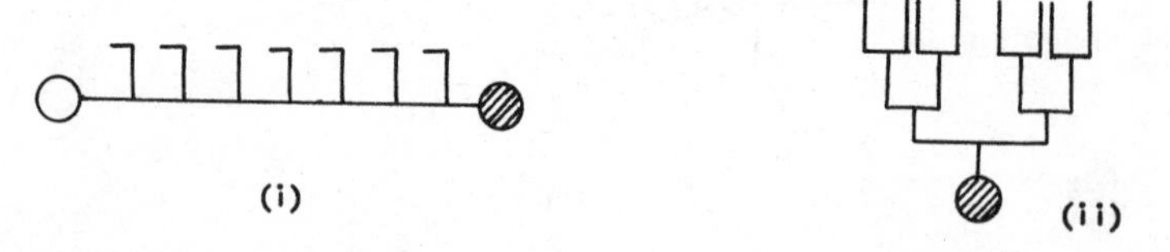

6. Velluz,L.; Nomine,G.; Amiard,G.; Torelli,V.; Cerede,J. Compt. Rend., 1963, 257, 3086.

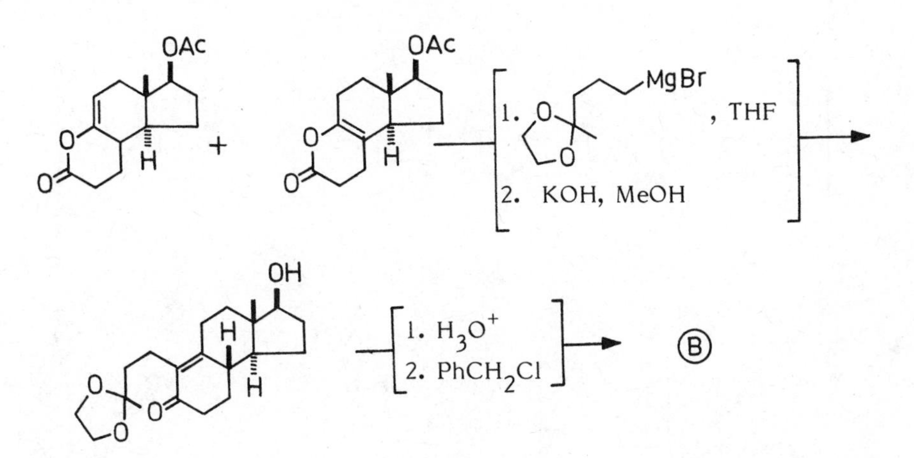

Recently, Stork et al.(7) have described a short and highly stereoselective synthesis of adrenosterone which is based on: (i) the construction of a trans-fused hydrindan by intramolecular Michael addition to control vicinal stereochemistry [transformation A→B] (8); (ii) obtaining the required stereochemistry at C-10 by construction of ring A on B-C-D tricyclic system carrying the C-10 Me and C-11 carbonyl (as enol ether) which direct the alkylation from the α-face (9).

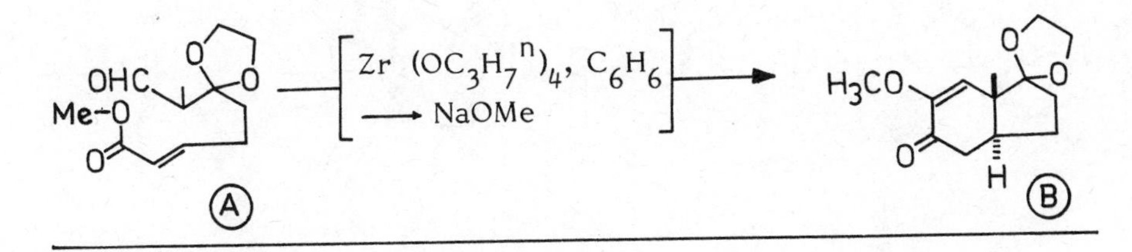

7. Stork, G., Winkler, J.D., Shiner, C.S.; J. Am. Chem. Soc., 1982 , 104, 3767.

8. Stork, G., Shiner, C.S. and Winkler, J., J. Am. Chem. Soc., 1982 , 104, 310.

9. Previous work (Stork, G. & Logusch, E. J. Am. Chem. Soc., 1980 , 102, 1218, 1219) suggested that the ring B must be twisted into a half-boat conformation to alleviate the severe interaction between the C-10 methyl and the methoxyl group at C-11, with increased accessibility of the α-face.

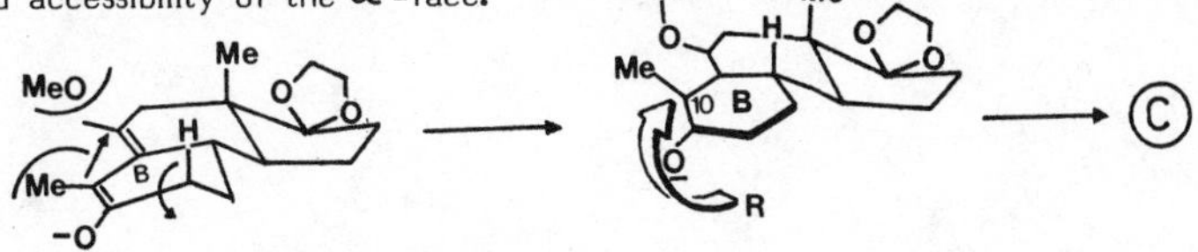

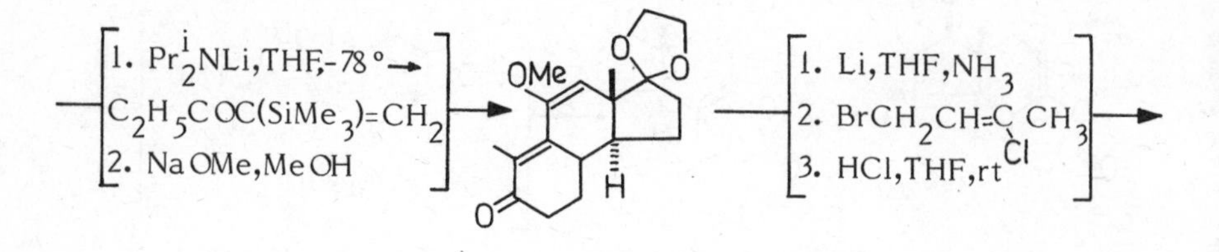

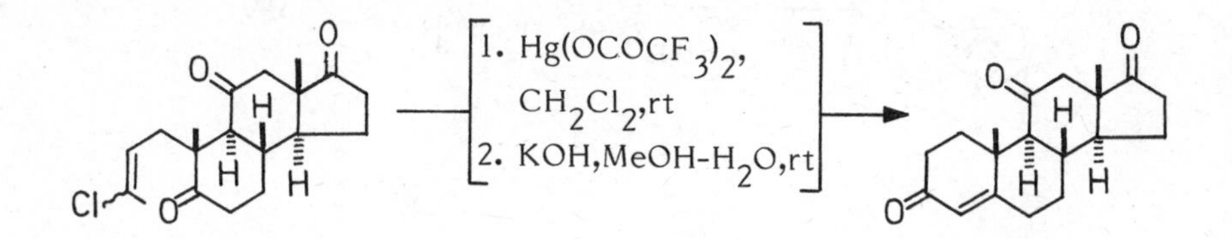

A novel D→BCD→ABCD route to 11-keto steroids described below, has been reported recently (10) which involves a high yield stereoselective intramolecular Diels-Alder reaction of furan-diene [A] as a key step.

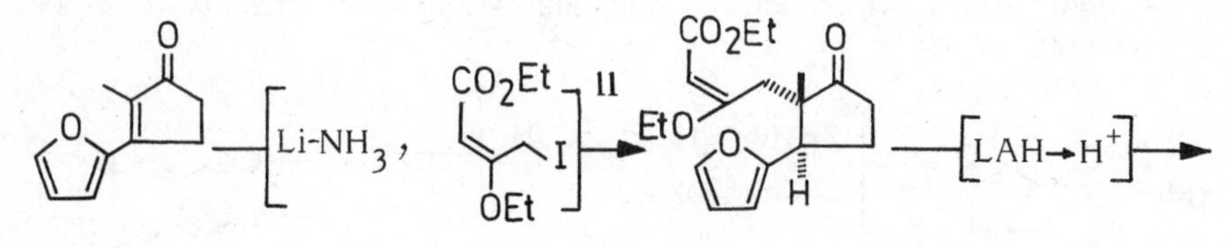

10. Van Royen, L.A., Mijngheer, R.; DeClarcq,P.J.; J. Tetrahedron, 1985, 41, 4667; and earlier references cited therein.

11. The product was obtained in 34% yield after HPLC purification along with 7% of 2,5-dialkylated product.

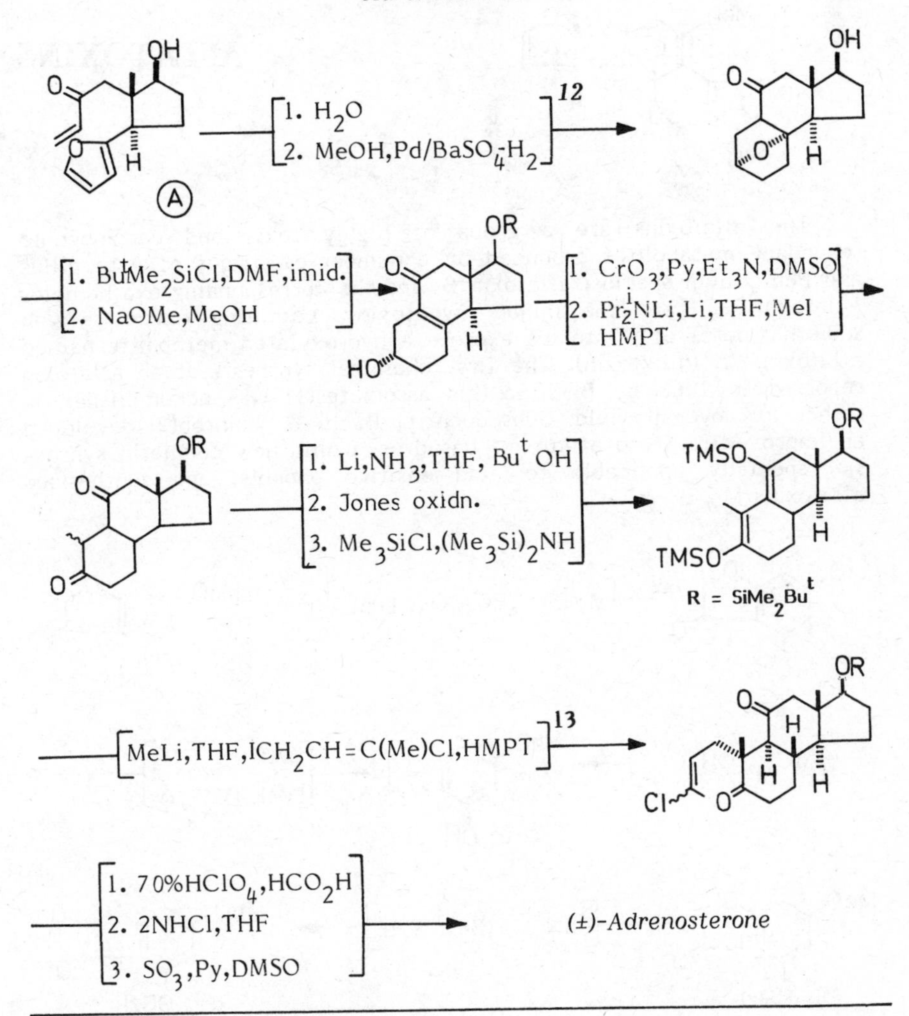

---

12. The cycloaddition did not proceed in non-polar solvents, presumably due to hydrogen bonding between the hydroxy group and the dienophilic side chain. The addition unexpectedly proceeded at a very fast rate in H_2O [c.f. Rideout, D.C. and Breslow,R. J. Amer. Chem. Soc., 1980, 102, 7816]. The adduct could easily revert back on isolation into organic phase, and it was found expedient to hydrogenate the aqueous emulsion of the adduct dissolved in methanol.

13. Stork et al had in their synthesis shown that 11-substituent because of steric bulk directs the introduction of the electrophile from the α-face (9).

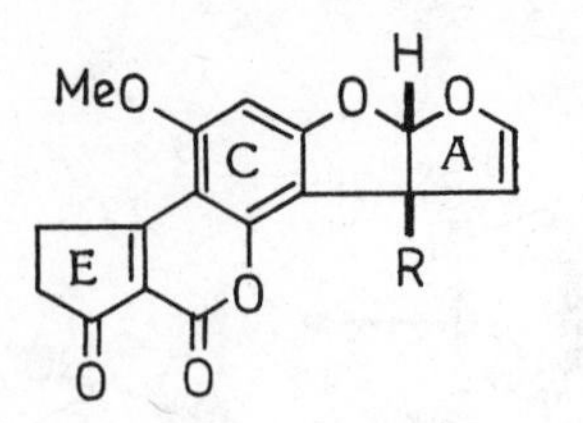

AFLATOXINS

The aflatoxins are a group of highly toxic and carcinogenic secondary metabolites produced by a number of strains of Aspergillus and Penicillium species; aflatoxin B and the corresponding oxa homolog Aflatoxin G are the major mycotoxins. Lactating cattle fed on sublethal doses of aflatoxins excrete a hydroxylated metabolite named aflatoxin M (milktoxin). The first chemical synthesis of an aflatoxin reported in 1966 by Büchi & his associates(1) was accomplished in rather low overall yield. Subsequently, Büchi & Weinreb(2) developed an improved route to aflatoxins based on a mild new coumarin synthesis, specially applicable to acid-sensitive phenols, and synthesised aflatoxins M , B & G .

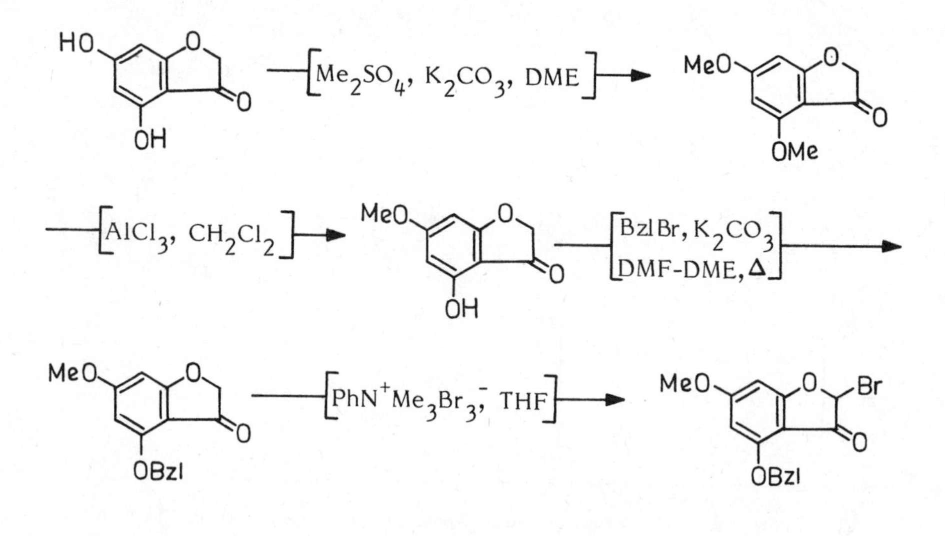

1. Büchi, G., Foulkes, D.M., Kurono, M., Mitchell, G.F., Schneider, R.S. J. Am. Chem. Soc., 1966, 88, 4534; 1967, 89, 6745.

2. Büchi, G., Weinreb, S.M. J. Am. Chem. Soc., 1969, 91, 5408; 1971, 93, 746.

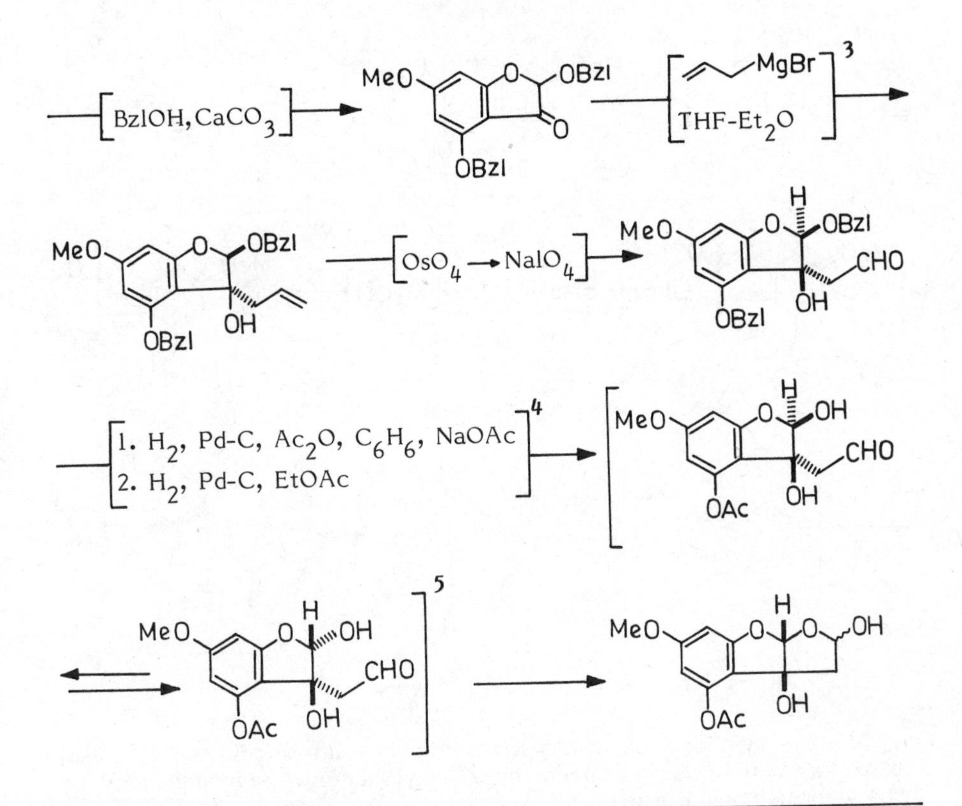

3. The Grignard reagent presumably attacks from the side opposite to that of the bulky benzyloxy group resulting in predominant formation of the trans isomer; see Cram, D.J., Wilson, D.R. J. Amer. Chem. Soc., 1963, 85, 1245, and the two isomers could be separated by chromatography. As both the isomers yielded the same tricyclic product, although the reaction is depicted with one isomer, it was found more practical to use the mixture of epimeric aldehydes for further transformations.

4. The first hydrogenation cleaves the benzyl group on the aromatic ring to give a monoacetate, which is further hydrogenated in a different solvent for removal of the second benzyl group.

5. The more stable cis-ring fused furobenzofuran is favored over the trans-fused system.

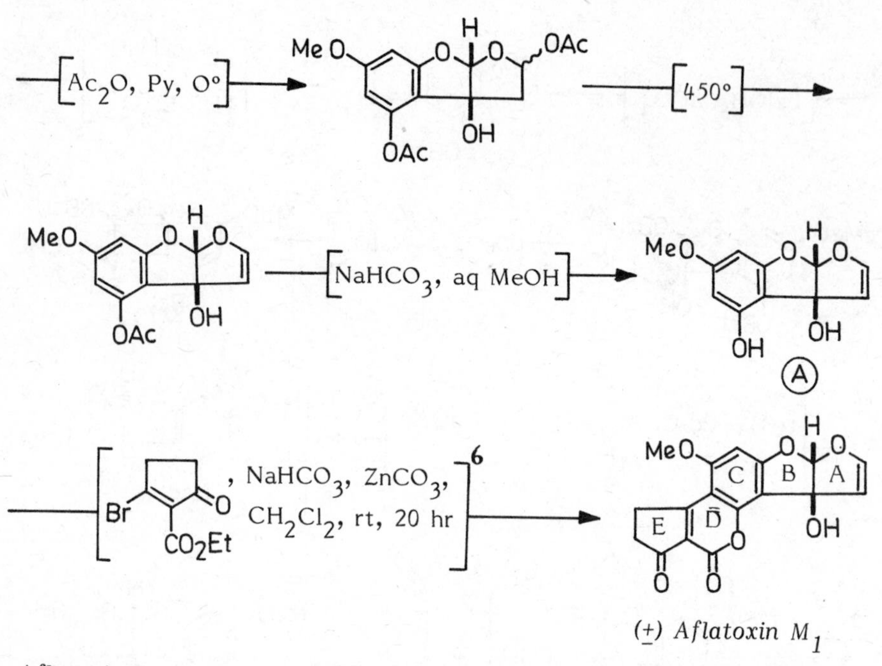

Aflatoxin B_1

For the synthesis of aflatoxin B_1 the key intermediate furo[2,3-b] benzofuran (C) was obtained by β-acyl lactone rearrangement of the 4-formylcoumarin (B) (1).

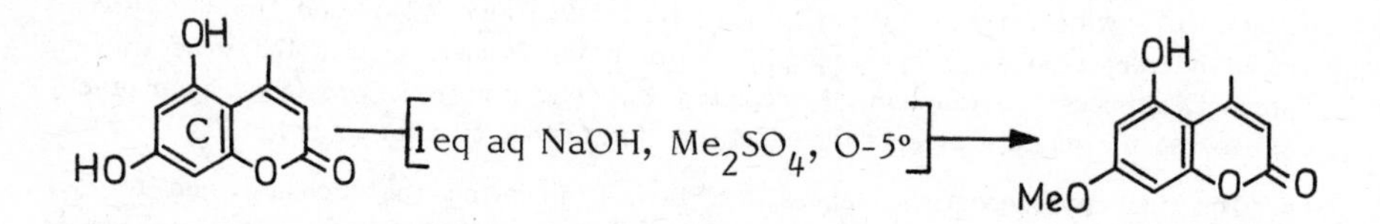

6. Owing to exceptional sensitivity of the tricyclic phenol (A) to acidic reagents, direct condensation with 2-carbethoxycyclopentane-1,3-dione under von Pechmann coumarin synthesis conditions proved impossible. However, 3-bromo-2-carbethoxycyclopent 2-enone readily underwent condensation to give the desired coumarin ring D when zinc carbonate was used both as catalyst and acid scavenger.

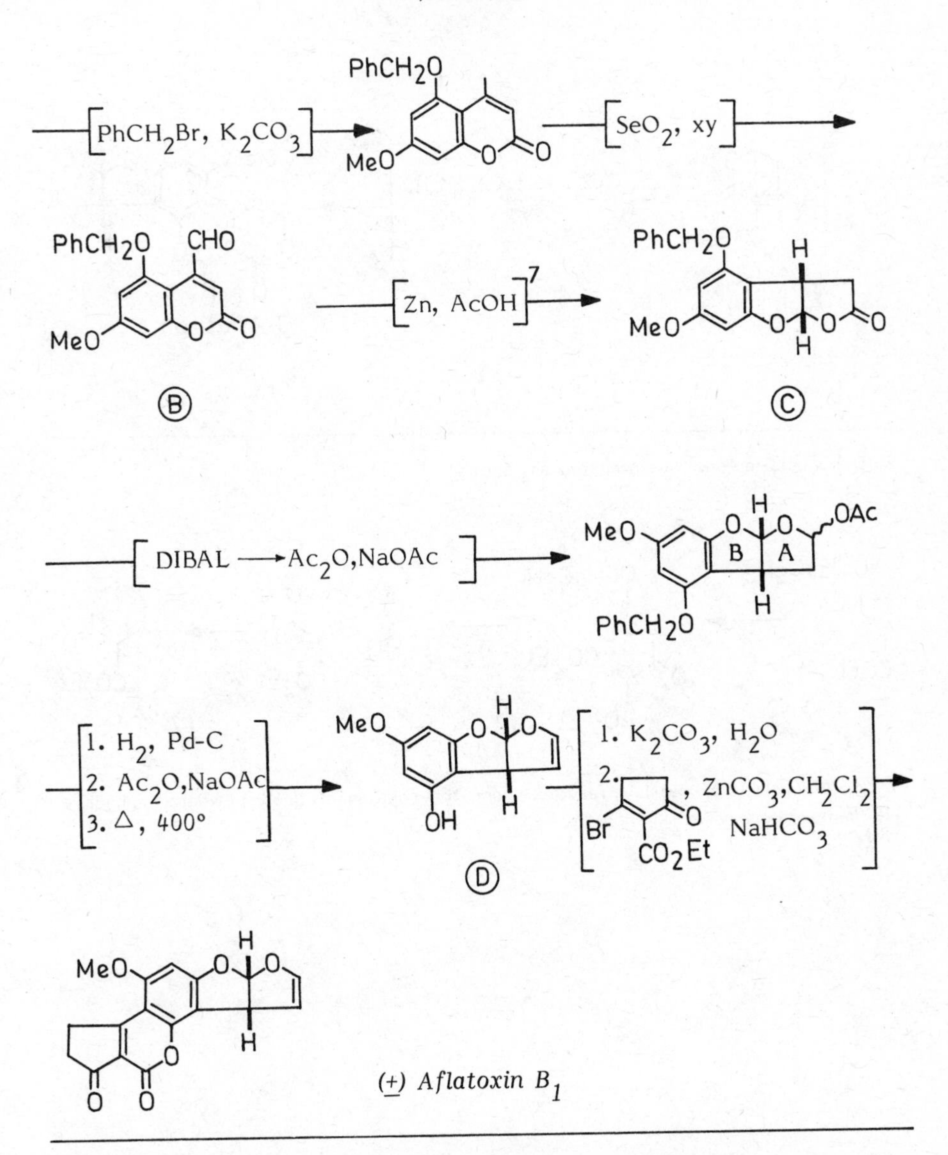

(±) *Aflatoxin* B_1

7. This rearrangement involves opening of the δ-lactone and formation of a new ring through the hydroxyl and the enolized carbonyl of the β-acyl group, Lawson, A. J. Chem. Soc., 1957, 144; Lange, C., Wamhoff, H., Korte, F. Ber. 1967, 100, 2312.

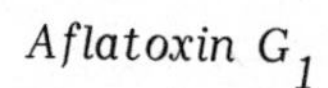

Aflatoxin G_1

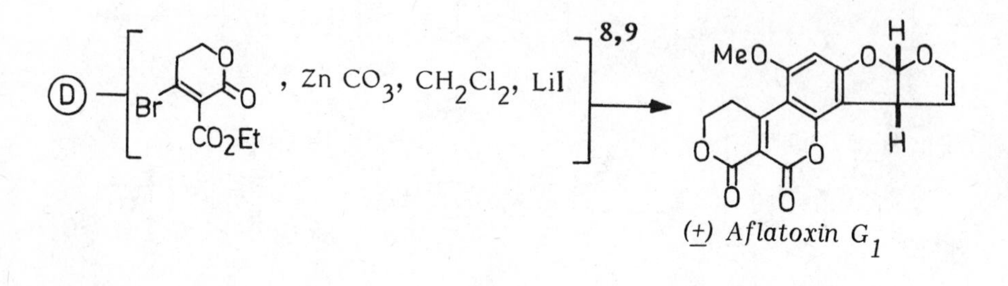

---

8. The bromo-lactone was prepared as follows :

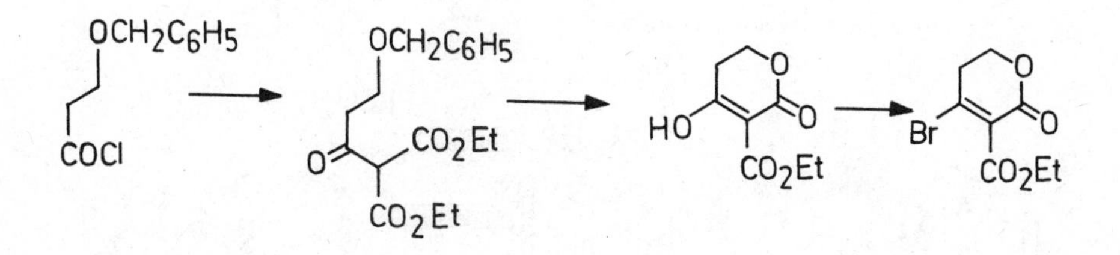

9. The bromolactone had low reactivity, and the addition of finely powdered anhydrous LiI improved the yield significantly.

13

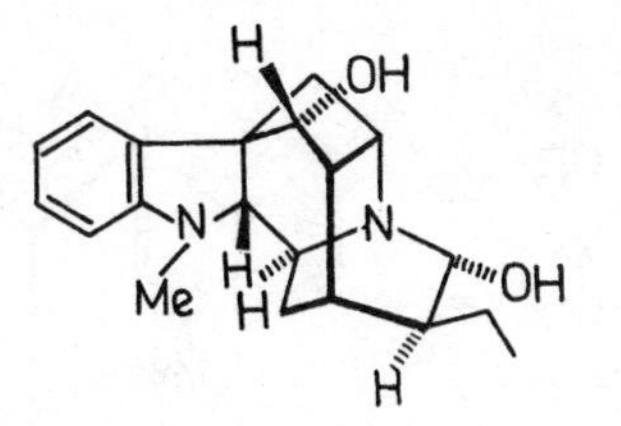

AJMALINE

Possessing six rings and nine asymmetric centres, ajmaline(1) presented a big challenge for total synthesis. In the first synthesis by Masamune et al(2) the characteristic quinuclidine ring of ajmaline was obtained by an oxidative scission of a cyclopentene-containing precursor [A] and cationoid cyclization of the resulting dialdehyde to give the key tetracyclic aldehyde [B].

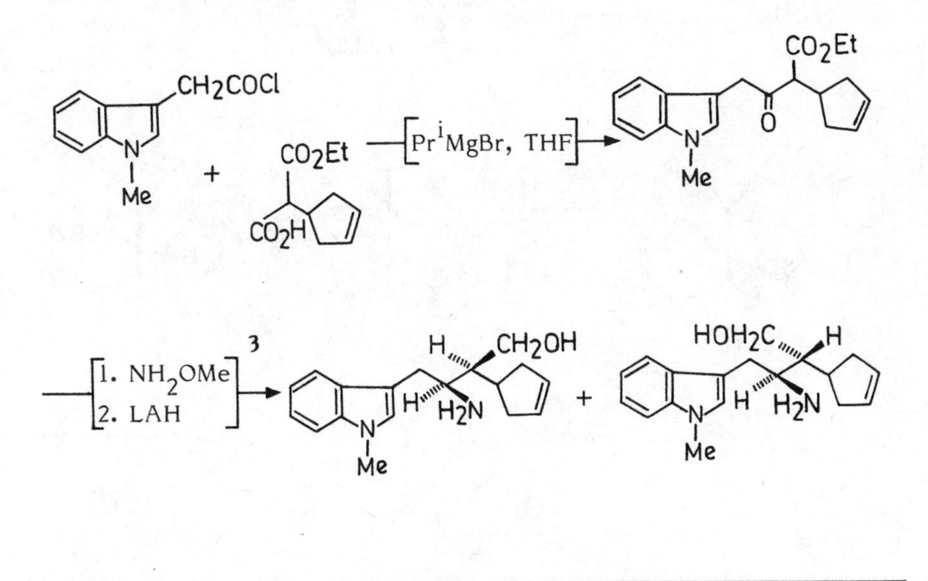

---

1. Named in memory of Hakim Ajmal Khan, a great physician of the Unani system in India in early this century; Siddiqui, S., Siddique, R.H. J. Ind. Chem. Soc., 1931, 8, 667.

2. Masamune, S., Ang, S.K., Egli, C., Nakatsuka, N., Sarkar, S.K., Yasunari, Y. J. Am. Chem. Soc., 1967, 89, 2506.

3. A 2:1 mixture of epimeric alcohols is produced in the reaction. However, both epimeric series are useful for synthesis, and are interconvertible at a later stage. The reaction sequence with only one epimeric amino alcohol is shown.

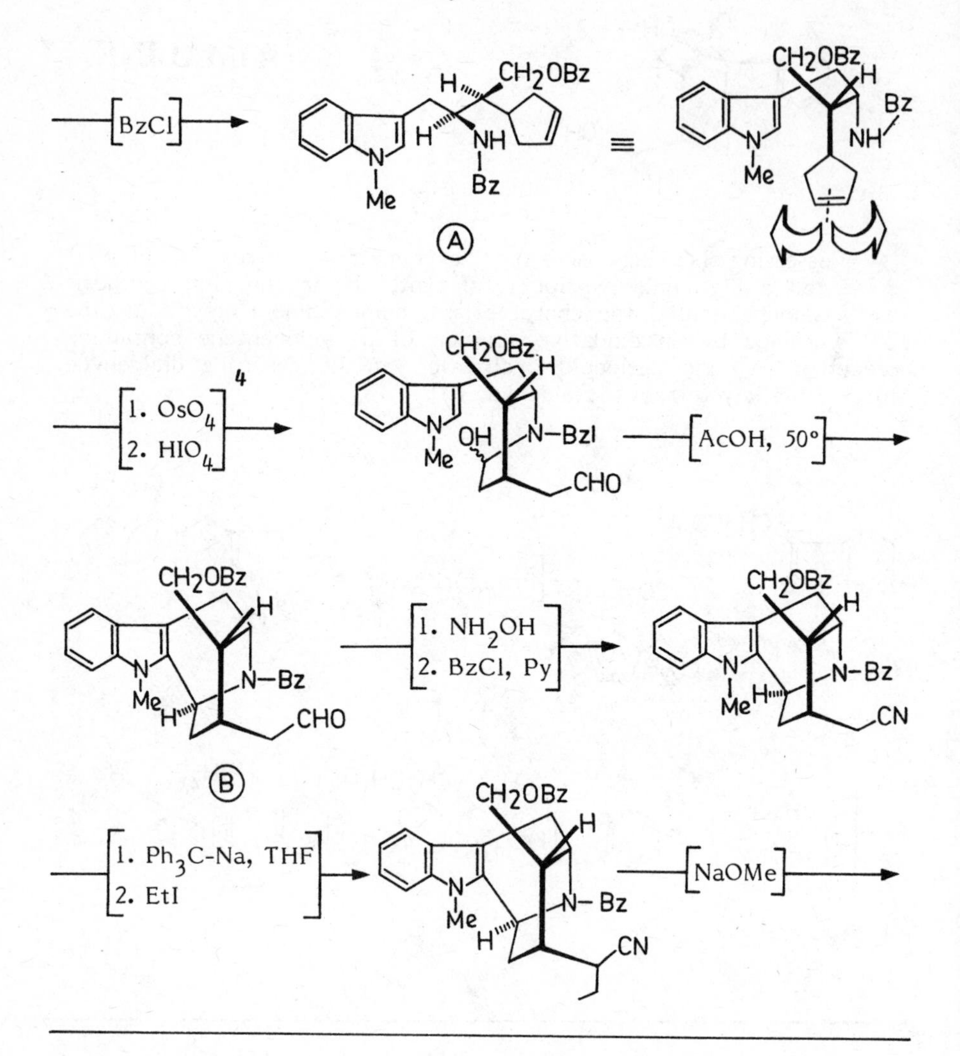

4. The preparation of dialdehydes by this procedure was utilised in earlier indole alkaloid synthesis: (a) van Tamelen, E.E., Shamma, M., Burgstahler, A.W., Wolinsky, J., Tamm, R., Aldrich, P.E. J. Am. Chem. Soc., 1958, 80, 5006; van Tamelen, E.E.,Dolby,L.J., Lawton, R.G. Tet. Letters, 1960, 19, 30.

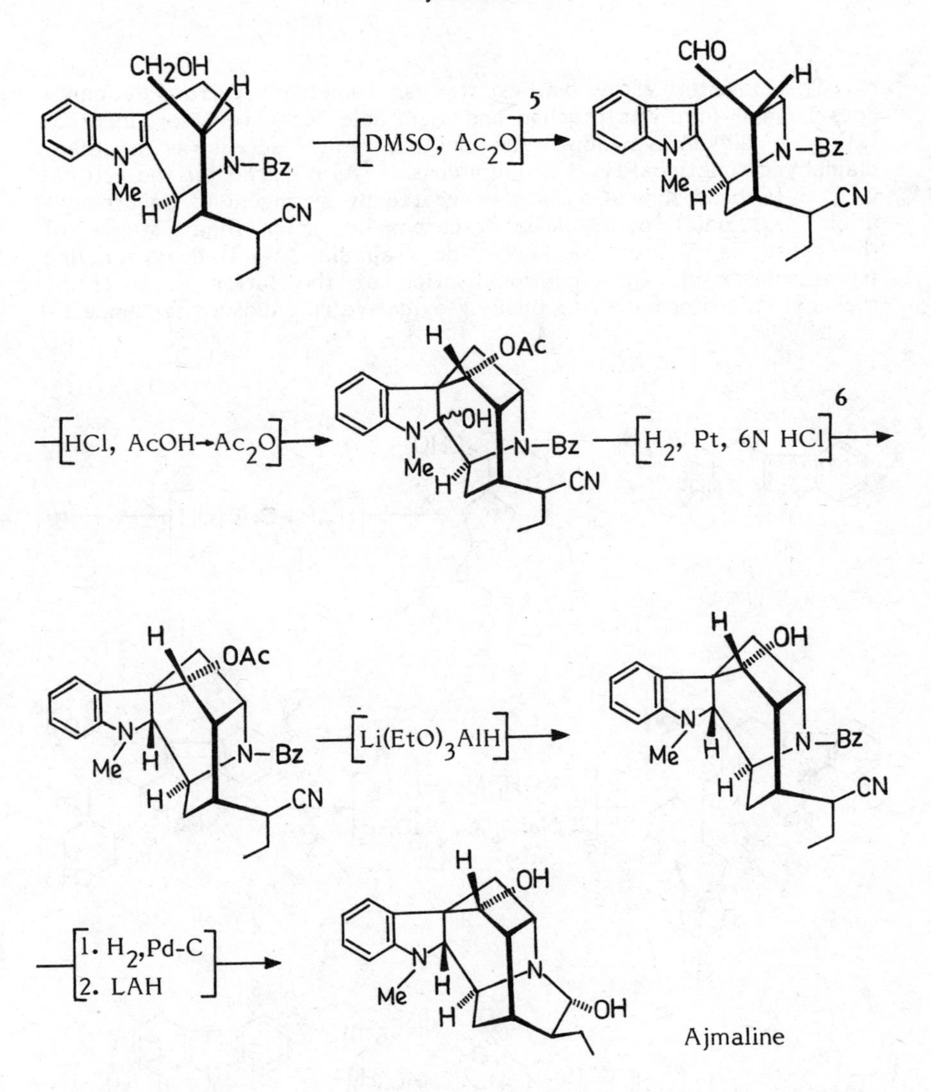

5. Fenselau, A.H., Moffat, J.G. J. Am. Chem. Soc., 1966, 88, 1762.

6. Since there is very little difference in steric hindrance between α and β sides of the molecule towards catalytic hydrogenation, reduction proceeds mainly from the α-side if the nitrogen is not protected (presence of a protonated nitrogen on α-side) giving the 2-epi series of ajmaline-type compounds. In order to encourage β-attack, therefore, reduction was carried out on the benzoylated amine, and this resulted in the exclusive formation of a compound of the normal series.

The biogenetic-type synthesis by van Tamelen & Oliver(7) commences from N-methyltryptophan and a suitable "C9" precursor, incorporating a dihydroxycyclopentene nucleus which serves as a latent dialdehyde functionality. The iminium salt [A] required for the critical C-5, C-16 bond formation was generated by an ingenious decarbonylation reaction(8) of β-carboline-carboxylic acid. Final stages of the synthesis utilise the known deoxyajmalal-A → 21-deoxyajmaline transformation(9), and functionalisation of the latter at C-21 by a phenyl chloroformate ring opening-oxidative ring closure sequence(10)

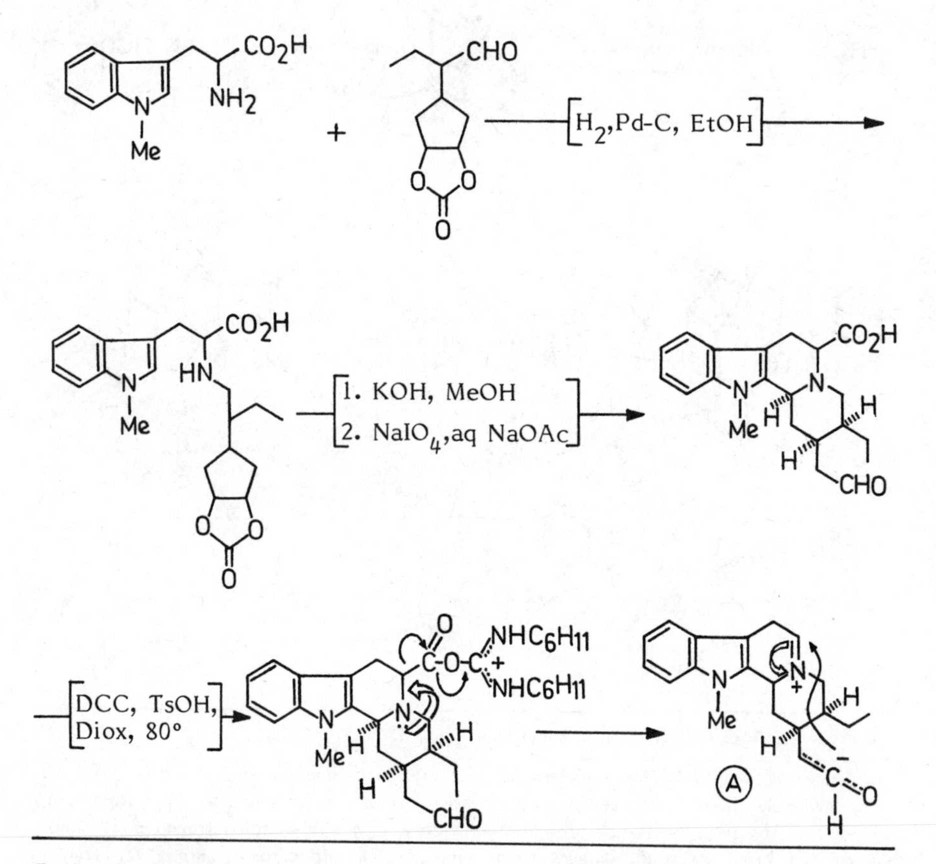

7. Van Tamelen,E.E., Oliver,L.K., J. Amer. Chem. Soc., 1970, 92, 2136; Bioorganic Chemistry, 1976, 5, 309.

8. Maksimov,V.I., Tetrahedron, 1965, 21, 687.

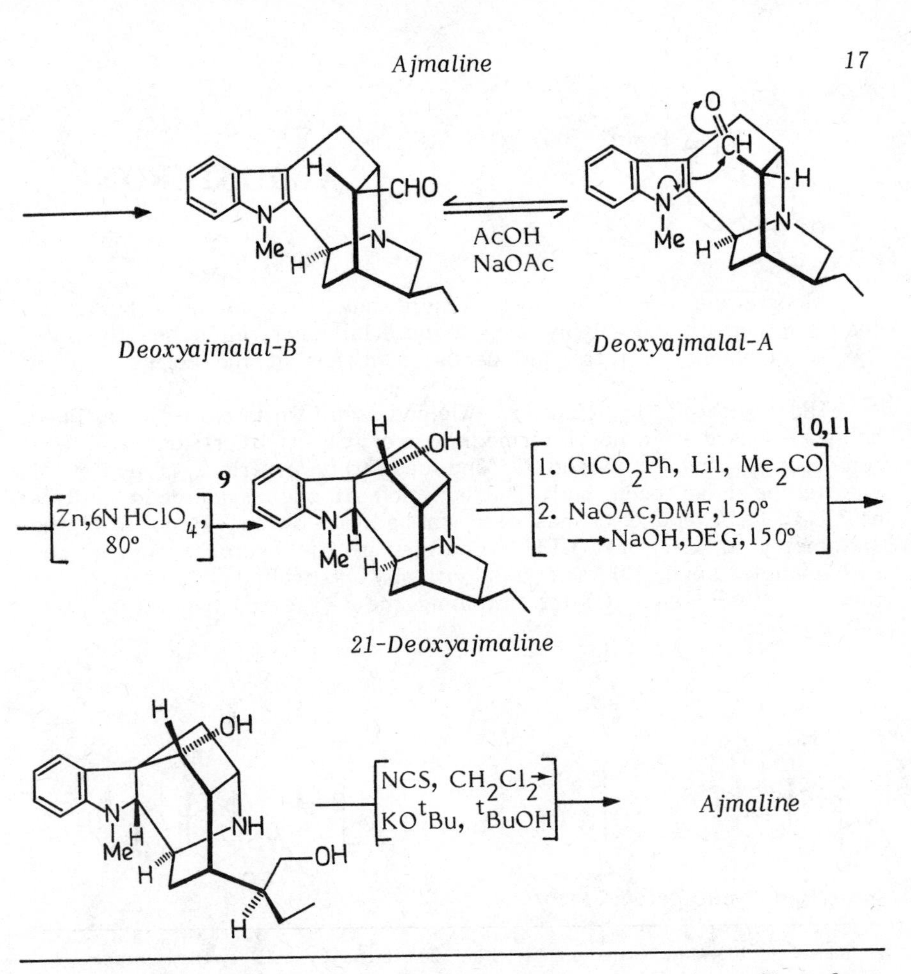

9. Bartlett, M.F., Lambert, B.F., Werblood, H.M., Taylor, W.I. J. Amer. Chem. Soc., 1963, 85, 475.

10. Hobson, J.D., McCluskey, J.G. J. Chem. Soc., 1967, 2015.

11. The reaction of phenyl chloroformate with tertiary aliphatic and alicyclic amines to give the corresponding N-carboxylates (pathway a) is an efficient general procedure for cleavage of tertiary amines under mild conditions, and provides a convenient alternative to the well-known von Braun cyanogen bromide cleavage. The choice of phenyl chloroformate over alkyl esters in this reaction probably results in suppression of the undesired competitive reaction pathway b.

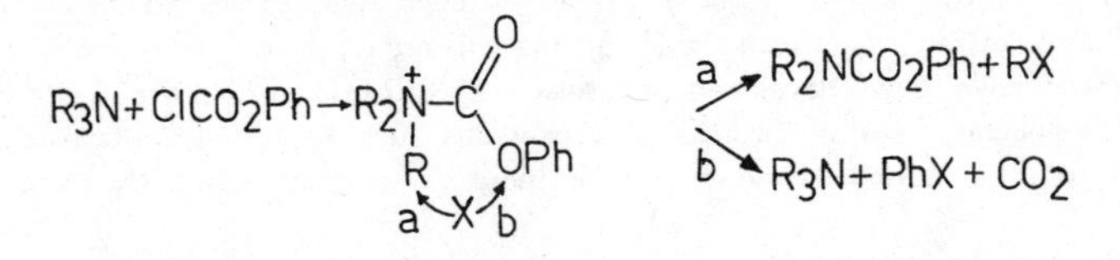

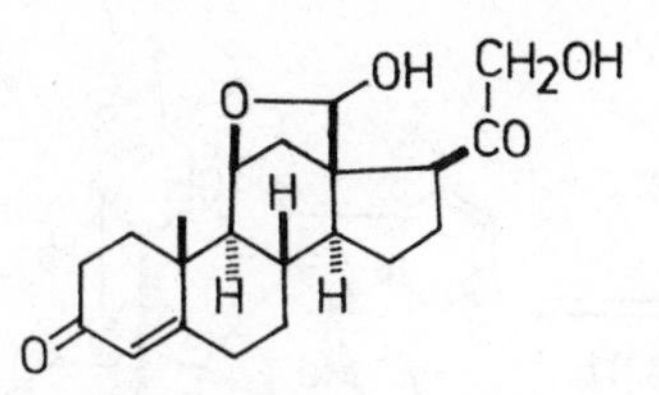

ALDOSTERONE

Aldosterone occurs in such minute quantities in natural sources that supplies of this highly active material have relied heavily upon the development of total and partial synthesis of the hormone (1,2).

The synthesis by Heusler, Wieland and Wettstein (3), outlined below, is based on a key intermediate from Sarett's cortisone synthesis representing ring A, B and C, and upon this the rings D and E of aldosterone have been built. Introduction of the asymmetric centres at C-13, C-14 and C-17 has been carried out by an extension of the asymmetry at C-11 to C-13 (A), stereospecific hydrogenation of the 14,15-double bond (B) and a kinetically controlled ketonization of the $\Delta^{17(20)}$ -enol (C) for obtaining the β-oriented hydroxy-acetone side chain at C-17.

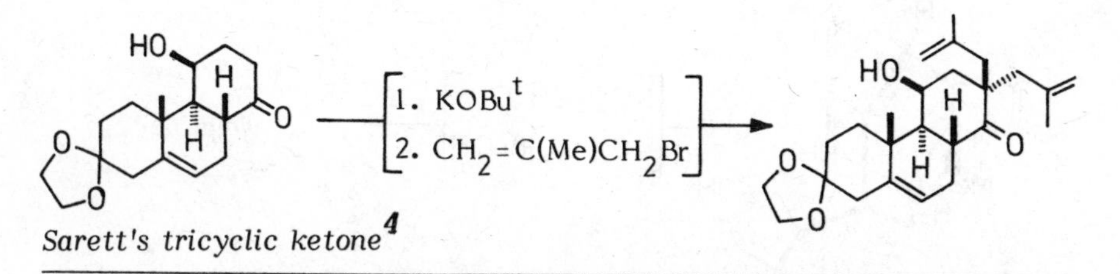

Sarett's tricyclic ketone[4]

---

1. Indeed, so urgent was the problem that aldosterone by total synthesis became the objective of a joint effort of no less than four notable group of workers of Basel, Zurich and Oss. The first synthesis was achieved by Wettstein,A.et al.at Ciba; the synthesis outlined above is a modification of the original sequence announced by these workers in 1955. For an engrossing account of the early history of aldosterone synthesis, see: Fieser,L.F.; Fieser,M. in "Steroids", Reinhold, New York (1959), p.713. Amongst other synthesis of aldosterone may be mentioned the synthesis of D-homoaldosterone, which could be converted to aldosterone; Szpilfogel,S.A.; Van Der Burg,W.J.; Siegmann,C.M.; Van Dorp,D.A. Rec. Trav. Chim., 1958, 77, 157, 171.

2. Aldosterone is one of those rare examples in which the partial practical synthesis of a complex natural product was realized long after the successful execution of a total synthesis. Since the partial synthesis of aldosterone from steroid precursors entails the functionalizing of an unactivated angular methyl group, many new and ingenious methods have been developed for selectively functionalizing the C-18 methyl group by intramolecular transfer substitution procedures such as of nitrile transfer (12); for a review of the earlier work see: Schaffner,K.; Arigoni,D.; Jeger,O. Experientia, 1960, 16, 169.

3. Heusler,K.; Wieland,P.; Wettstein,A. Helv. Chim. Acta, 1959, 42, 1586.

4. See under synthesis of cortisone.

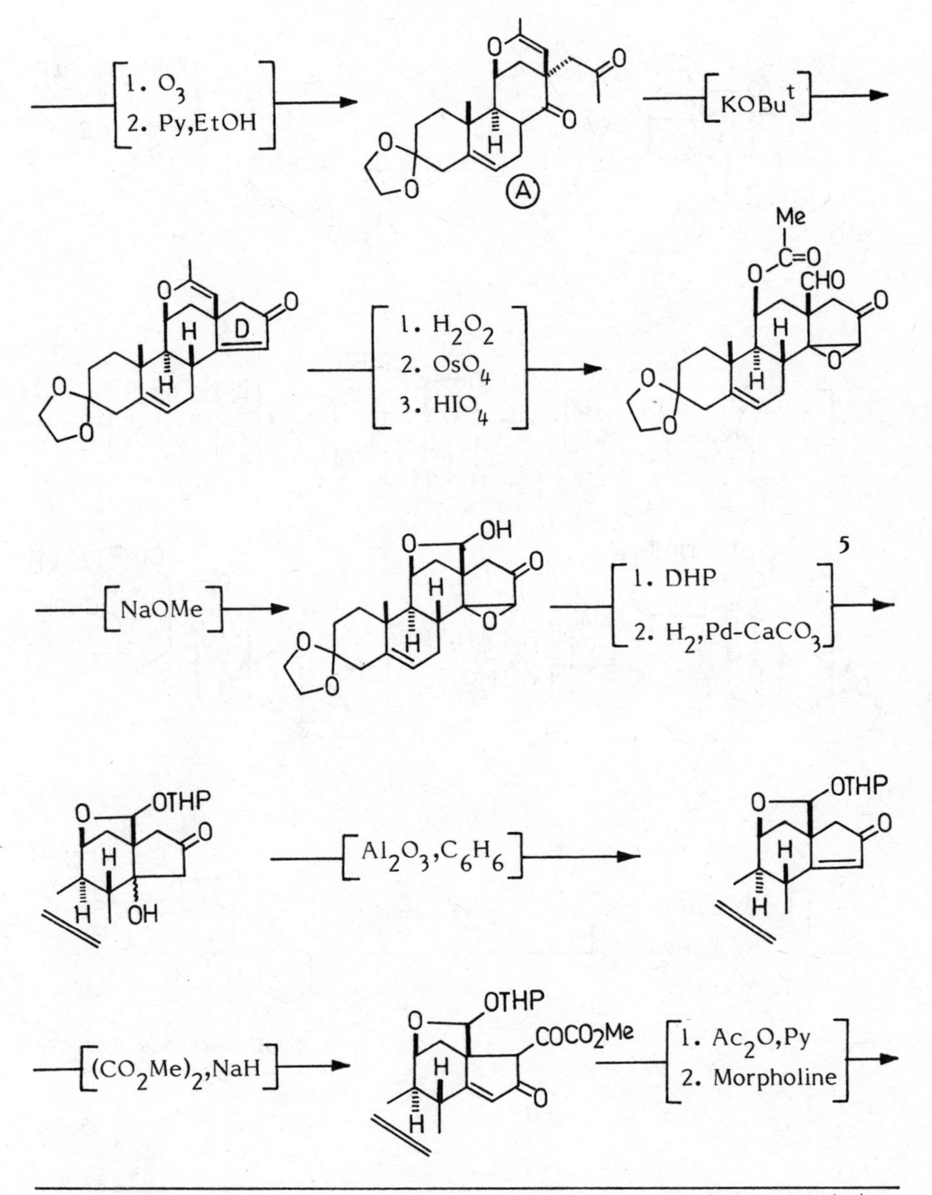

5. The 18,11β-cyclohemiacetal was preserved throughout the synthesis as a tetrahydropyranyl ether.

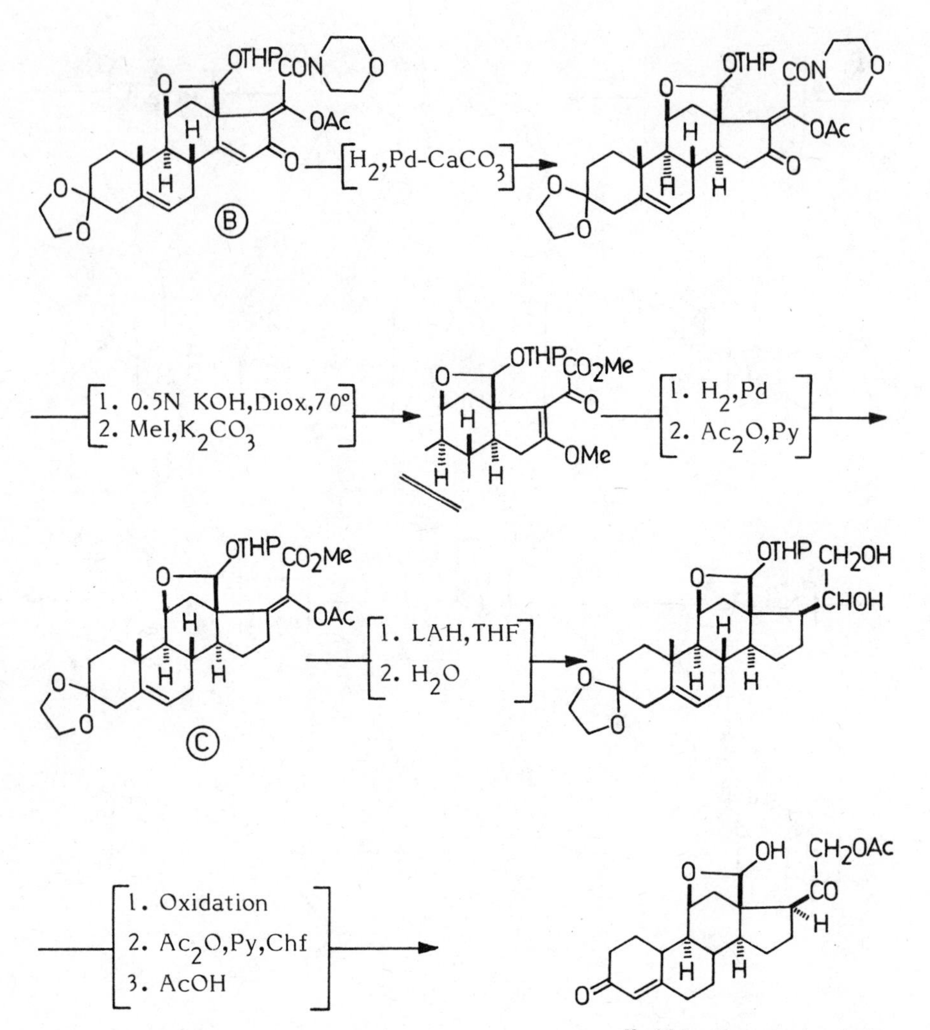

dl-Aldosterone Acetate

Johnson and his colleagues(6-10)described a highly stereoselective total synthesis of aldosterone using the "hydrochrysene approach". Introduction of the 11β-hydroxyl in (A) is based on a method developed earlier at Wisconsin for the production of 11-oxygenated steroids (10). Material for completing the hemiacetal ring has been obtained by an oxidative fission of ring D in the homosteroid (B).

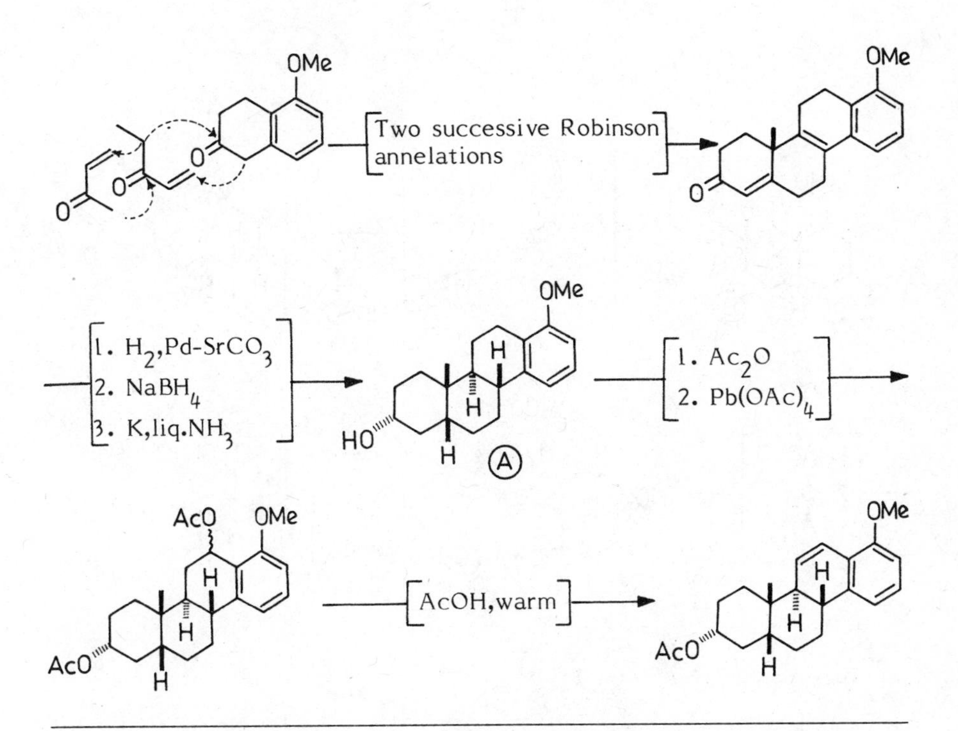

6. Johnson,W.S.; Collins,J.C.; Pappo,R.; Rubin,M.B. J. Am. Chem. Soc., 1958, 80, 2585; Johnson,W.S.; Collins,J.C.; Pappo,R.; Rubin,M.B.; Kropp,P.J.; Johns,W.F.; Pike,J.E.; Bartmann,W. ibid, 1963, 85, 1409.

7. Johnson,W.S.; Szmuszkovicz,J.; Rogier,E.R.; Hadler,H.I.; Wynberg,H. J. Am. Chem. Soc., 1956, 78, 6285.

8. Johnson,W.S.; Rogier,E.R.; Szmuszkovicz,J.; Hadler,H.I.; Ackerman,J.; Bhattacharyya,B.K.; Bloom,B.M.; Stalmann,L.; Clement,R.A.: Bannister,B.; Wynberg,H. J. Am. Chem. Soc., 1956, 78, 6289.

9. Johnson,W.S.; Kemp,A.D.; Pappo,R.; Ackermann,J.; Johns,W.F. J. Am. Chem. Soc., 1956, 78, 6312.

10. Johnson,W.S.; Pappo,R.; Johns,W.F. J. Am. Chem. Soc., 1956, 78, 6339.

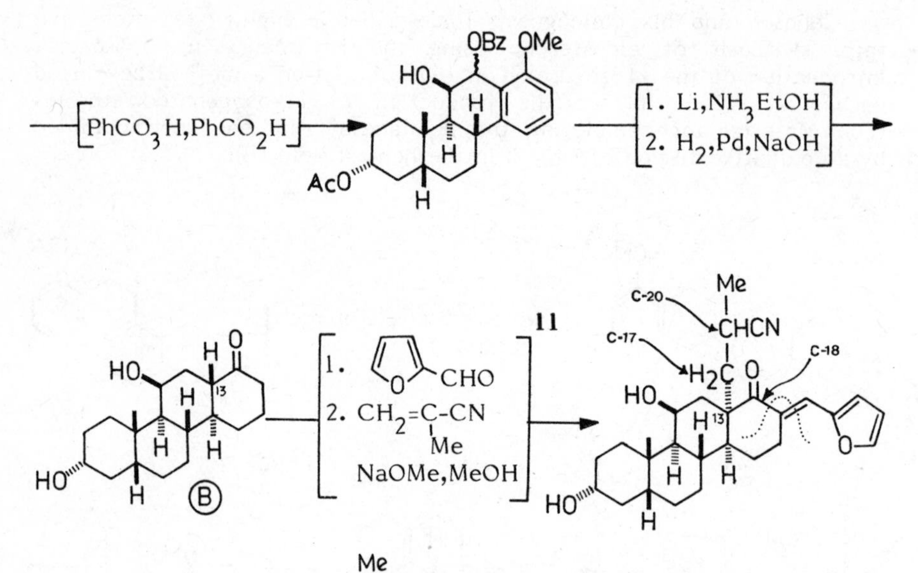

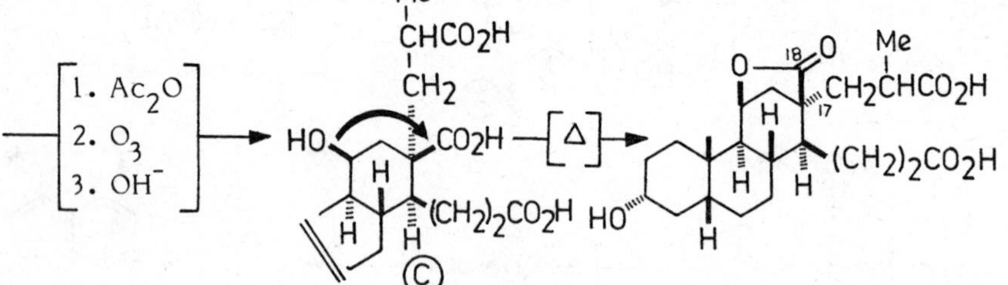

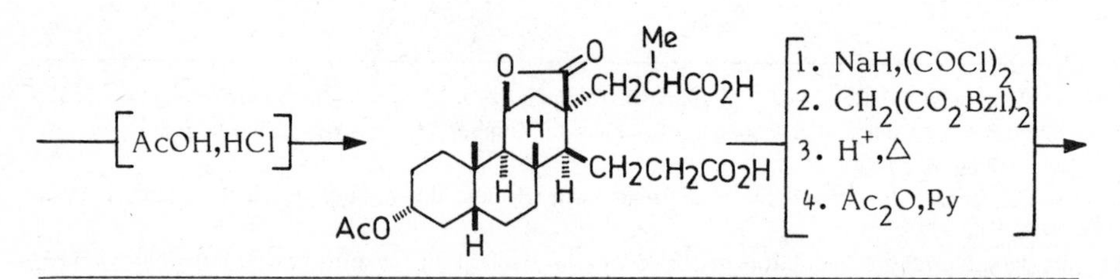

11. Introduction of the angular substituent in the tetracyclic D-homosteroid (B) results in a C/D-cis ring fusion with the alkylating group attached to C-13 in an α-orientation, thus totally unsuited for incorporation as the 11,13α-oriented ring E. The incoming α-alkylating group was therefore destined to serve as the C-17 hydroxyacetone side chain and it befell the newly formed β-oriented C-13 carboxyl group, obtained during oxidative fission of ring D, to serve as C-18 of aldosterone.

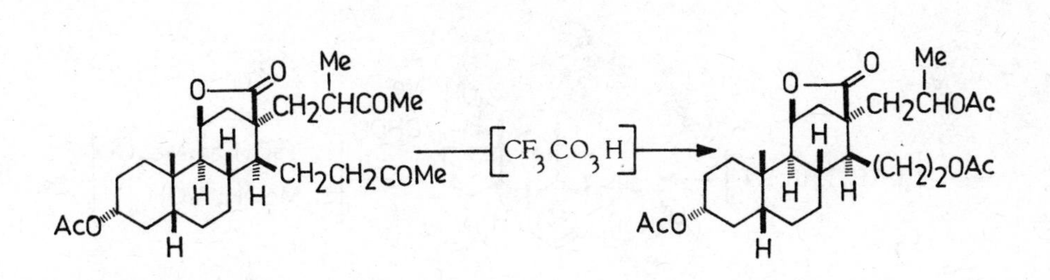

Me
O
CH2CHCOMe
H
CH2CH2COMe
H H
AcO
H
CF3 CO3 H
Me
O
CH2CHOAc
H
(CH2)2OAc
H H
AcO
H

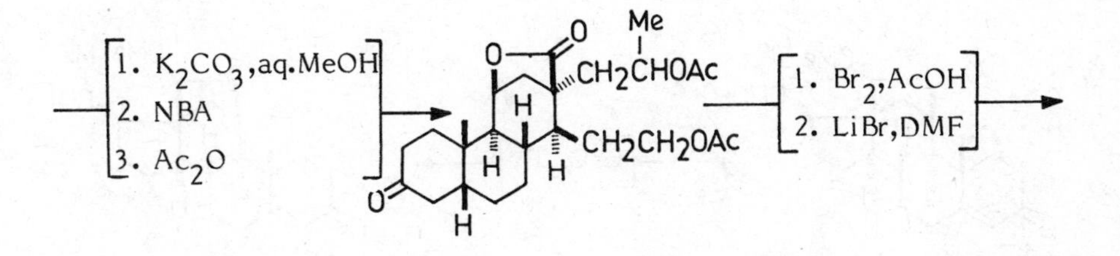

1. K2CO3,aq.MeOH
2. NBA
3. Ac2O
Me
O
CH2CHOAc
H
CH2CH2OAc
H H
O
H
1. Br2,AcOH
2. LiBr,DMF

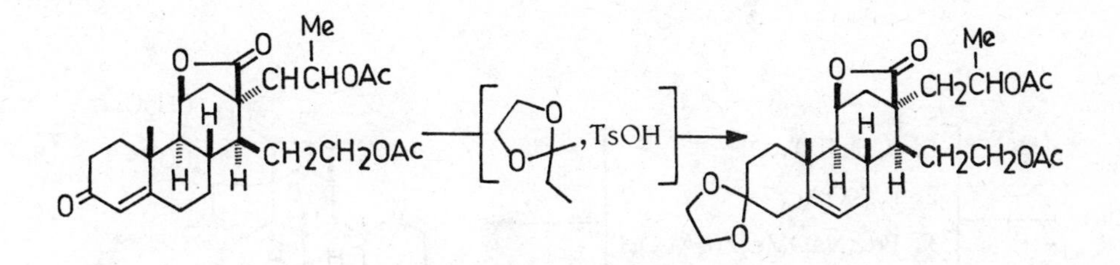

Me
O
CHCHOAc
H
CH2CH2OAc
H H
O
,TsOH
Me
O
CH2CHOAc
H
CH2CH2OAc
H H

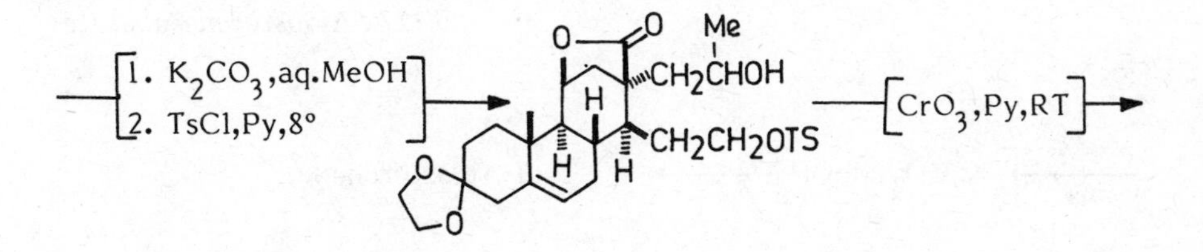

1. K2CO3,aq.MeOH
2. TsCl,Py,8°
Me
O
CH2CHOH
H
CH2CH2OTS
H H
CrO3,Py,RT

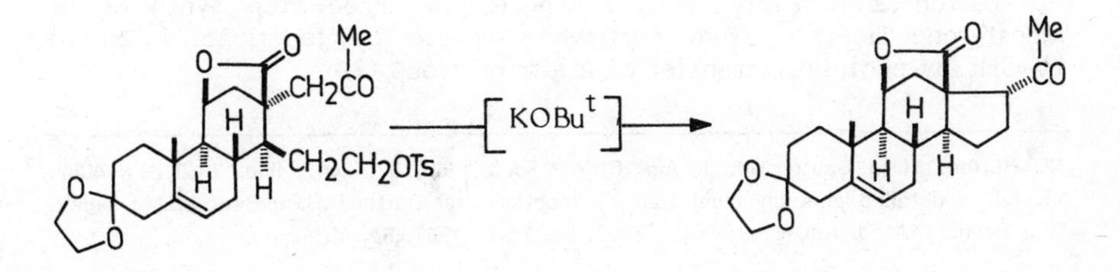

Me
O
CH2CO
H
CH2CH2OTs
H H
KOBut
Me
O
CO
H
H H

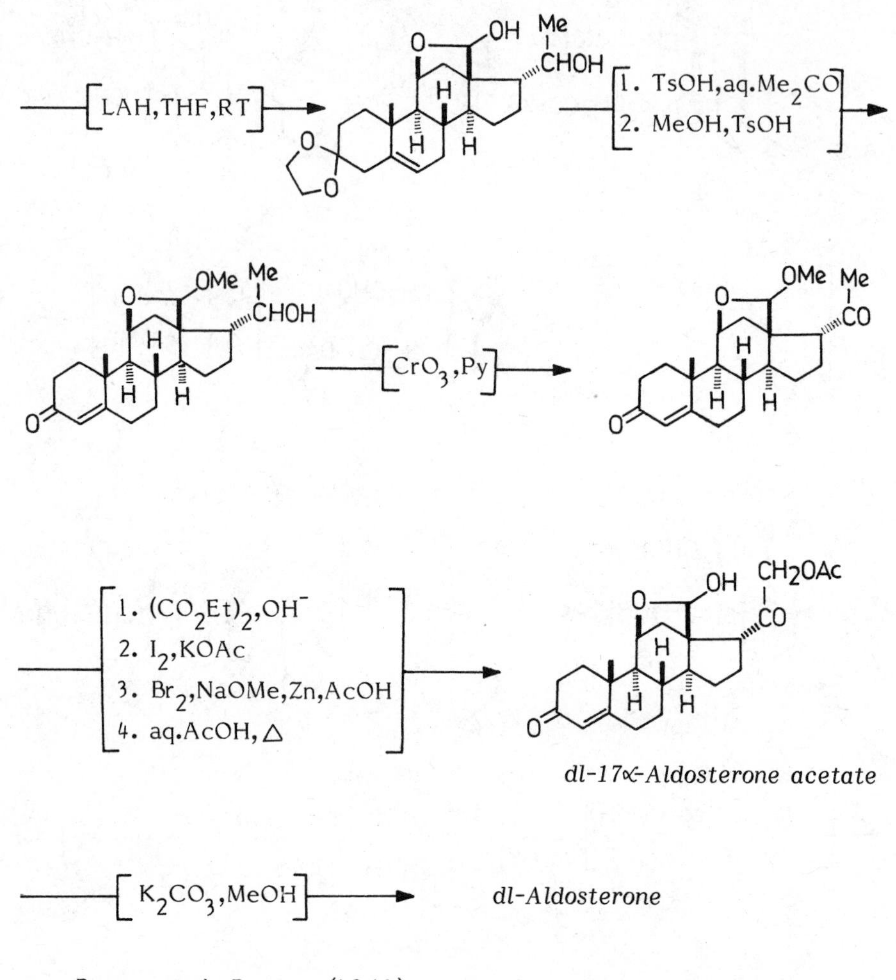

Barton and Beaton (12,13) reported a three step synthesis of aldosterone acetate from cortisone acetate by functionalisation of 18-CH_3 by photolytic transfer of a nitrite group (2).

---

12. Barton,D.H.R.; Beaton,J.M. J. Am. Chem. Soc., 1960, 82, 2641; ibid, 1961, 83, 4083.
13. For a detailed photochemical transfer reaction see: Barton,D.H.R.; Beaton,J.M.; Geller, L.E.; Pechet,M.M. J. Am. Chem. Soc., 1960, 82, 2640; 1961, 83, 4076.

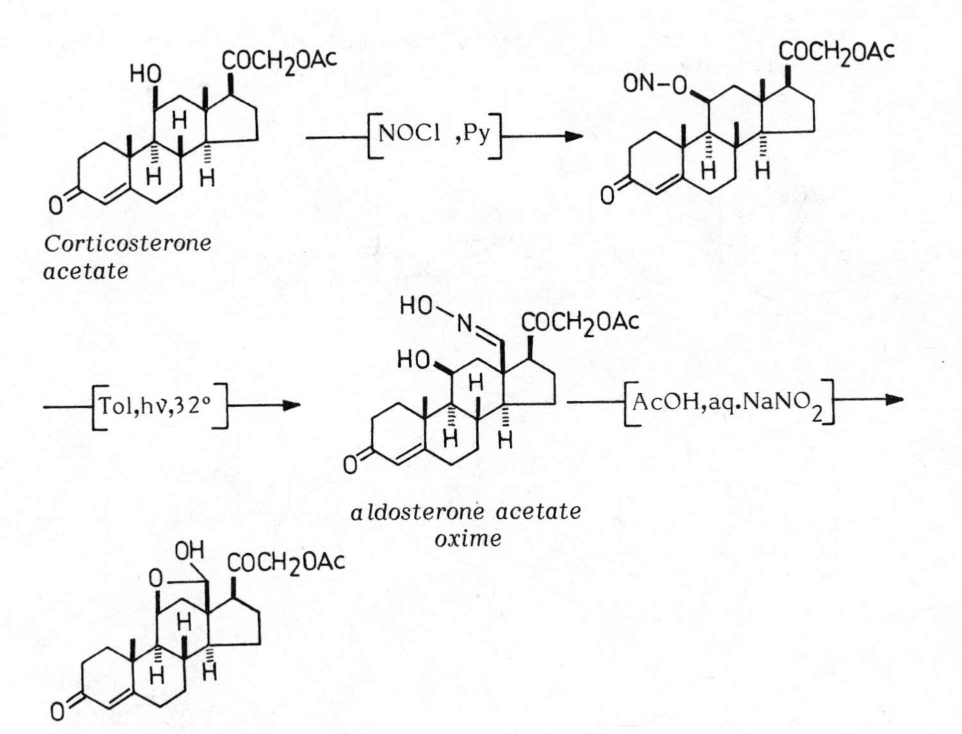

Corticosterone acetate

aldosterone acetate oxime

aldosterone 21-acetate

The overall yield in this reaction was rather low, on account of the competition between C-19 and C-18 functionalisation. In a further refinement of this method, it was observed that photolysis of the nitrites of the corresponding 1,2-dehydro derivatives resulted in almost exclusive C-18 functionalisation, and thus afforded greatly improved yields of aldosterone through 1,2-dehydro-(I) and of 1,2,6,7-didehydro-(II) aldosterone 21-acetates (14).

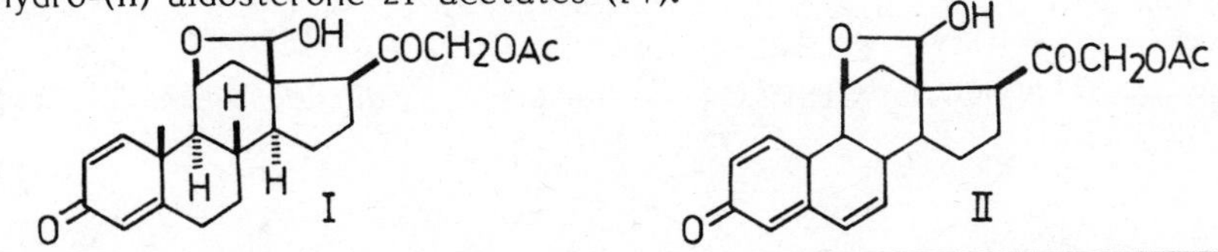

14. Barton,D.H.R.; Basu,N.K.; Day,M.J; Hesse,R.H.; Pechet,M.M.; Starratt,A.N. J. Chem. Soc. Perkin Trans., 1975, 2243.

Miyano (15) has more recently described an efficient synthesis of aldosterone from 20-hydroxy-18,20-cyclosteroids.

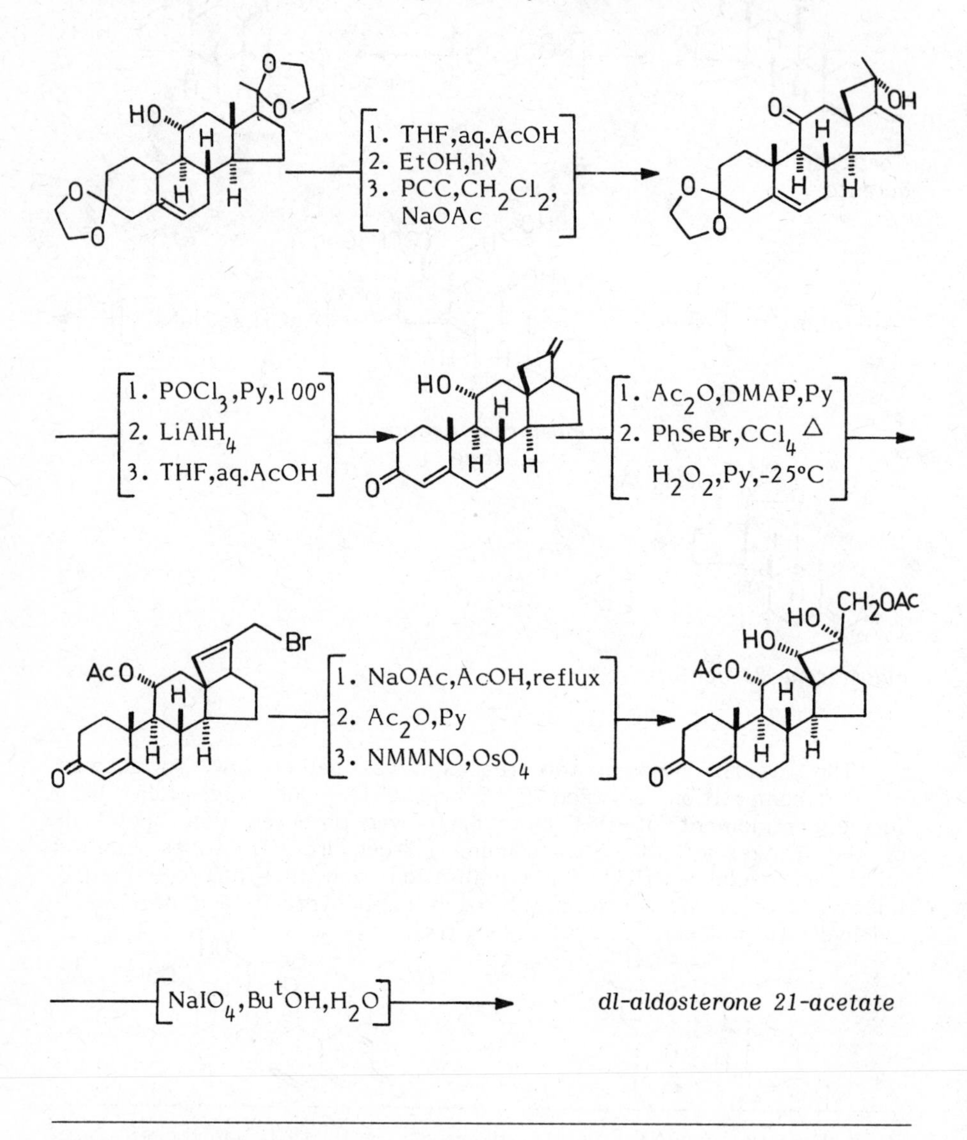

dl-aldosterone 21-acetate

---

15. Miyano,M. J. Org. Chem., 1981, 46, 1846.

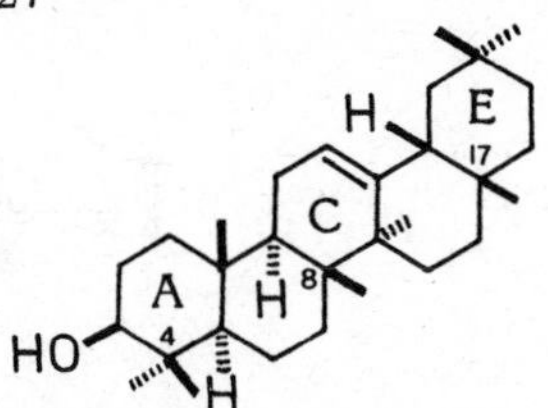

AMYRIN

A general approach to the oleanane skeleton has been the construction of a tetracyclic (AB-DE) system (1) in which ring C is completed by an internal cationic cyclization reaction (3-5). The synthesis of 18α-olean-12-ene (C) based on this approach is described below (2). The remaining task of introducing a β-hydrogen at C-18 and a 3β-hydroxyl into the pentacyclic system (C) has been accomplished by Barton and his colleagues (7), thus completing a total synthesis of β-amyrin. Introduction of the oxygen function in ring A has been accomplished by a novel photolytic exchange process (8) involving a functional derivative based on the ring C olefinic linkage.

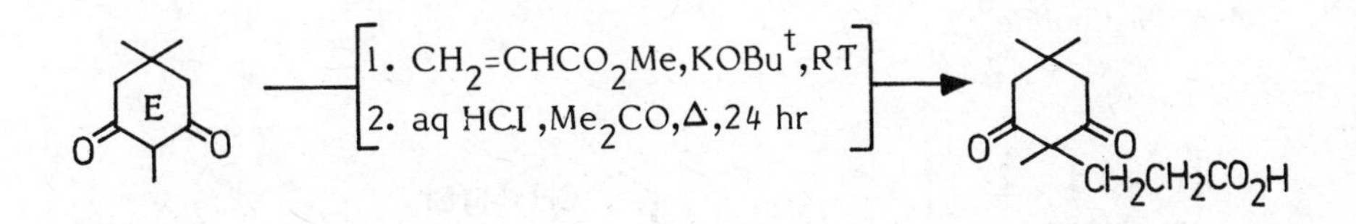

---

1. This approach, which was first suggested by Halsall and Thomas, provides an admirable solution to the problem of introducing the vicinal C-8 and C-14 methyl groups, an otherwise difficultly accessible feature of ring C. The desired anti-trans arrangement of rings B,C and D is obtained during cyclization when a carbonium ion is generated at C-8, Halsall,T.G.; Thomas,D.B. J. Chem. Soc., 1956, 2431.

2. The outlined synthesis is a composite of the efforts of groups of workers; the steps leading to the key tetracyclic ketone (B) are taken from Corey's synthesis of olean-11,12,13,18-diene (ref.3) while subsequent steps leading to the pentacyclic skeleton are from the synthesis of Barltrop et al (ref.4); Ghera and Sondheimer (5) have also synthesised this intermediate by a similar approach.

3. Corey,E.J.; Hess,H.J.; Proskow,S. J. Am. Chem. Soc., 1959, 81, 5258; 1963, 85, 3979.

4. Barltrop,J.A.; Littlehailes,J.D.; Rushton,J.D.; Rogers,N.A.J. Tetrahedron Lett., 1962, 429.

5. Ghera,E.; Sondheimer,F. Tetrahedron Lett., 1964, 3887.

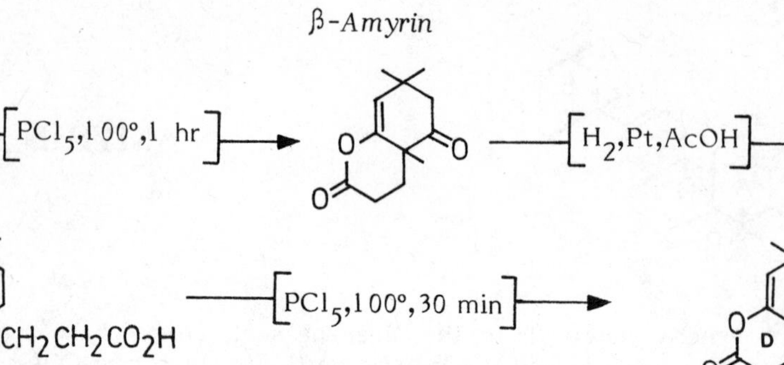

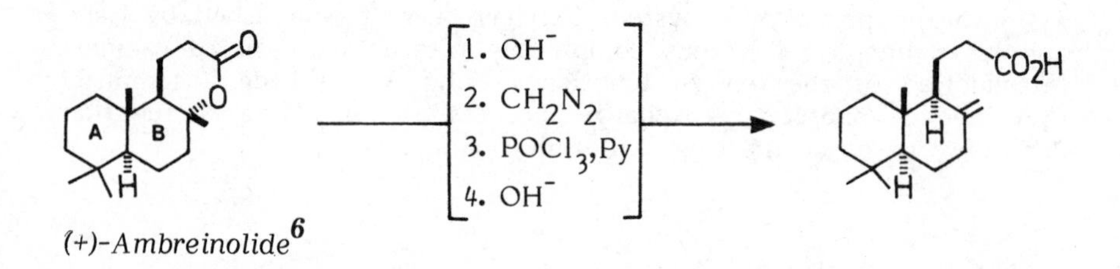

(+)-Ambreinolide[6]

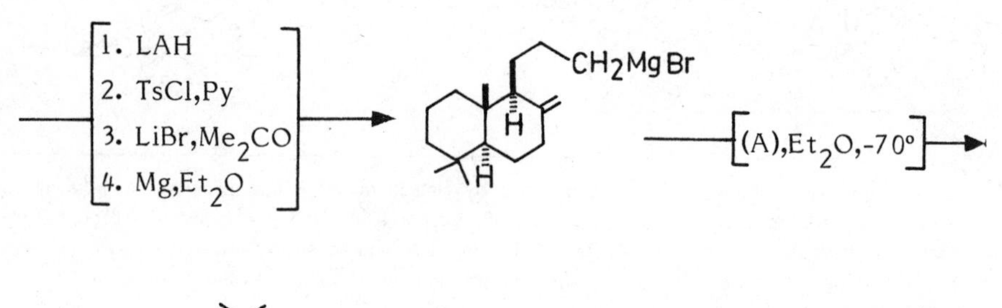

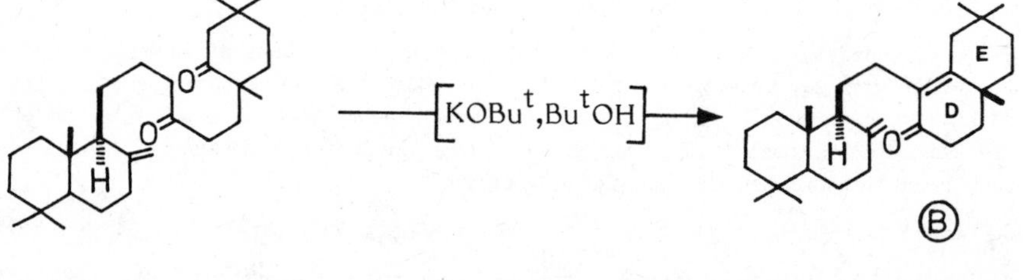

6. For a total synthesis of (±)-ambreinolide, see Dietrich,P., Lederer,E., Helv. Chim. Acta, 1952, 35, 1148.

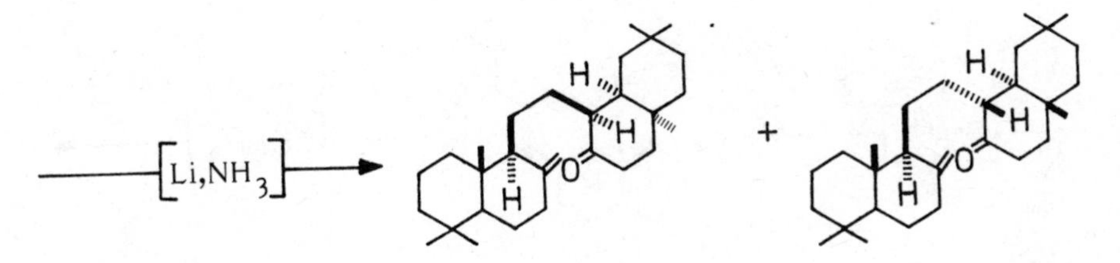

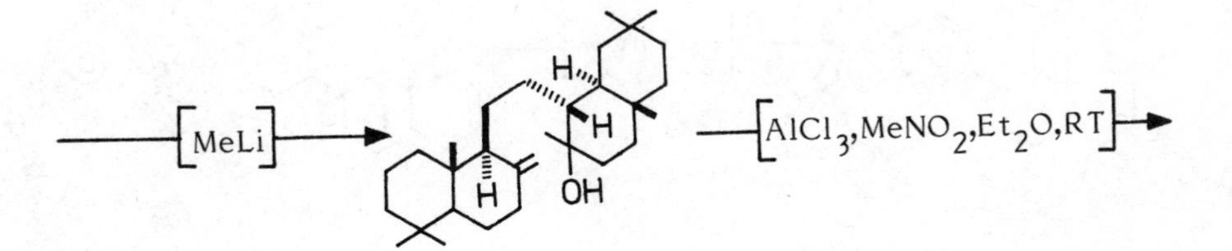

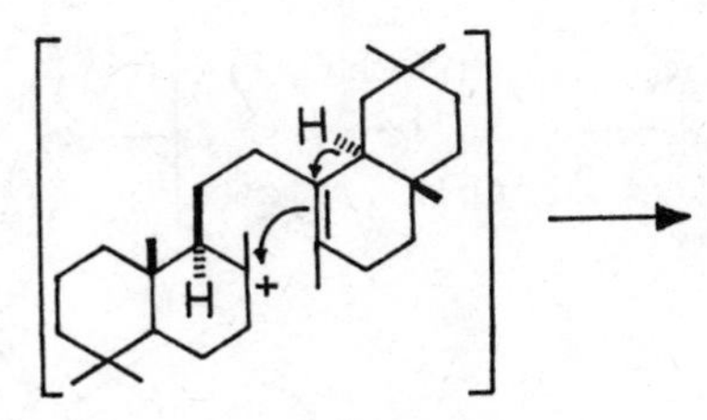

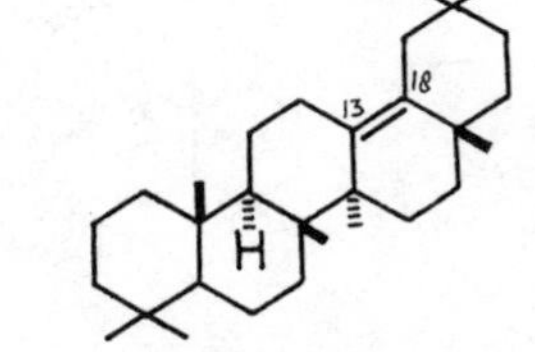

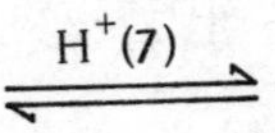

Olean-13(18)-ene
(δ-amyrene)

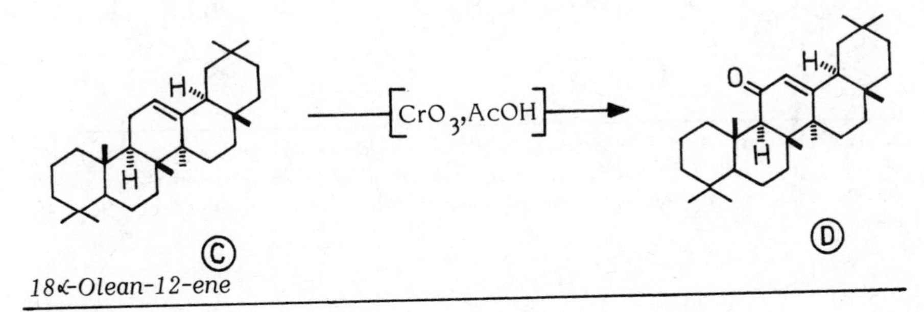

18α-Olean-12-ene

---

7. Barton,D.H.R., Lier,E.F., McGhie,J.F., J. Chem. Soc.(C), 1968, 1031.

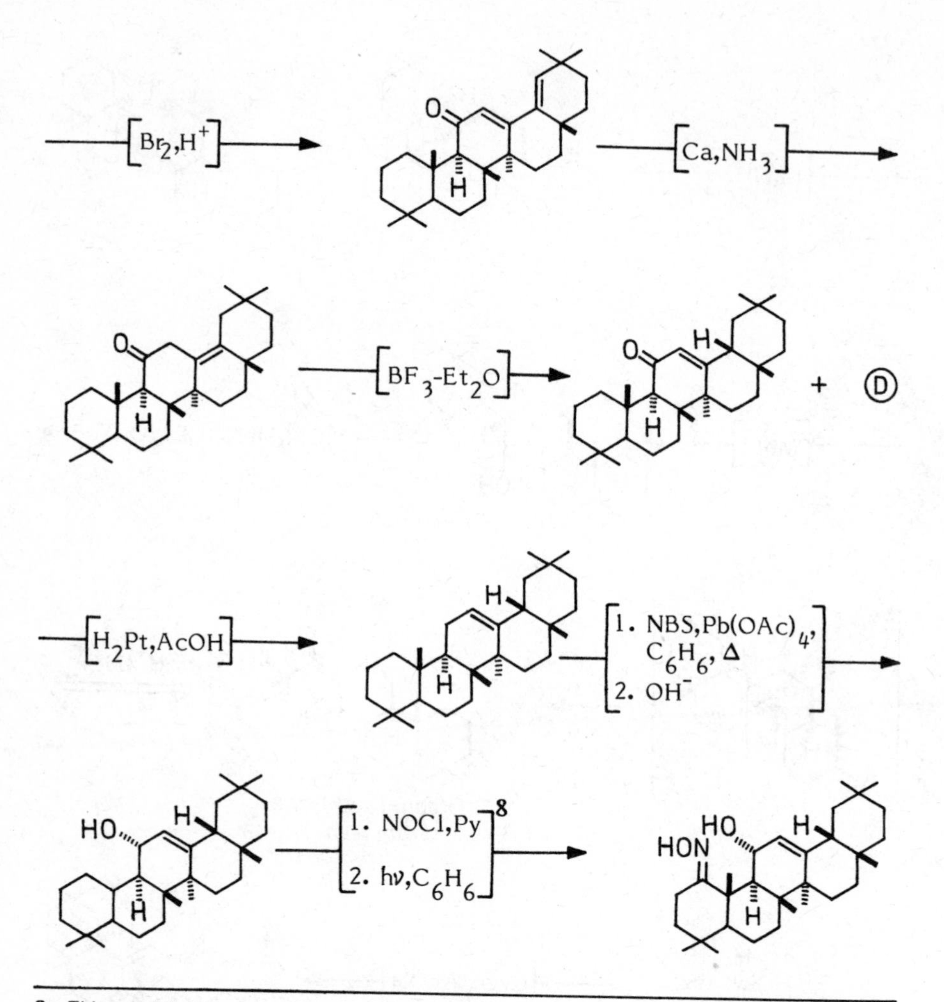

8. This photolytic exchange process involving the intramolecular transfer of the NO of the organic nitrite to a γ-carbon atom probably involves an activated alkoxy radical (i), Barton,D.H.R., Beaton,J.M., Geller,L.E., Pechet,M.M., J. Am. Chem. Soc., 1961, 83, 4076.

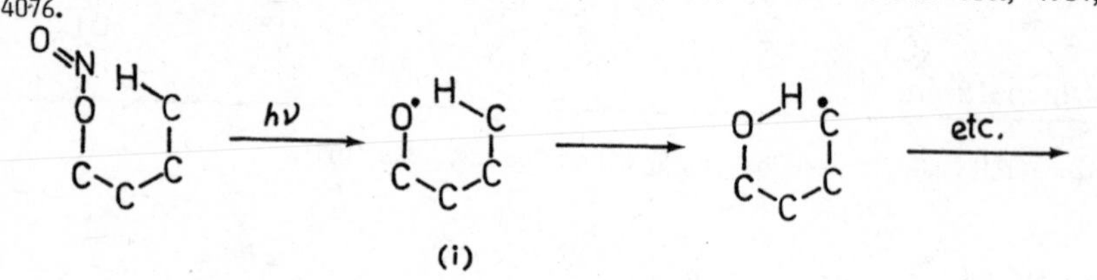

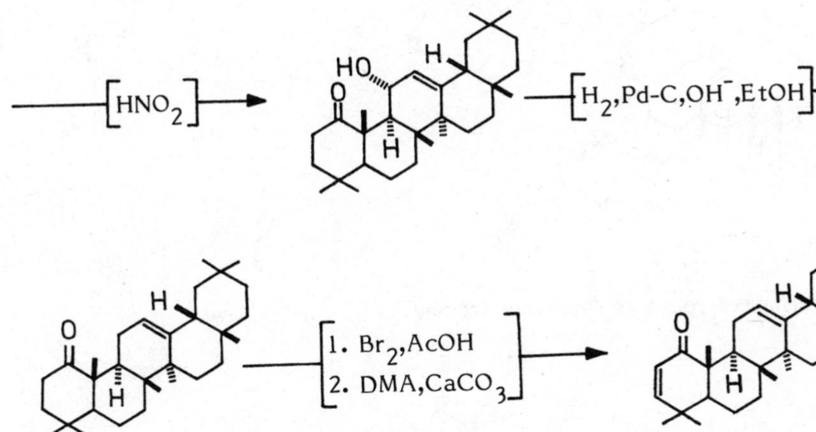
HNO2
HO
H
O
H
H2,Pd-C,OH⁻,EtOH
H
O
H
1. Br2,AcOH
2. DMA,CaCO3

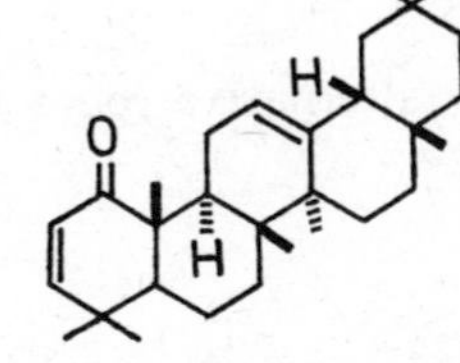
H
O
H

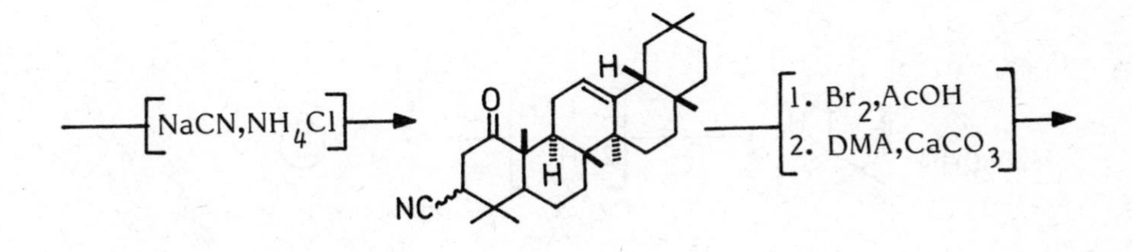
NaCN,NH4Cl
H
O
H
NC
1. Br2,AcOH
2. DMA,CaCO3

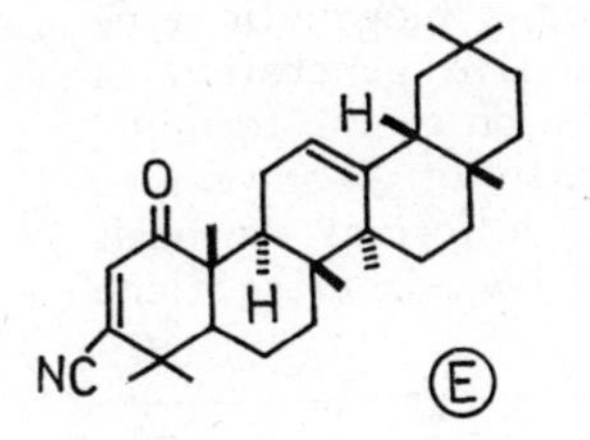
H
O
H
NC
E

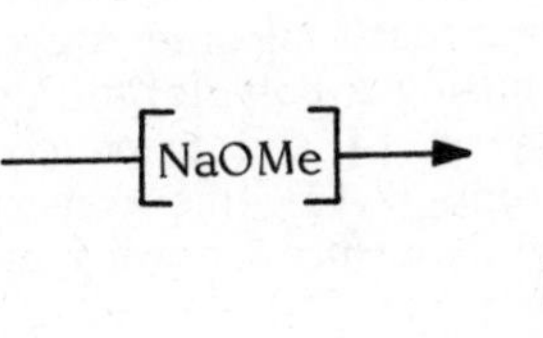
NaOMe

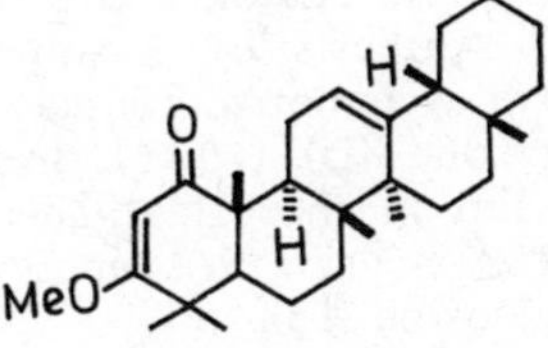
H
O
H
MeO

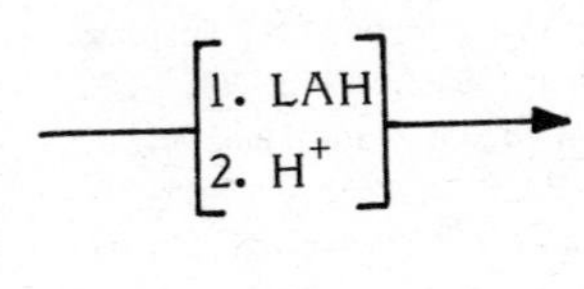
1. LAH
2. H⁺

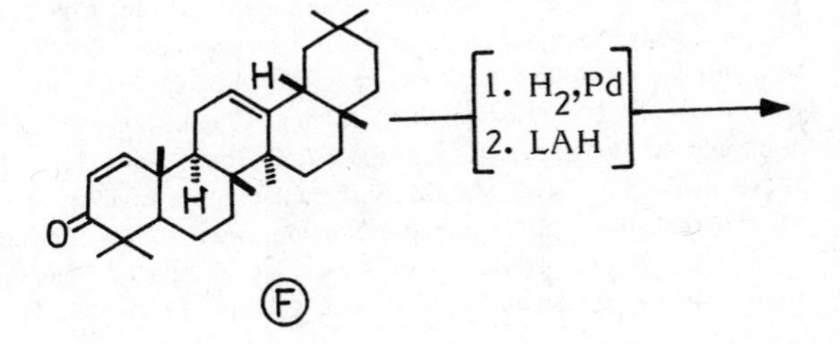
H
H
O
F
1. H2,Pd
2. LAH

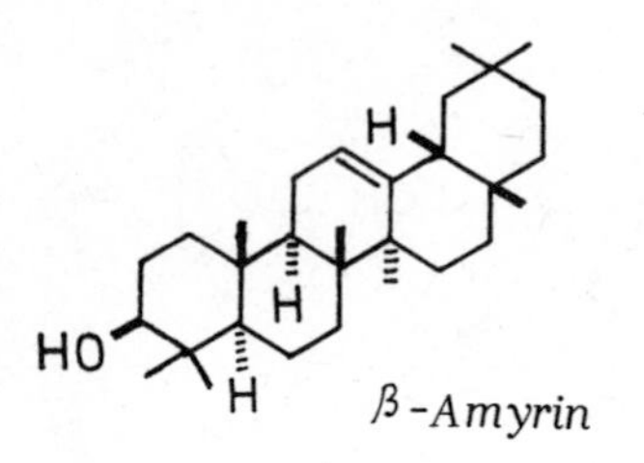

β-Amyrin

An alternative route to β-amyrin from (E) (9) is as follows:

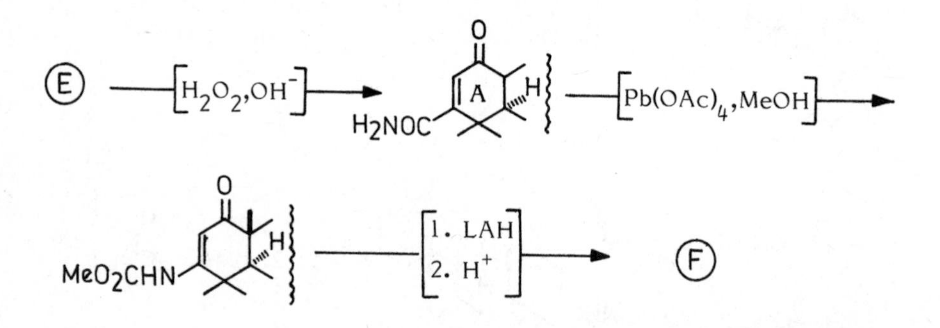

Van Tamelen and his associates have described a biogenetic type of synthesis of δ-amyrin featuring the stereoselective generation of five asymmetric centres during a polyolefinic cyclization of the terminal epoxide (D) (10,11). In view of the prior conversion of δ-amyrin to β-amyrin through β-amyrene (7,12) this constitutes a formal synthesis of β-amyrin. Presumably in nature amyrins are derived from squalene 2,3-oxide (13).

---

9. Involves the intermediate formation of an isocyanate.which adds on MeOH.

10. Van Tamelen,E.E.; Seiler,M.P.; Wierenga,W. J. Am. Chem. Soc., 1972, 94, 8229.

11. For related biogenetic-type total syntheses, see Van Tamelen,E.E.; Milne,G.M.; Suffness, M.I.; Rudler Chauvin,M.C.; Anderson,R.J.; Achini,R.S. J. Am. Chem. Soc., 1970, 92, 7202; van Tamelen,E.E.; Anderson,R.J. J. Am. Chem. Soc., 1972, 94, 8225; van Tamelen,E.E.; Holton,R.A.; Hopla,R.E.; Konz,W.E. J. Am. Chem. Soc., 1972, 94, 8229.

12. Brownlie,G.; Fayez,M.B.E.; Spring,F.S.; Stevenson,R.; Strachan,W.S. J.Chem.Soc,1956,1377.

13. Subsequently (S)-squalene-2,3-oxide was shown to be the exclusive precursor of β-amyrin in plant system: Barton,D.H.R. et al, Chem. Commun., 1974, 861; for biosynthesis from squalene see:Suga,K.et al, Chem. Letters, 1972, 129, 313.

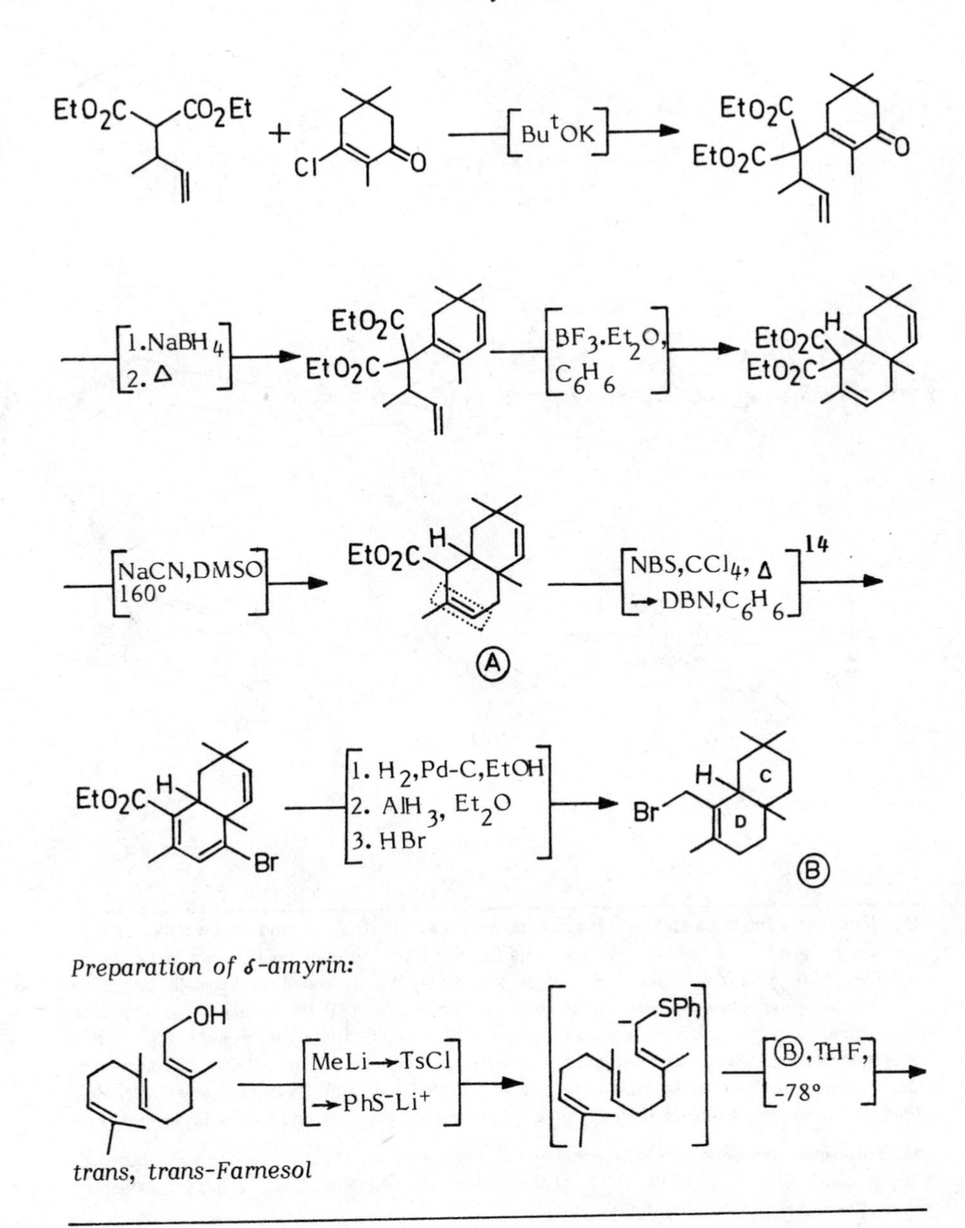

14. The β,γ-unsaturated ester (A) could not be converted directly by base or acid to the α,β isomer.

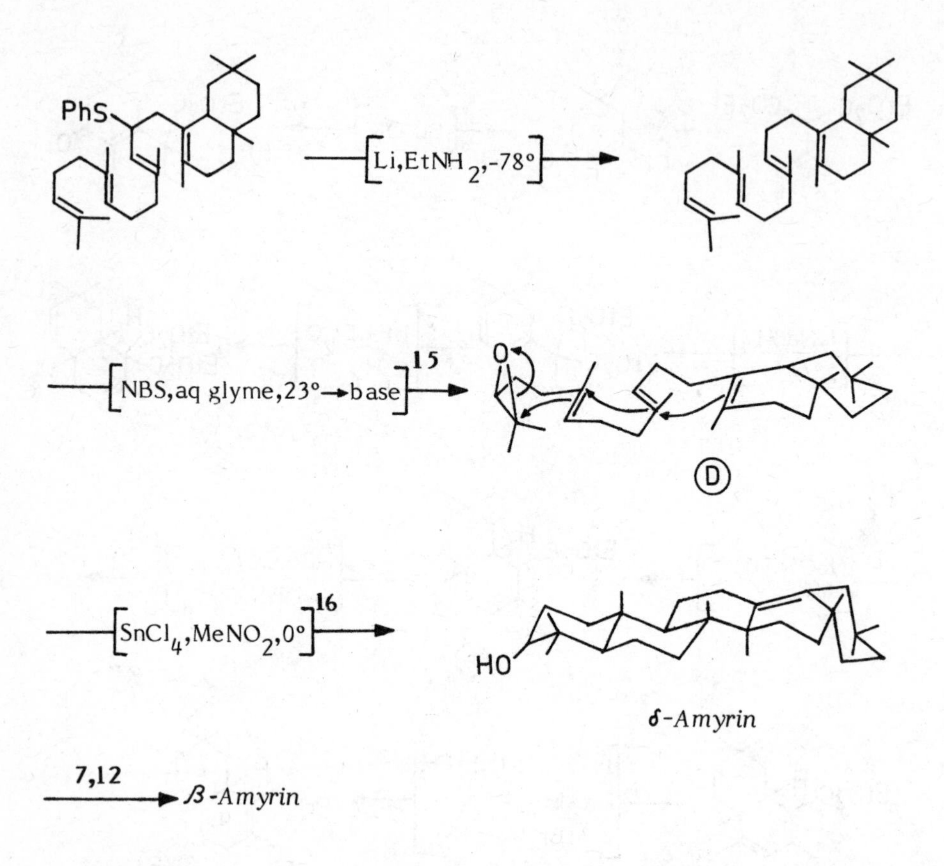

15. Selective oxidation of the terminal double bond in the starting hydrocarbon should not be possible in a solvent of low polarity because the molecule would exist in an uncoiled, fully extended state, with all of the double bonds equally vulnerable to attack by the oxidizing agent. However, in a more polar medium the molecule is expected to assume a highly coiled, compact conformation such that the system of internal hydrogen bonds would be disrupted as little as possible. Under these circumstances the internal double bonds would be sterically shielded and therefore chemically less reactive, while the terminal olefinic links would remain exposed and available for reaction.

16. For papers on definitive stereochemistry of polyene cyclization, see Stork,G.; Burgstahler,A.W. J. Am. Chem. Soc., 1955, 77, 5068; Eschenmoser,A.; Ruzicka,L.; Jeger,O.; Arigoni,D. Helv. Chim. Acta, 1955, 38, 1890.

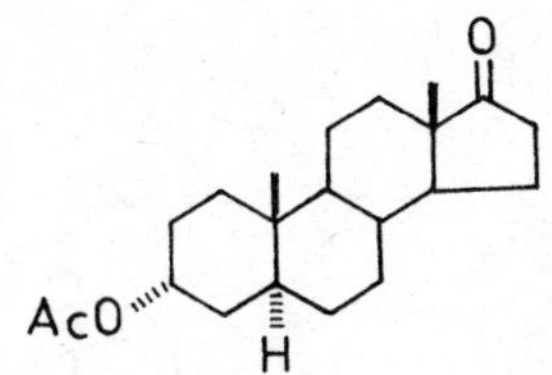

ANDROSTERONE

An emerging and useful facet of the art in organic synthesis is the deployment of distal control elements to effect the functionalisation of unactivated locations, at will, at prespecified sites. This approach would enable the realization of seemingly impossible transformations and also lead to a better understanding as to how similar changes may be carried out by enzymes in diverse biologically important processes. The pioneering work of Breslow in this domain can best be illustrated with the transformation of the readily available cholesterol to androsterone (1), a transformation earlier carried out by classical methods (2).

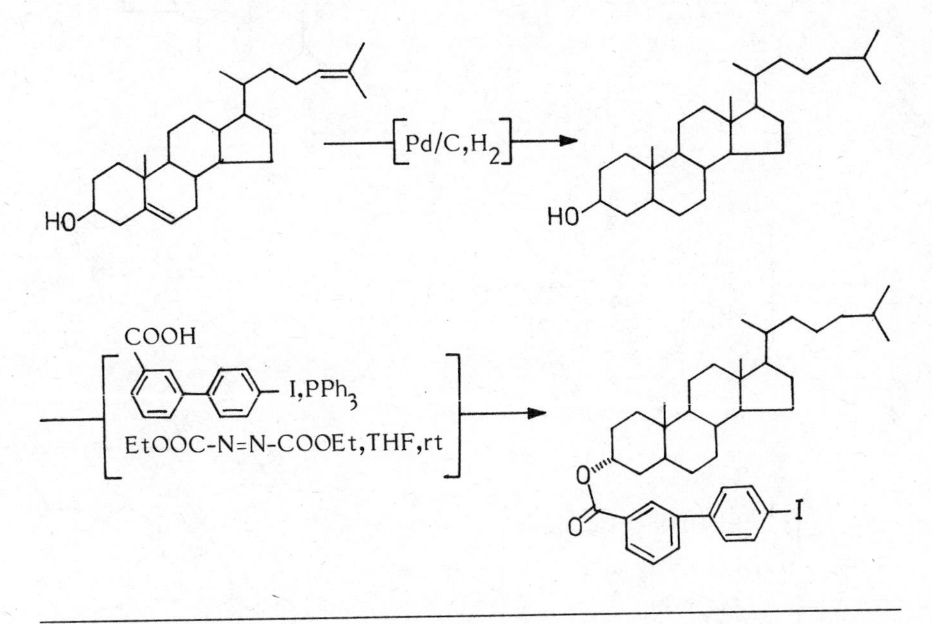

1. Breslow,R.; Corcoran,R.J.; Snider,B.B.; Doll,R.J.; Khanna,P.L.; Kaleya,R. J. Am. Chem. Soc., 1977, 99, 905.

2. Ruzicka,L.; Goldberg,M.W.; Brunger,H. Helv. Chem. Acta, 1934, 17, 1389; Ruzicka,L. Wirz,H.; Meyer,J. ibid, 1935, 18, 998; Marker,R.E.J. Am. Chem. Soc., 1935, 57, 1755.

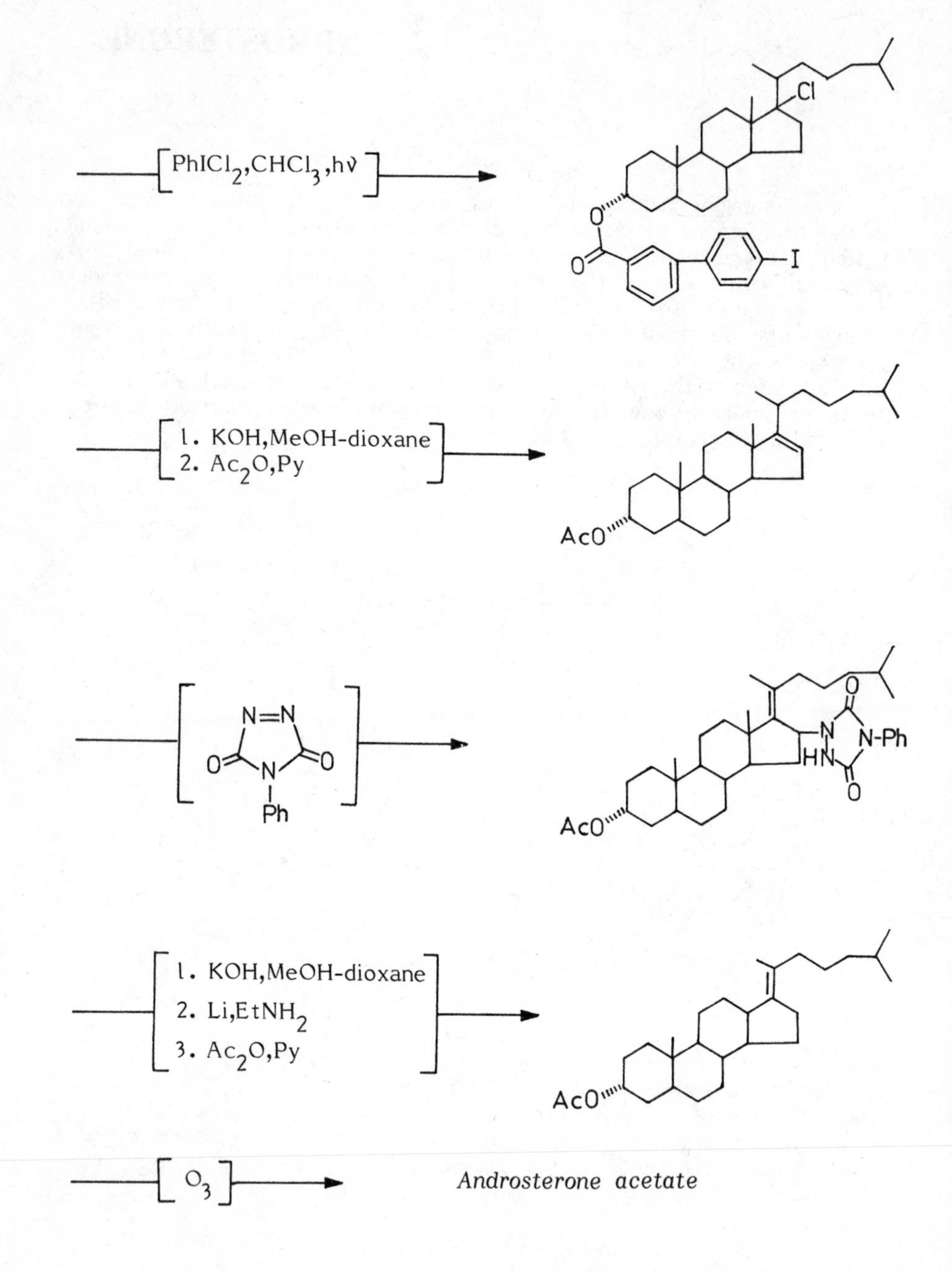

Androsterone acetate

ANNULENES

Annulenes (1) are systems that are circumscribed by a periphery. The resulting annular space can be either void or contain a variety of atoms or substructures to provide viability for the π array, without becoming part of the aromatic system. There is evidence from NMR spectroscopy that annulenes having 4n+2 π electrons do possess a ring current whilst in 4n π electron systems there is no ring current (2).

Sondheimer and co-workers made a series of annulenes, from [14]-annulene onwards, and examined their properties in the light of Huckel rule (2,3). The general principle followed in the synthesis of these compounds was the cyclization of 1,ω-acetylenes by oxidative coupling, followed by base catalyzed prototropic rearrangement of the linear 1,5-enynes to conjugated ene-ynes, and catalytic reduction of the latter over Lindlar catalyst to the annulenes. In general these compounds were found to be unstable to prolonged keeping.

[14]Annulene

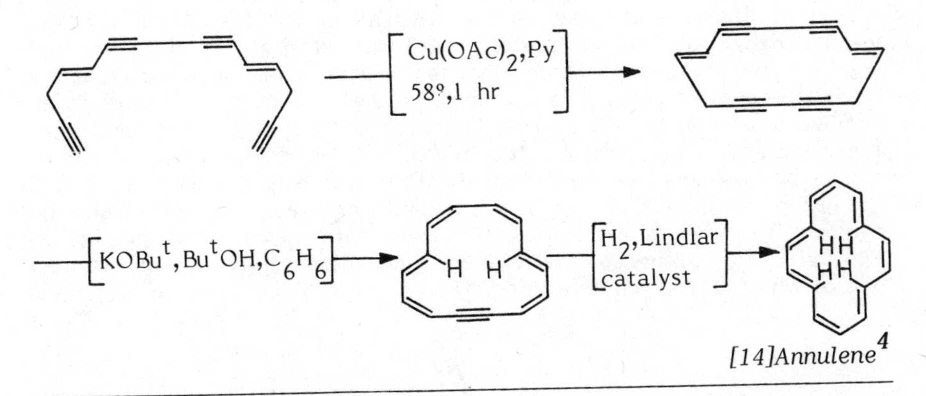

[14]Annulene[4]

---

1. The name annulenes has been given to completely conjugated cyclic polyolefins. Individual compounds are described by adding a number in square brackets indicating the number of atoms in the ring.
2. Sondheimer,F. Pure & Applied Chem., 1963, 7, 3639.
3. Sondheimer,F.; Calder,I.C.; Elix,J.A.; Gaoni,Y.; Garratte,P.J.; Grohmann,K.; DiMaio,G.; Mayer,J.; Sargent,M.V.; Wolovsky,R. Chem. Soc. Spl. Publn., 1967, 21, 75.
4. Sondheimer,F.; Gaoni,Y. J. Am. Chem. Soc., 1960, 82, 5765.

[18]Annulene

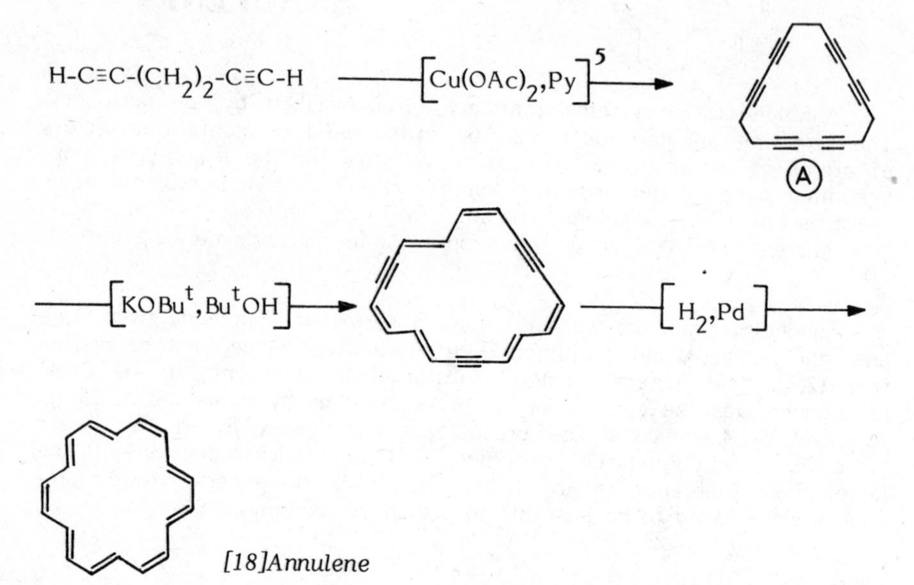

[18]Annulene

In [18]annulenes there exists a dynamic process involving the in-out change of the 18 protons as 3 degenerate sets of 6. In the case of substituted annulenes, the substituent being necessarily larger than hydrogen, would opt to stay out. In the case of a 1,2-disubstituted annulenes, it can be shown that in order to keep the substituent out, the same set of 6 protons has to be in and consequently the nmr would be temperature independent. The synthesis of such a system has been accomplished in a novel and ingenious manner from the cyclooctatetraene dimer Ⓐ (6). Cyclooctatetraene when heated at 100° gives a mixture of two dimers formed in almost equal quantity one of which is the required dimer Ⓐ.

---

5. In this oxidative coupling of 1,5-hexadiynes, in addition to the trimer (A), tetramer, pentamer, hexamer and heptamer were also formed and could be made to undergo the same reactions to form the corresponding higher annulenes.

6. Schroder,G.; Neuberg,R.; Oth,J.F.M. Angew. Int., 1972, 11, 51.

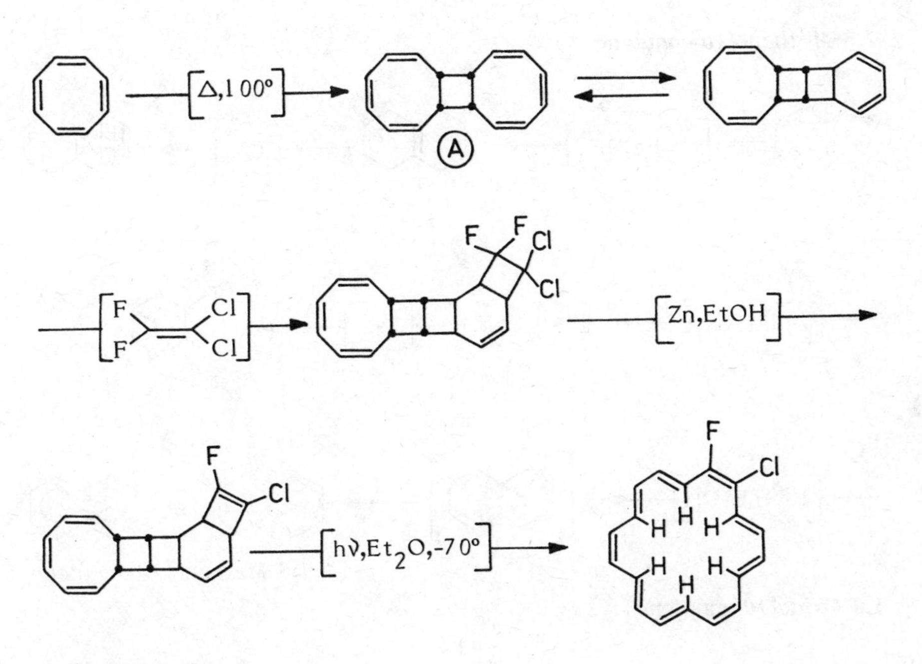

Bridged annulenes

Severe H-H interactions prevent the immediate 4n+2 homologs of benzene, starting from cyclodecapentaene, assuming coplanarity, a condition that is essential for gainful overlap and therefore aromaticity. Investigation of the possibility that the replacement of these hydrogens by bridging would alleviate the unfavourable interactions has led to the preparation of bridged annulenes; where the bridge could be CH_2/O/NH/CO, although apparently violating the "Bredt rule" by possessing double bonds at the bridge-head, these have been shown to be truly aromatic compounds on the basis of physical and chemical probes. The syntheses are based on the increased susceptibility of the central bond in 1,4,6,8-tetrahydronaphthalene to electrophilic reactions; the remaining double bonds are converted to a tetraene by sequence of bromine addition and dehydrohalogenation (7-10).

---

7. Vogel,E.; Roth,H.D. Angew. Chem. Int.Ed, 1964, 3, 228; Vogel,E.; Boll,W.A. ibid, 1964, 3, 642.

8. Vogel,E.; Biskup,M.; Pretzer,W.; Boll.W.A. Angew. Chem. Int., 1964, 3, 642; Sondheimer,F. Shani,A. J. Am. Chem. Soc., 1964, 86, 3168.

9. Vogel,E.; Pretzer,W.; Boll,W.A. Tetrahedron Lett., 1965, 3613.

10. Homologs corresponding to cyclopentadiene anion and tropylium cation have been made; for an extensive and illuminating review of the subject, see: Vogel,E. Aromaticity, Chem. Soc. Spl. Publn., 1967, 21, 113.

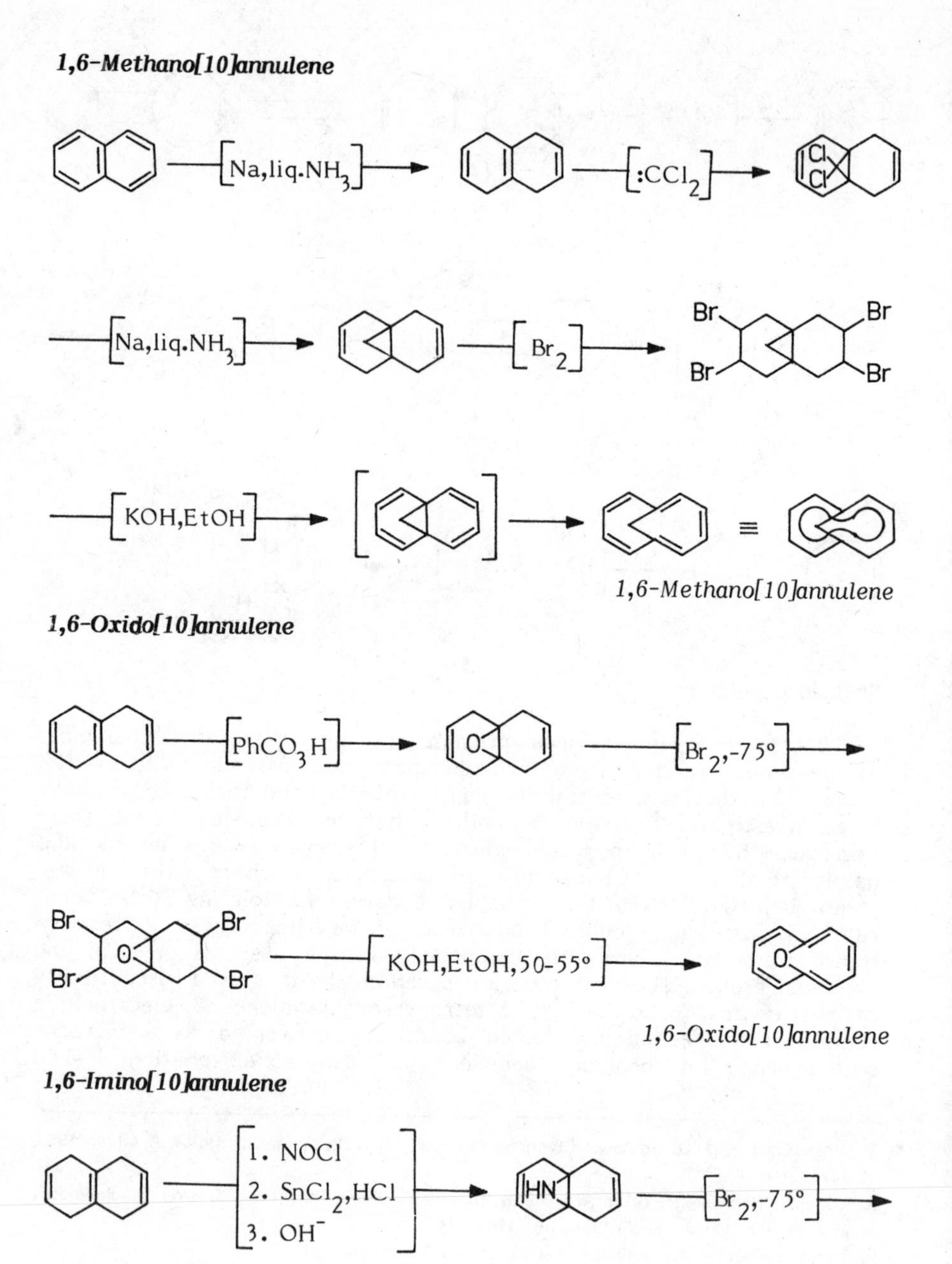
1,6-Methano[10]annulene
Na,liq.NH3
:CCl2
Cl
Cl
Na,liq.NH3
Br2
Br
Br
Br
Br
KOH,EtOH
≡
1,6-Methano[10]annulene
1,6-Oxido[10]annulene
PhCO3H
O
Br2,-75°
Br
Br
Br
Br
O
KOH,EtOH,50-55°
O
1,6-Oxido[10]annulene
1,6-Imino[10]annulene
1. NOCl
2. SnCl2,HCl
3. OH⁻
HN
Br2,-75°

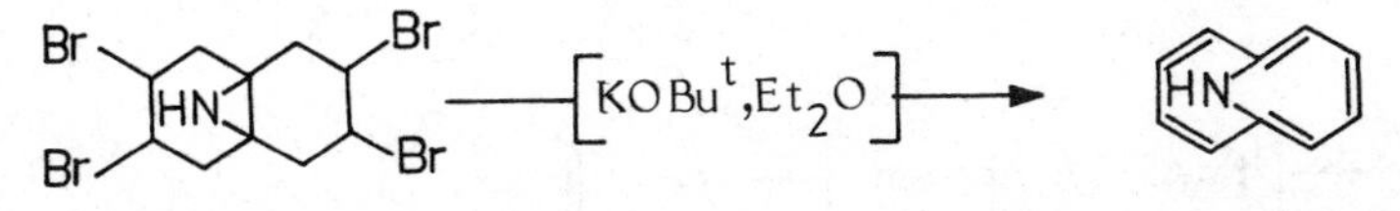

1,6-Imino[10]annulene

In another approach, the unfavourable H-H interactions are overcome by means of an ethane bridge. Molecular models show that in this two-carbon bridged annulene the peripheral atoms lie in a plane (11). The crucial tetracyclic intermediate (A) was synthesized by an elegant intramolecular oxidative coupling of a dimethoxymetacyclophane (12).

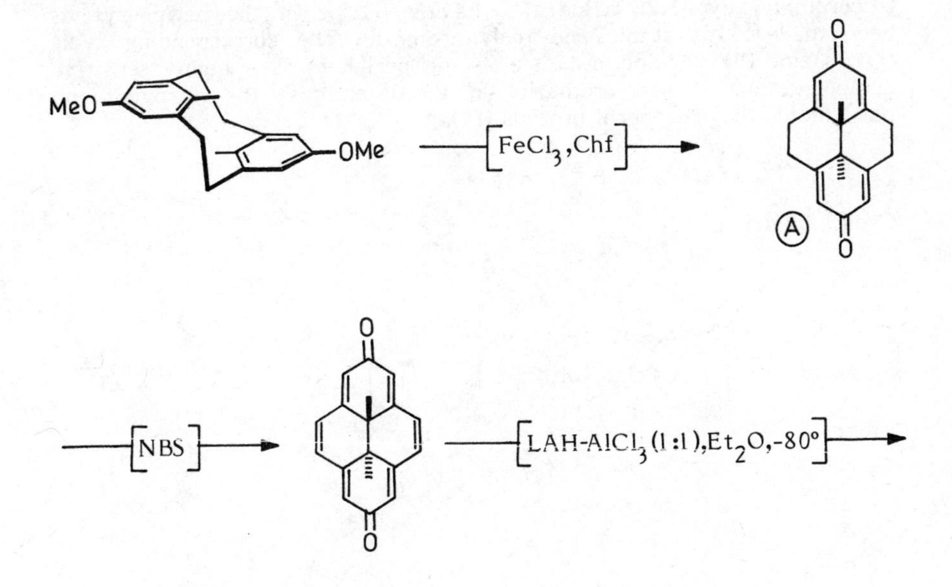

---

11. The molecule fulfills the criteria for aromaticity in terms of ring current and reactivity, thus showing that systems with cavities in the middle and having 4n+2 electrons in the periphery can be truly aromatic. The existence of ring current has been dramatically shown by the heavy shielding of the methyl groups (δ = -4.25) by the ring current. In accordance with predictions based on Woodward-Hoffman rule, the molecule undergoes conrotatory photolytic opening.

12. Boekelheide,V.; Phillips,J.B. J. Am. Chem. Soc., 1963, 85, 1545; idem, Proc. Natl. Acad. Sci. (U.S.A.), 1964, 51, 550.

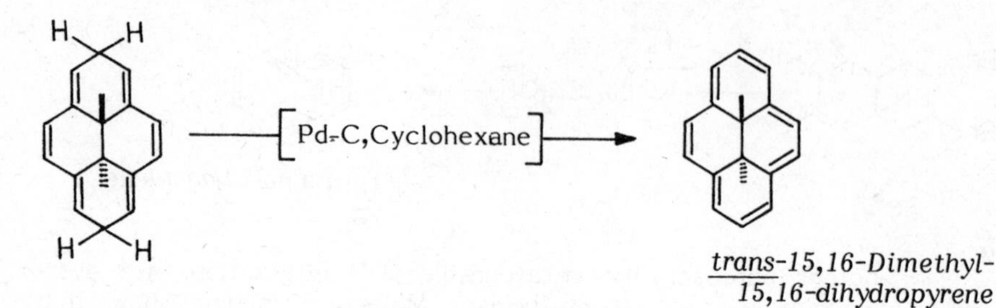

trans-15,16-Dimethyl-15,16-dihydropyrene

A particularly interesting case is where a single nitrogen serves as the bridge, without the lone pair being a part of the periphery. Accordingly, cycl[3.2.2]azine (I) having 10 π e in the periphery has been made (13) is stable and truly aromatic. The corresponding cycl-[3.3.3]azine (II) having a 12 π e in the periphery is a highly sensitive compound and is non aromatic. In the later case the nitrogen lone pair shields the peripheral protons (14).

Cycl[3.2.2]azine

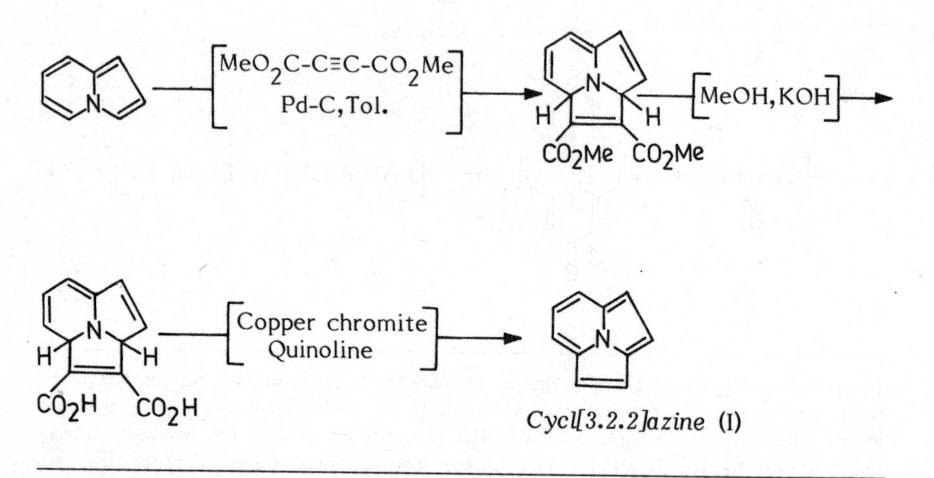

Cycl[3.2.2]azine (I)

---

13. Galbraith,A.; Small,T.; Barnes,R.A.; Boekelheide,V. J. Am. Chem. Soc., 1961, 83, 453.
14. Farquhar,D.; Leaver,D. Chem. Commun., 1969, 24.

Cycl[3.3.3]azine

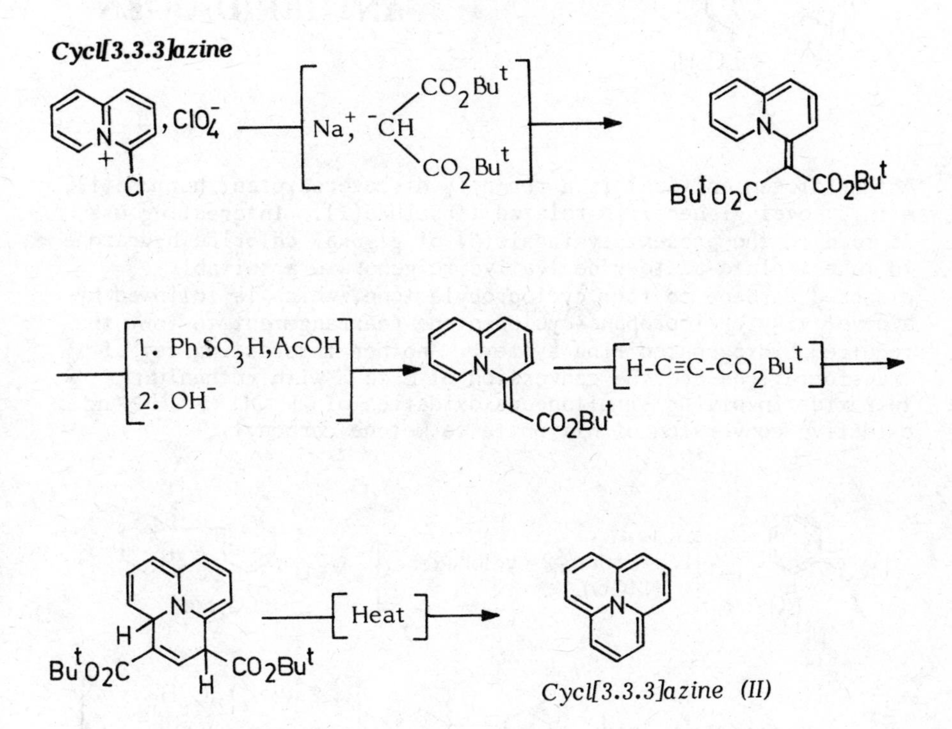

Cycl[3.3.3]azine (II)

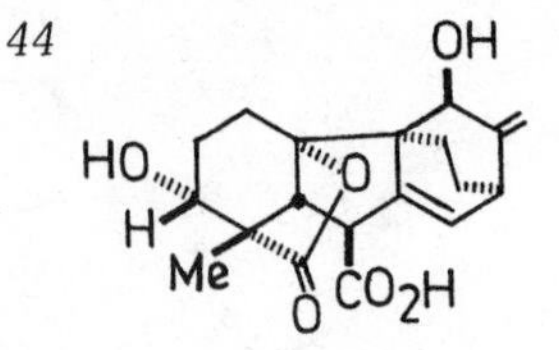

ANTHERIDIOGEN

Antheridiogen An (Aan) is a recently discovered plant hormone(1), with a novel gibberellin-related structure(2). Interesting use is made in the present synthesis(3) of glyoxal chloride hydrazone to make a diazo-acetoxy derivative to generate a suitably disposed carbene to form cyclopropylactone, which is followed by a novel vinylcyclopropane-cyclopentene rearrangement to form the required hydrogenated ring system. Another interesting set of transformations are the conversion of B to C with ruthenium tetroxide involving simultaneous oxidation of CH_2OH to CHO and oxidative conversion of nitronate to ketone carbonyl.

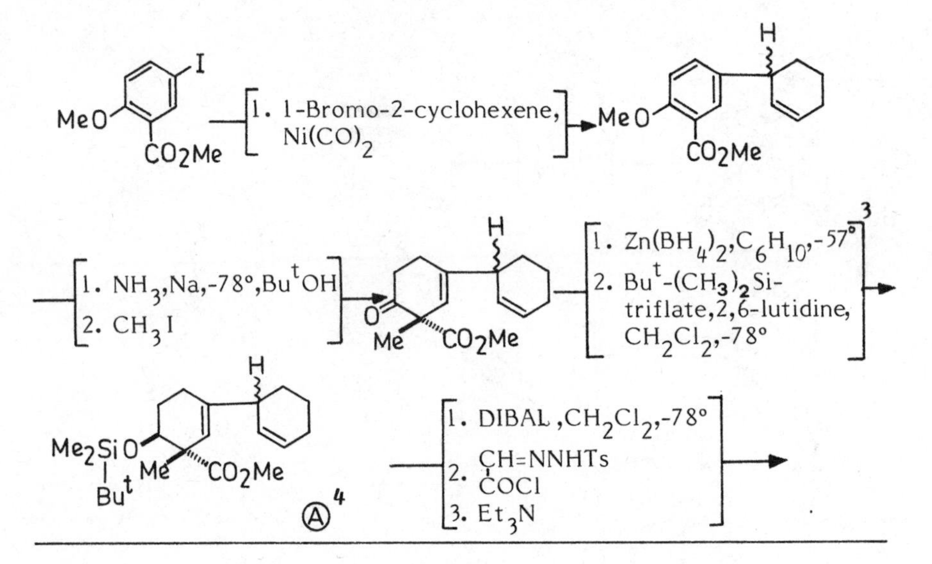

1. Naf,U,Nakanishi,K.,Endo,M.;Bot.Rev., 1975, 41, 315.

2. Nakanishi,K.,Endo,M.,Naf,U.,Johnson,L.F.,J.Am.Chem.Soc., 1971, 93, 5579.

3. Corey,E.J. and Myers,A.G.,J.Am.Chem.Soc., 1985, 107, 5574-5576.

4. Relative stereochemistry at the adjacent stereocenters in (A) is indicated by expected hydride attack at the less-screened face of the keto group in a Zn-related β-Keto ester.

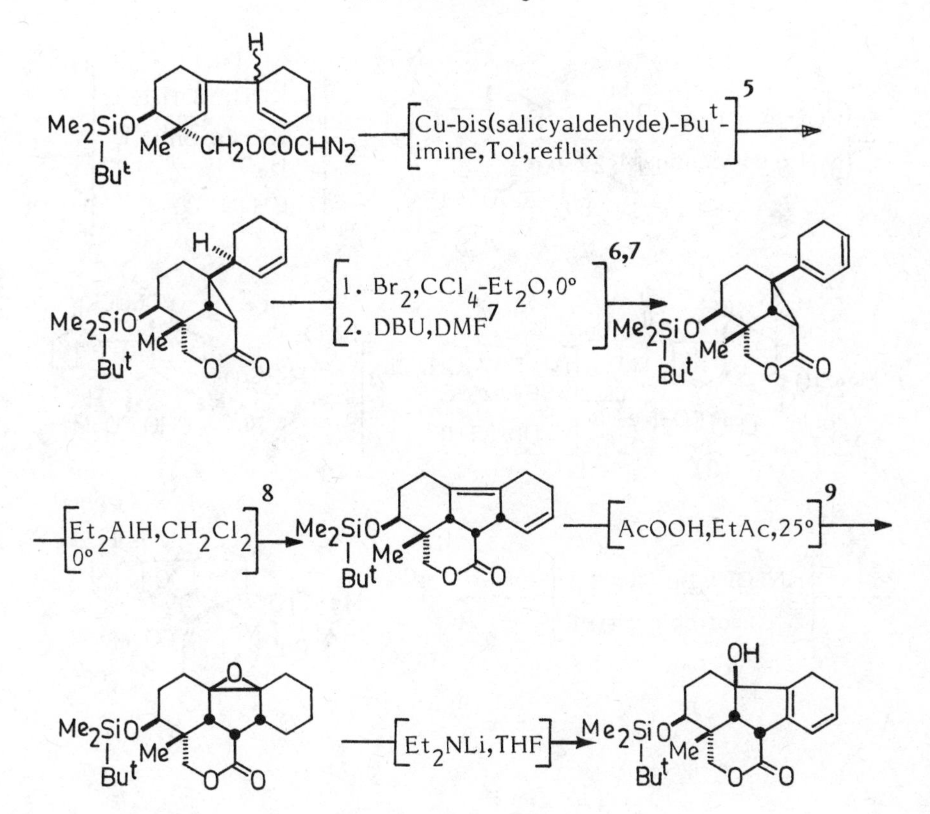

---

5. This internal carbenoid addition generates the cyclopropyl lactone stereospecifically.

6. This reaction formed the four isomeric trans-dibromides, and the mixture was used for next reaction.

7. DBU converted two of the dibromides to diene and left the others unchanged, which could be converted to the starting material by Zn-AcOH.

8. This is a novel version of the vinyl cyclopropane-cyclopentene rearrangement.

9. Epoxidation takes place from the less screened face of the double bond.

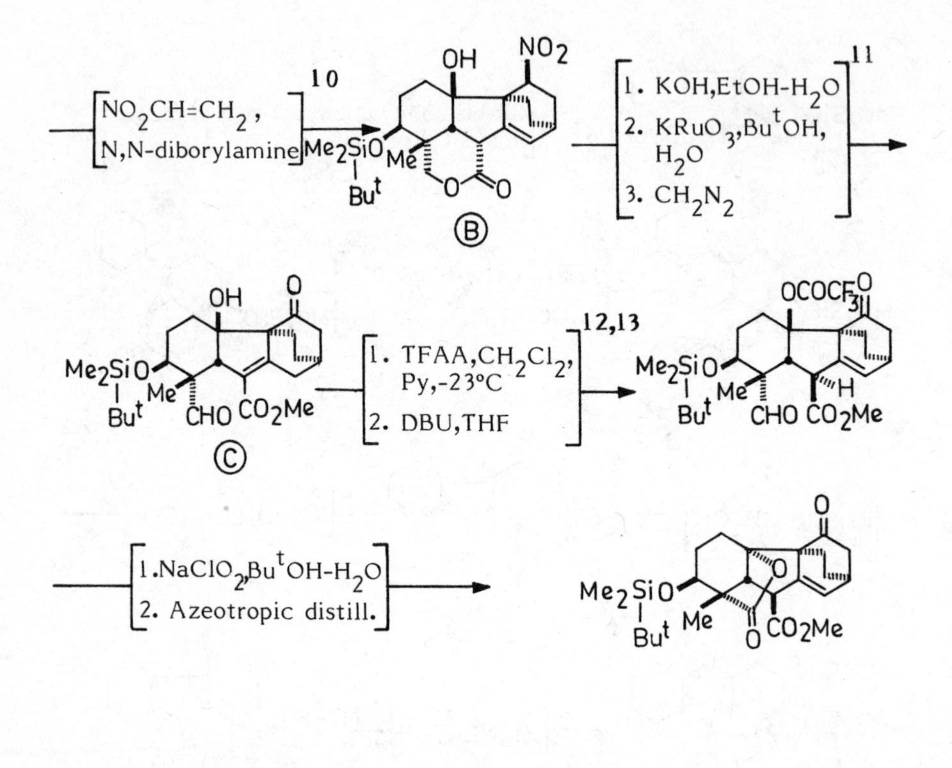

10. This reaction proceeds with orientational- and stereo-specificity.

11. This conversion was unusual for the number of structural changes which included olefinic transposition, δ-lactone hydrolysis, RuO_4 oxidation of CH_2OH to CHO and oxidative Nef conversion by Ru(VI) oxidation.

12. In addition to trifluoracetylation of angular hydroxyl concomitant $\alpha,\beta \rightarrow \beta,\gamma$ migration of the double bond.

13. This led to isomerisation to the more stable β-(methoxycarbonyl)epimer.

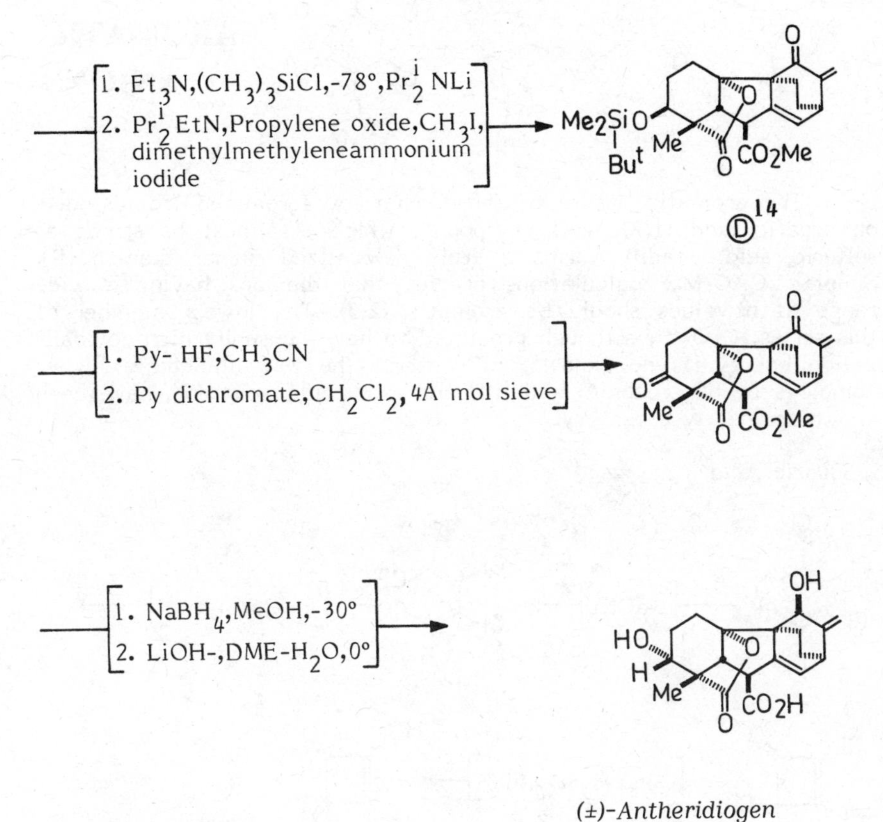

(±)-Antheridiogen

---

14. Reduction of Ⓓ with $NaBH_4$ in MeOH gave stereo-selectively the β-alcohol, which on desil ylation yielded the product with stereochemistry assigned to antheridiogen, but its nmr spectrum proved to be very different from that reported for the latter and therefore its 3-epimer was then synthesised.

AROMATIC ANIONS

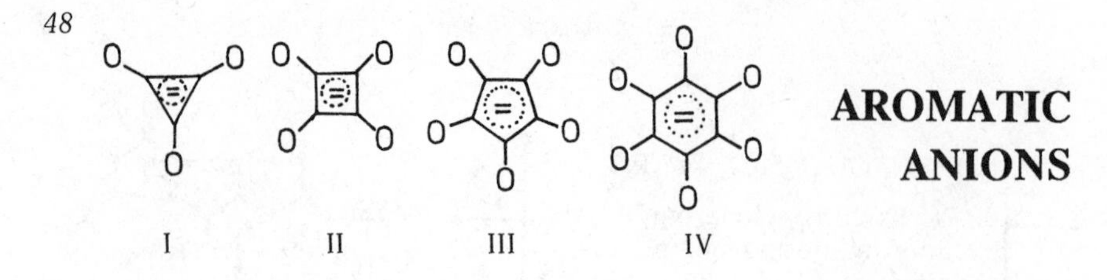

The aromatic nature of these anions was realized from studies on squaric acid (1,2). This compound, which is almost as strong as sulfuric acid, readily forms a truly delocalized planar dianion [II]. Simple LCAO-MO calculations predict that dianions having a wide range of n values should be aromatic (2,3). The lowest member of this series, $C_3O_3^{-2}$, although predicted to have unusually high delocalization energy, is not known; in contrast the two immediate higher homologs of II, croconic acid [III] and rhodizonic acid [IV], have been known for nearly a century.

Squaric Acid[4]

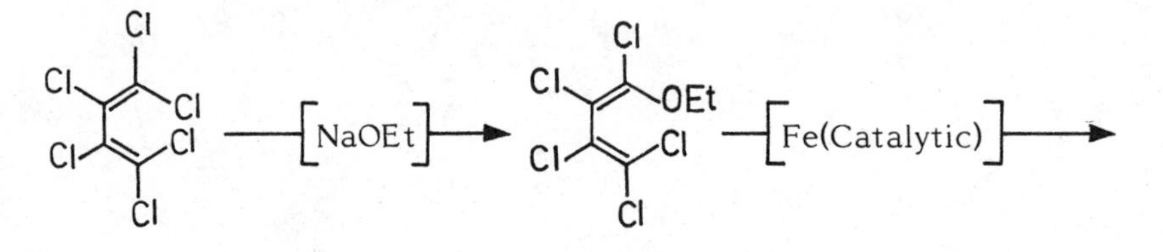

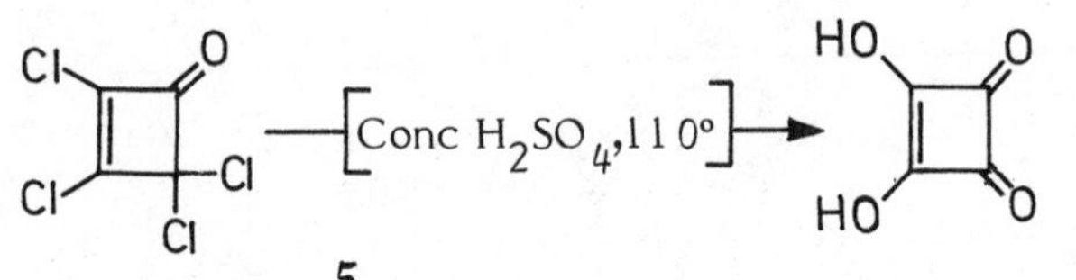

Squaric Acid

Croconic Acid[5]

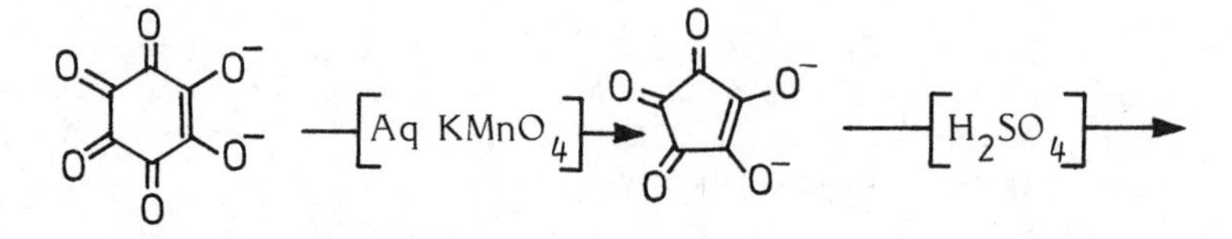

---

1. Maahs,G, Hegenberg,P., Angew. Chem. Internat. Ed., 1966, 5, 888.

2. West,R., Powell,D.L., J. Am. Chem. Soc., 1963, 85, 2577.

3. LCAO-MO calculations show that at least in the case of small rings the -2 ions are more stable than those having smaller or larger charge.

4. Maahs,G., Annalen, 1965, 686, 55.

5. Yamada,K., Hirata,Y., Bull. Chem. Soc. (Japan), 1958, 31, 551.

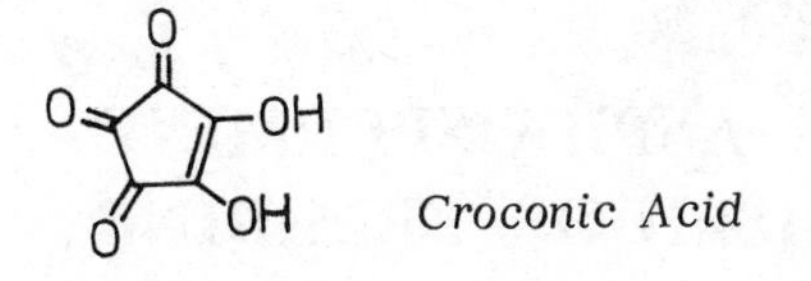

Croconic Acid

Rhodizonic Acid[6]

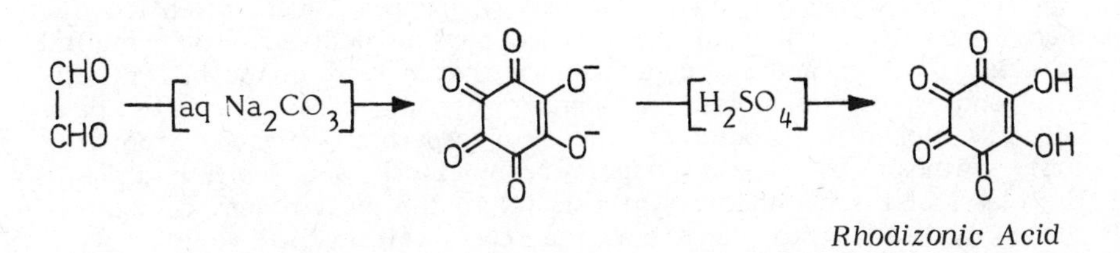

Rhodizonic Acid

An ingenious transformation of diethyl squarate to diethyl dreieck-saure (diethyl three cornered acid) has been accomplished, although its further transformation to free acid could not be achieved (7).

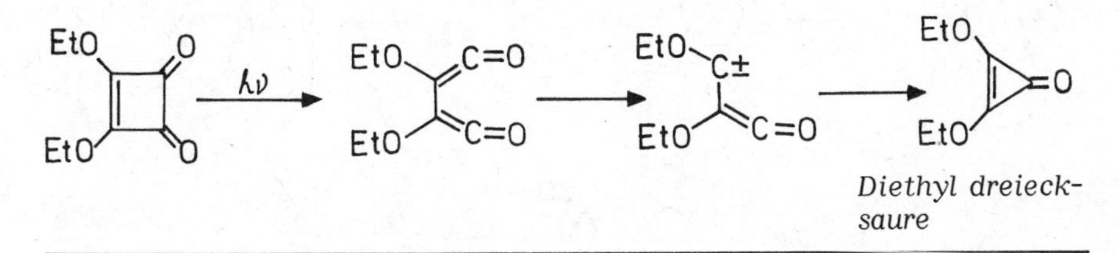

Diethyl dreieck-saure

---

6. Homolka,B., Ber., 1921, 54, 1393.

7. Dehmlow,E.V., Tetrahedron Lett., 1972, 1271.

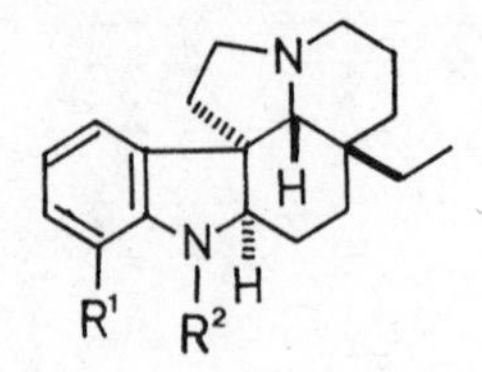

ASPIDOSPERMINE
ASPIDOSPERMIDINE

The Aspidosperma alkaloids have attracted much attention for many years on account of the marked physiological activity exhibited by many of them and the central place they occupy in the biosynthesis of monoterpene alkaloids. A number of ingenious strategies have been developed to construct the Aspidosperma skeleton, from the first synthesis of (±)-**aspidospermine** by Stork and Dolfini utilising the classical Fischer-indole synthesis (1) to the most recent quinodimethane cyclisation to construct the core tetracyclic skeleton (12).

Stork and Dolfini (1) in their synthesis of aspidospermine, utilized the arylhydrazone-carbazolenine rearrangement on a ring C-D-E tricyclic ketone (B) for construction of the characteristic pentacyclic indoline nucleus of aspidospermine.

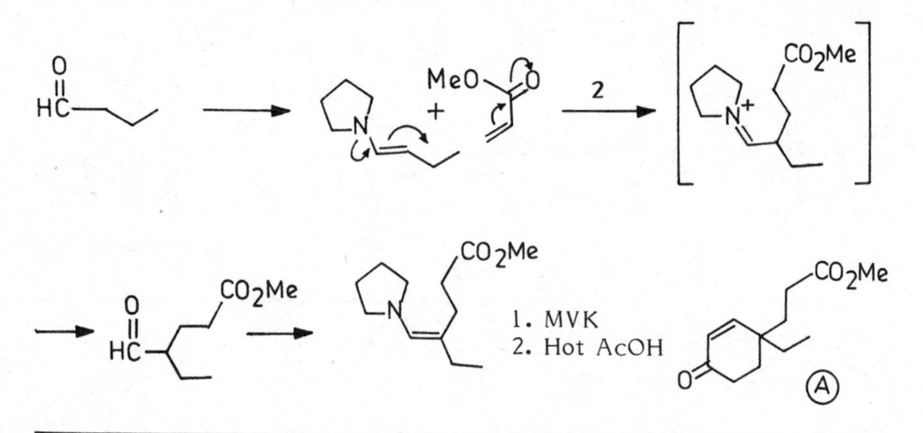

1. Stork,G.; Dolfini,J.E. J. Am. Chem. Soc., 1963, 85, 2872.
2. Enamine alkylation procedure, Stork,G.; Brizzolara,A.; Landesman,H.; Szmuszkovicz,J.; Terrell,R. J. Am. Chem. Soc., 1963, 85, 207.
3. Reaction of (A) with ammonia gave a mixture of cis and trans keto lactams (i) which formed the unwanted linear system (ii) on Fischer indole synthesis, indicating that enolization at C-2 occurs away from the ring junction. Construction of ring E was therefore undertaken at this stage in order to induce enolization in the desired direction.

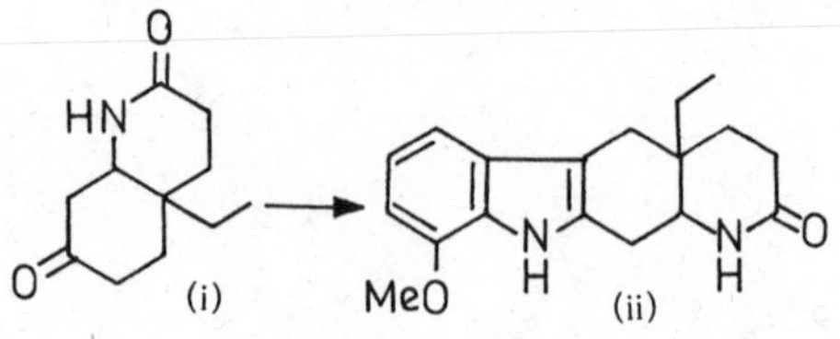

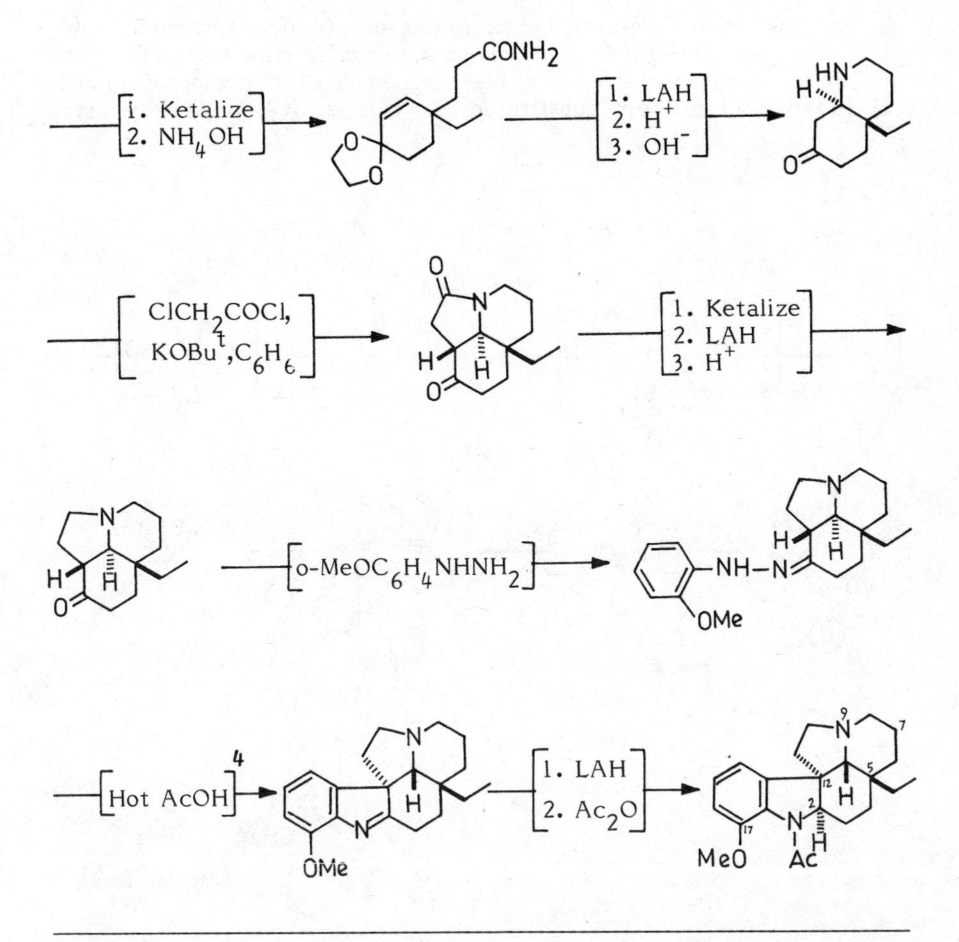

---

4. Although all four stereoisomers of (B) have been synthesized (refs.1,5 and 6) and Stork's tricyclic ketone has been shown to possess the stereochemistry depicted in (iv), the stereochemistry of these intermediates has little effect on final outcome of the synthesis. During the Fischer cyclization there is equilibration at C-12 and C-19 asymmetric centres involving a retro-Mannich reaction which results in the formation of the most stable relative arrangement of the three asymmetric centres in rings C,D and E. Therefore, in principle, all the four isomers [(i)-(iv)] are capable of conversion to the natural product.

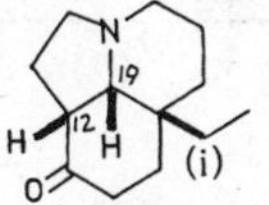
(i)

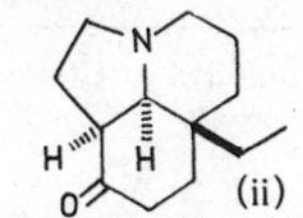
(ii)

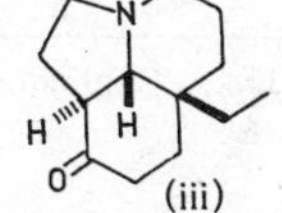
(iii)

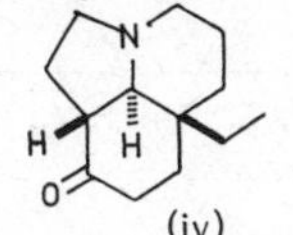
(iv)

5. Ban,Y.; Sato,Y.; Inoue,I.; Nagai,M.; Oishi,T. Terashima,M.; Yonemitsu,O.; Kanaoka,Y.; Tetrahedron Lett., 1965, 2261.

6. Kuehne,M.E; Baylia,C. Tetrahedron Lett., 1966, 1311.

Stevens et al. described the synthesis of the hydrolulidone (D) of Stork and Dolfini by a fundamentally different approach which involves acid-catalysed thermal rearrangement of a cyclopropylimine, (B) to an appropriately substituted 2-pyrroline, (C), as a key step (7).

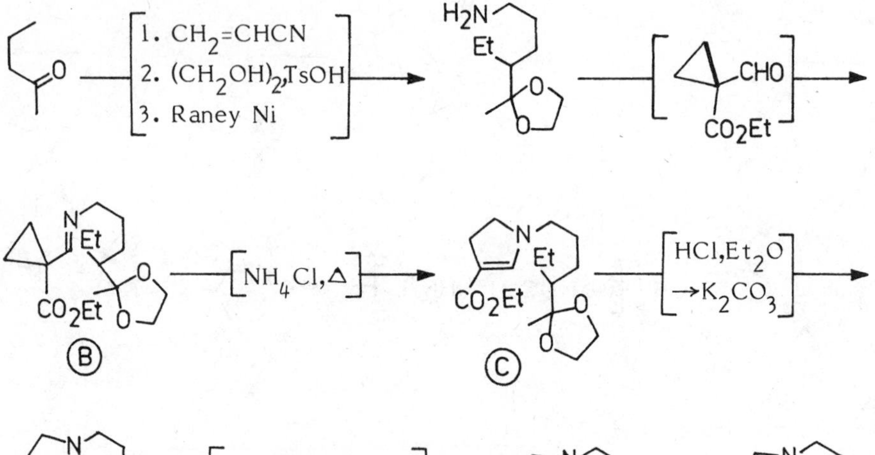

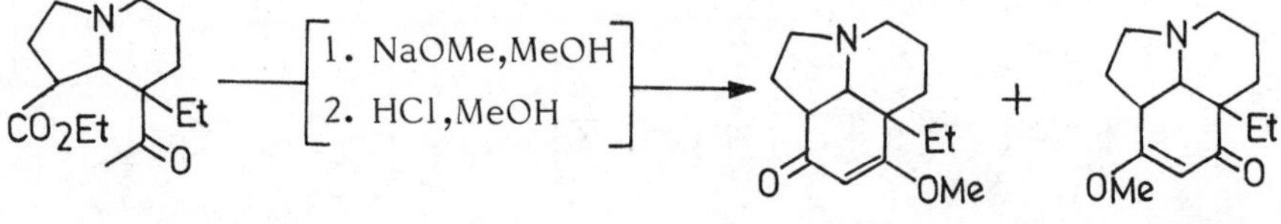

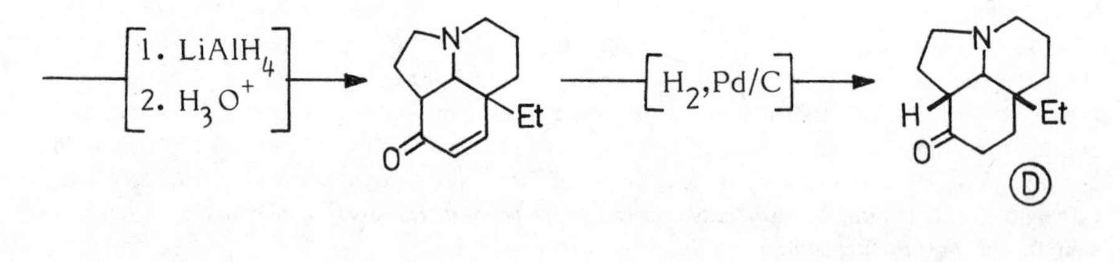

7. Stevens,R.V.; Fitzpatrick,J.M.; Kapla,M.; Zimmerman,R.L. Chem. Comm., 1971, 857.

Martin et al. have used yet another approach to the synthesis of the key hydrolulidone system by an intramolecular [4+2] cycloaddition reaction of enamides (8,9).

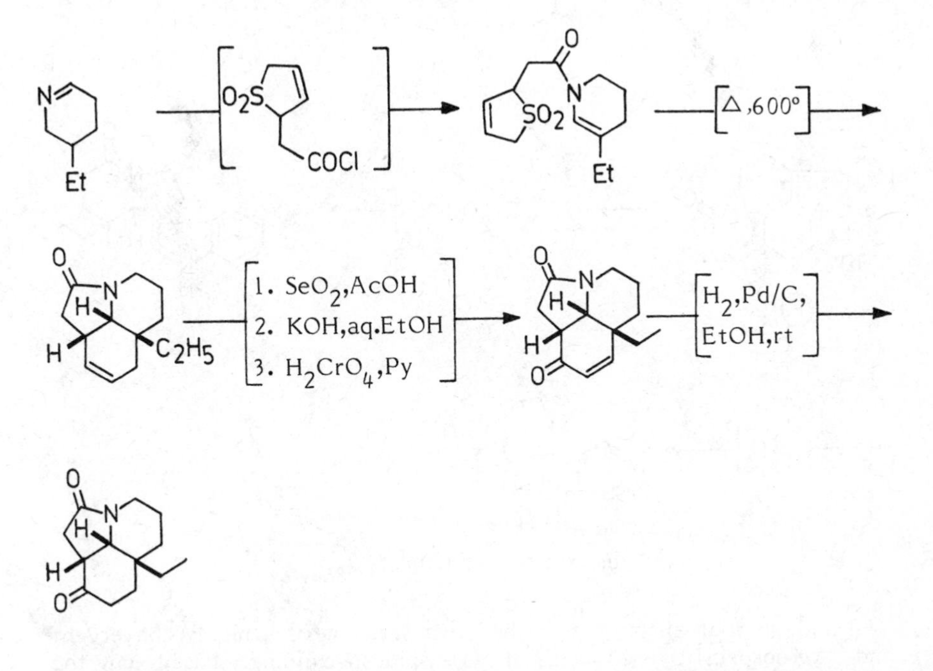

In the synthesis by Harley-Mason and Kaplan the pentacyclic system with the correct stereochemistry was obtained through a deep seated rearrangement of the octahydro-pyridocarboline intermediate (E) (10,11).

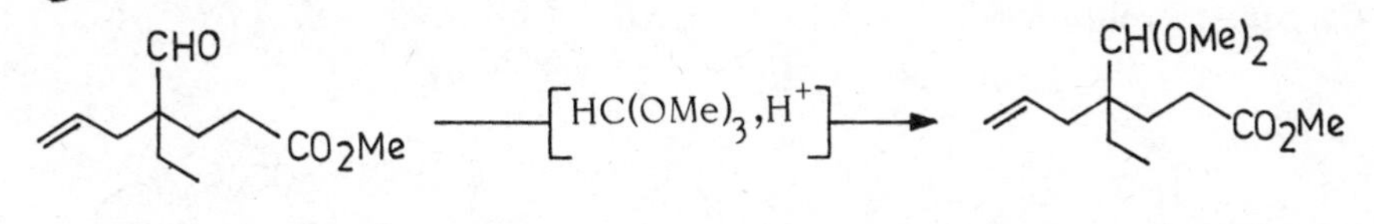

---

8. Martin,S.F.; Desai,S.R.; Phillips,G.W.; Miller,A.C. J. Am. Chem. Soc., 1980, 102, 3294.
9. For some of the other synthesis of the hydrolulidone; see: Ban,Y.; Akagi,M.; Oishi,T.; Tetrahedron Lett., 1969, 2057; Ban,Y.; Lijima,I.; Inone,I.; Akagi,M.; Oishi,T. ibid, 1969, 2067; Klioze,S.S.; Darmory,F.P. J. Org. Chem., 1975, 40, 1588.
10. Harley-Mason,J.; Kaplan,M. Chem. Comm., 1967, 915.
11. It appears that this remarkable stereospecific skeletal rearrangement is initiated by the formation of a carbonium ion at C-16 of the tetracyclic lactam (E).

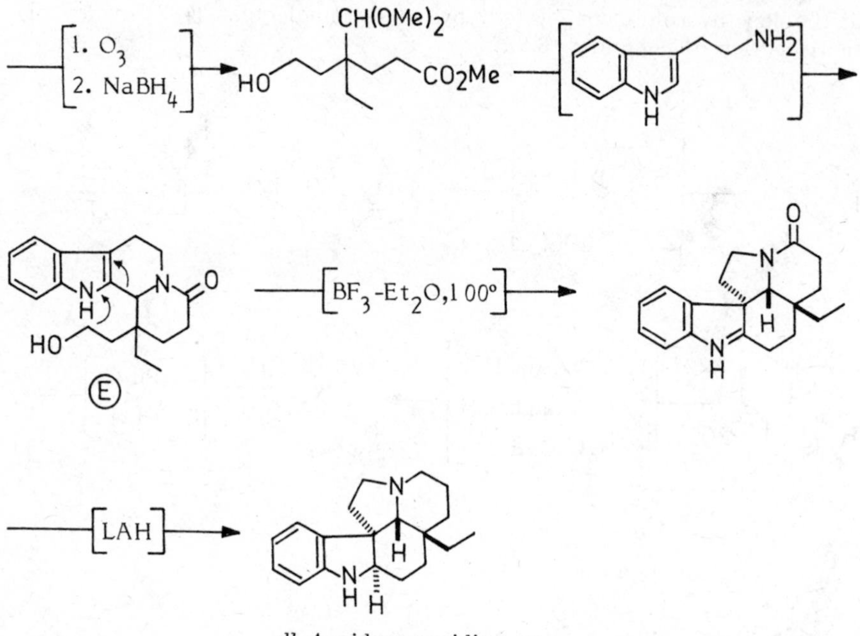

dl-Aspidospermidine

Gallagher et al. have described two versions of a highly convergent and stereospecific synthesis of (±)-aspidospermidine, based on the utility of 2,3-quinodimethane systems to construct polycyclic systems by [4+2] cycloaddition with defined stereochemistry at a number of contiguous asymmetric centres (12).

Carbamate route

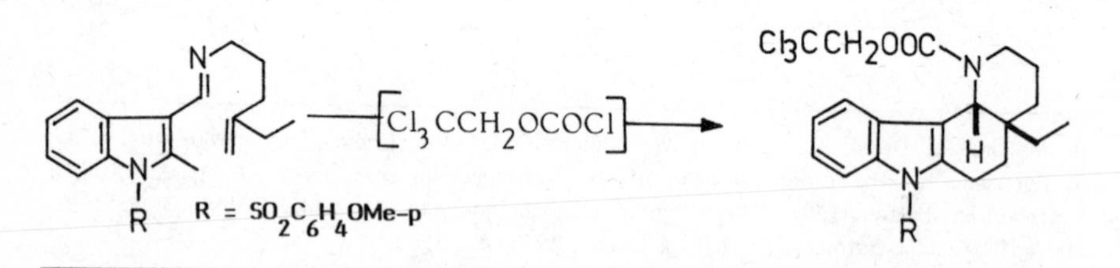

12. Gallagher,T.; Magnus,P.; Huffman,J.C. J. Am. Chem. Soc., 1982, 104, 1140; 1983, 105, 4750.

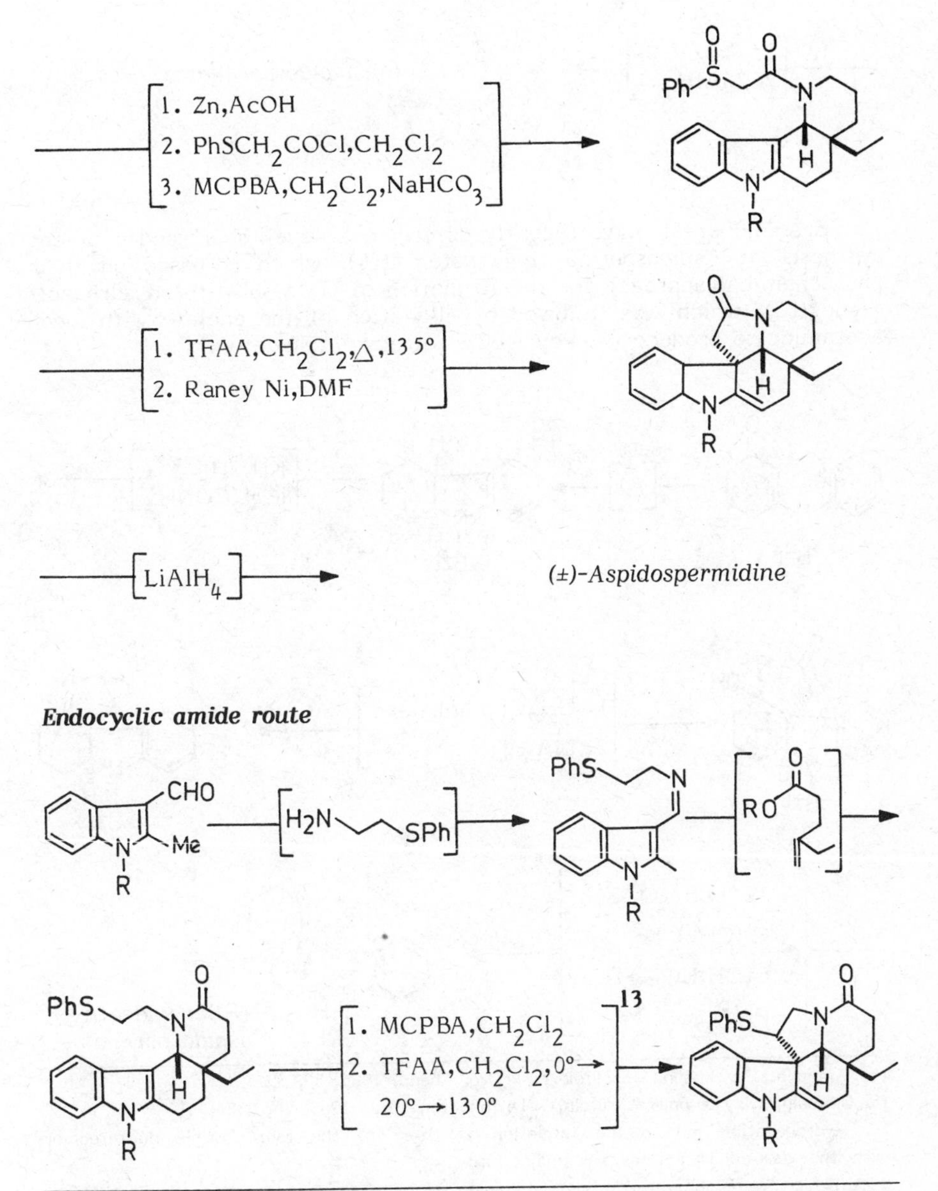

13. The assignment of stereochemistry at C-12 is based upon mechanistic consideration that align the sulfonium ion trans to the indole 2,3-double bond as shown below, and confirmed by obtaining the same product from Vincadiffermine; Gallagher,T.; Magnus,P.; Huffman,J.C. J. Am. Chem. Soc., 1983, 105, 4750.

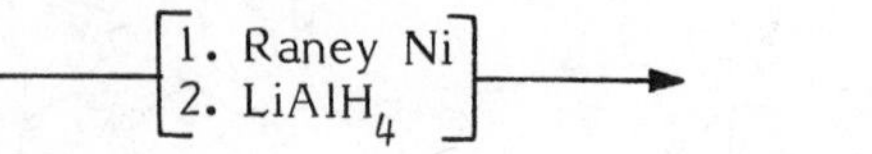

(±)-Aspidospermidine

Gramain et al. have recently described a novel, short and efficient synthesis of Aspidosperma ring system (14), which is based on their photochemical approach for the formation of C-4a-substituted carbazol-4-one (15), which was followed by alkylation of the enolate with iodoacetamide to produce the key ABCE tetracyclic skeleton.

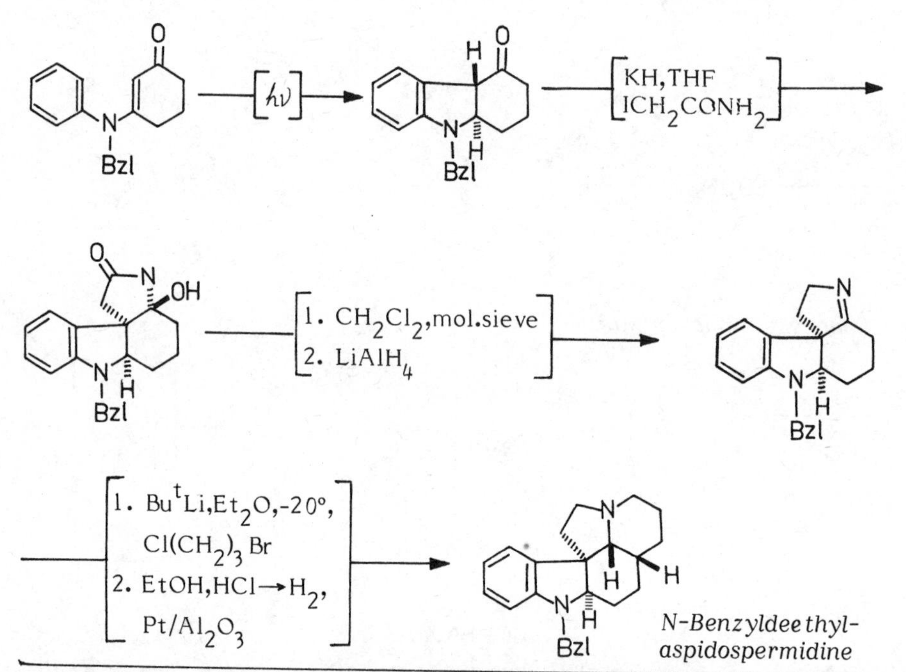

14. Gramain,J.C.; Husson,H.P.; Troin,Y. J. Org. Chem., 1985, 50, 5517.
15. Gramain,J.C.; Husson,H.P.; Troin,Y. Tetrahedron Lett., 1985, 26, 2323.
16. Hydrogenation led to the formation of three of the four possible diastereomers with the desired isomer as the major product, which were separated easily by column chromatography; the other two isomers were assigned the stereochemistry shown below on the basis of spectroscopic data.

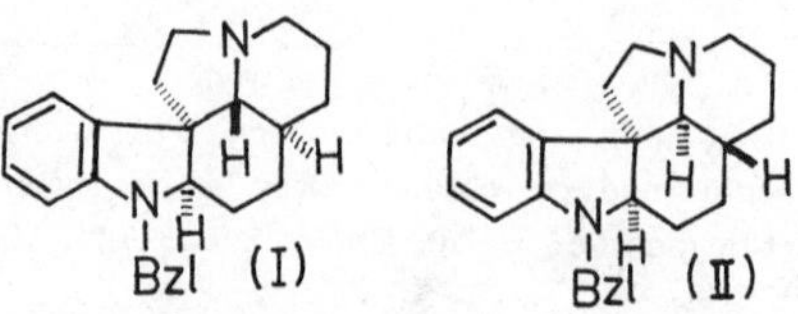

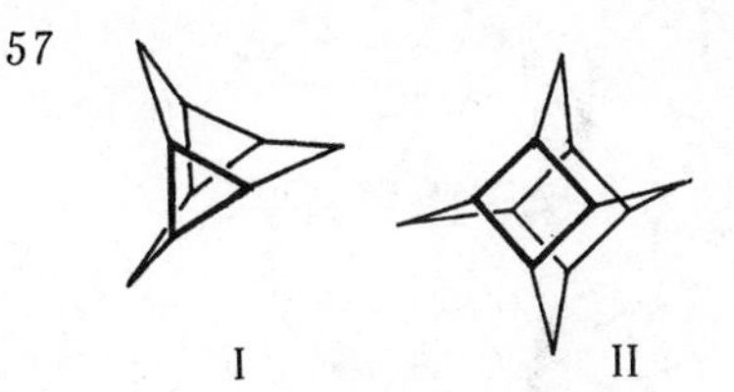

ASTERANES

Methylene bridging of parallelly stacked cyclic duplexes results in a periphery of rigidly held cyclohexane boats with a pleasing design that has been appropriately named as asteranes. The tri-asterane (I) has been prepared from cyclohexa[1,4]diene (1).

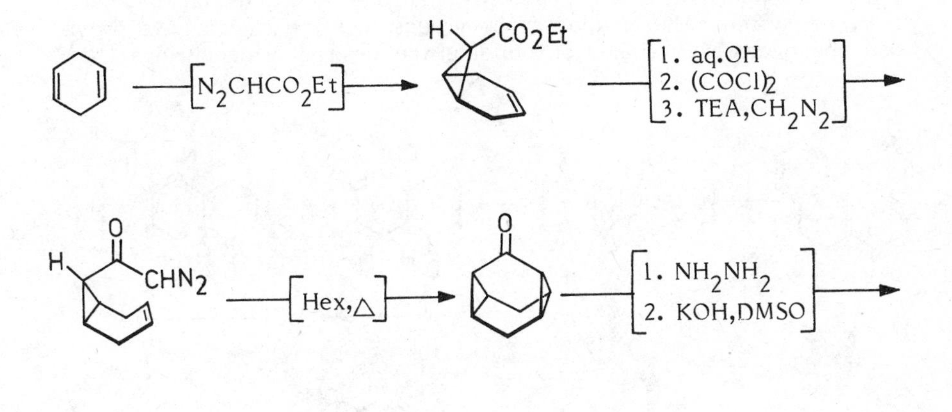

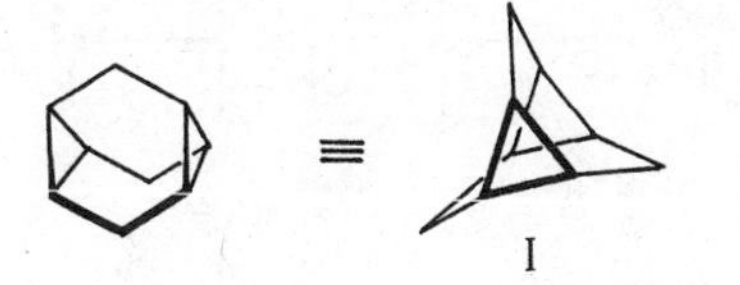

Tetra-asterane (II) has been prepared involving a key 4+4 photo-cyclo addition (2).

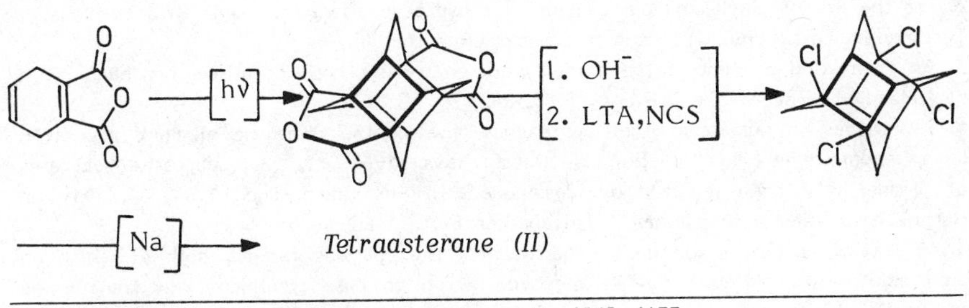

Na → *Tetraasterane (II)*

1. Biethan,U.; Gizycki,U.; Musso,H. Tetrahedron Lett., 1965, 1477.

2. Musso,H. Angew. Chem. Int. Ed., 1975, 14, 180.

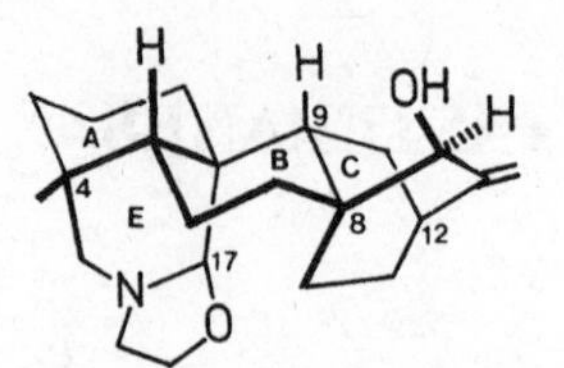

ATISINE

A major obstacle in the synthesis of diterpene alkaloids is the problem of constructing the A-E and C-D bridge ring systems, each of which have one end located at an angular position. The key to the first successful synthesis of atisine (1) was the application of a versatile new hydrocyanation procedure (2) for functionalizing the otherwise inaccessible C-10 and C-8 angular positions on the ABC tricyclic system. The angular cyano groups thus introduced have provided the necessary handle for building the desired bridged rings (3,4).

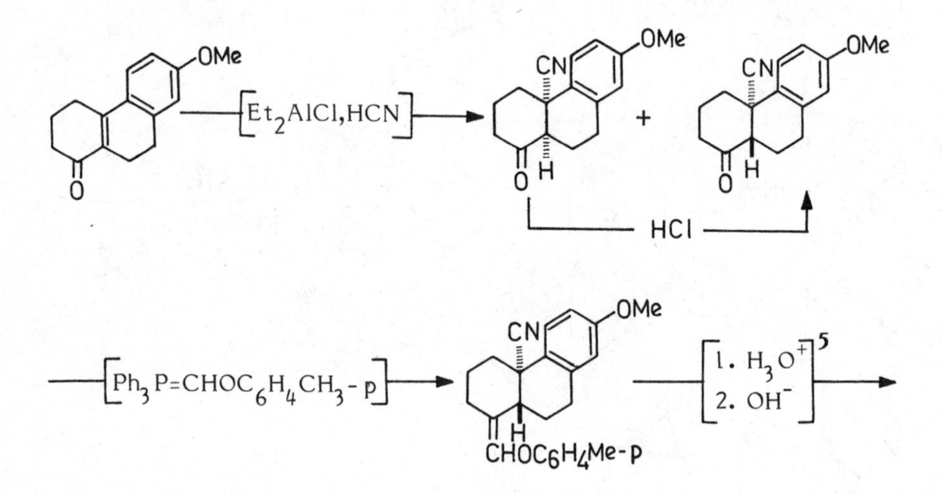

1. Nagata,W.; Sugasawa,T.; Narisada,M.; Wakabayashi,T.; Hayase,Y. J. Am. Chem. Soc., 1963, 85, 2342; 1967, 89, 1499.

2. For the use of alkylaluminium cyanides for hydrocyanation, see: Nagata,W.; Yoshioka,M. Tetrahedron Lett., 1966, 1913 and references cited therein.

3. An interesting account of the earlier work on the synthesis of atisine has been given by Ireland,R.E. Record Chem. Progr., 1963, 24, 225.

4. References to numerous other approaches towards the diterpene alkaloids are given in a recent review by Pelletier,S.W. Quart. Revs., 1967, 21, 525; successful syntheses of atisine have been reported by Masamune,S. J. Am. Chem. Soc., 1964, 86, 291 and Guthrie,R.W., Valenta,Z.; Wiesner,K. Tetrahedron Lett., 1966, 4645.

5. In Wittig reaction a mixture of geometrical isomers was formed, both of which on mild acid treatment gave the 9α-aldehyde which on base treatment was transformed to the 9β-epimer.

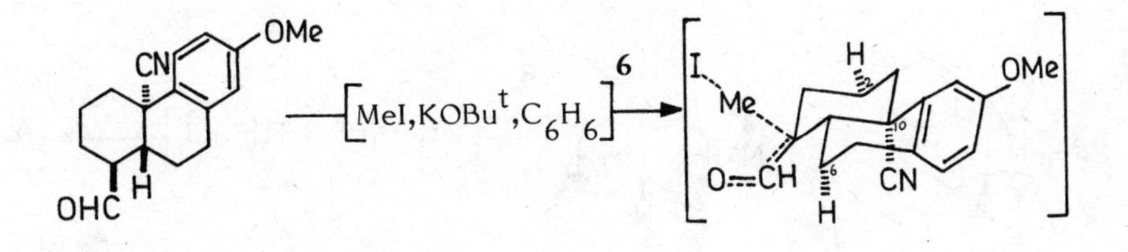

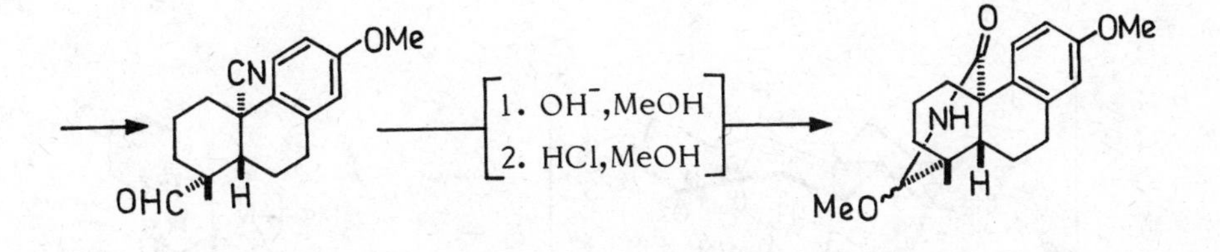

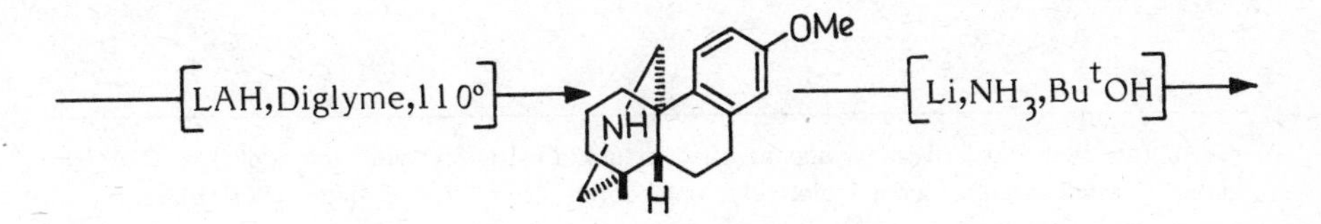

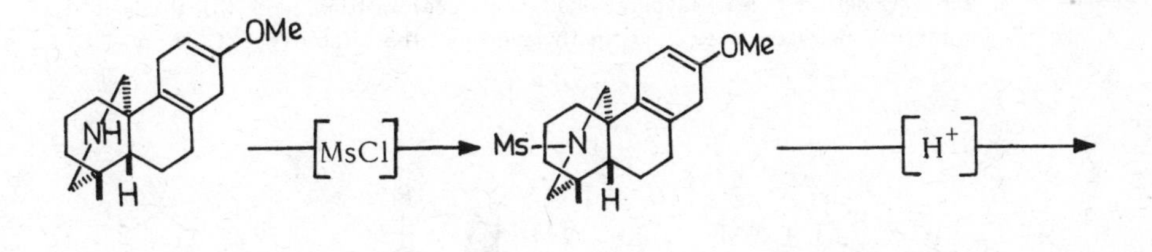

---

6. The approaching electrophile enters equatorially, avoiding axial substituents at C-2, C-6 and C-10.

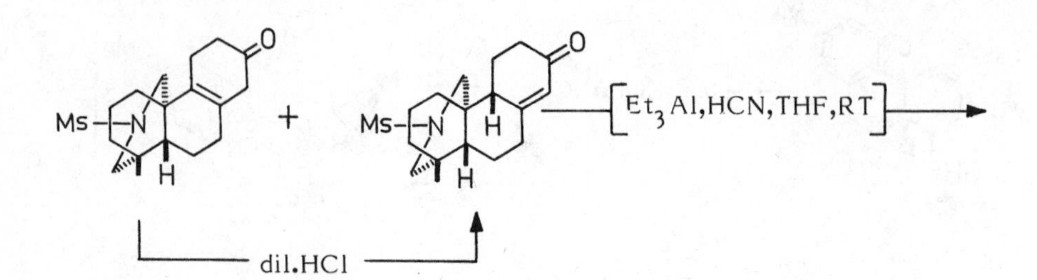

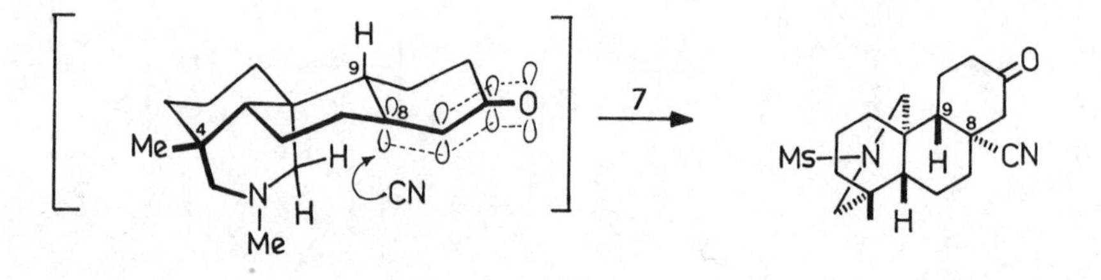

7. At the first sight it may appear that attack of the cyanide anion at the C-8 trigonal C-atom would occur relatively unhindered from the β-face [cf.(i)]resulting in a preferential formation of the B/C-cis fused product (ii). However, since we are dealing with transition states in which there must be a maximum overlap between the cyanide anion and the developing p-orbital at C-8 (i.e., C-8 is more nearly tetragonal), geometry of the transition state will resemble more the final product. In such an event two factors serve to militate against (i) as the more probable transition state: (a) ring C is in the energetically less favoured half-boat conformation, and (b) there is 1,3-diaxial interaction between the C-17 methylene and the protons at C-14 and C-12.

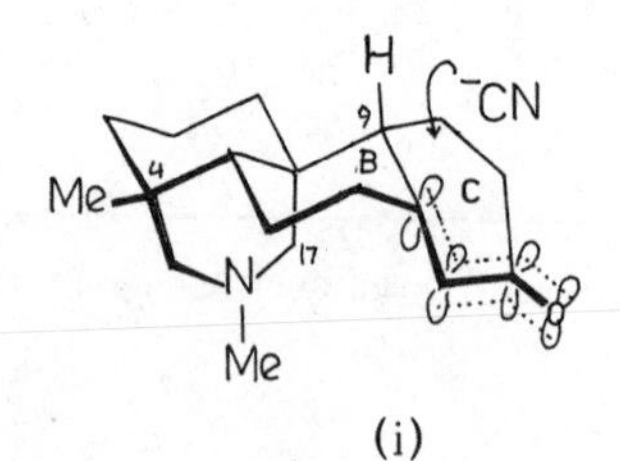

(i)

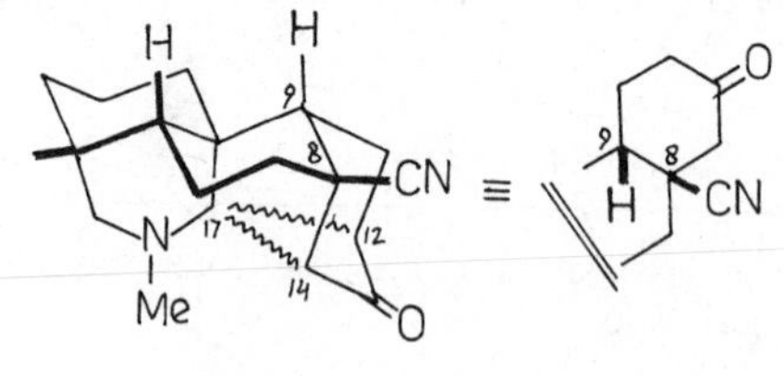

(ii)

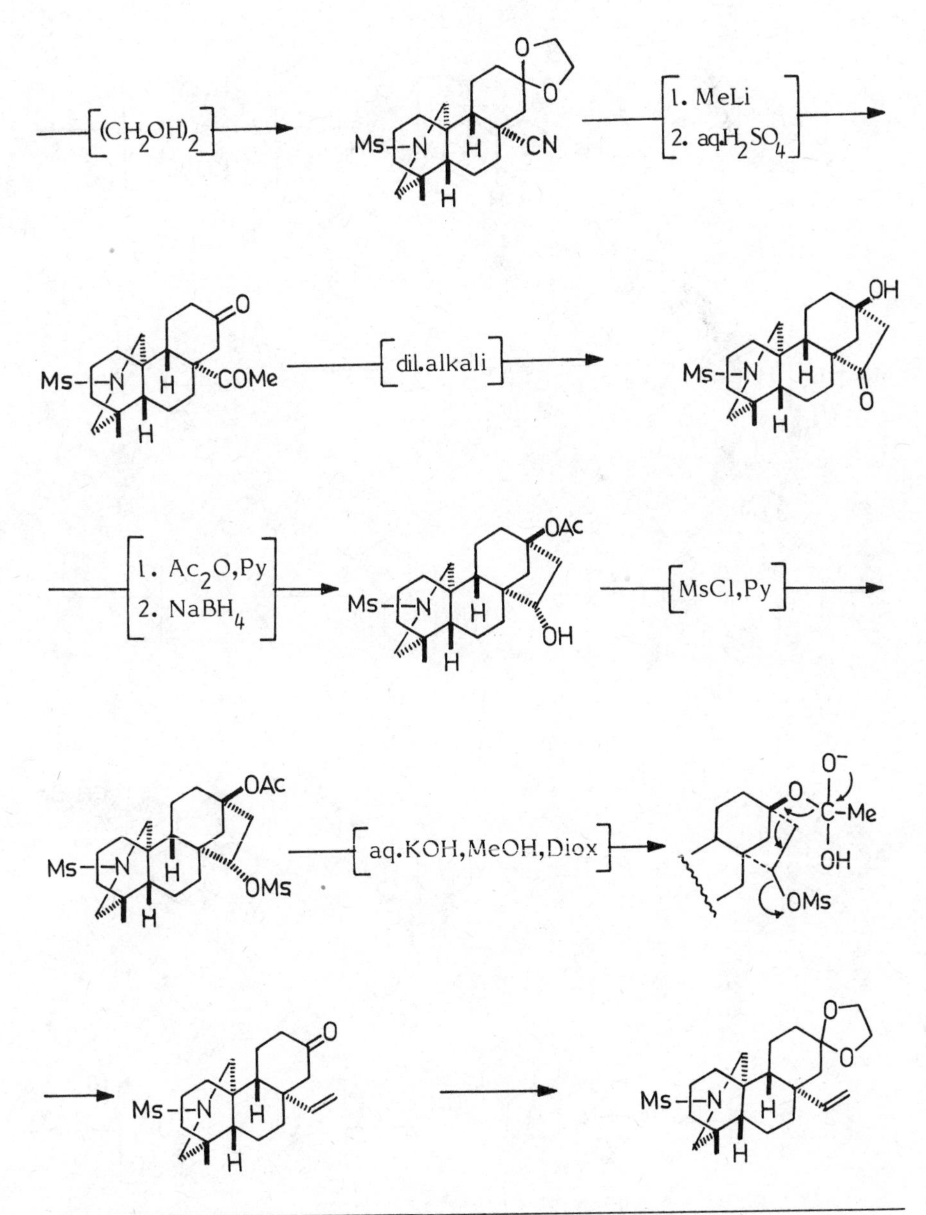

8. cf. Grob,C.A.; Schiess,P. Angew. Chem. Internat. Ed., 1967, 6, 1.

9. Employing this intermediate Nagata *et al.*(1) have also carried out the synthesis of dl-Veatchine and dl-garryine using approaches similar to that employed for atisine.

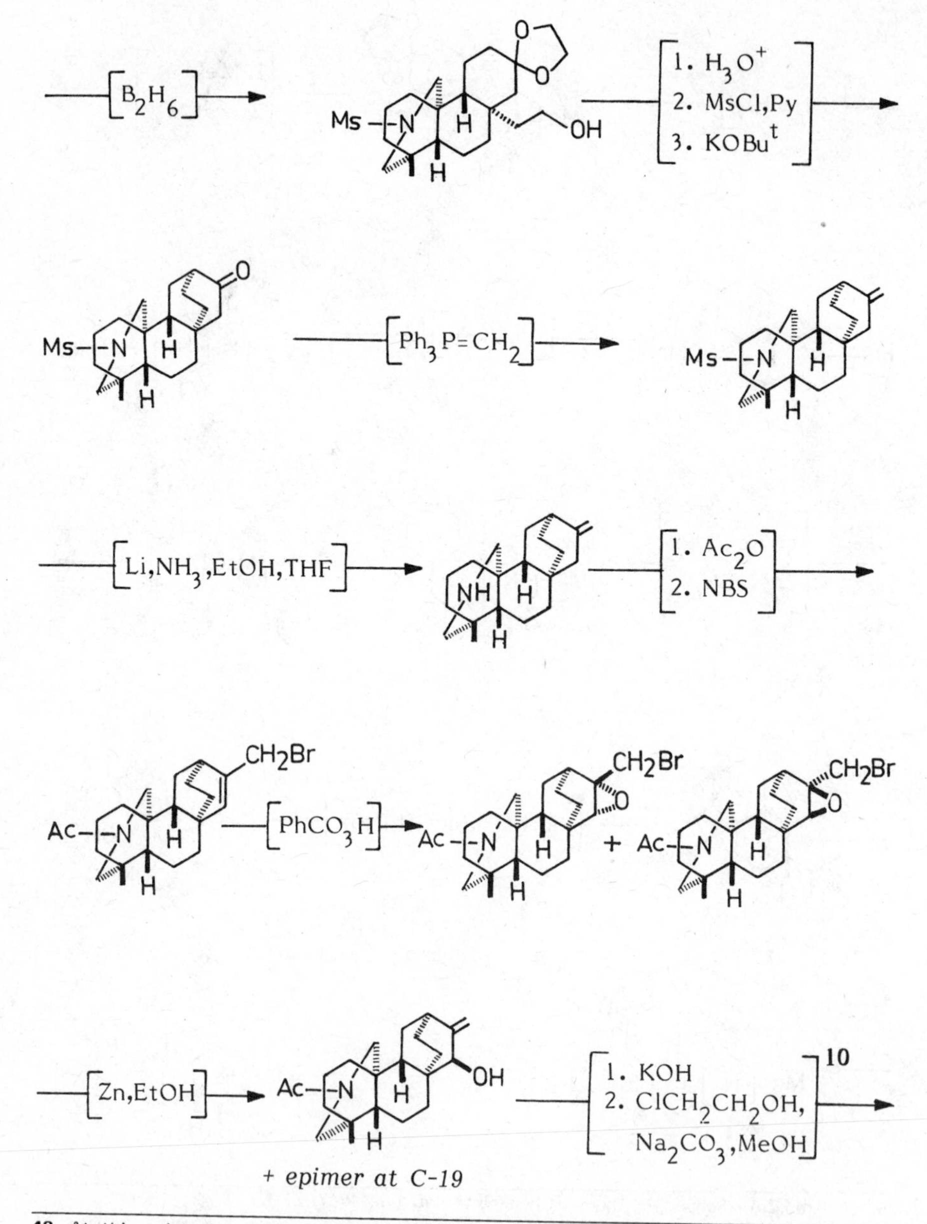

+ *epimer at C-19*

10. At this point the synthetic scheme intersects the partial synthesis of atisine reported by Pelletier,S.W.; Jacobs,W.A. J. Am. Chem. Soc., 1956, 78, 4144.

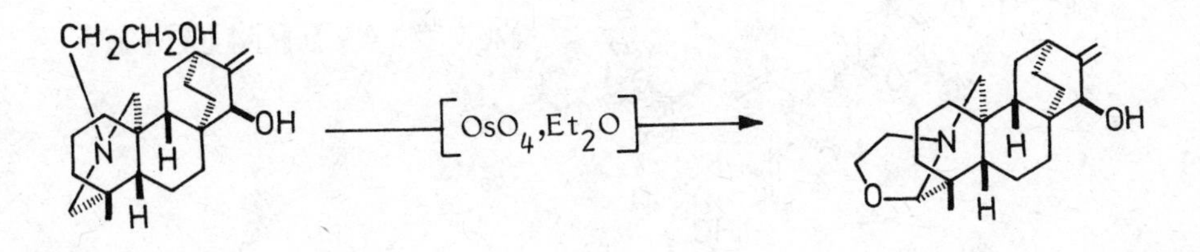
CH2CH2OH
OH
N
H
H
OsO4,Et2O
O
N
H
OH
H

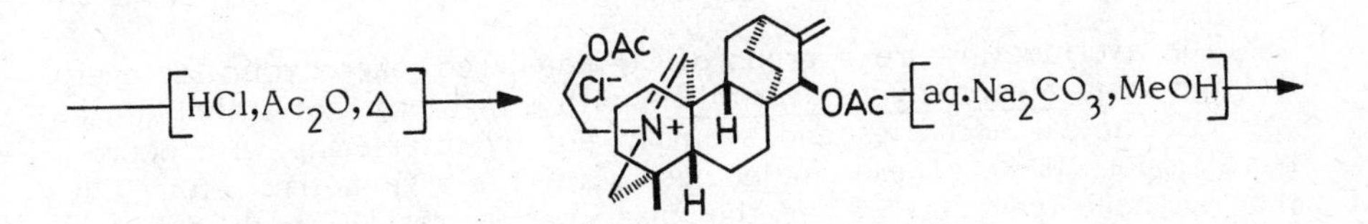
HCl,Ac2O,Δ
OAc
Cl⁻
N+
H
H
OAc
aq.Na2CO3,MeOH

dl-Atisine

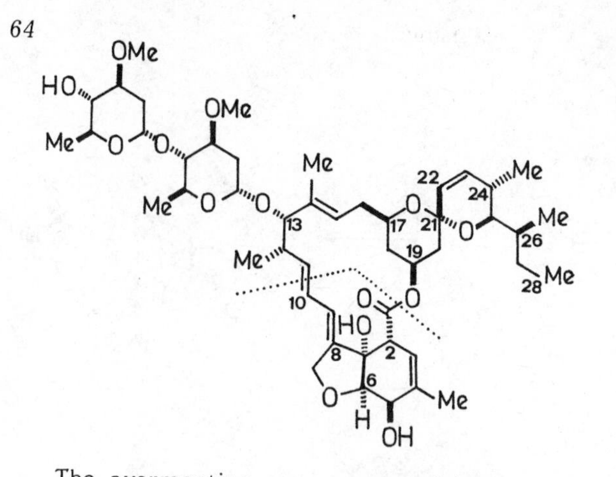

AVERMECTIN

The avermectins are a group of 16-membered macrocyclic lactones isolated from Streptomyces avermitilis with a high order of anthelmintic and insecticidal activities and seem to act by interfering with neuro-transmission (1-3). These molecules with 12 asymmetric atoms in the aglycone part and 8 in the disaccharide residue with delicate functionalities offered a veritable challenge for synthesis. The first synthesis of avermectin B_{1a} described below is based on the strategy of constructing the molecule in two units, the "northern" segment (C_{11} -C_{28}) utilising chirons derived from (S)-malic acid and L-isoleucine, and the "southern" segment (C_1 -C_{10}), followed by coupling the two, macrolactonisation, stereocontrolled glycosylation and adjustment of functionalities (4).

"Northern" Segment (C_{11}-C_{28} fragment)

C_{11}-C_{14} Unit :

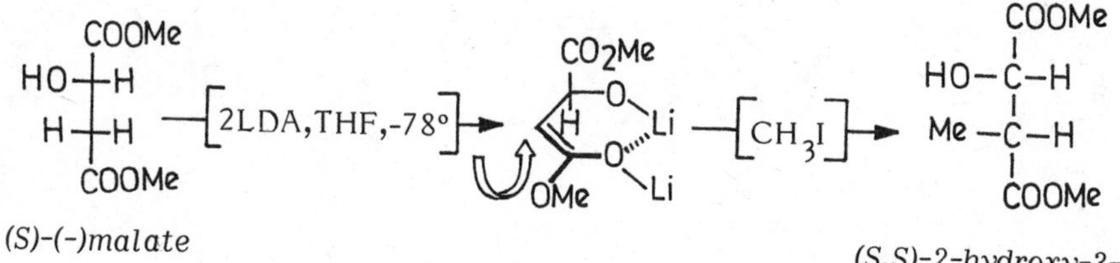

(S)-(-)malate

(S,S)-2-hydroxy-3-methylsuccinate

---

1. Burg,R.W.; Miller,B.M.; Baker,E.E. Antimicrob. Agents & Chemother., 1979, 15,361.
2. Miller,T.W.; Chaiet,L.; Cole,D.J. Antimicrob. Agents & Chemother., 1979, 15, 368.
3. Chabbale,J.C.; Mrozik,H.; Tolman,R.L. et al, J. Med. Chem., 1980, 23, 1134.
4. Hanessian,S.; Ugolini,A.; Dube,D.; Hodges Paul,J.; Andre,C. J. Am. Chem. Soc., 1986, 108, 2776, and references cited therein of their earlier work on the synthesis of some fragments.
5. Seebach,D.; Wasmuth,D. Helv. Chim. Acta, 1980, 63, 197.

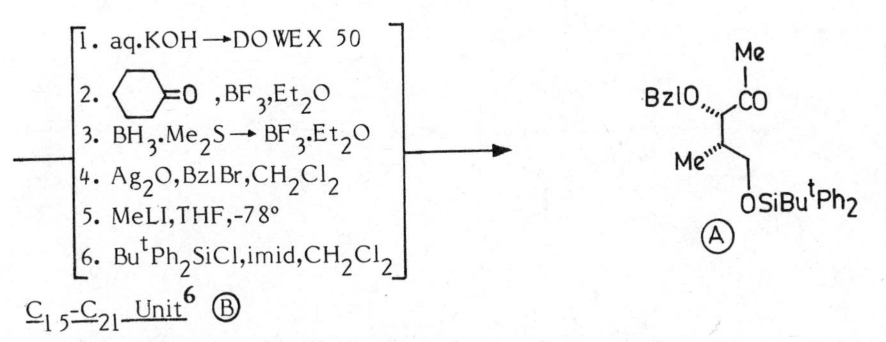

C_{15}-C_{21} Unit[6] (B)

This was synthesised both from D-glucose and (S)-(-)-malic acid as follows:

From D-Glucose

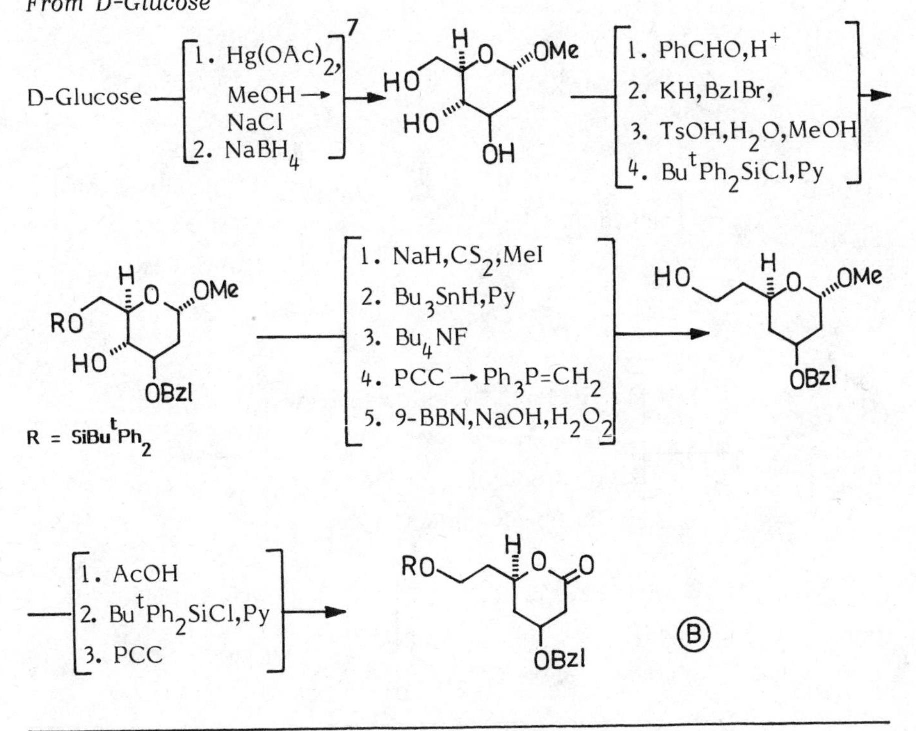

6. Hanessian,S.; Ugolini,A.; Therien,M. J. Org. Chem., 1983, 48, 4427.

7. Corey,E.J.; Fuchs,P.L. Tetrahedron Lett., 1972, 3769.

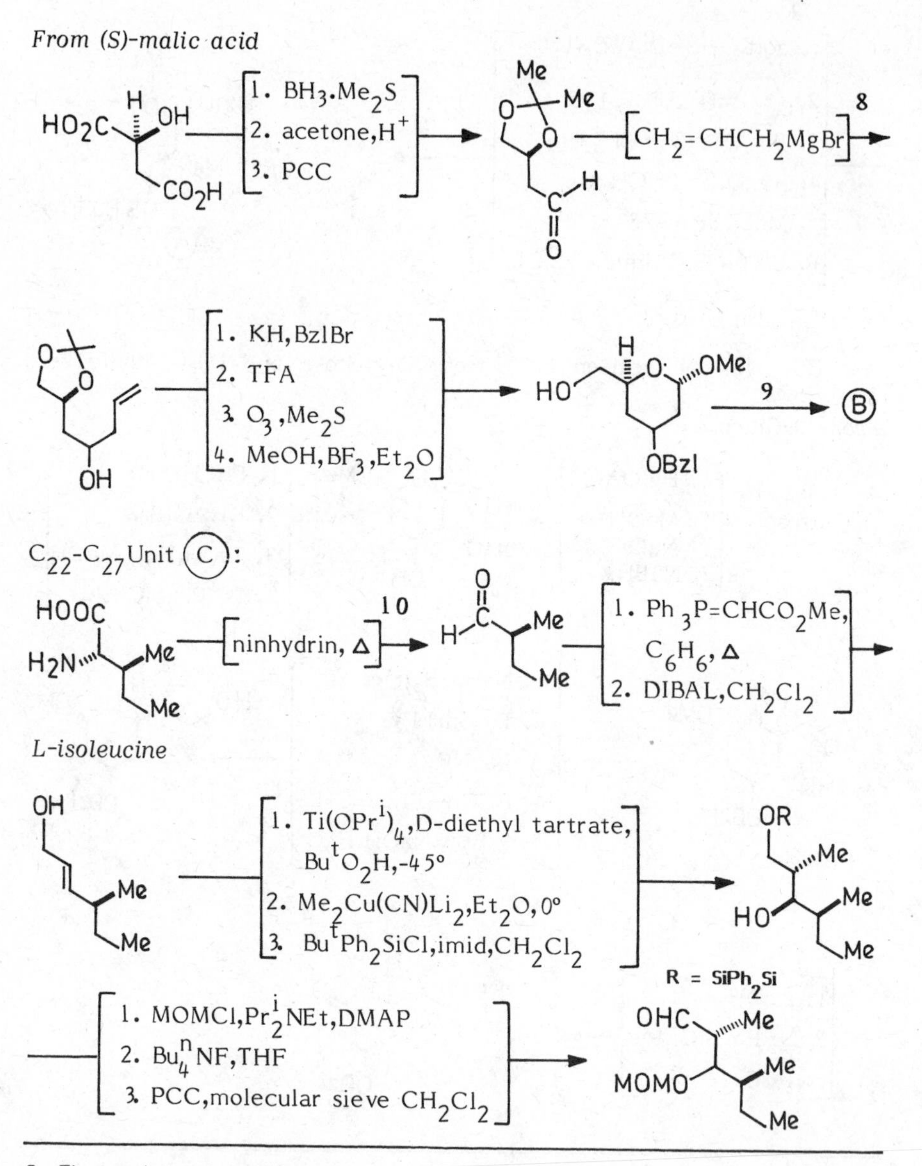

8. The product was obtained as a 1:1 mixture of homoallylic alcohols, which could be separated by chromatography.

9. Chain was extended and (B) obtained as in the case of (D)-glucose.

10. Fores, W.S. J. Am. Chem. Soc., 1954, 76, 1377.

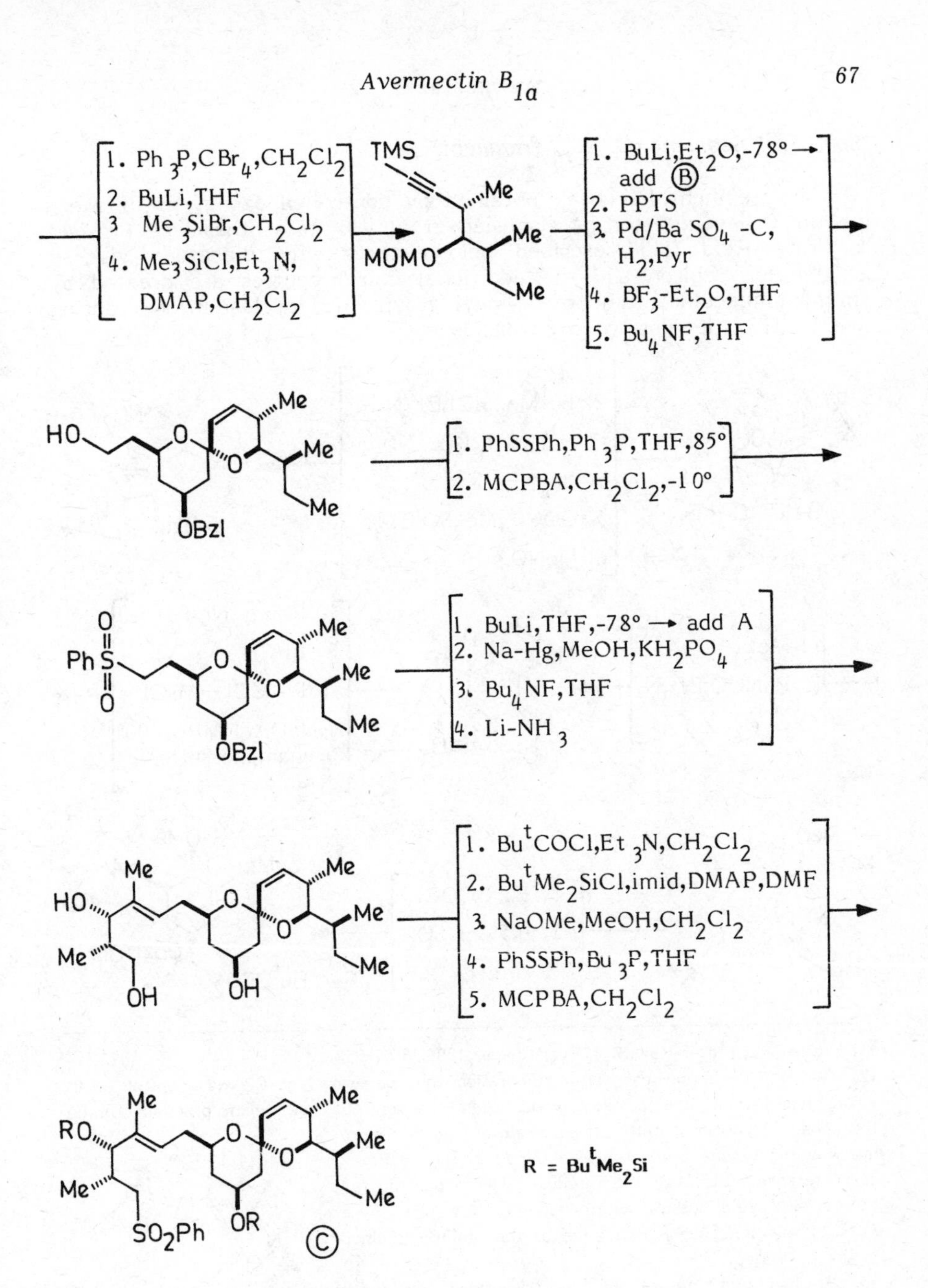
1. Ph_3P,CBr_4,CH_2Cl_2
2. BuLi,THF
3 Me_3SiBr,CH_2Cl_2
4. Me_3SiCl,Et_3N, DMAP,CH_2Cl_2
TMS
MOMO
Me
1. BuLi,Et_2O,-78° → add B
2. PPTS
3. Pd/Ba SO_4 -C, H_2,Pyr
4. BF_3-Et_2O,THF
5. Bu_4NF,THF
HO
OBzl
1. PhSSPh,Ph_3P,THF,85°
2. MCPBA,CH_2Cl_2,-10°
PhS
1. BuLi,THF,-78° → add A
2. Na-Hg,MeOH,KH_2PO_4
3. Bu_4NF,THF
4. Li-NH_3
OH
1. Bu^tCOCl,Et_3N,CH_2Cl_2
2. Bu^tMe_2SiCl,imid,DMAP,DMF
3. NaOMe,MeOH,CH_2Cl_2
4. PhSSPh,Bu_3P,THF
5. MCPBA,CH_2Cl_2
RO
SO_2Ph
OR
C
R = Bu^tMe_2Si

"Southern" Segment (C_1-C_{10} fragment)

This segment has been obtained by controlled ozonolysis of avermectin B seco-ester by Hanesian et al (4). Its synthesis by Prashad & Fraser-Reid (11) described below starts with diacetone glucose as the core chiron and the additional chiral centres are created by an intramolecular nitrile oxide-vinyl group (3+2) cycloaddition reaction (INOC) (12), leading to oxahydrindene.

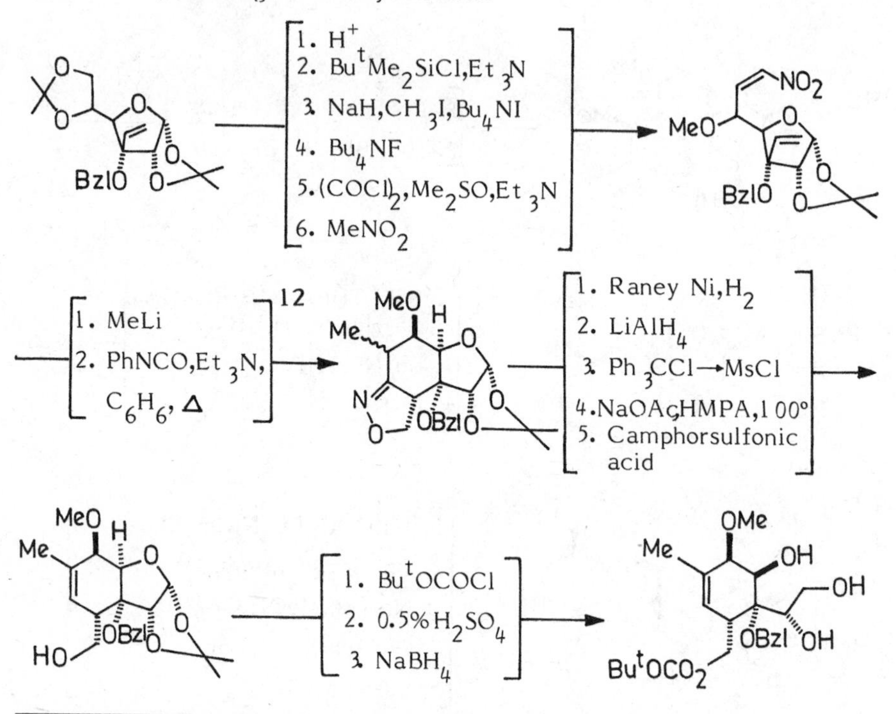

11. Prashad,M.; Fraser-Reid,B. J. Org. Chem., 1985, 50, 1564.

12. The INOC reaction and the configuration thus generated at C_2 was central to the whole strategy; of the possible modes of cyclisation the one which proceeds through the chair transition state shown below seemed more likely and only one product was obtained in this reaction. The stereochemistry of the product was supported by the H1-H2 relationship determined by the 250 MHz PMR of the benzylidene derivative of the olefin obtained in the next step.

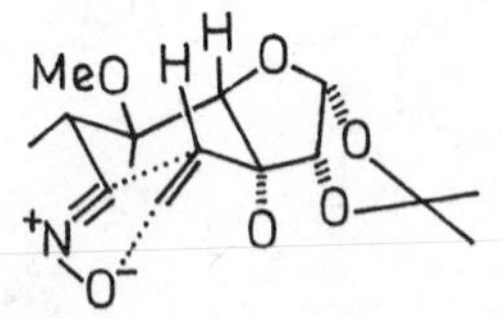

13. Prashad and Fraser-Reid have described this synthesis upto the primary alcohol, while Hanessian, et al (4) have stated that the synthesis of (D) will be reported later.

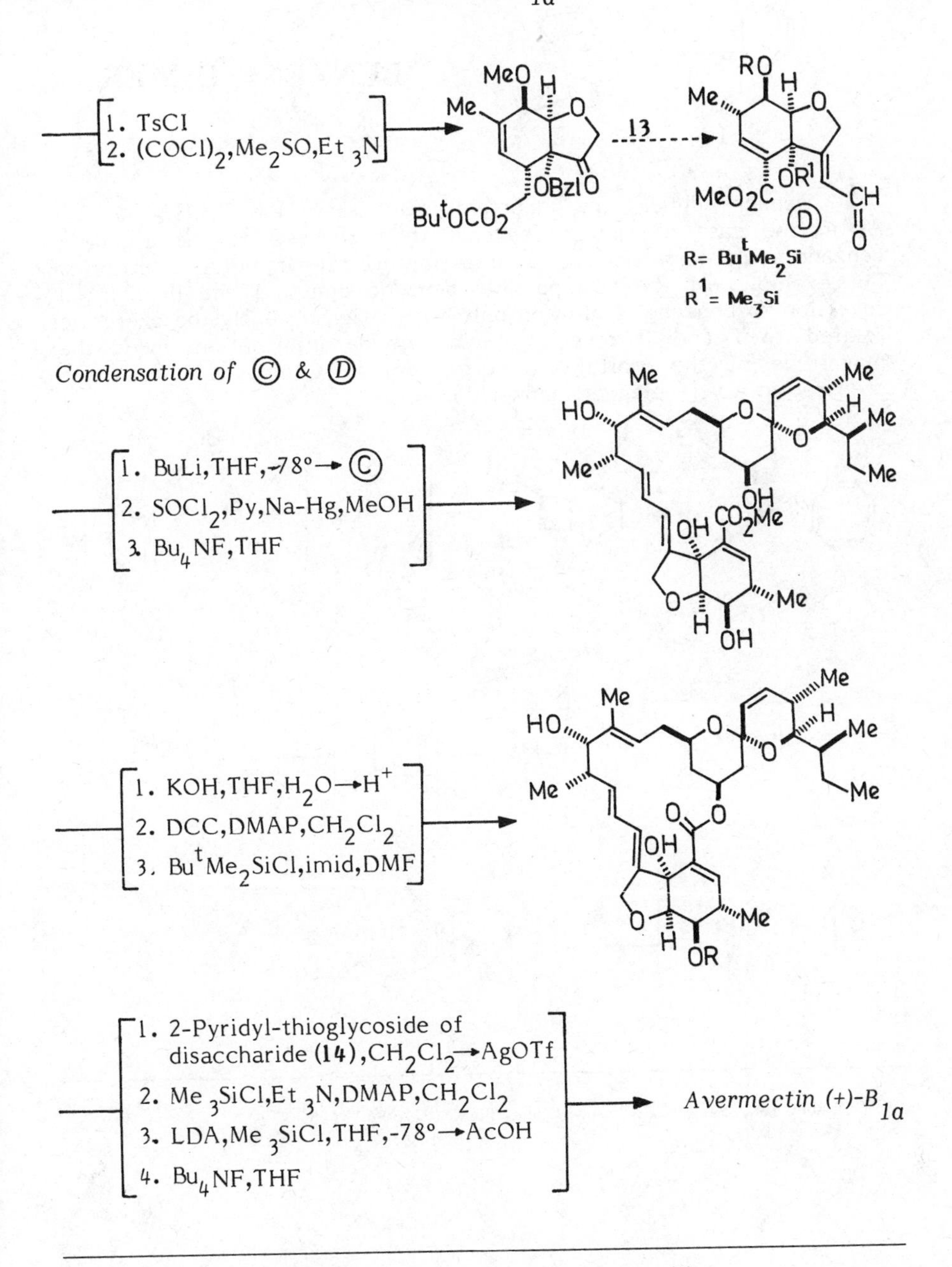

[1. 2-Pyridyl-thioglycoside of disaccharide (**14**),CH_2Cl_2→AgOTf
2. Me_3SiCl,Et_3N,DMAP,CH_2Cl_2
3. LDA,Me_3SiCl,THF,-78°→AcOH
4. Bu_4NF,THF] → *Avermectin (+)-B_{1a}*

---

14. The disaccharide subunit was obtained from avermectin B_{1a} (4).

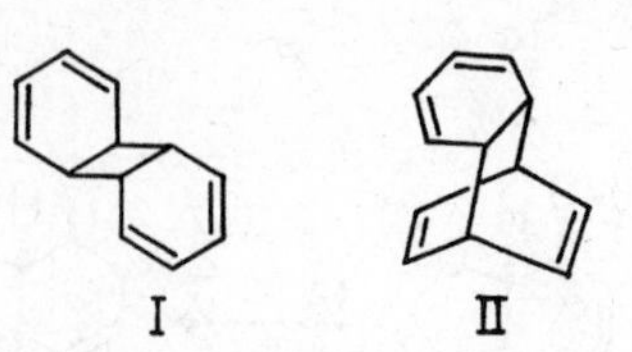

BENZENE DIMER

Benzene dimers are important members of the $C_{12}H_{12}$ family. As in the case of Dewar benzene, their reversal to the aromatic benzene could be restrained by imposing orbital symmetry constraints. For example, of the two possible benzene dimers I and II, thermal reversion to benzene is allowed only for II. Consequently the energetics related to I and II reversal could provide information about the magnitude of the orbital symmetry control. Compound I has been prepared in a very elegant manner (1).

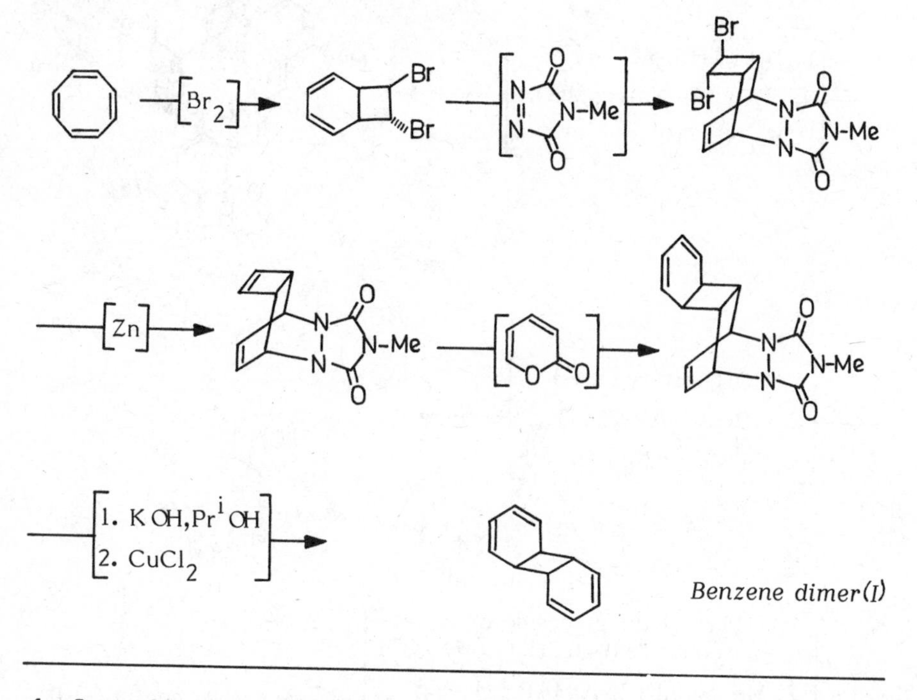

Benzene dimer(I)

1. Berson, J.A., Davis, R.F. J. Am. Chem. Soc., 1972, 94, 3658.

BENZENE OXIDES

The exploitation of fragile sigma bonds in pericyclic reactions have led to many novel systems, the most notable being Doering's bullvalene (1). The inspiration for bullvalene came from the understanding of the facile I--II change. Thus, benzene oxide (I,X=O) gives

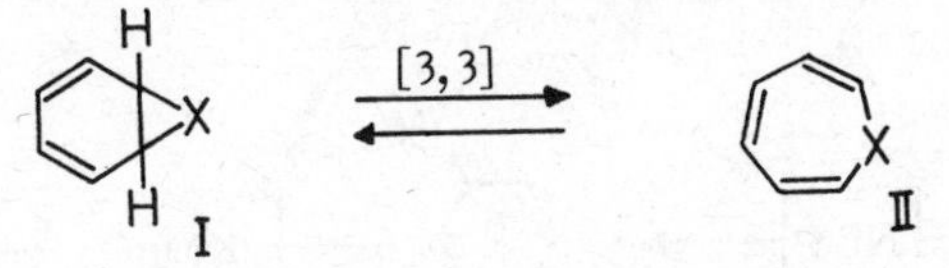

oxepin (II,X=O) and it could be predicted that Z,Z benzene dioxide [III] should give 1,4-dioxocin [IV] and finally that all Z benzene trioxide [V] should give VI. These expectations have been realized (2). The interrelationships that exist in the benzene oxide series is illustrated below:

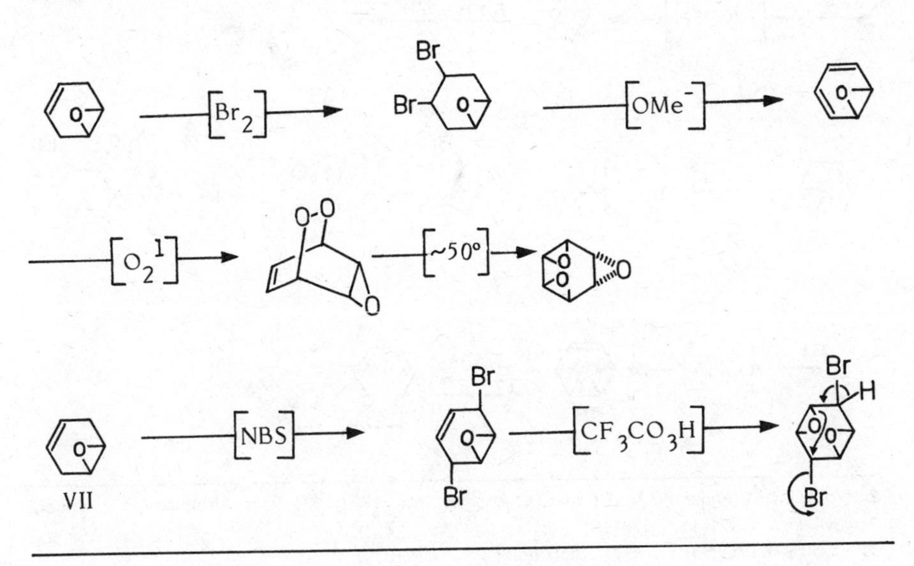

1. Von Doering, W., Ferrier, B.M., Fossel, E.T., Hartenstein, J.H., Jones Jr., M., Klumpp, G., Rubin, R.M., Saunders, M., Tetrahedron, 1967, 23, 3943.

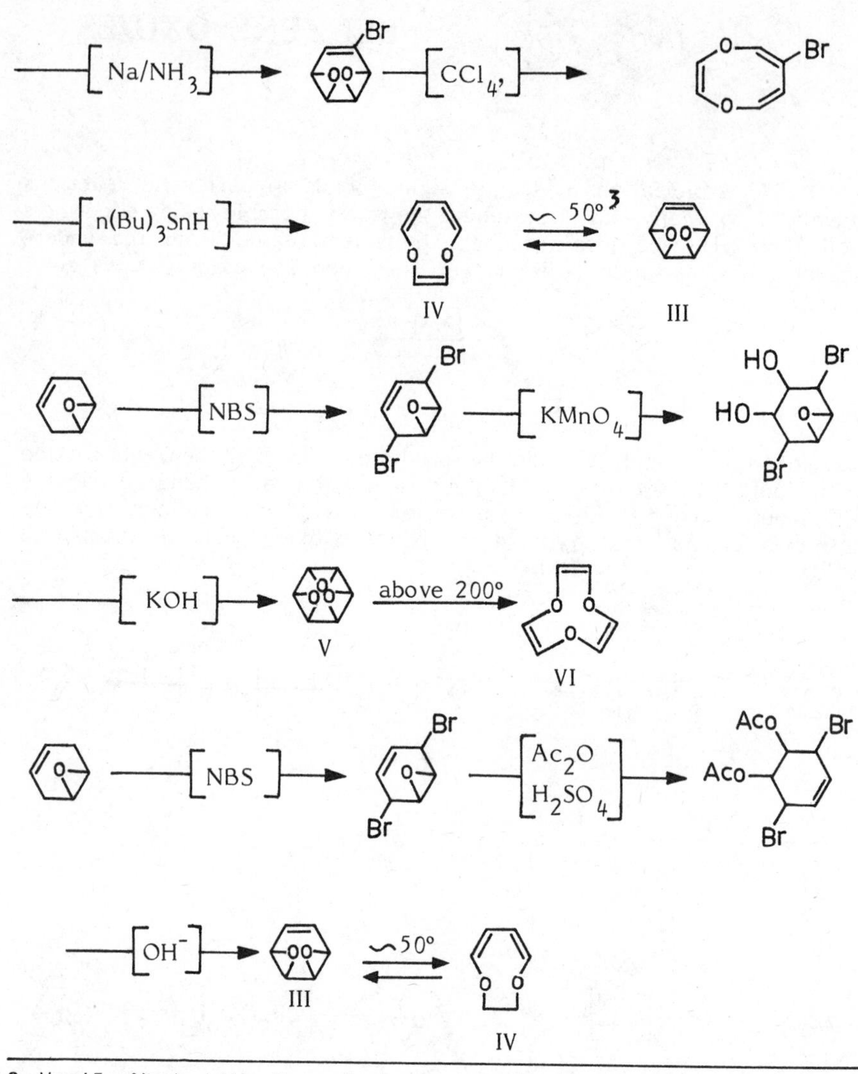

2. Vogel,E.; Altenbach,H.J.; Cremer,D. Angew. Int., 1972, 11, 935; Altenbach,H.J.; Vogel,E. Angew. Int., 1972, 11, 937.

3. At 50°, III:IV::5:95; Ea = 27 kcal/mole; A = 7.1×10^{13}.

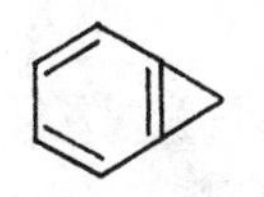

BENZOCYCLOPROPENE

Benzocyclopropene is an intriguing compound in which two of the sigma bonds of benzene are bent to form a three membered ring(1). The compound, prepared(2) in an ingenious manner by a retro Diels-Alder reaction, was found to be a reasonably stable one and spectral data showed no noteworthy disturbance of delocalization of the π system(3).

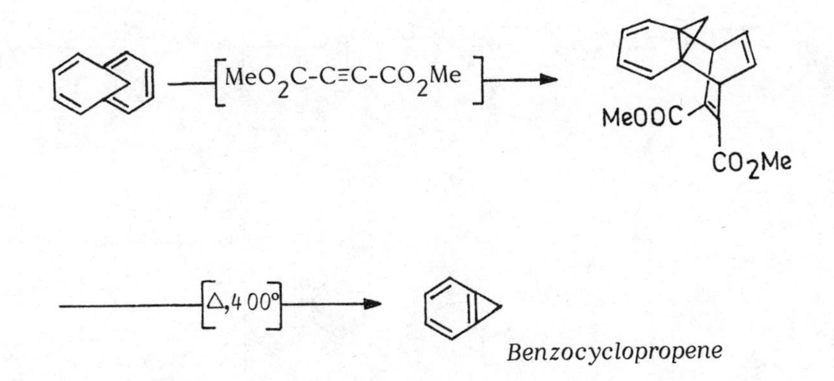

Benzocyclopropene

1. E. Fullman and E. Buncel, J. Am. Chem. Soc., 85, 2106 (1963).
2. E. Vogel, W. Grimme and S. Korte, Tetrahedron Letters, 3625 (1965).

3. The bridge double bond, however, is highly reactive:

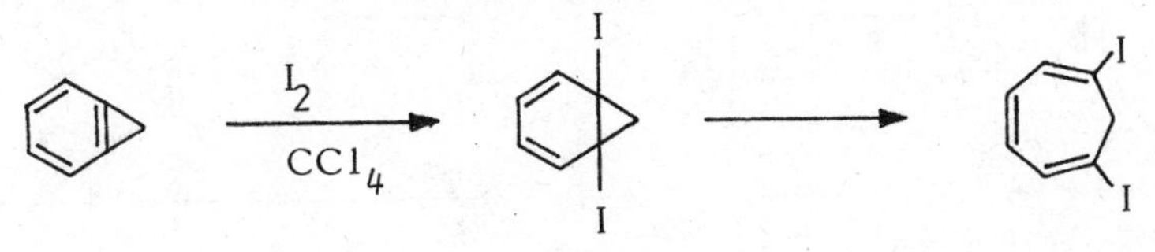

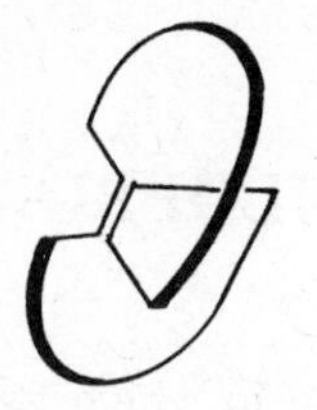

BETWEENANENES

Betweenanenes represent a particularly aesthetically pleasing carbon constellation. These chiral compounds are π systems, necessarily sandwiched "between" the alkyl chains. The first ingenious and selective synthesis of [10,10]betweenanene was accomplished from cyclododecane-1,2-dione (1).

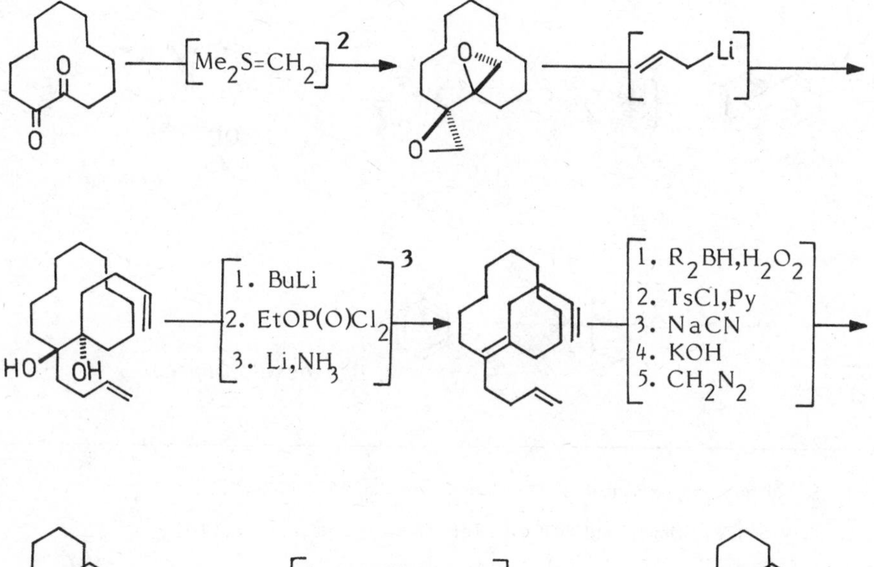

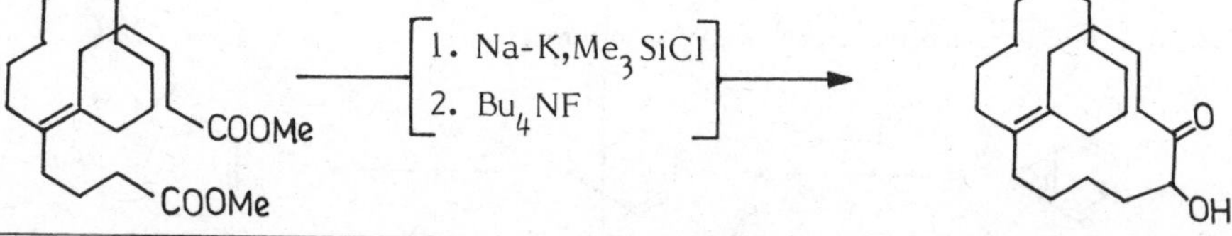

---

1. Marshall,J.A.; Lewellyn,M.E.J. Am. Chem. Soc., 1977, 99, 3508.

2. A mixture of trans and cis bisepoxides formed in a ratio of 1.5:1 was obtained, separable by column chromatography. The subsequent reaction sequence when carried out on the cis bisepoxide afforded bicyclo [10.10.0]docos-1(12)-ene.

Bicyclo[10.10.0]docos-1(12)-ene

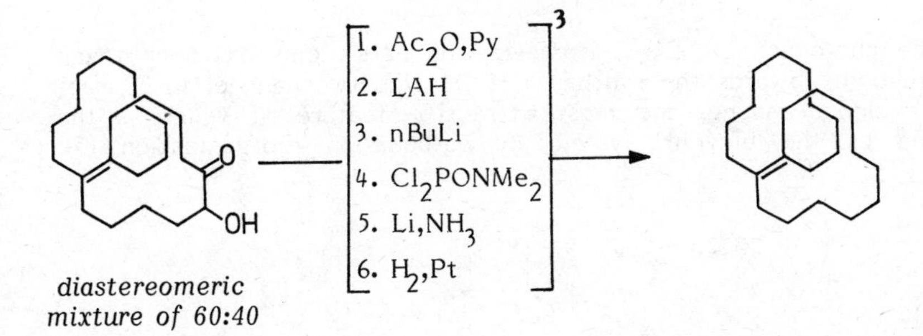

diastereomeric mixture of 60:40

A variant of the above strategy was to start with cyclododecanone (4).

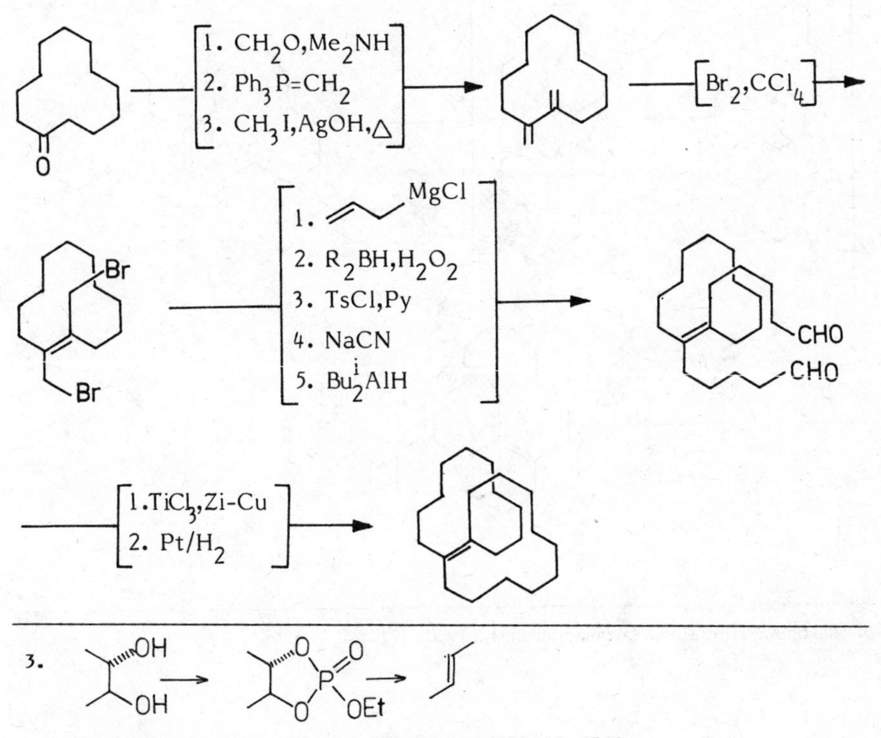

4. Marshall, J.A.; Chung, Kyoo-Hyun, J. Org. Chem., 1979, 44, 1566.

The photochemical E→Z isomerization of systems has been taken advantage of towards the synthesis of [11,11]betweenanene, in 3 steps from cyclododecanone, the most attractive feature of which is the creation of the bicyclic system by carbocation reorganization (5).

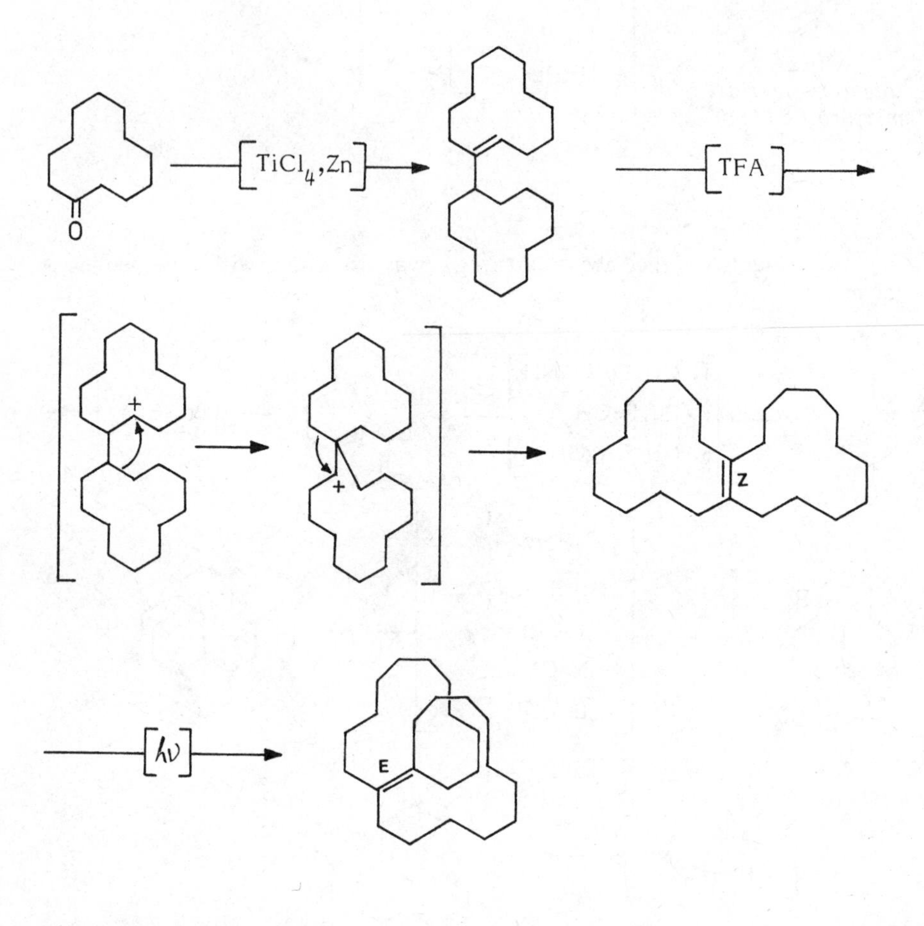

5. Nickon,A.; Zurer,P.St.J.Tetrahedron Lett., 1980, 3527.

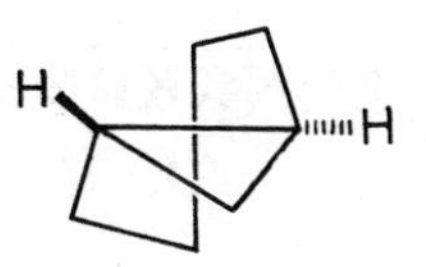

E-BICYCLO(4,1,0)HEPTANE

A cyclopropane unit would considerably enhance the strain relative to the corresponding olefins in alicyclic systems. On this basis, E-bicyclo(4,1,0)heptane [I] should be considerably more strained than E-cyclohexene. Yet, such systems have been made involving a highly stereospecific ring contraction step (1).

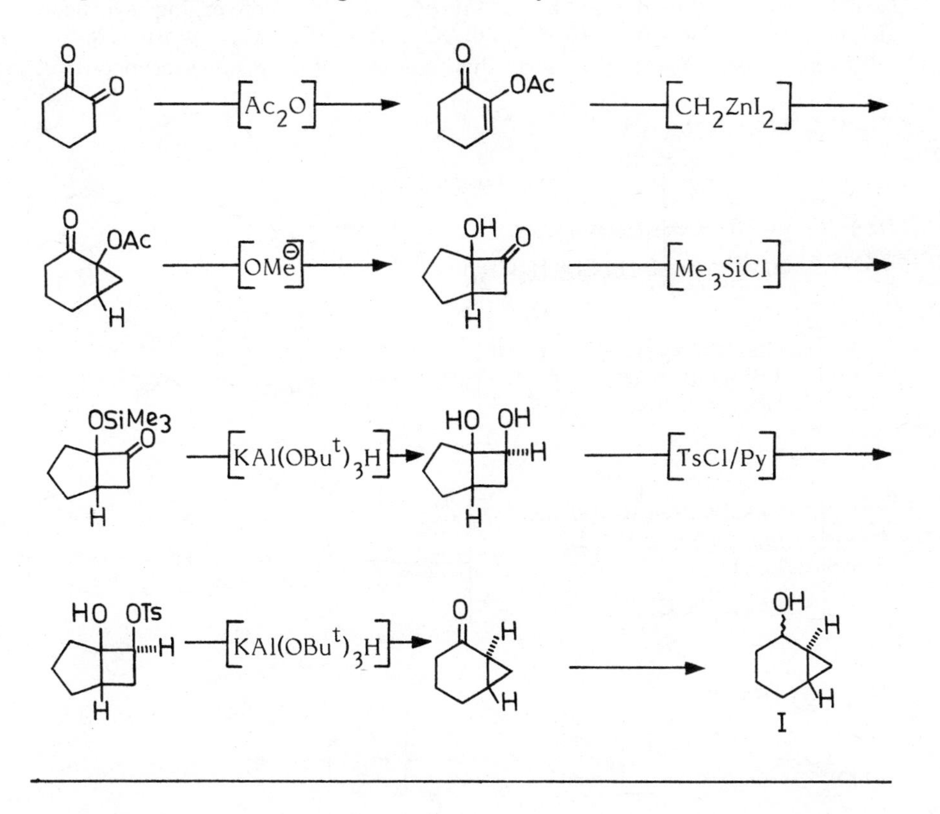

1. Paukstelis, J.V., Kao, J. J. Am. Chem. Soc., 1972, 94, 4783.

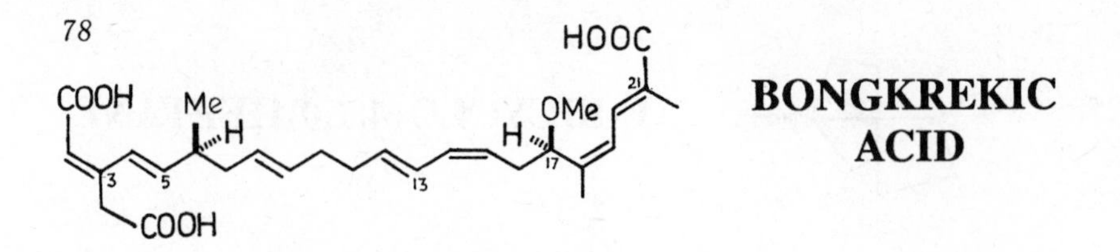

BONGKREKIC ACID

Bongkrekic acid, a toxin produced by Pseudomonas cocovenenans, has a high affinity for the mitochondrial translocator protein necessary for ATP function in eukaryotic cells (1). Its synthesis by Corey and his associates (2) is based on a retrosynthetic recognition of two β-methyl glutaconate units in two end segments of the molecule, which led to the design of a stereocontrolled converging synthesis from three fragments, C_1-C_4, C_5-C_{13} and C_{14}-C_{22} with required absolute stereochemistry, and this allowed for good stereocontrol.

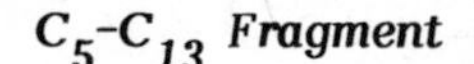

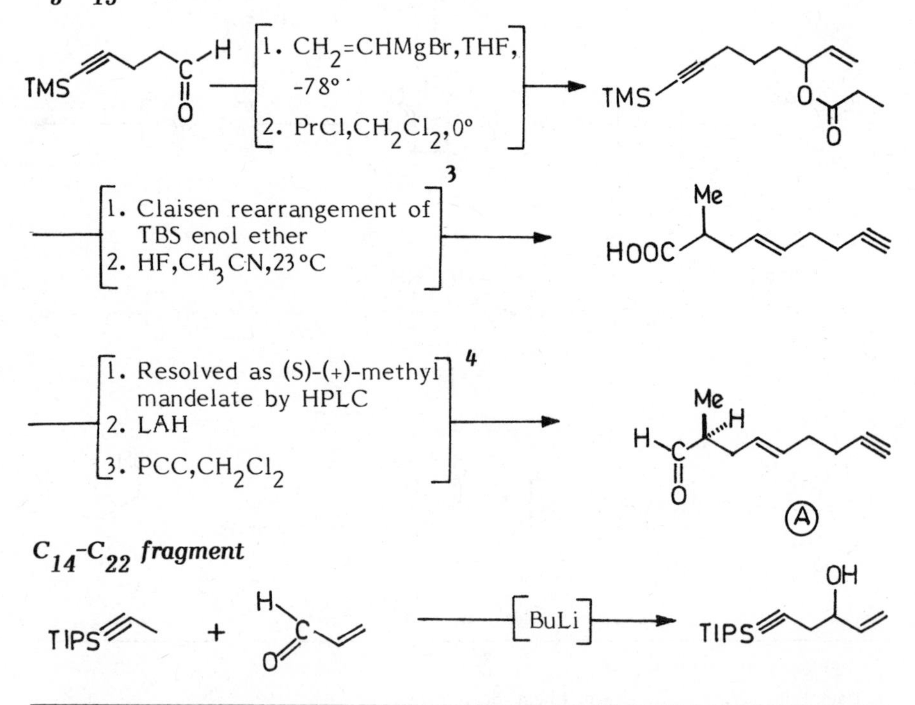

1. Khugensberg,M. Trends Biochem. Sci., 1979, 4, 249, and references cited therein.
2. Corey,E.J.; Tromantano,A. J. Am. Chem. Soc., 1984, 106, 462.
3. Ireland,R.E.; Mueller,R.H.; Willart,A.K. J. Am. Chem. Soc., 1976, 98, 2868.
4. The product was obtained as a mixture of (S,S) and (R,S) diastereomers.

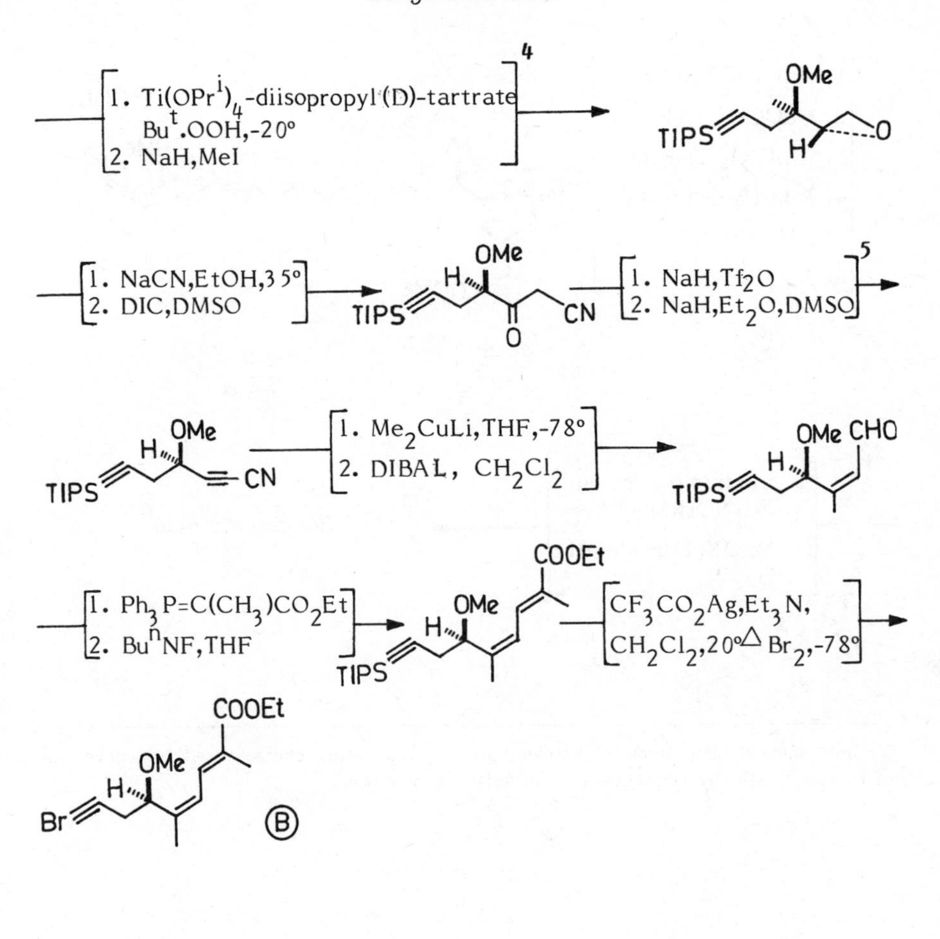

C_1-C_{13} fragment

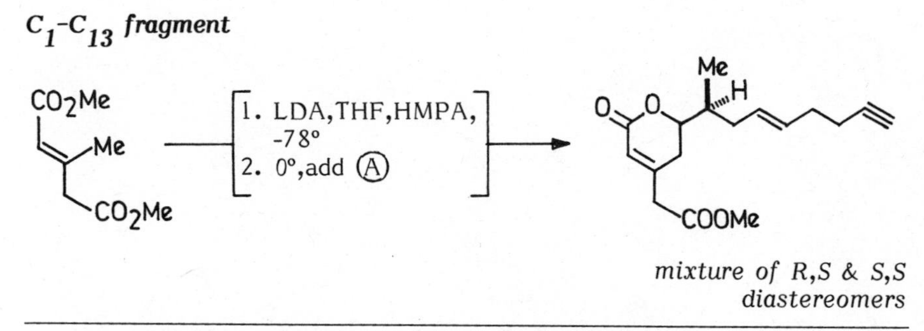

mixture of R,S & S,S diastereomers

---

5. This is a novel method of enyne generation through enol-triflate.

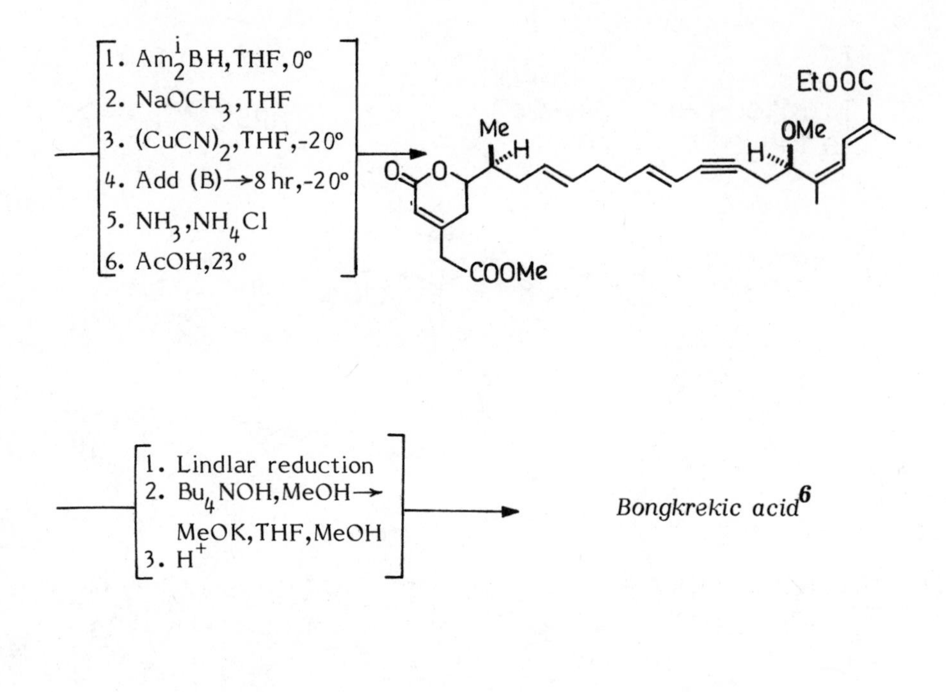

---

6. Free acid in neat form is unstable, and it was better characterised by conversion to trimethyl ester by treatment with etherial diazomethane.

81

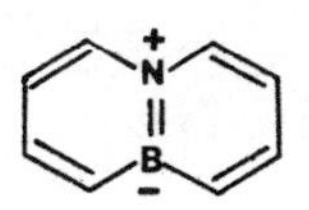

10,9-BORAZARONAPHTHALENE

Dewar's extensive work on heterocyclic compounds containing boron and nitrogen has demonstrated that an aromatic C-C bond could effectively be replaced by a B-N bond. In the synthesis of 10, 9-borazaronaphthalene (1) the final step involved Pd/C dehydrogenation of a perhydro precursor.[1] The compound looks and smells like naphthalene and nuclear magnetic resonance studies show that it is truly aromatic.

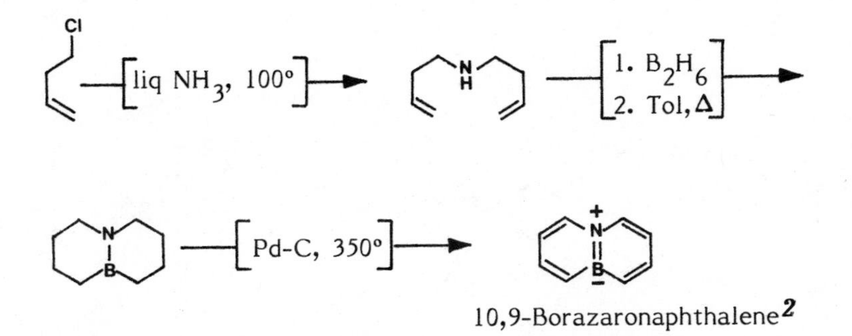

10,9-Borazaronaphthalene[2]

---

1. M.J.S. Dewar, G.J. Gleicher and B.P. Robinson, J. Am. Chem. Soc., 86, 5098 (1964); M.J.S.Dewar and R. Jones, J. Am. Chem. Soc., 90, 2137 (1968); F.A. Davis, M.J.S. Dewar, R. Jones and S.D. Worley, J.Am. Chem. Soc., 91, 2094 (1969).

2. The most interesting reaction of this compound thus far reported is the one leading to another B-N aromatic system on treatment with 1,1,3,3-tetraethoxypropane and CF_3CO_2H: M.J.S. Dewar and R. Jones, Tetrahedron Letters, 2707 (1968).

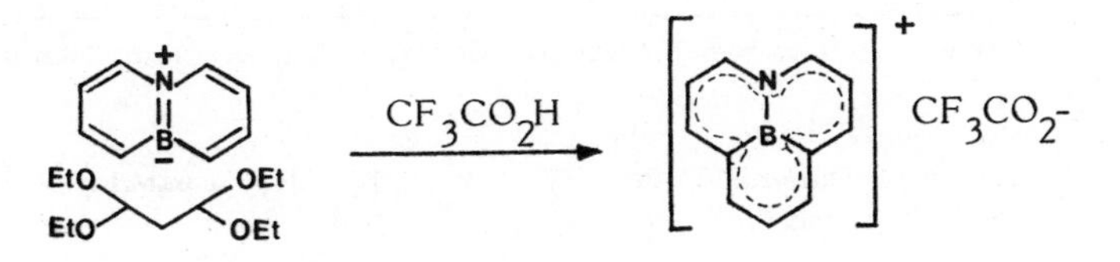

BULLVALENE

The synthesis of bullvalene(1-3), which undergoes rapidly reversible and structurally degenerate isomerization, demonstrated in a remarkable manner the predictive powers of organic chemistry.

Schroder's 'spectacular' synthesis[2]

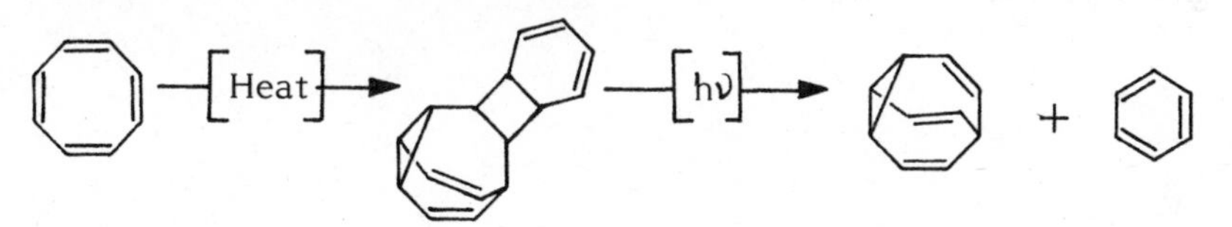

Doering's 'rational' synthesis[3]

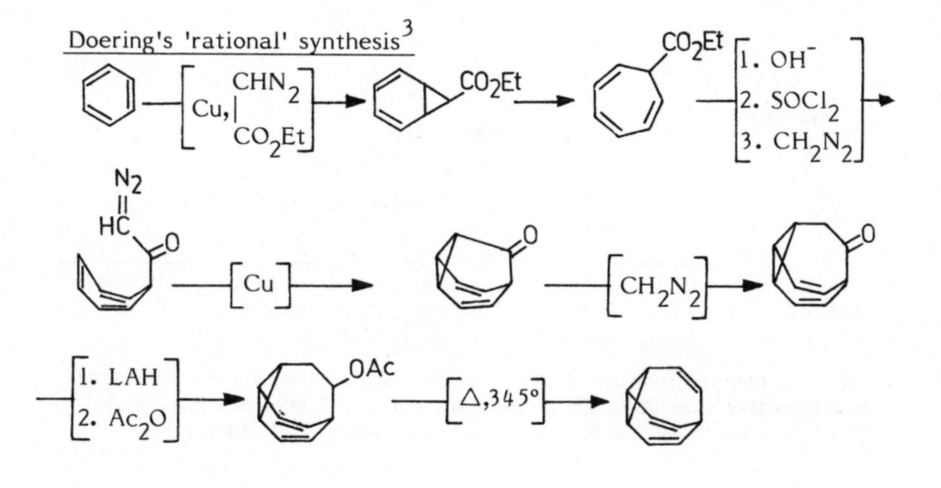

---

1. For a recent review, see: Schroder,G.; Oth,J.F.M. Angew. Chem. Internat. Ed., 1967, 6, 414.
2. Schroder,G. Angew. Chem. Internat. Ed., 1963, 2, 481.
3. Von Doering,W.; Ferrier,B.M.; Fossel,E.T.; Hartenstein,J.H.; Jones,M.Jr.; Klumpp,G.; Rubin, R.M.; Saunders,M. Tetrahedron, 1967, 23, 3943.

83

CANTHARIDIN

Although the possibility of introducing the structural features of cantharidin, virtually all in one step, by a diene addition attracted the attention of many workers, much of the earlier efforts directed along these lines were infructuous (1). The earlier stereospecific synthesis by Stork and his associates (2) which involved two successive diene reactions, and that by Schenk & Wirtz (3) which involved a diene-type photochemical peroxidation, were based on classical lines. It is only recently that Dauben, Kessel and Takemura achieved a short and efficient synthesis based on Diels-Alder reaction by carrying out the reaction under high pressure (4).

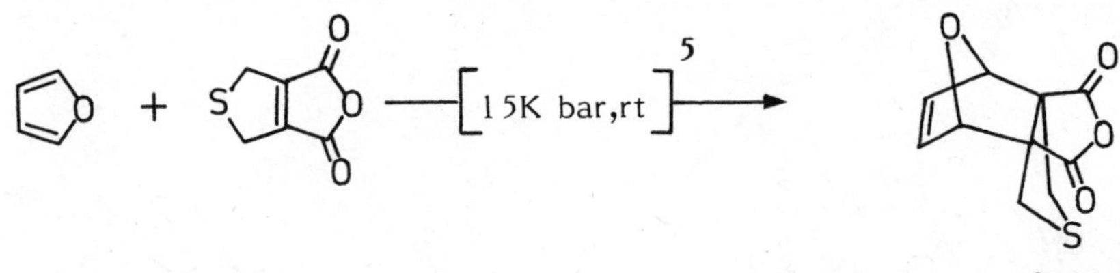

85:15 mixture of this and the endo-anhydride

—[Raney Ni]→ *Cantharidin*

---

1. For a review of the previous synthetic efforts, see: Alder,K.; Schumacher,M. Fortschr. Chem. Org. Naturstoffe, 1953, 10, 87.

2. Stork,G.; van Tamelen,E.E.; Friedman,L.J.; Burgstahler,A.W. J. Am. Chem. Soc., 1951, 73, 4501; 1953, 75, 384.

3. Schenck,G.O.; Wirtz,R. Naturwiss, 1953, 40, 581.

4. Dauben,W.G.; Kessel,C.R.; Takemura,K.H. J. Am. Chem. Soc., 1980, 102, 6893

5. Furan, on account of its aromaticity is a poor diene, and high temperature could not be used due to thermal cycloreversion; however, pressure in the range of 10-20K bars facilitates Diels-Alder reaction of furan. Dimethylmaleic anhydride is a poor dienophile on account of unfavourable electronic (electron donation by Me group) and steric factors (crowding by Me groups). It was anticipated that a sulfur-containing methylene bridge in place of the Me groups would reduce the electron donating character of the Me groups and also reduce the steric demands of the disubstituted maleic anhydride.

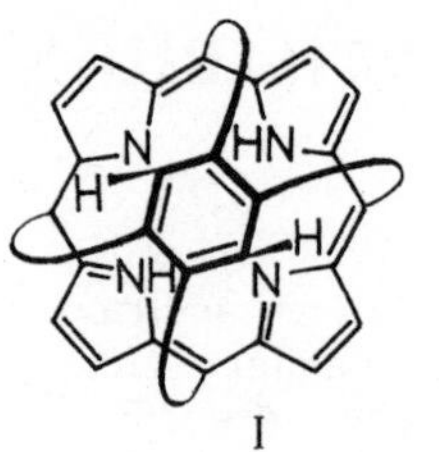

CAPPED PORPHYRINS

The "Capped" porphyrin I has been made in a very elegant manner by condensing a suitably substituted tetraldehyde with four molecules of pyrrole to provide stereoselectively the tetrasubstituted meso-porphyrin. It not only forms complexes with oxygen which are indefinitely stable at -20°C, but in the presence of imidazole it effectively carries oxygen (2,3).

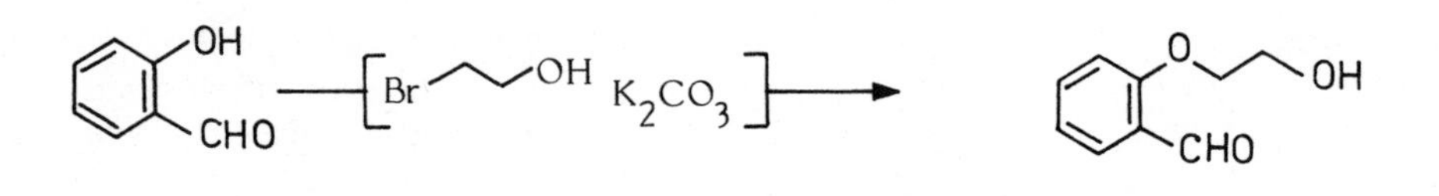

1. Almog,J.; Baldwin,J.E.; Dyer,R.L.; Peters,M. J. Am. Chem. Soc., 1975, 97, 226.
2. The porphyrin group in haemoglobin is remarkably different from the unencumbered parent system in the sense that the haemoglobin-oxygen complex is more stable and does not exhibit tendency to oxidize to the Fe^{3+} system. Further, most fascinating is the observation that the Fe^{2+} in haemoglobin can be replaced by Co^{2+} without significantly affecting its oxygen carrying properties; in contrast, the parent Co^{2+}-porphyrin shows no tendency to bind oxygen. Evidentally, the protein plays a dominant role in stabilizing the oxygen complexes and this could be achieved by shielding one of the porphyrin faces, thus making the formation of the stable octahedral bi-liganded complexes impossible, whilst permitting the oxygen coordination. The stability of the dioxygen adduct is largely dependent on the nature and concentration of the coordinating base.

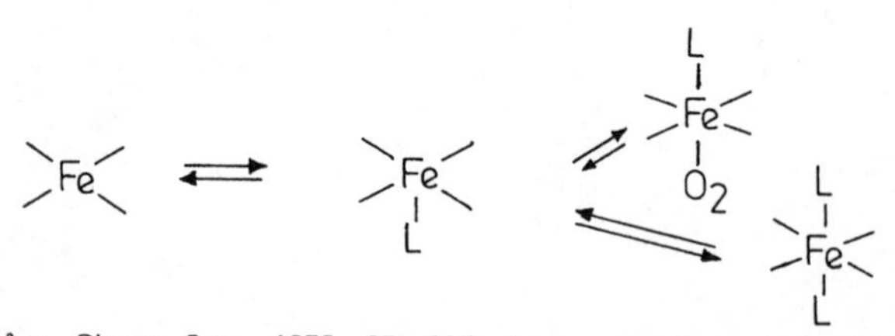

3. Baldwin,J. J. Am. Chem. Soc., 1975, 97, 227; for a recent review relating to blood substitutes, see: Riess,J.G.; LeBlanc,M. Angew. Chem. Int. Ed., 1978, 17, 621.

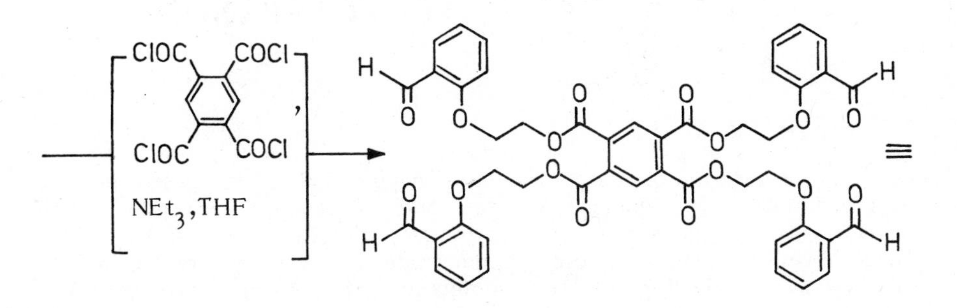

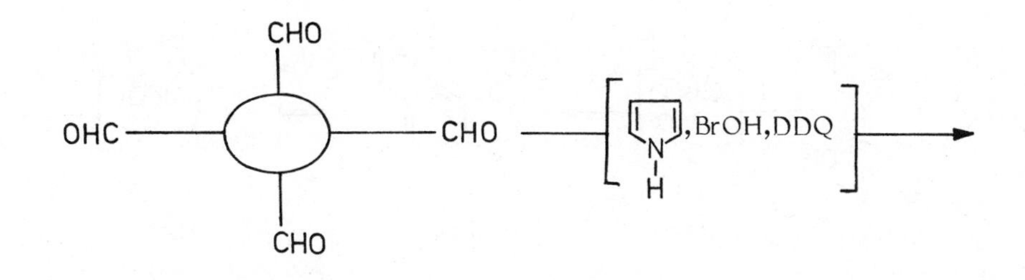

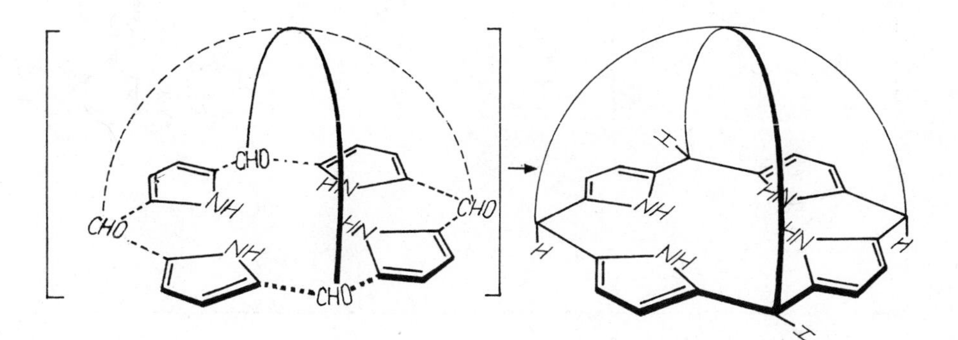

——[$FeCl_3$,THF,2.Cr^{II}]——► *Capped Porphyrin*

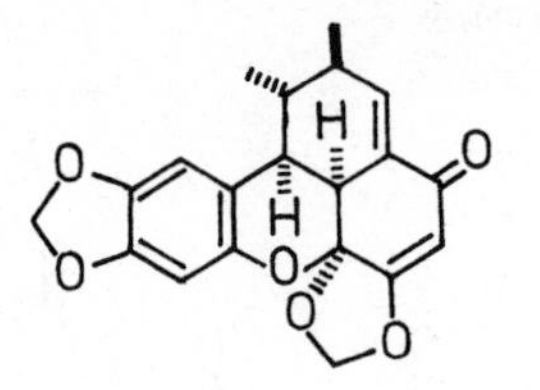

CARPANONE

The synthesis of carpanone by Chapman et al is based on the cycloaddition of o-quinonemethides and nucleophilic olefins which generated contiguous multiple assymetric centres in one step, in this case five. The required o-quinonemethide (A) was generated in situ by phenolic coupling of two molecules of 2-(trans-1-propenyl)-4,5-methylenedioxyphenol.

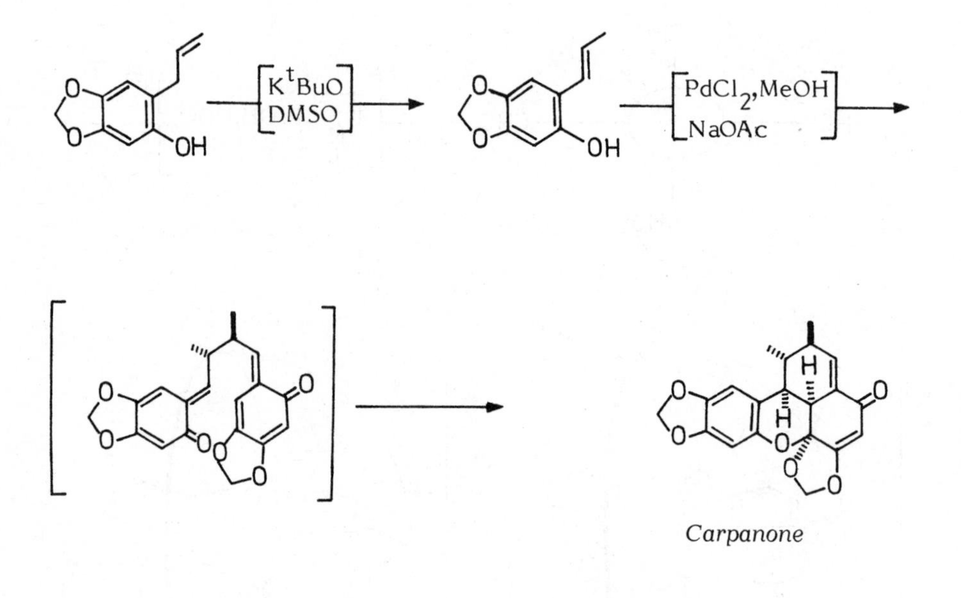

Carpanone

1. Chapman,O.L., Engel,M.R., Springer,J.P., Clardy,J.C.;J. Am. Chem. Soc., 1971, 93, 6696.

2. Chapman,O.L., McIntosh,C.L., Chem. Commun., 1971, 383.

87

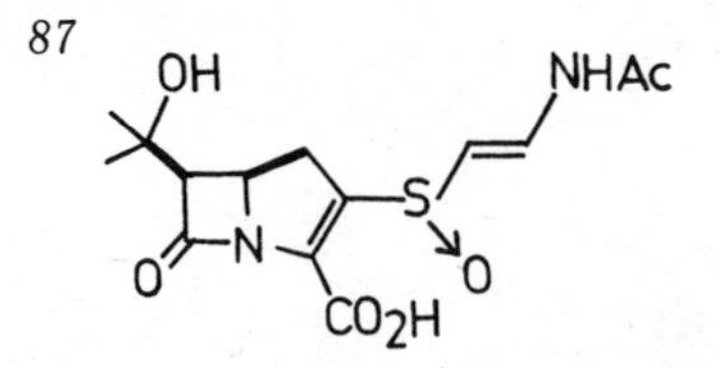

CARPETIMYCIN A

Carpetimycin is an important representative of cis-substituted carbapenem antibiotics. Although a number of synthesis of trans-substituted carbapenems have been reported (1), this is the first chiral synthesis of a cis-substituted carbapenem antibiotic (2). The synthesis involves the enzymatic conversion of prochiral diester (A) to a chiral acid-ester and an efficient novel conversion of a trans-substituted δ-lactone (B) to a cis-substituted β-lactam.

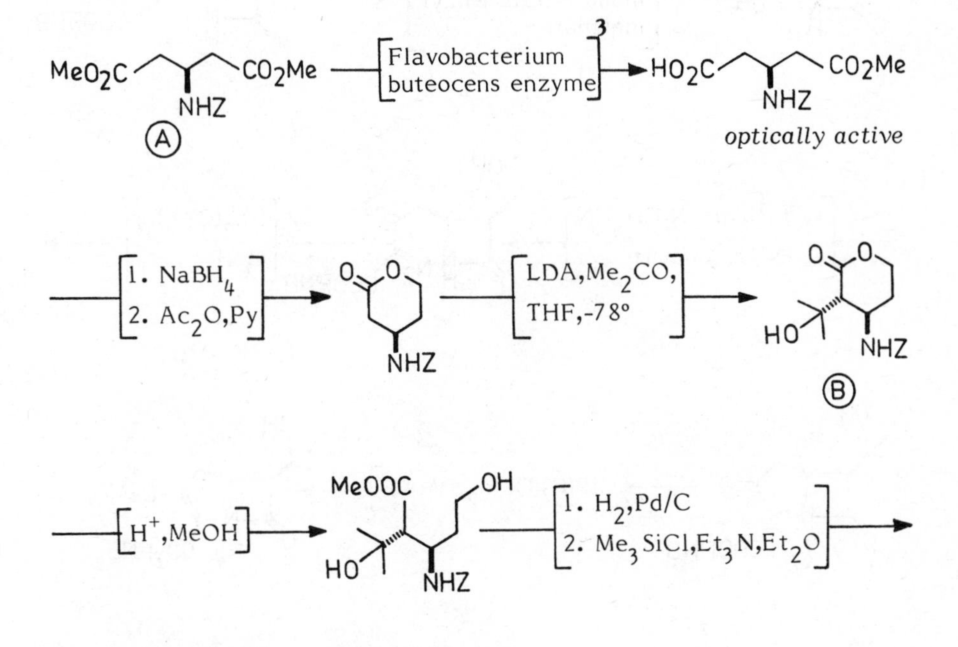

1. see Thienomycin, p. 351.

2. Limori,T.; Takahashi,Y.; Izawa,T.; Kobayashi,S.; Ohno,M. J. Am. Chem. Soc., 1983, 105, 1659.

3. Ohno,M.; Kobayashi,S.; Limori,T.; Wang,Y.F.; Izawa,T. J. Am. Chem. Soc., 1981, 103, 2405.

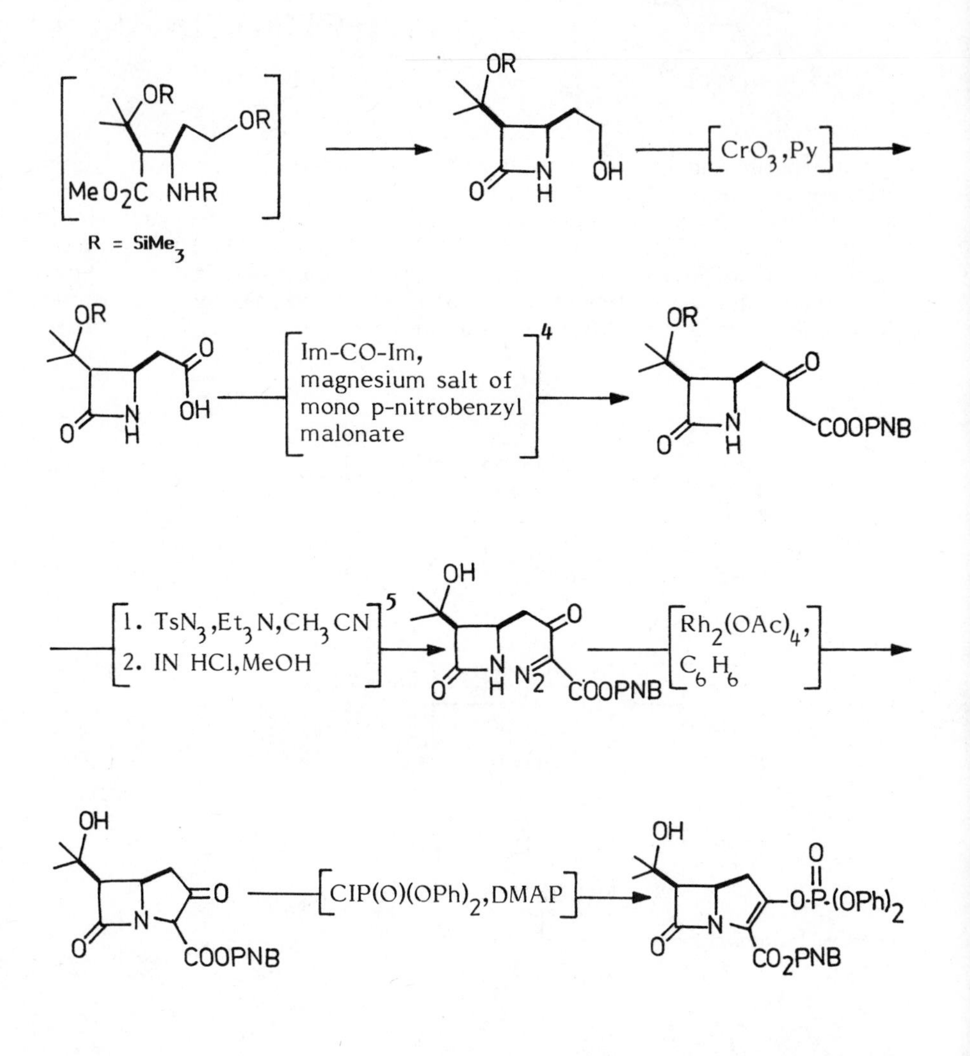

4. Brooks,D.W.; Lu,L.D.L.; Masamune,S. Angew. Chem. Int. Edn.,1979, 18, 72.

5. Salzmann,T.N.; Ratcliffe,R.W.; Christensen,B.G.; Bouffard,F.A. J. Am. Chem. Soc., 1980, 102, 6161.

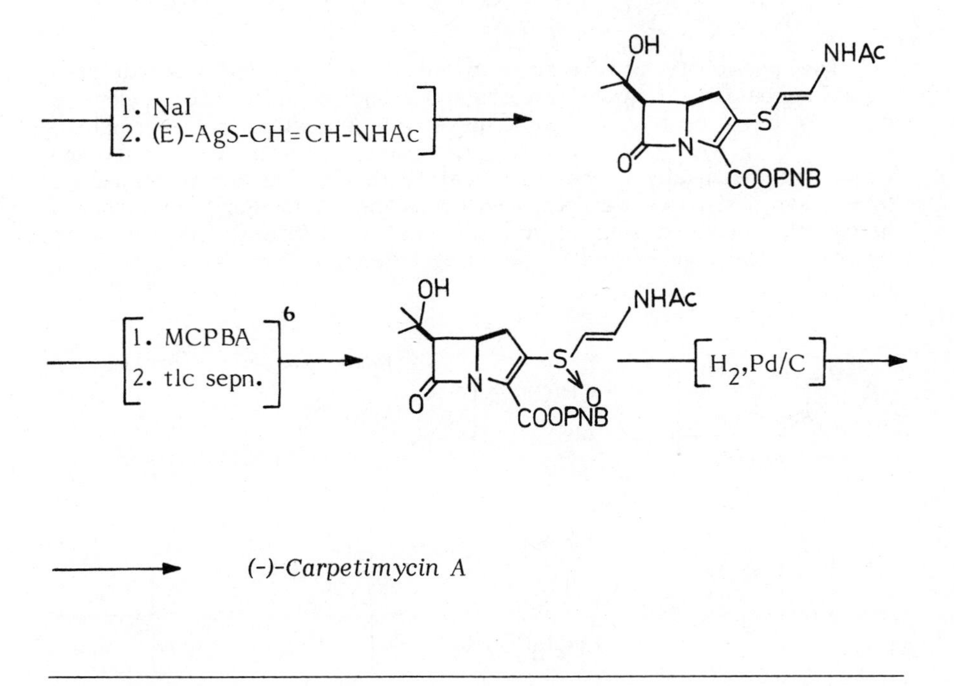

---

6. In oxidation a mixture of (R)- and (S)-sulfoxides was obtained, which were separated by preparative tlc in 45 and 47% yield respectively; hydrogenolysis of (R)-sulfoxide yielded (-)-carpetimycin A.

CATENANES

The possibility of creating an interlocked system has captured the imagination of many investigators and a variety of ingenious methods have been tried to achieve this objective(1). Lüttringhaus and Schill(2) were the first to effect the synthesis of a catenane employing essentially a non-statistical approach. The key intermediate in this synthesis, (A), was prepared by an intramolecular N-dialkylation across the benzene ring of a bis-ω -haloalkyl acetal (B), which on hydrolysis easily generated the required ketone.

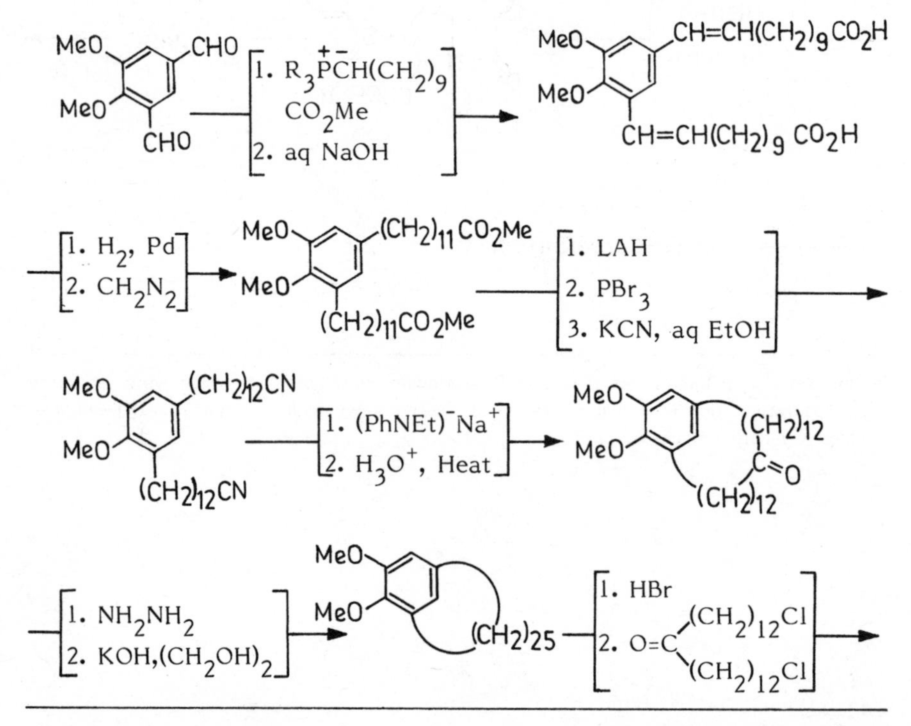

1. Frisch, H.D., Wasserman, E. [J. Am. Chem. Soc., 1961, 83, 3789], have dealt with chemical topology, examining in some detail several model systems and also possible ways by which interlocked systems could be created.

2. Schill, G., Lüttringhaus, A. Angew. Chem. Internat. Ed., 1964, 3, 546; Lüttringhaus,A., Isele, G. ibid, 1967, 6, 956; Schill, G. Ber. 1967, 100, 2021. Schill, C., Logemann, E., Zurcher, C. Angew. Int., 1972, 11, 1089; Schill, G., Logemann, E., Vetter, W. Angew Int., 1972, 11, 1089.

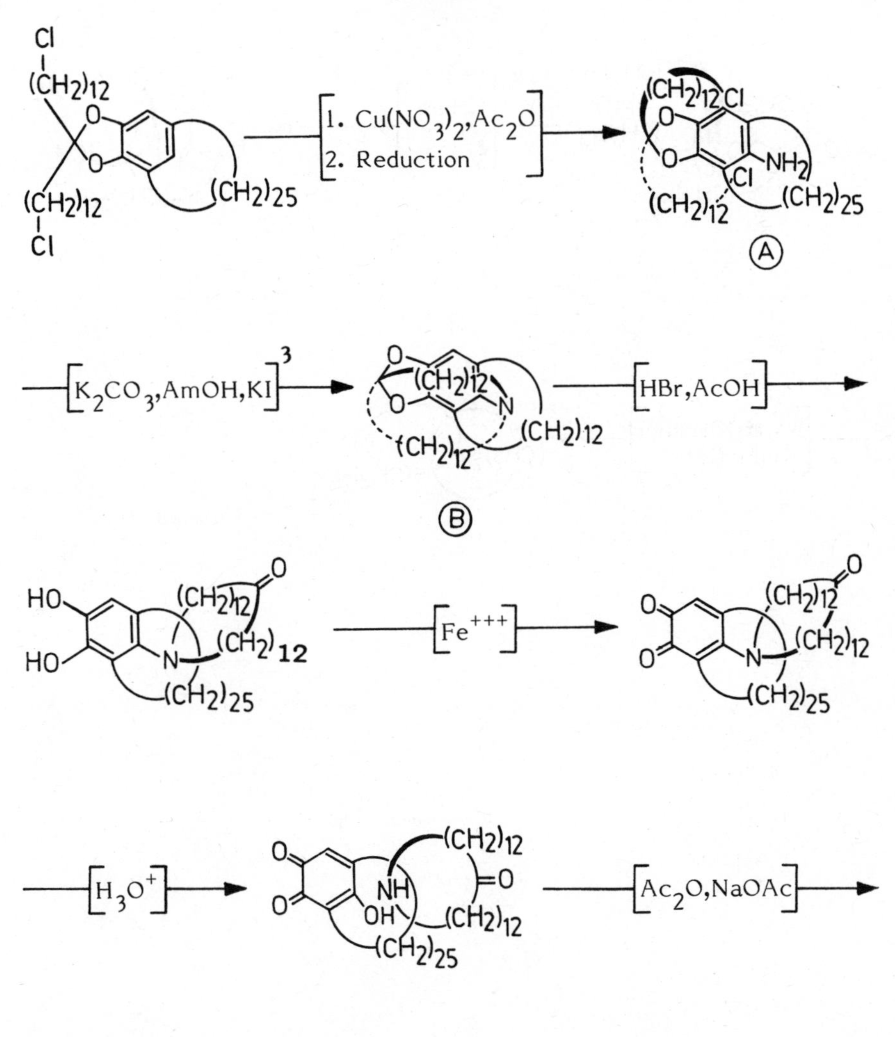

---

3. The possible alternative mode of cyclization of the "ansa" compound (not involving interlocking) is not sterically favoured.

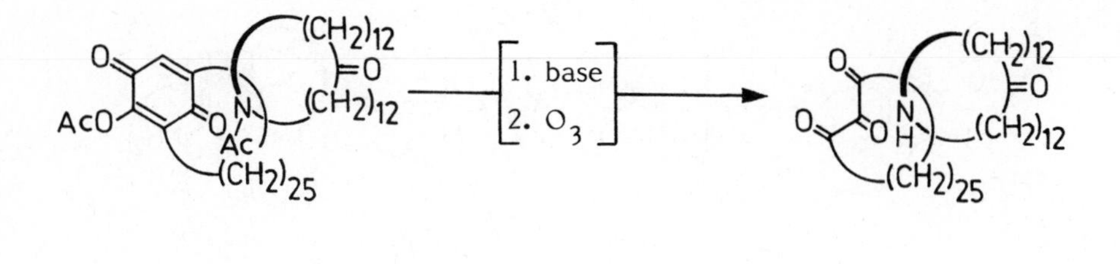
(CH2)12
O
AcO
N
O
Ac
(CH2)12
(CH2)25
1. base
2. O3
O
O
O
N
H
(CH2)12
O
(CH2)12
(CH2)25

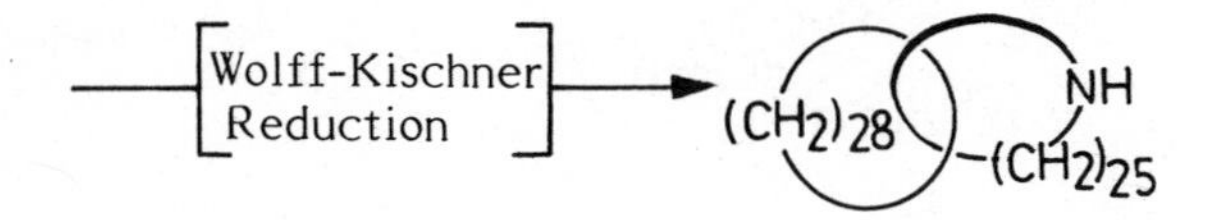
Wolff-Kischner
Reduction
(CH2)28
NH
(CH2)25

Catenanes

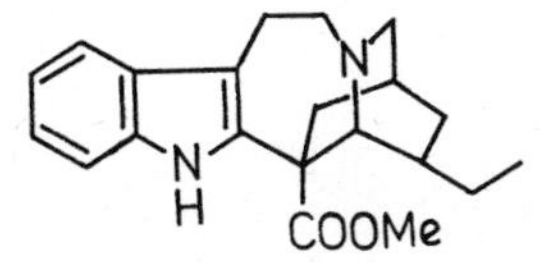

CATHARANTHINE

Catharanthine is only a minor constituent of the alkaloids of Catharanthus roseus, and therefore its practical synthesis is of critical importance for the preparation of semisynthetic vinblastine, vincristine and related dimeric alkaloids. A number of syntheses have been described for (±)-catharanthine (1). Raucher and his associates have very recently described an efficient total synthesis of (±)-catharanthine which employs a Diels-Alder reaction between a dihydropyridine and α-chloroacryloyl chloride to assemble the appropriately substituted isoquinuclidine intermediate (A).

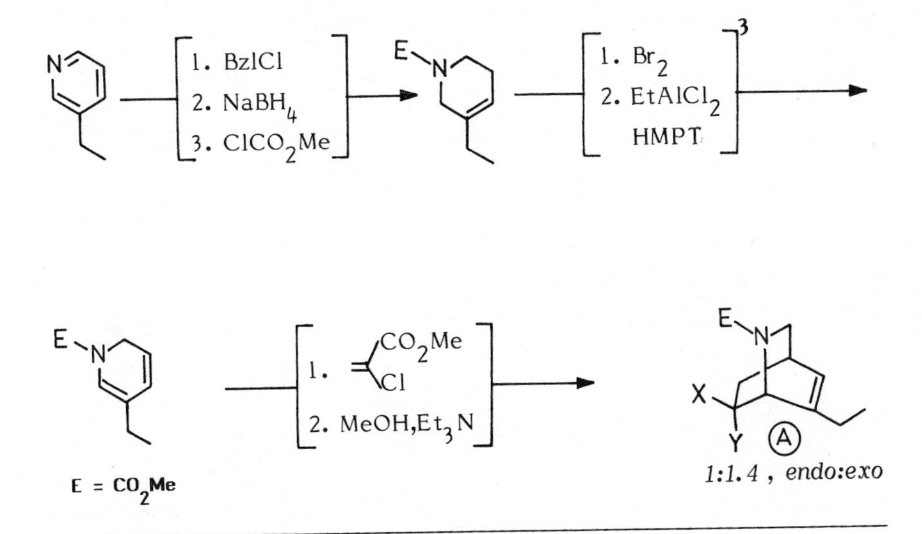

1:1.4 , endo:exo

1. (a) Buchi,G.; Kulsa,P.; Ogasawara,K.; Rosati,R. J. Am. Chem. Soc., 1969, 92, 999; (b) Marazano,C.; LeGoff,M.T.; Fourrey,J.L.; Das,B. J. Chem. Soc., Chem. Commun., 1981, 389; (c) Keuhne,M.E.; Bornmann,W.G.; Earley,W.G.; Mark,I. J. Org. Chem., 1986, 51, 2913.
2. Raucher,S.; Bray,B.L. J. Org. Chem., 1985, 50, 3236; Raucher,S.; Bray,B.L.; Lawrence,R.F. J. Am. Chem. Soc., 1987, 109, 442.
3. This dehydrobromination procedure is crucial for this preparation since other reagents such as DBU, DBN, quinoline were not successful.

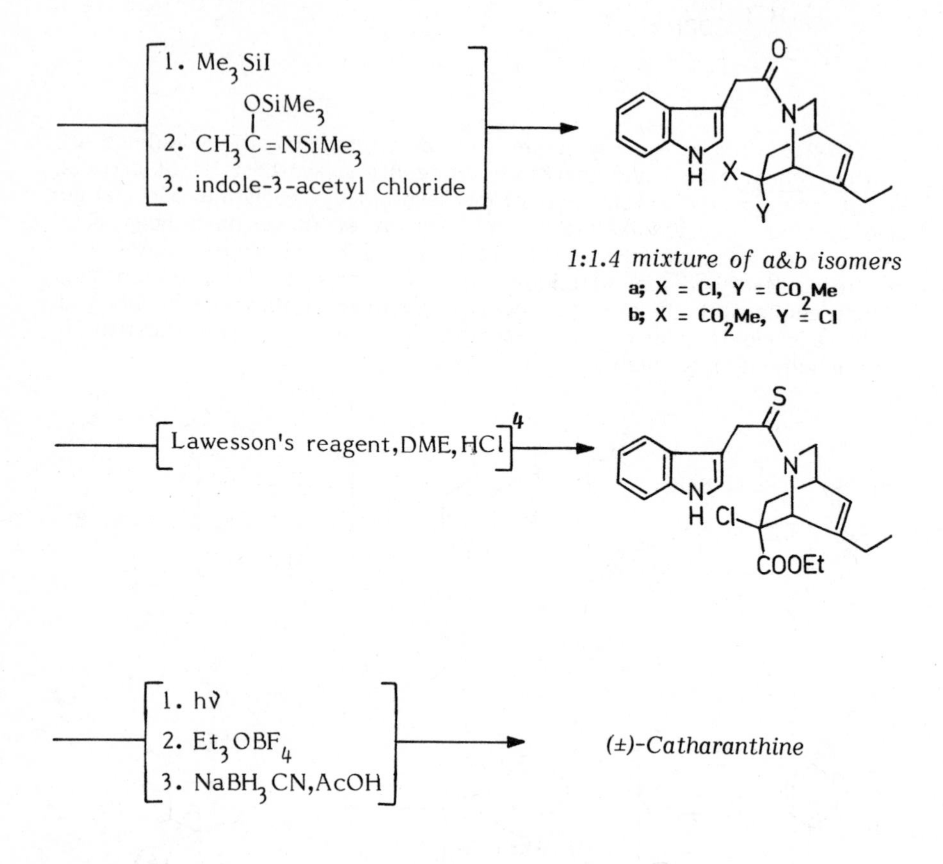

4. This thioamide formation did not proceed well without catalytic amount of anhydrous HCl, which presumably causes isomerisation of isomer b to a, which undergoes thionation readily.

CAVITANDS

An emerging facet in the art of organic synthesis is the creation of tailor made cavities on a rigid frame. Such structures, called cavitands, are molecular vessels which can harbor substrates and reagents. They could be used to simulate biological processes, since, in such enzyme controlled changes the acceptor sites possess rigid cavities whose internal surface are compatible with that of the substrates and reagents. A range of cavitands have been made (1) and cavitands II and III, arising from the resorcinol-acetaldehyde tetramer I, would be a good illustration. Molecular models show that the cavity dimensions of II can accommodate single molecule of CH_2Cl_2, $CHCl_3$, THF or 4 molecules of water. In III the void is big enough to engulf one molecule of [2.2]paracyclophane or 10 of water (2).

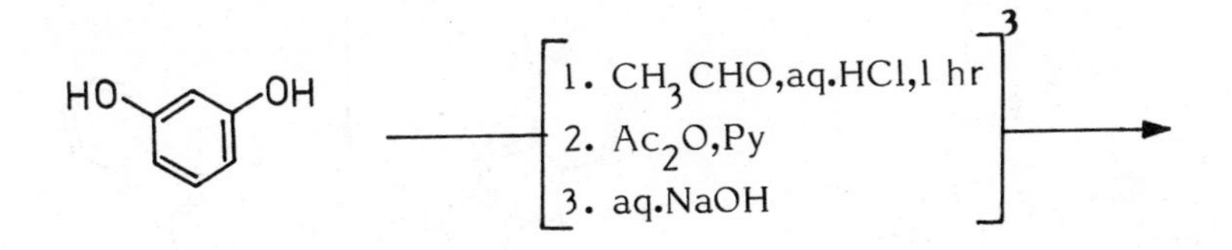

1. Cram,D.J. Science, 1983, 219, 1177; Cram,D.J. Trueblood Top. Curr. Chem., 1981, 98, 43.

2. Moran,J.R.; Karbach,S.; Cram,D.J. J. Am. Chem. Soc., 1982, 104, 5826.

3. The reaction gave, in addition to the all cis I, the cis, trans, cis isomer in smaller amounts, which could be separated by acetylation.

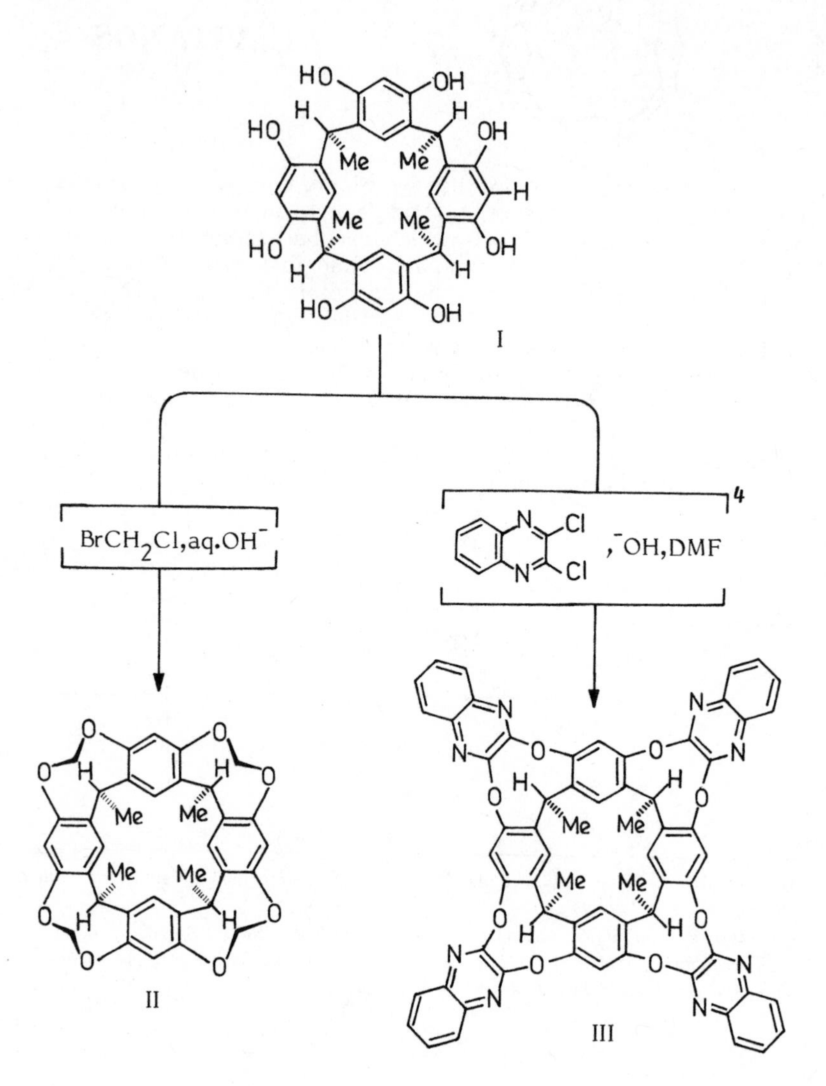

4. Compound III always crystallized with solvents (DMF,$CHCl_3$) which could not be freed even at 100°C, 10^{-5} torr.

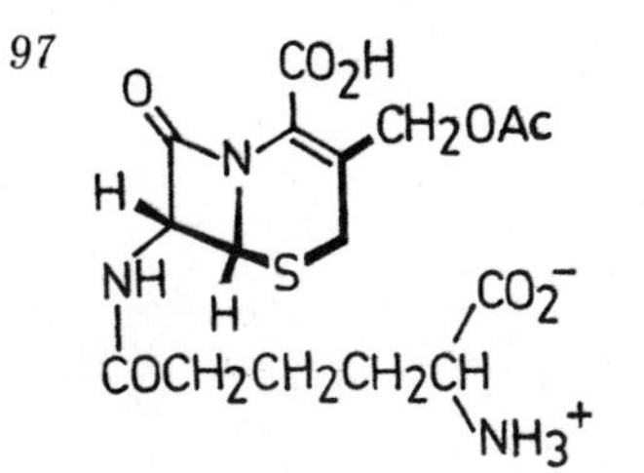

CEPHALOSPORIN-C

Beset by the same problems that had made difficult the synthesis of penicillin, attempts for the synthesis of cephalosporin (1) projected through 7-acylaminocephalosporanic acid--the transitory dihydrothiazine counterpart of penicilloic acid--made little progress (2,3). In an ingenious solution of the synthetical problem, Woodward and his colleagues (4) have obtained the fused β-lactam-dihydrothiazine ring by Michael addition of a dialdehyde, followed by cyclization, on the key β-lactam intermediate (A) which represents the common structural feature of both penicillins and cephalosporins.

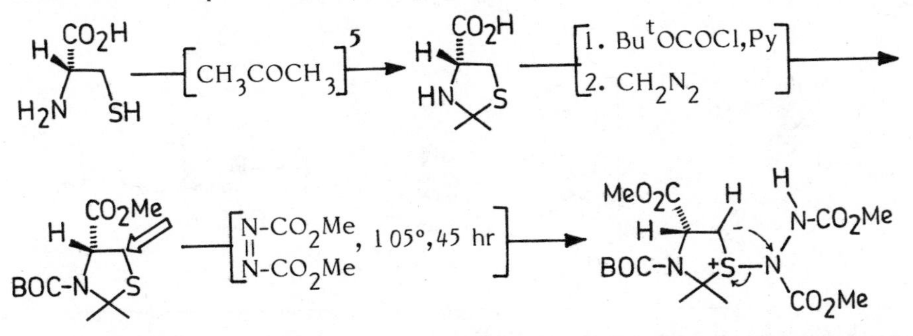

1. For a review, see Abraham,E.P., Quart. Rev., 1967, 21, 231.

2. Galantay,E., Engl,H., Szabo,A., Fried,J., J. Org. Chem., 1964, 29, 3560; Sheehan,J.C., Schneider,J.A., ibid, 1966, 31, 1635. The closest approach to cephalosporins, using a route similar to Sheehan's penicillin synthesis is the synthesis of a derivative of cephalosporin C_C by Heymes,R., Amiard,G., Nomine,G., Compt. Rend., 1966, 263, 170.

3. The interesting possibility of realizing the desired dihydrothiazine nucleus by ring expansion of thiazolidines has been explored with some success, Morin,R.B., Jackson,B.G., Mueller,R.A., Lavagnino,E.R., Scanlon,W.B., Andrews,S.L., J. Am. Chem. Soc., 1963, 85, 1896; Stork,G., Cheung,H.T., J. Am. Chem. Soc., 1965, 87, 3783.

4. Woodward,R.B., Heusler,K., Gosteli,J., Naegeli,P., Oppolzer,W., Ramage,R., Ranganathan,S., Vorburggen,H., J. Am. Chem. Soc., 1966, 88, 852; Woodward,R.B., Science, 1966, 153, 487.

5. Binding of the reactive amino and sulfhydryl groups of L(+)-cysteine into a thiazolidine ring followed by introduction of tert-butoxycarbonyl group on the nitrogen was carried out to enhance reactivity of the methylene group.

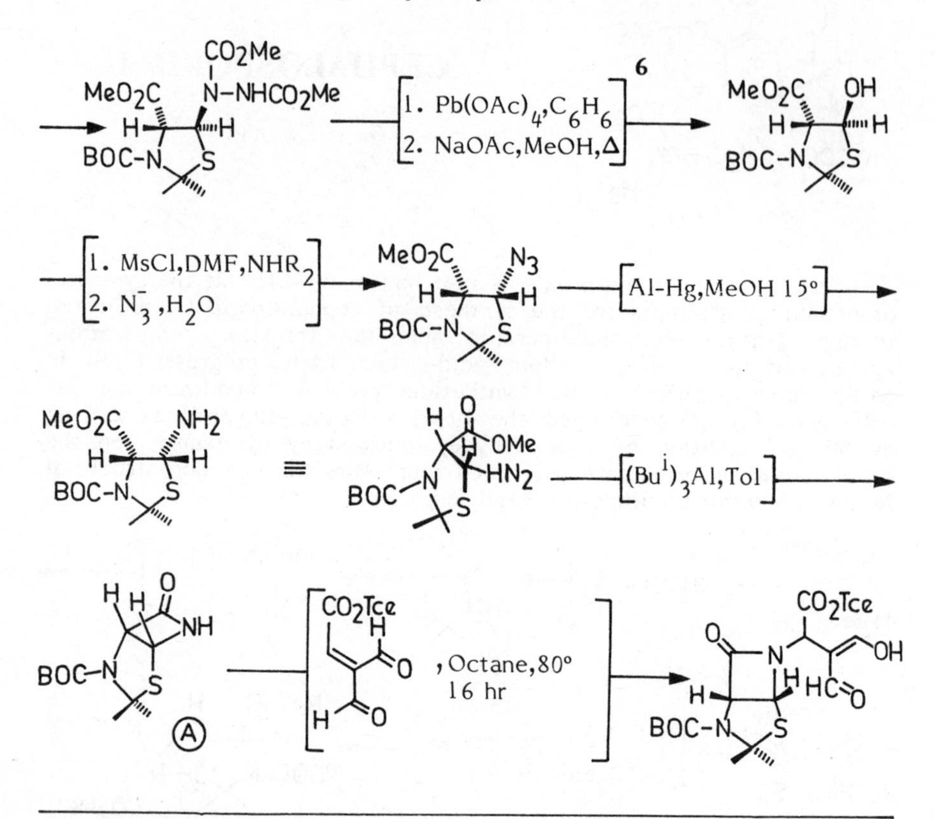

6. Introduction of the trans- acetoxy group occurs under directive influence of the bulky carbomethoxy on the adjacent carbon and is probably initiated by oxidation of the substituted hydrazine part:

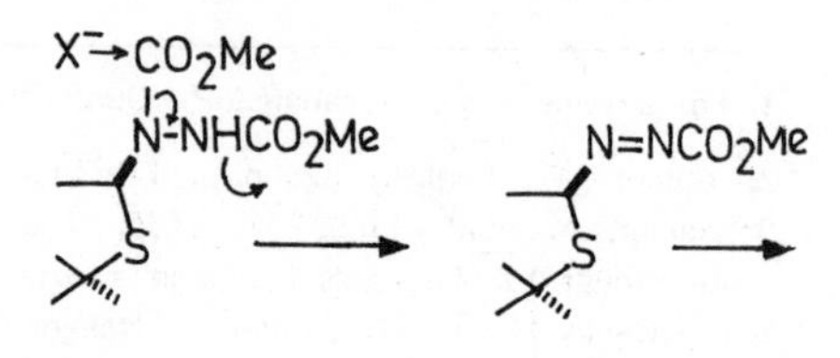

MeO2C N-NHCO2Me
H
BOC-N S

OAc
N=NCO2Me
S
B⁻

OAc
H
S

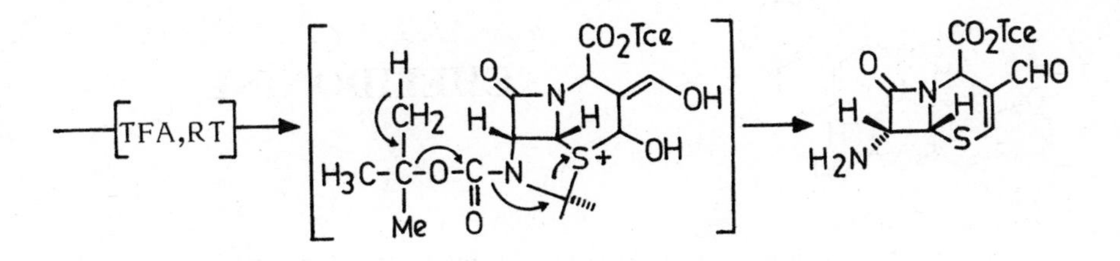

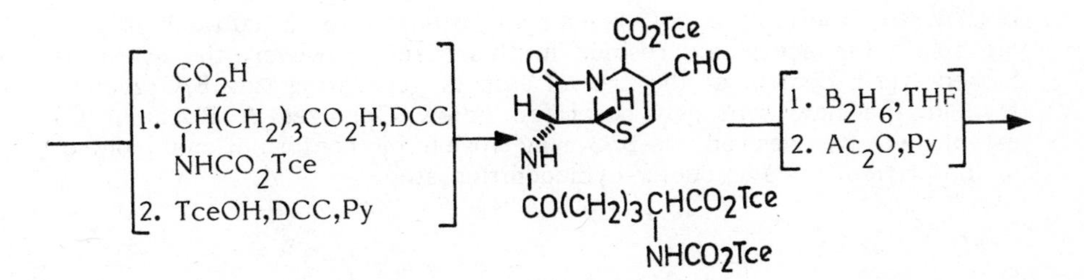

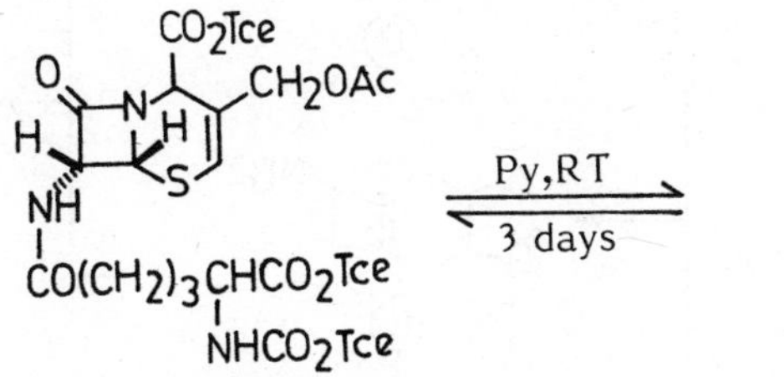

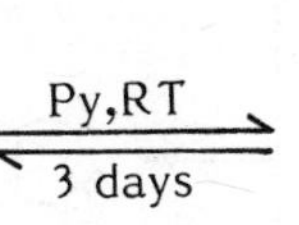

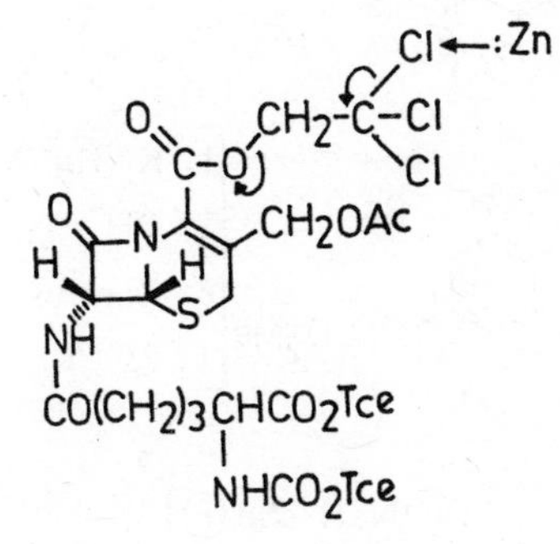

—[Zn,aq AcOH,0°]→ *Cephalosporin-C*

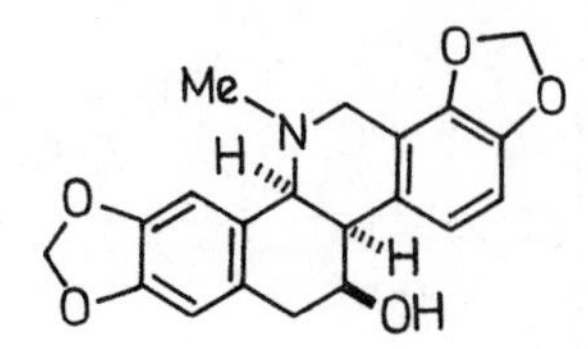

CHELIDONINE

The first synthesis of chelidonine by Oppolzer and Keller (1) exploited a versatile approach to annellated polycyclic heterocycles based on intramolecular cycloaddition of o-quinodimethanes, generated readily by thermolysis of benzocyclobutenes, to a suitably placed dienophile (an acetylenic residue in this case). However, the synthesis lacked steric control at the crucial step of generating ring B/C geometry. In the new synthesis described below Oppolzer & Robbiani (2) established the desired cis-B/C ring fusion by conformational control in the critical intramolecular cycloaddition step.

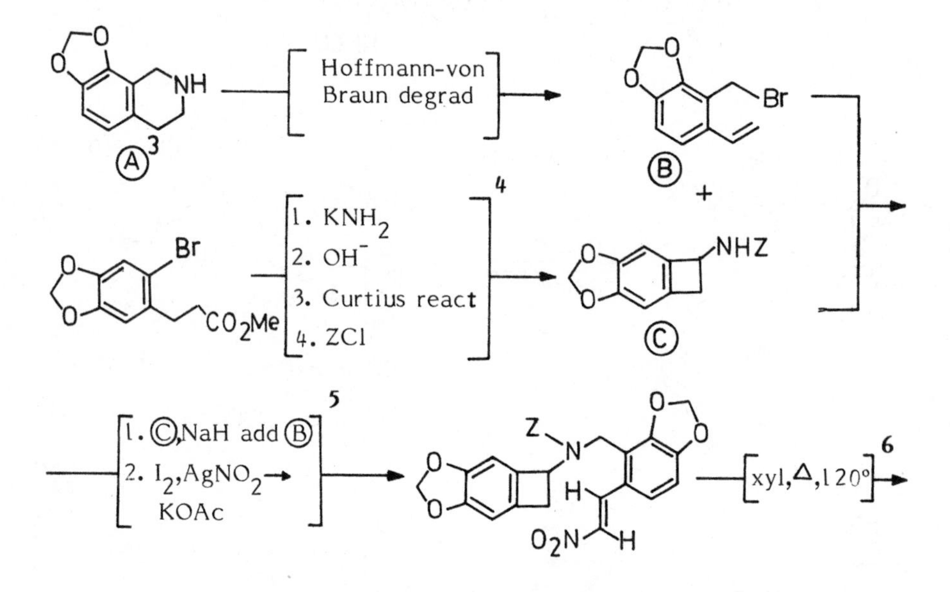

1. Oppolzer,W.; Keller,K. J. Am. Chem. Soc., 1971, 93, 3836.

2. Oppolzer,W.; Robbiani,C. Helv. Chim. Acta, 1983, 66, 1119.

3. (A) was obtained from 2,3-methylenedioxybenzaldehyde by standard reactions.

4. The cyano benzocyclobutene intermediate was also prepared by flash pyrolysis of 4,5-methylenedioxy-2-methyl-α-chlorobenzylcyanide.

5. In the nitro displacement reaction (E)-nitrostyrene was the only isolable product.

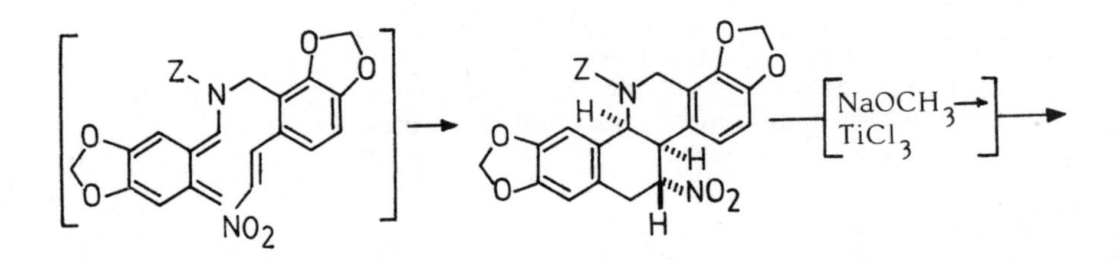

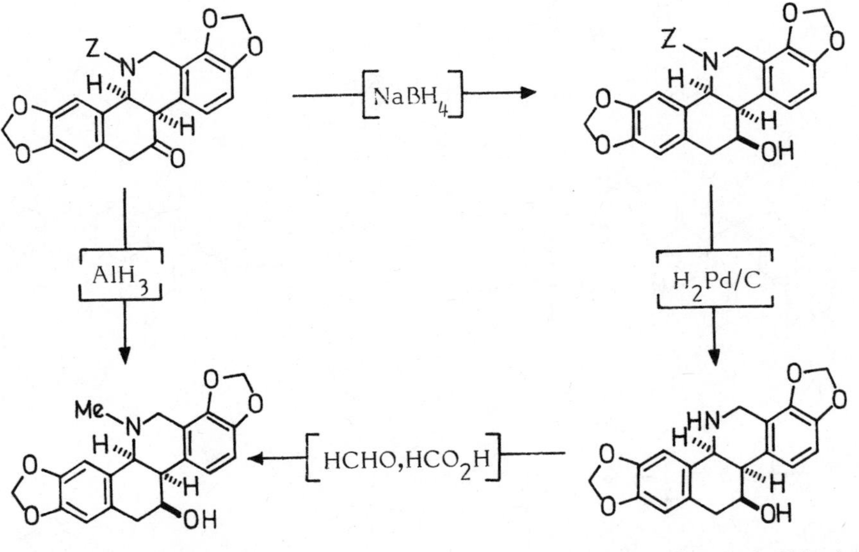

(±) Chelidonine (±) Norchelidonine

---

6. There would be a strong preference for exo-nitro transition state (D) over endo-nitro-transition state (E), which would explain the observed stereoselectivity. The addition of benzocyclobutene to ω-nitrostyrene took place in the opposite direction and also gave a 2:1 mixture of epimers at C-carbamate position, thus emphasising the advantages of intra- vs inter-molecular cycloadditions for regio- and stereo-chemical control.

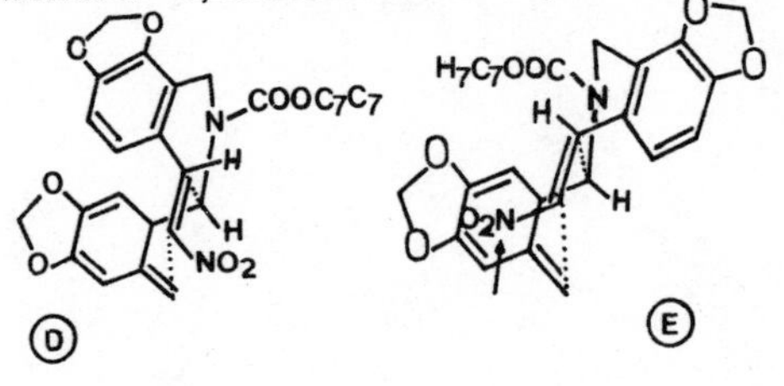

Cushman _et al_ (7) developed a synthesis of benzophenanthridine alkaloids, including chelidonine, based on addition of homophthallic anhydride to a Schiff's base and optimised the reaction conditions for the desired thermodynamically less stable cis-diastereoisomer.

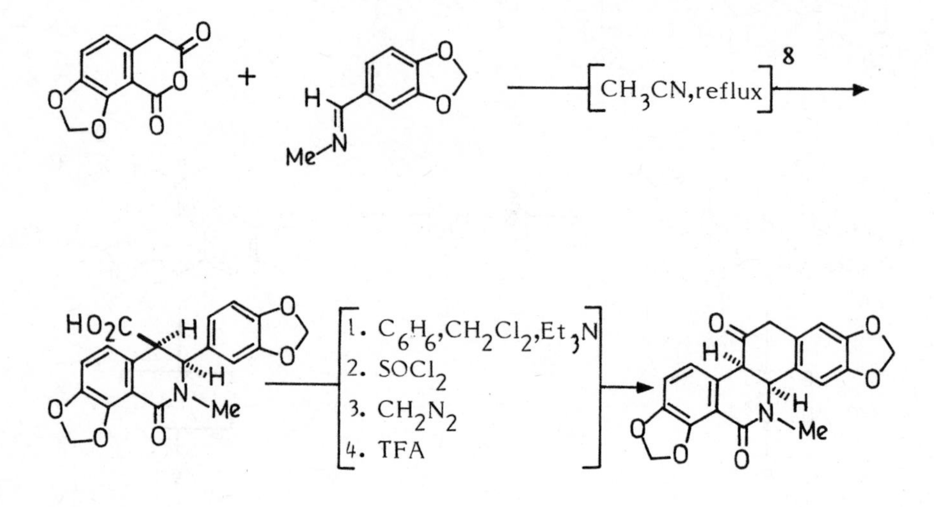

—[$LiAlH_4$]→ *(±) Chelidonine*

---

7. Cushman,M.; Choong,Tung-Chung; Valko,J.T.; Koleck,M.P. J. Org. Chem.,1980, 45, 5067; Tetrahedron Lett., 1980, 21, 3845.

8. Of a number of solvents and reaction condition tried this gave the most favourable ratio of the required cis to trans isomers, 67:33; the cis diastereomer is thermodynamically less stable.

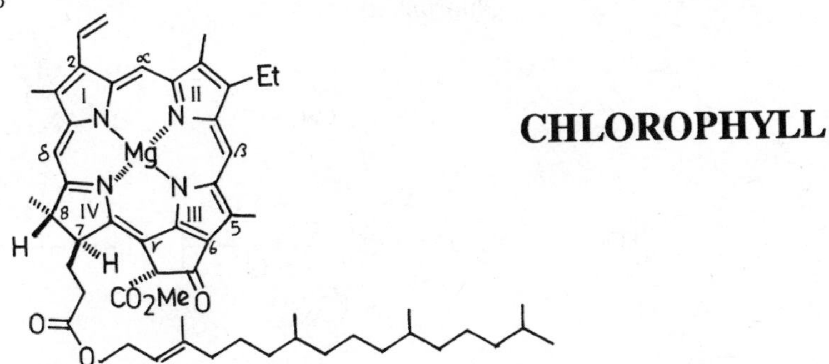

CHLOROPHYLL

The total synthesis of so complex a substance as chlorophyll-α accomplished by Woodward and his colleagues (1) stands as a masterpiece of synthetic organic chemistry. A consideration of the stereochemical aspects and the interatomic distances in the environs of ring III and IV in chlorophyll reveals that the porphyrin→purpurin equilibration (B) is clearly a favoured transformation (conversion of trigonal C-7 and C-8 to tetrahedral configuration), resulting in release of steric compression due to molecular overcrowding along the γ-periphery (2). The creation of the isocyclic ring and selective delivery of the hydrogen at C-8 [transformation (B)] are an outcome of this understanding. Other noteworthy features ofthe synthesis are generation of the vinyl group (4) and fashioning of the γ -substituent by a novel oxidation of the isocyclic ring to give chlorin e_6 from which the sequence of reactions to chlorophyll-α had already been established by Willstatter and Fischer (8).

---

1. Woodward,R.B. et al. J. Am. Chem. Soc., 1960, 82, 3800; Woodward,R.B. Angew. Chem., 1960, 72, 651.

2. The release in steric compression along the crowded γ-periphery that can occur in going from a porphyrin to a purpurin structure is depicted below:

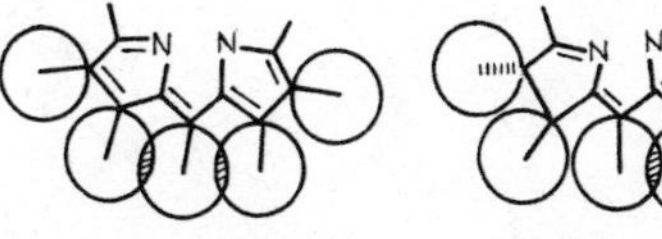

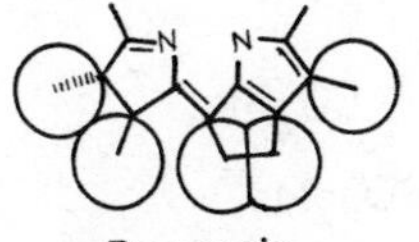

Porphyrin *Chlorin* *Purpurin*

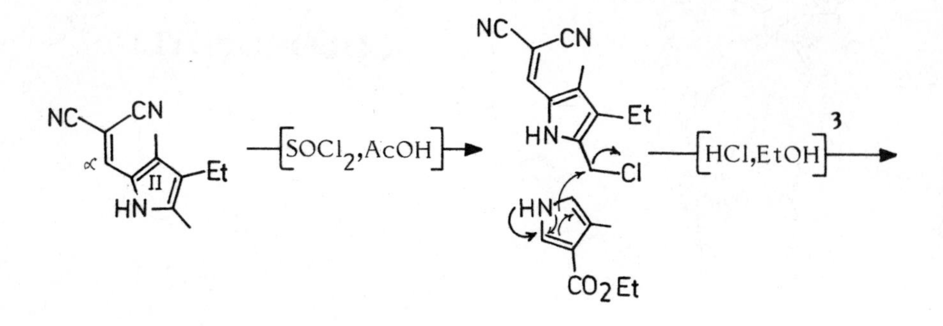

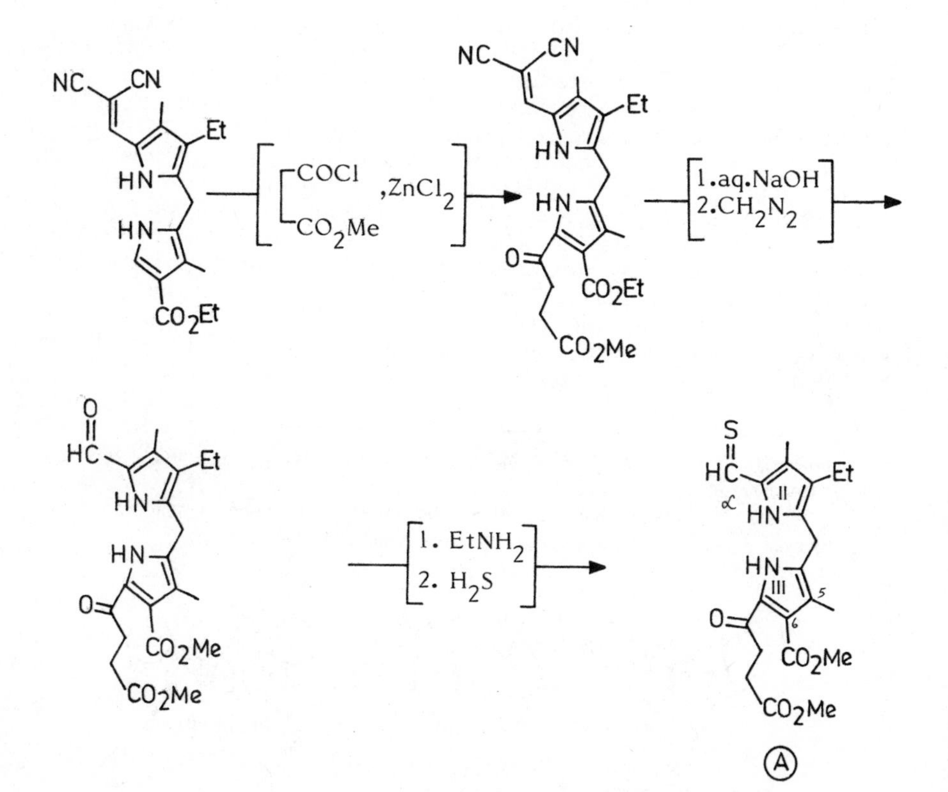

---

3. Malonodinitrile serves as an acid-stable aldehyde protecting group.

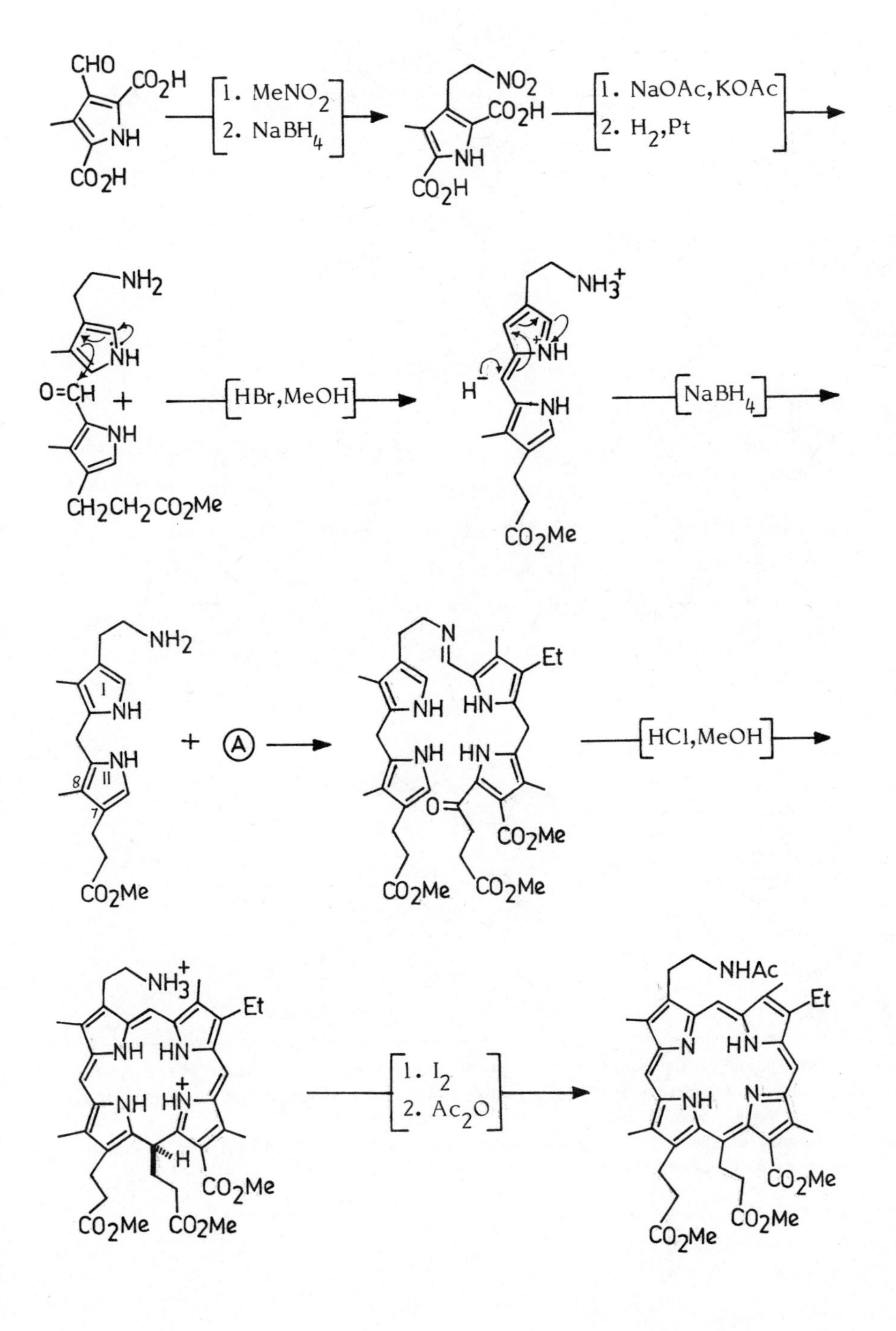

CHO
CO_2H
NH
CO_2H
1. $MeNO_2$
2. $NaBH_4$
NO_2
CO_2H
NH
CO_2H
1. NaOAc, KOAc
2. H_2, Pt
NH_2
NH
O=CH +
NH
$CH_2CH_2CO_2Me$
HBr, MeOH
NH_3^+
NH
H
NH
CO_2Me
$NaBH_4$
NH_2
I
NH
NH
II
8
7
CO_2Me
+ A
N
Et
NH
HN
NH
HN
O
CO_2Me
CO_2Me
CO_2Me
HCl, MeOH
NH_3^+
Et
NH
HN
NH
HN
H
CO_2Me
CO_2Me
CO_2Me
1. I_2
2. Ac_2O
NHAc
Et
N
HN
NH
N
CO_2Me
CO_2Me
CO_2Me

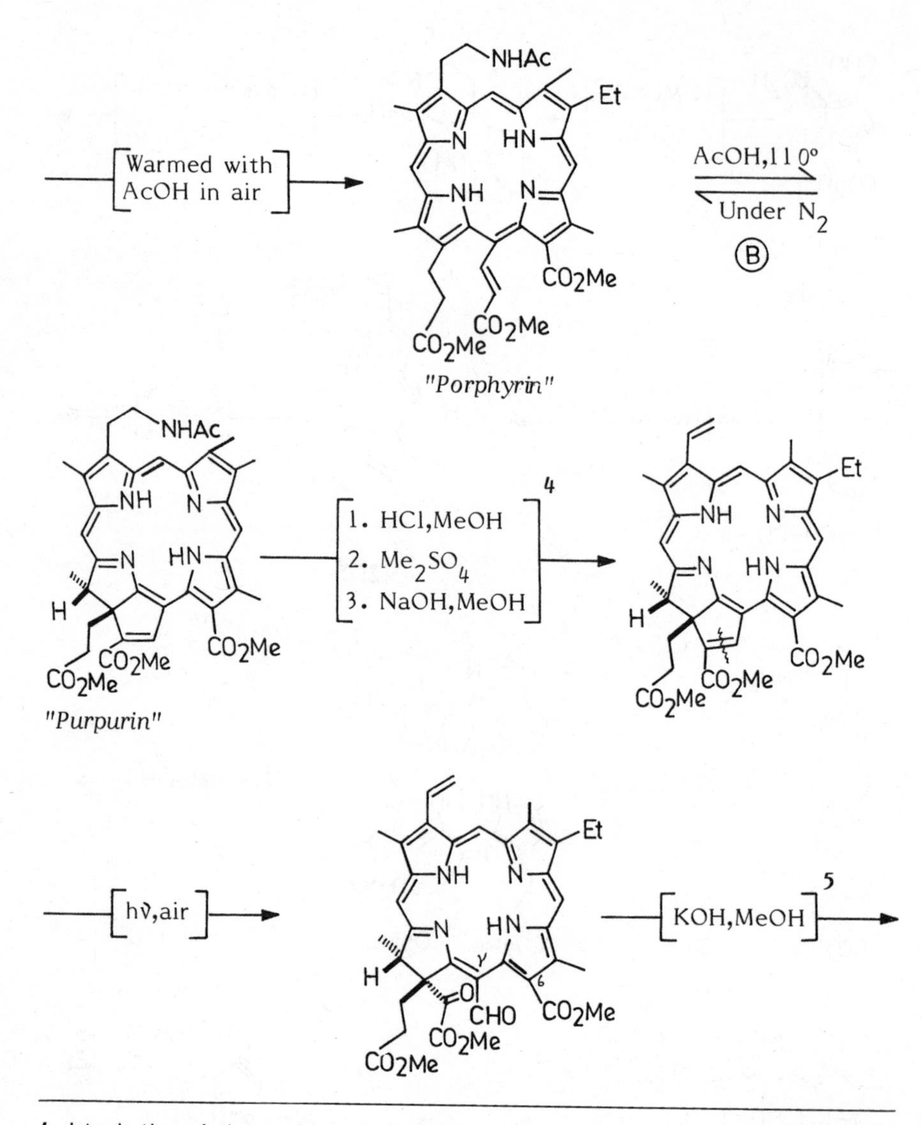

---

4. Introduction of the sensitive vinyl group, deferred until a late stage in the synthesis, by Hofmann degradation of a β-aminoethyl group was a strategy first evolved by Woodward in the synthesis of quinine.

5. Removal of the oxalyl group and simultaneous cyclization of formyl and 6-carbethoxy group occurs in this treatment.

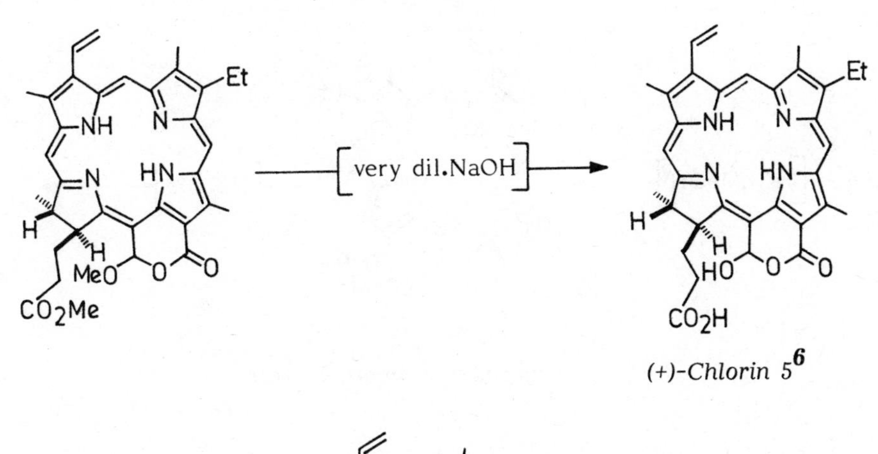

(+)-Chlorin 5[6]

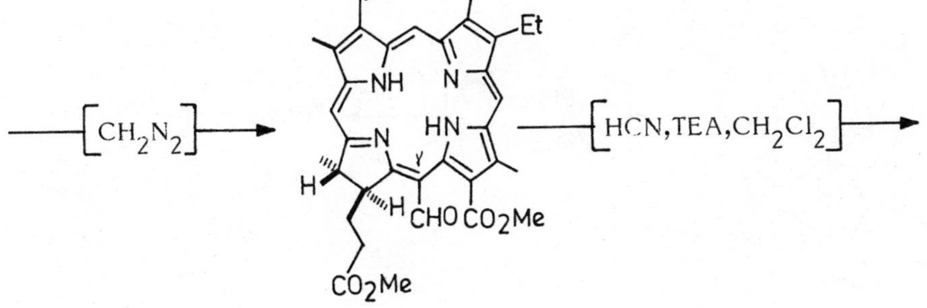

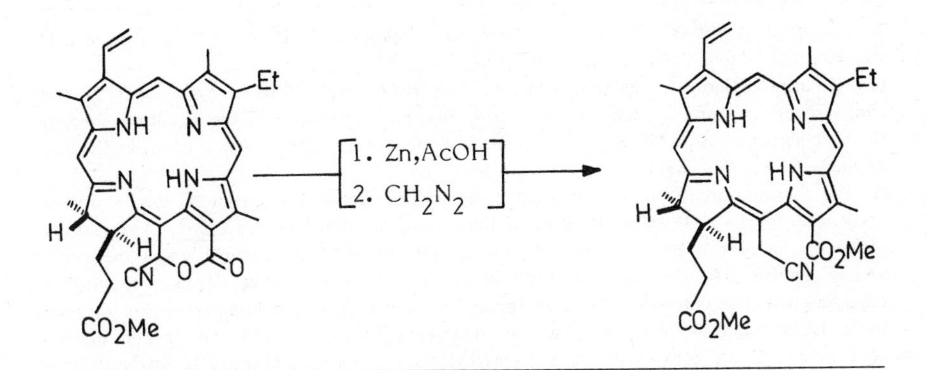

---

6. Resolved via quinine salt.

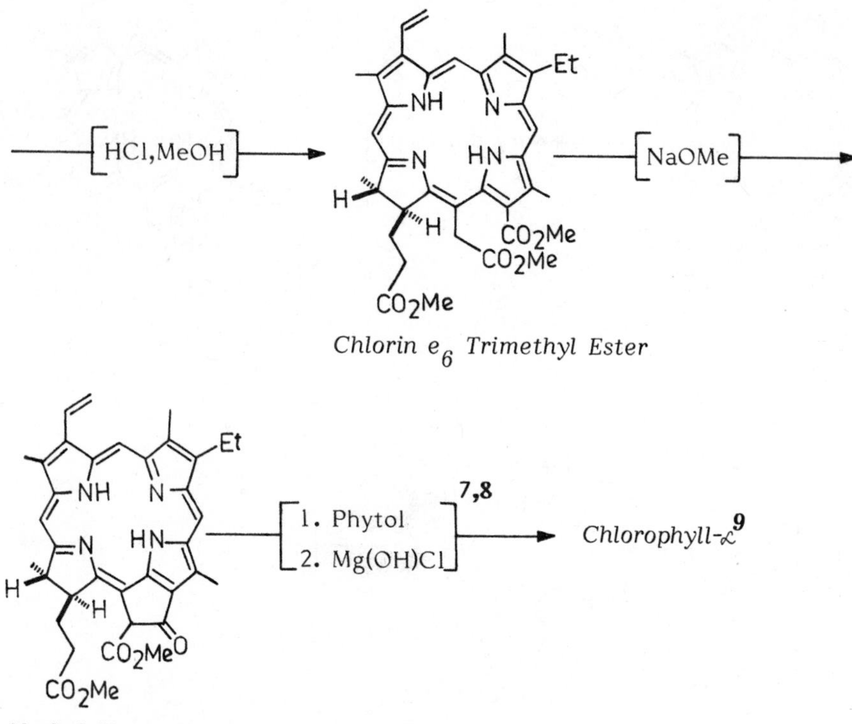

---

7. The total synthesis of phytol has been accomplished by Burrell,J.W.K.; Jackman,L.M.; Weedon,B.C.L. Proc. Chem. Soc., 1959, 263.

8. For introduction of the phytyl group, see: Willstater,R.; Stoll,A. Annalen, 1911, 380, 148; Fischer,H.; Stern,A. Annalen, 1935, 519, 244. For conversion of the resulting pheophytin to chlorophyll, see: Willstater,R.; Forsen,L. Annalen, 1913, 396, 188; Fischer,H.; Goebel,S. Annalen, 1936, 524, 269.

9. The closest approach to chlorophyll by Hans Fischer had been a total synthesis, of phaeporphyrin-α_5 [Fischer,H.; Stier,E.; Kanngiesser,W. Annalen, 1940, 543, 258]. Fischer's attempts, cut short by an untimely death, have been finally brought to a close in a second synthesis of the plant pigment by a German team led by Strell. The synthesis proceeds with the porphyrin → chlorin reduction using sodium and isoamyl alcohol (although there is no rigorous proof of the two hydrogens having entered the desired positions at C-7 and C-8), conversion of a γ-methyl to an acetic acid residue, introduction of a 2-acetyl as the vinyl group progenitor and conversion to pheophorbide-α, which has been previously transformed into chlorophyll, Strell,M.; Kalojanoff,A.; Koller,H. Angew. Chem., 1960, 72, 169. See also: Johnson,A.W. Science Progr., 1961, 49, 77.

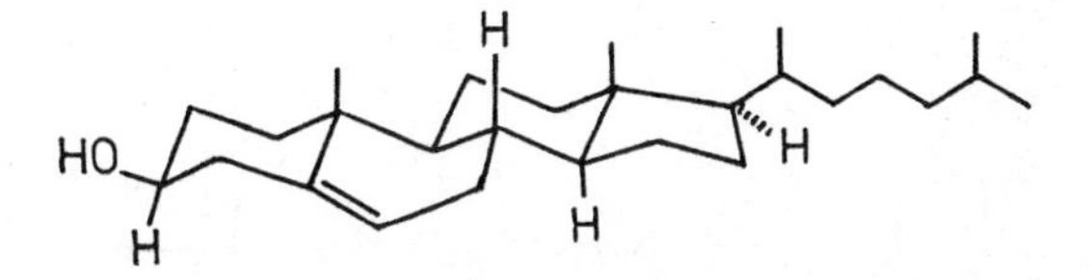

CHOLESTEROL

Two total syntheses (1) of cholesterol were reported almost simultaneously in 1951 (2). The synthesis developed by Woodward and his colleagues (3) is outlined below. It commences with introduction of the future ring D as a six-membered ring in a cis-fused bicyclic adduct (A) representing the C/D ring unit. Elaboration of ring B gives the key tricyclic ketone (B) in which the desired anti-trans structure has been ensured through enol mediated equilibrations at C-8 and C-14 on two separate occasions. Completion of ring A followed by the ring contraction sequence furnishes Woodward's steroid (C) which can be transformed into cholesterol (4).

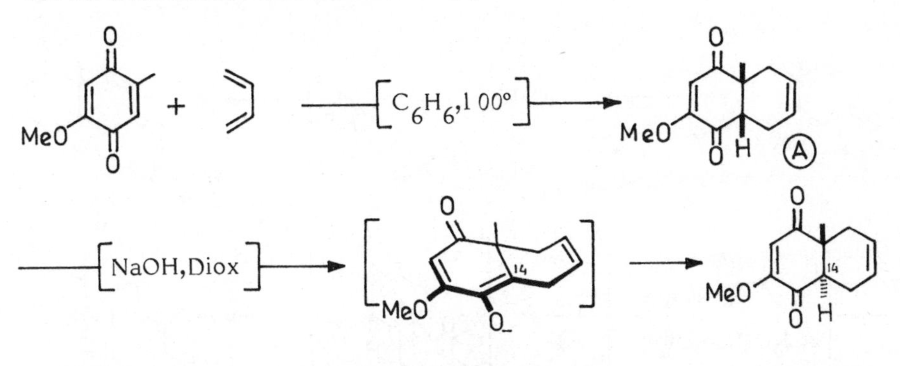

1. One of these syntheses was carried out at Harvard (ref.2) and the second in Robinson's Laboratory at Oxford. The classical approach of Robinson and his coworkers which resulted in a non-stereospecific synthesis of Reich's ketone, was based on a scheme devised as early as 1941; see: Cornforth,J.W. Prog. Org. Chem., 1955, 3, 21; it consists essentially in the addition of ring D (cf. equilenin) onto a saturated A-B-C tricyclic diketone derived from 1-methyl-5-methoxy-2-tetralone; Cardwell,H.M.E.; Cornforth,J.W.; Duff,S.R.; Holtermann,H.; Robinson,R. Chem. Ind., 1951, 389; J. Chem. Soc., 1953, 361.

2. Cholesterol has also been synthesised from dihydroprogesterone by adding the isohexyl side chain; Keana,J.F.W.; Johnson,W.S. Steroids, 1964, 4, 457.

3. Woodward,R.B.; Sondheimer,F.; Taub,D.; Heusler,K.; McLamore,W.M. J. Am. Chem. Soc., 1951, 73, 2403, 3547, 3548; 1952, 74, 4223.

4. Woodward's synthetic steroid (C) can be transformed into intermediates from which the paths to both androgenic and progestational hormones are available. If the $\Delta^{9(11)}$-double bond in (C) is utilized for introduction of the 11-oxygen function, the route to cortisone also lies open.

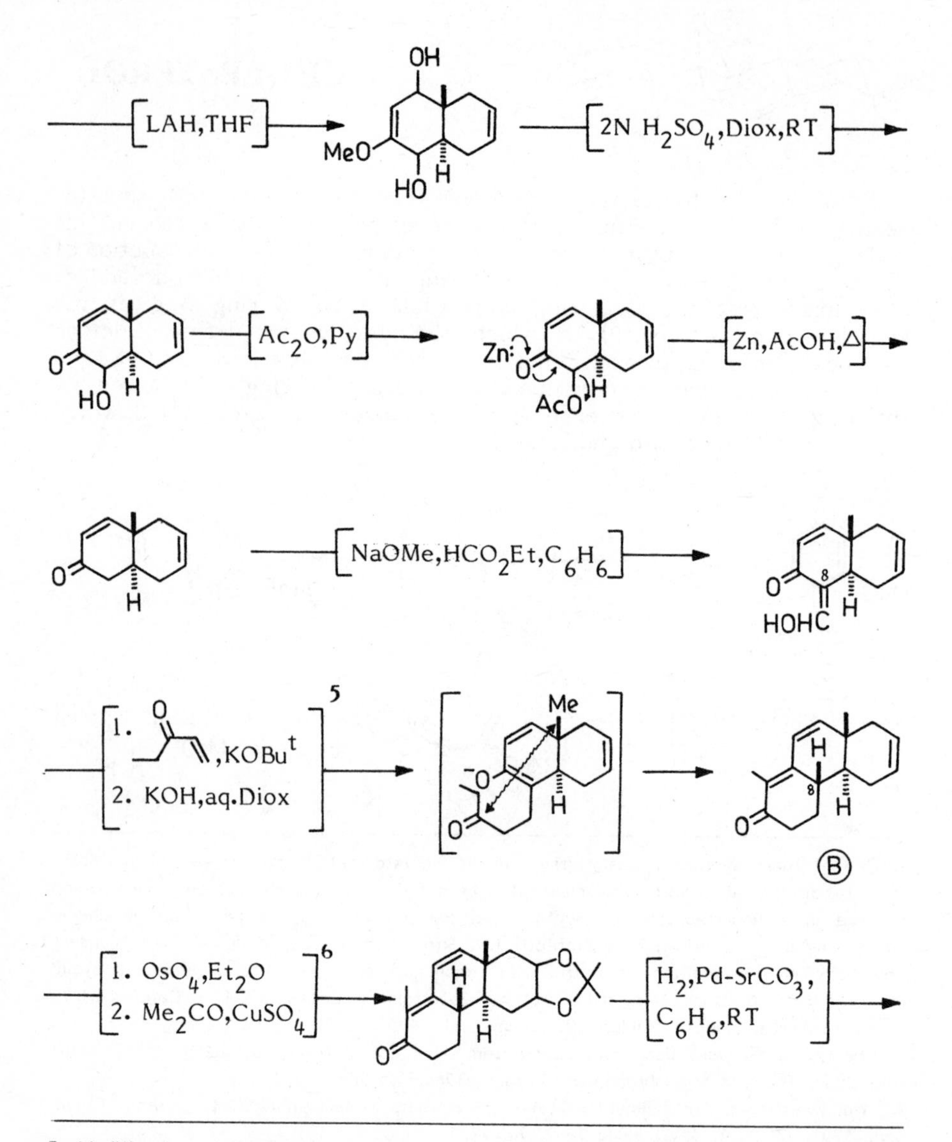

5. Modification of Robinson's annelation reaction by Shunk,C.H.; Wilds,A.L. J. Am. Chem. Soc., 1949, 71, 3946, for preparation of polycyclic cyclohexenones.

6. Iodine, silver acetate and wet acetic acid may be employed with advantage for oxidation of the double bond to a β-cis glycol; Woodword,R.B.; Brutcher,F.V. J. Am. Chem. Soc., 1958, 80, 209.

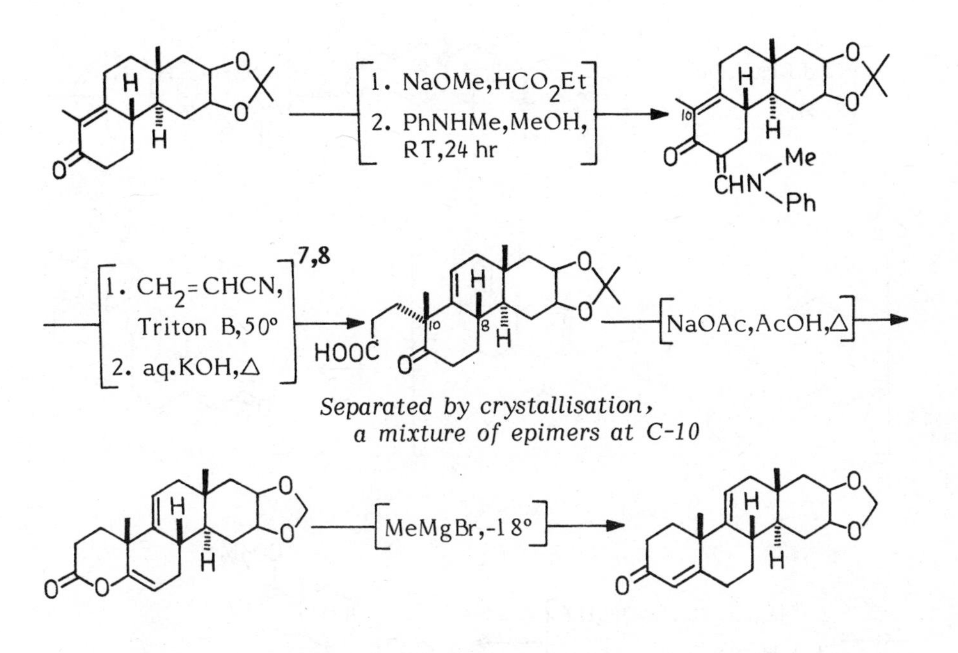

---

7. cf. Turner,R.B. J. Am. Chem. Soc., 1950, 72, 579; Fujimoto,G.I. ibid, 1951, 73, 1856.

8. Introduction of the asymmetric centre at C-10 results in the formation of two isomers in the ratio of 2:1, the incoming group preferentially occupying the axial position (attack from β-face). In the case at hand, where there is little difference in accessibility to the delocalized carbanion either from the α- or the β-side, the steric course of this addition appears to be determined largely by energy of the intermediate transition state. If the latter resembles the final ketonic product during rehybridization and establishment of the new C-C bond, then approach of the electrophilic attacking reagent from the β-face occurs in an intermediate resembling the pre-chair form (i). On the other hand α-attack (which results in formation of the desired 10-β-methyl isomer) occurs in an energetically less favoured pre-twist intermediate (ii). This probably explains the unfavourable stereochemical outcome of this reaction, cf. Velluz,L.; Valls,J.; Nomine,G. Angew. Chem. Internat. Ed., 1965, 4, 181.

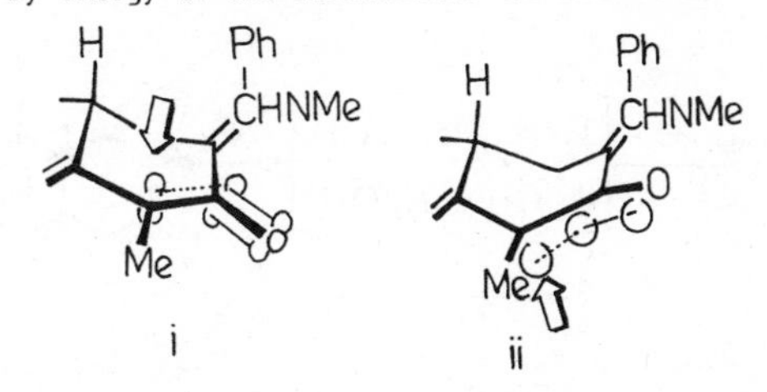

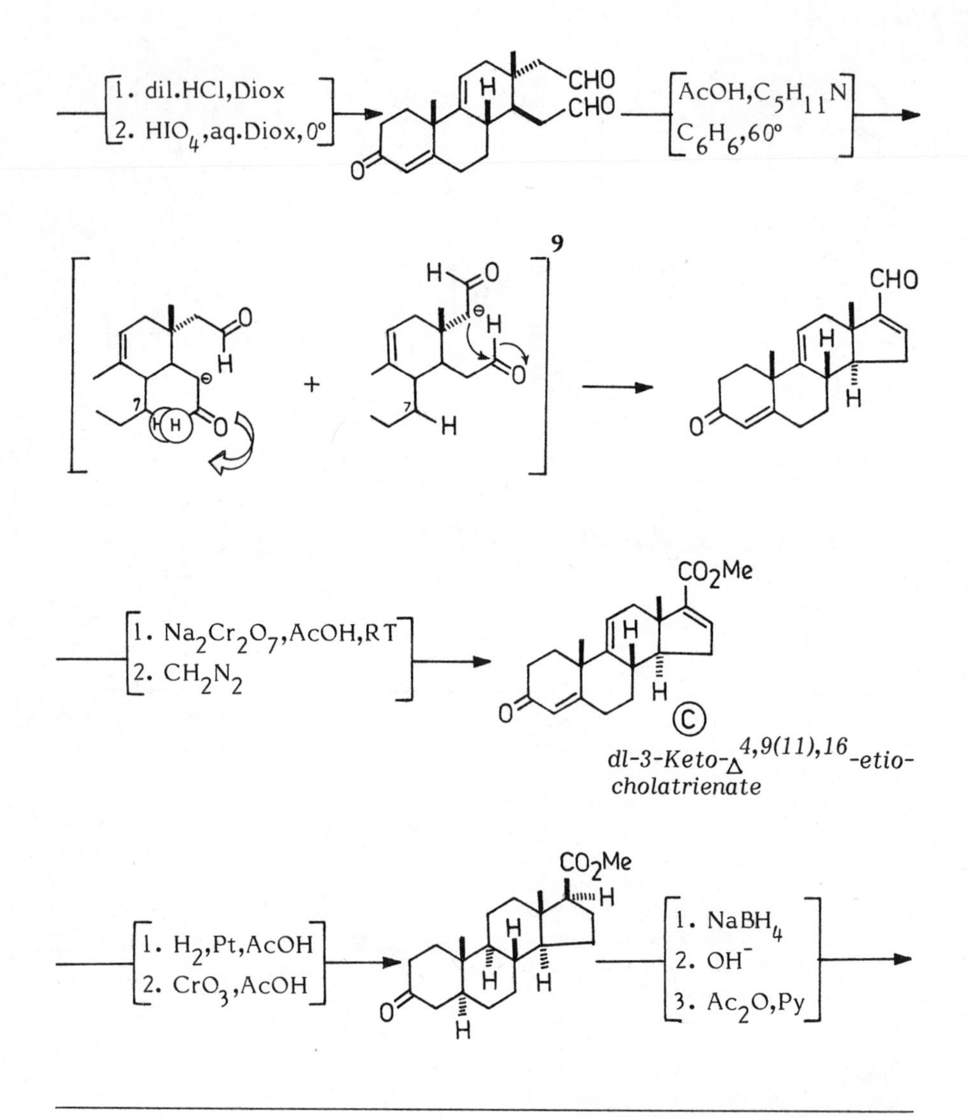

9. Aldol condensation occurs preferentially on the upper methylene group. This is attributed to the relatively less crowded environment of the upper methylene group, whereas either the catalyst fails to gain access to the lower methylene group or the anion, if formed, cannot be restrained with sufficient facility in a suitable orientation for the cyclization to take place.

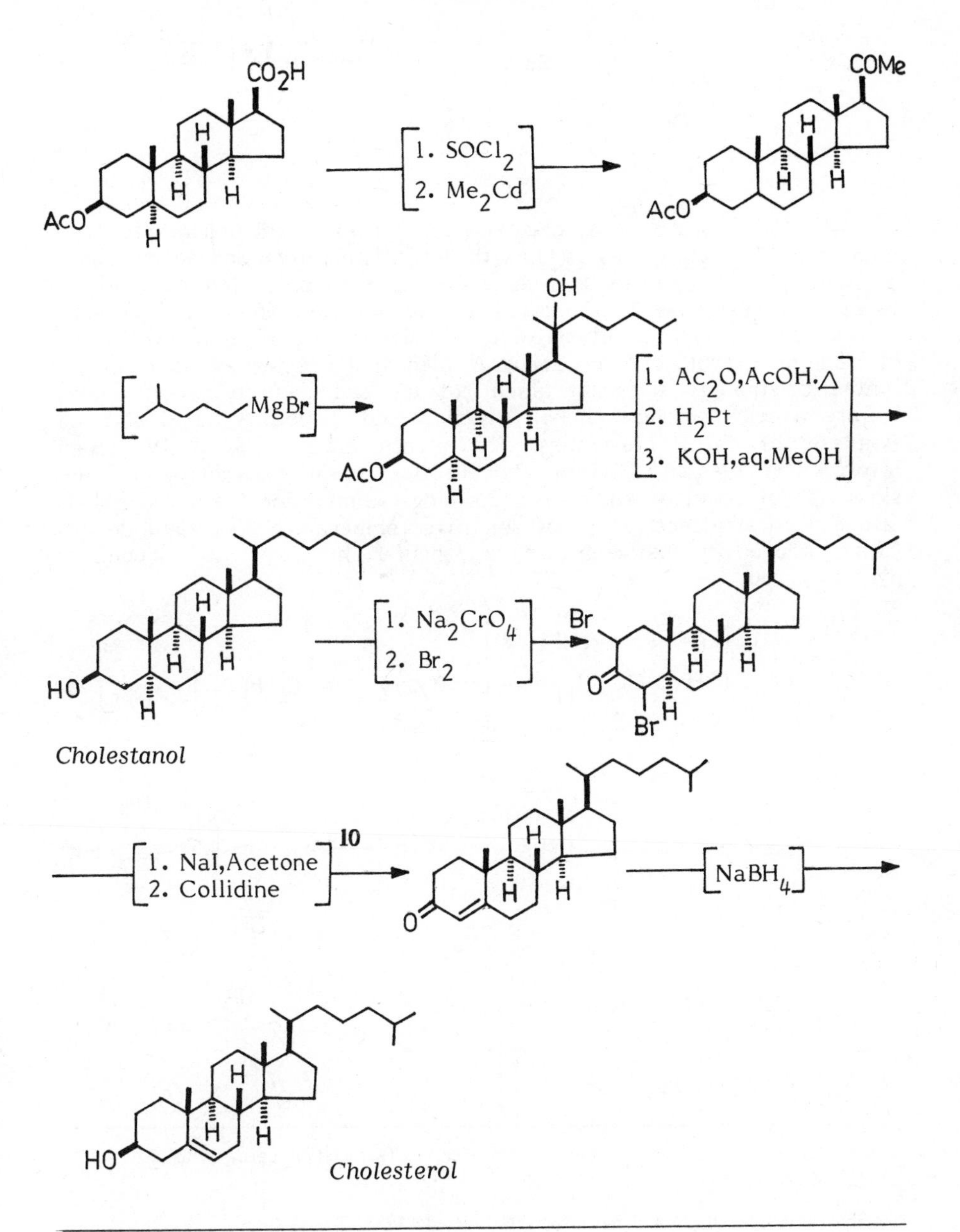

10. Collidine causes 4-dehydrobromination of the axial 4-bromo and hydrogenolysis of the presumably equatorial 2-iodo via an enolate anion in the 2-iodo-4-bromo intermediate, cf. Rosenkranz,G.; Mancera,O.; Gatica,J.; Djerassi,C. J. Am. Chem. Soc., 1950, 72, 4077.

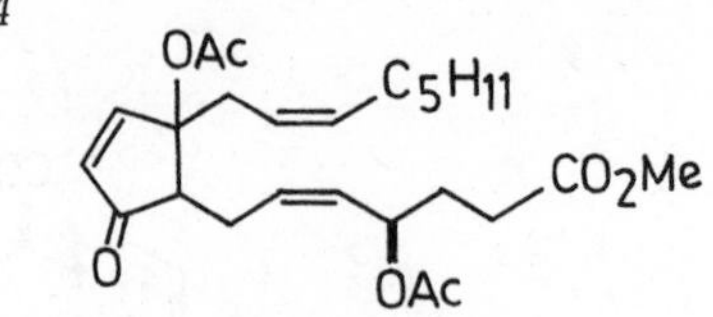

CLAVULONES

Clavulones are a newly discovered family of novel marine eicosanoids from Clavularis viridis(1), with antiinflammatory and antitumour activities(1,2). Clavulone II is the 5,6-E isomer of I, while Clavulone III is the 5,6-E, 7,8-Z isomer of I. Clavulone I can be isomerised to clavulone II & III by acid catalysis. The present synthesis of racemic clavulone I by Corey & Mehrotra (3) has as its essential synthetic strategy the attachment of ω- and then α-carbon chains to a preformed cyclopentadiene, which would be easily applicable to synthesis of chiral clavulones. The attachment of the α-side chain involves an elegant Claisen condensation taking advantage of the slowness of cyclopentadienone forming elimination. The synthesis also makes an effective use of selective temperature dependent desilylating propensity between tertiary/angular hydroxyl and secondary hydroxyl group.

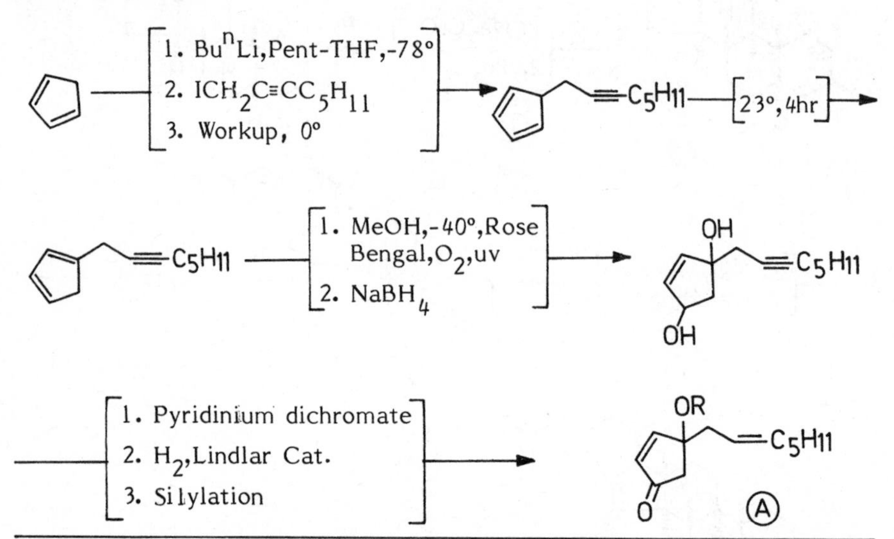

1. (a) Kikuchi,H., Tsukitani,Y., Iguchi,K., Yamada,Y., Tetrahedron Lett., 1982, 23, 5171; (b) ibid, 1983, 24, 1549.

2. Kabayashi,M., et al, 26th Symp. Chem. Nat. Prod., 1983, 228.

3. Corey,E.J., Mehrotra, Mukund M., J. Am. Chem. Soc., 1984, 106, 3384.

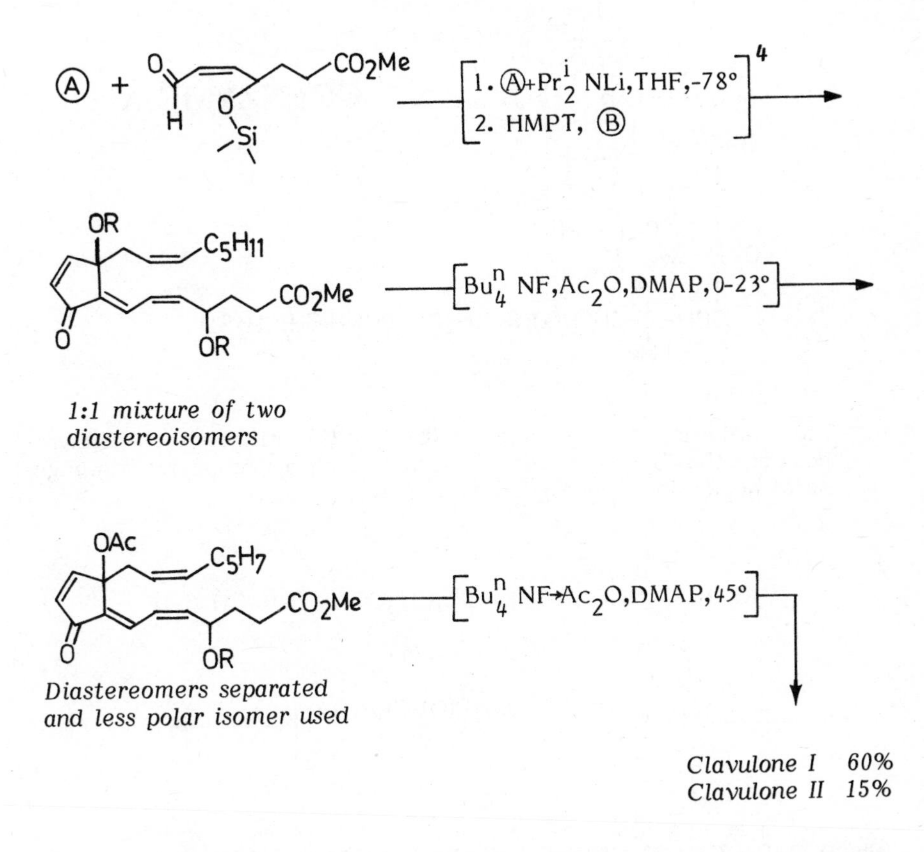

4. The success of this remarkable aldolisation has been attributed (3) to the relative slowness of cyclopentadienone forming elimination.

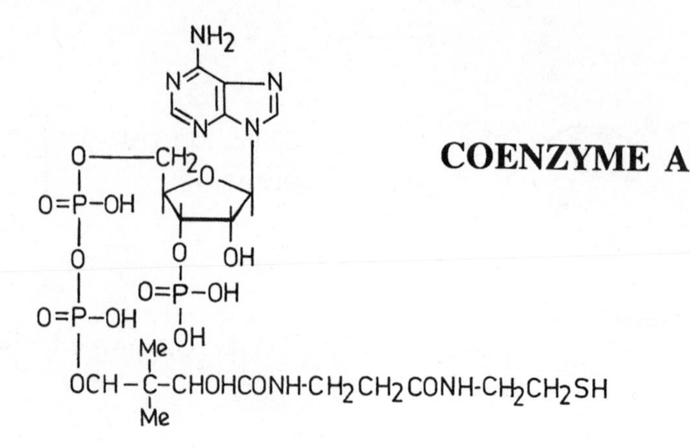

COENZYME A

The synthesis of coenzyme A (Co A) (1) utilizes the efficient method for the synthesis of nucleoside-5' pyrophosphates through the use of nucleoside-5' phosphoramidates (2).

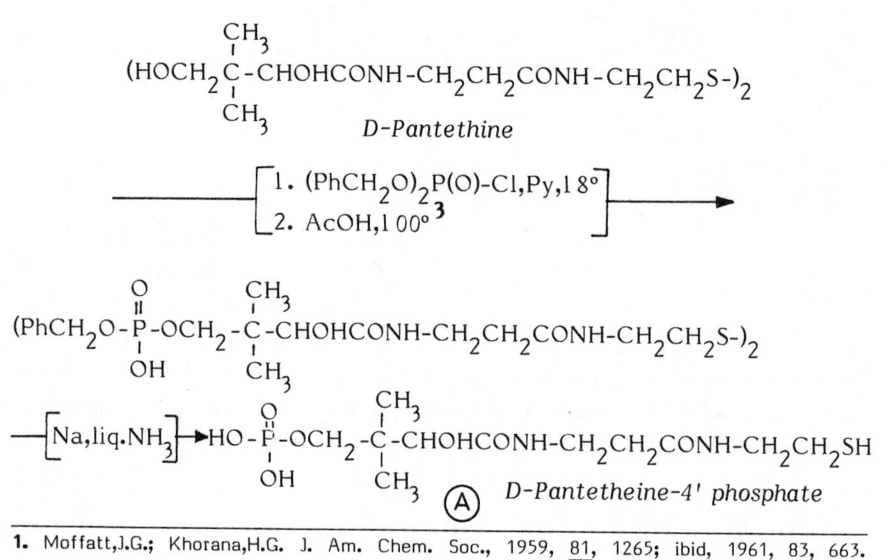

1. Moffatt,J.G.; Khorana,H.G. J. Am. Chem. Soc., 1959, 81, 1265; ibid, 1961, 83, 663.
2. Phosphoramidates are reactive intermediates and are very useful for the specific synthesis of the pyrophosphate bond, cf. Clark,V.M.; Kirby,G.W.; Todd,A.R. J. Chem. Soc., 1957, 1497; Moffatt,J.G.; Khorana,H.G. J. Am. Chem. Soc., 1958, 80, 3756; Moffatt,J.G. Khorana,H.G., ibid, 1961, 83, 649. Using this method the synthesis of a number of nucleotide coenzymes such as UDPG and FAD has been carried out, Moffatt,J.G.; Khorana,H.G. J. Am. Chem. Soc., 1958, 80, 3756; Roseman,S.; Distler,J.J.; Mofatt,J.G.; Khorana,H.G. ibid, 1961, 83, 659.
3. If AcOH treatment was omitted the product was very heavily contaminated with 2',4'-cyclic phosphate. The formation of the cyclic phosphate presumably occurred as a transesterification reaction of the tertiary phosphate; AcOH treatment presumably converts it into monophosphate, which is not a good transesterificating agent.

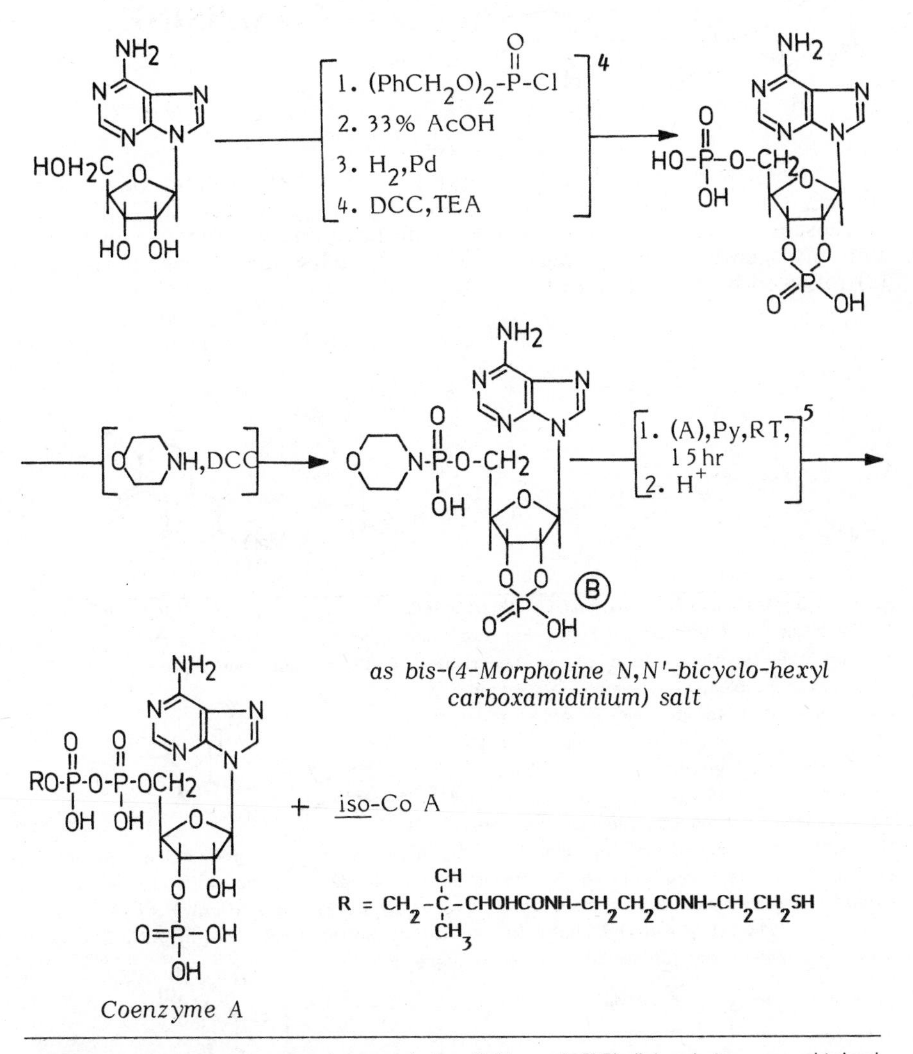

4. During phosphorylation a mixture of the 5',2'- and 5',3'-diphosphates was obtained. Treatment with AcOH caused partial debenzylation. The mixture of diphosphates thus obtained after hydrogenation could also be treated directly with morpholine and DCC, when morpholidate (B) was obtained in one step in an overall yield of 90–95%.

5. After the reaction the products were separated by chromatography on ion exchange resin and ECTEOLA column.

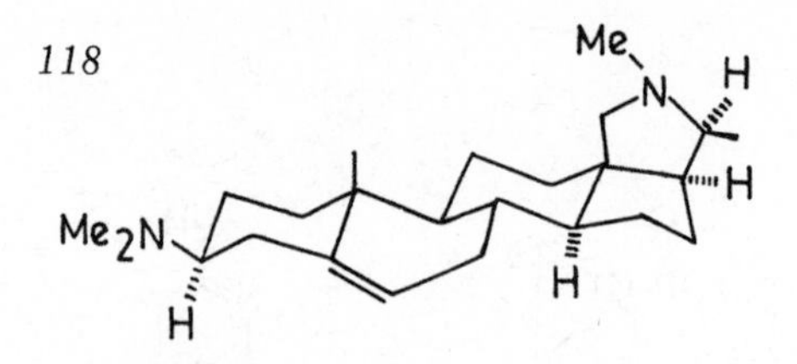

CONESSINE

Conessine is marked by the presence of a heterocyclic ring fused to ring D of the steroid nucleus. Stork and his colleagues (1), in their stereospecific total synthesis of conessine have chosen first to construct this nitrogen-containing ring on a B-C-D tricyclic intermediate Ⓐ (2) followed by completion of ring A.

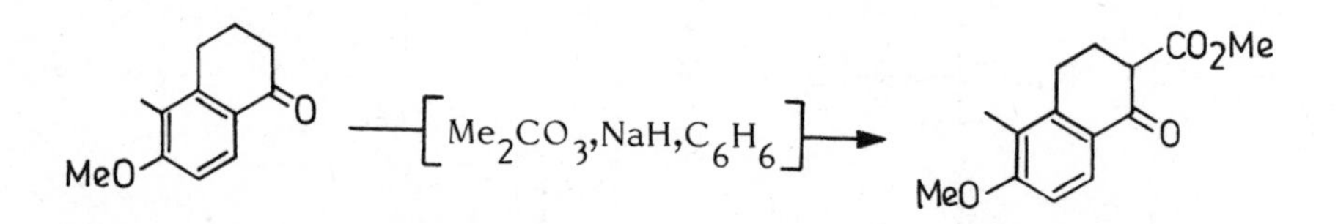

1. Stork,G.; Darling,S.D.; Harrison,I.T.; Wharton,P.S. J. Am. Chem. Soc., 1962, 84, 2018.
2. Synthesis of the nitrogen-containing conessine ring system from steroid precursors requires the functionalization of an unactivated C-18 angular methyl group. At this point the problems involved are shared with the related steroid hormone, aldosterone; and the ingenious use of proximity effects for selective introduction of reactive functions at C-18 [cf.(i)] are discussed by Schaffner,K.; Arigoni,D.; Jeger,O. Experientia, 1960, 16, 169. Special mention, however, must be made of the use of Hofmann-Loffler-Freytag reaction involving the free-radical chain decomposition of a C_{20}-N-chloroamine in the synthesis of dihydroconessine (ii→v) by Corey,E.J.; Hertler,W.R. J. Am.Chem. Soc., 1959, 81, 5209, and the photolytic method of Barton,D.H.R.; Starratt,A.N. J. Chem. Soc., 1965, 2444, used for partial synthesis of conessine.

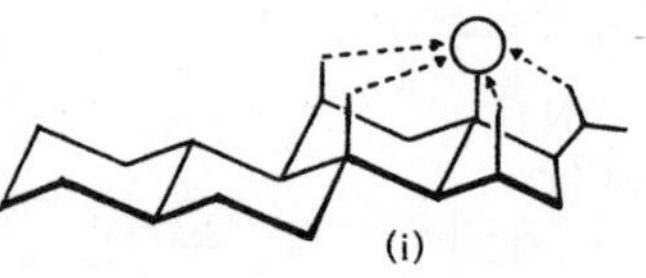
(i)

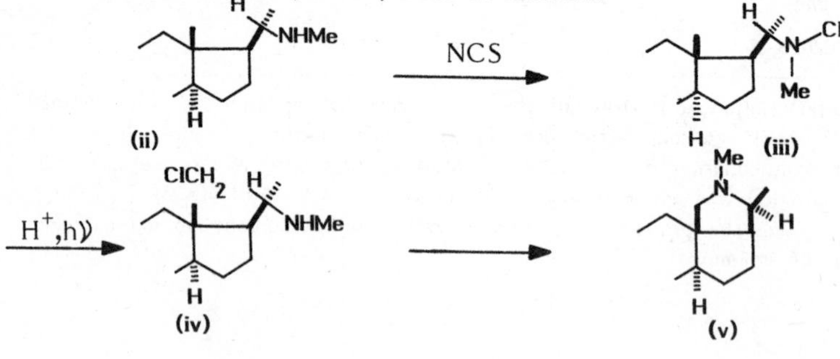

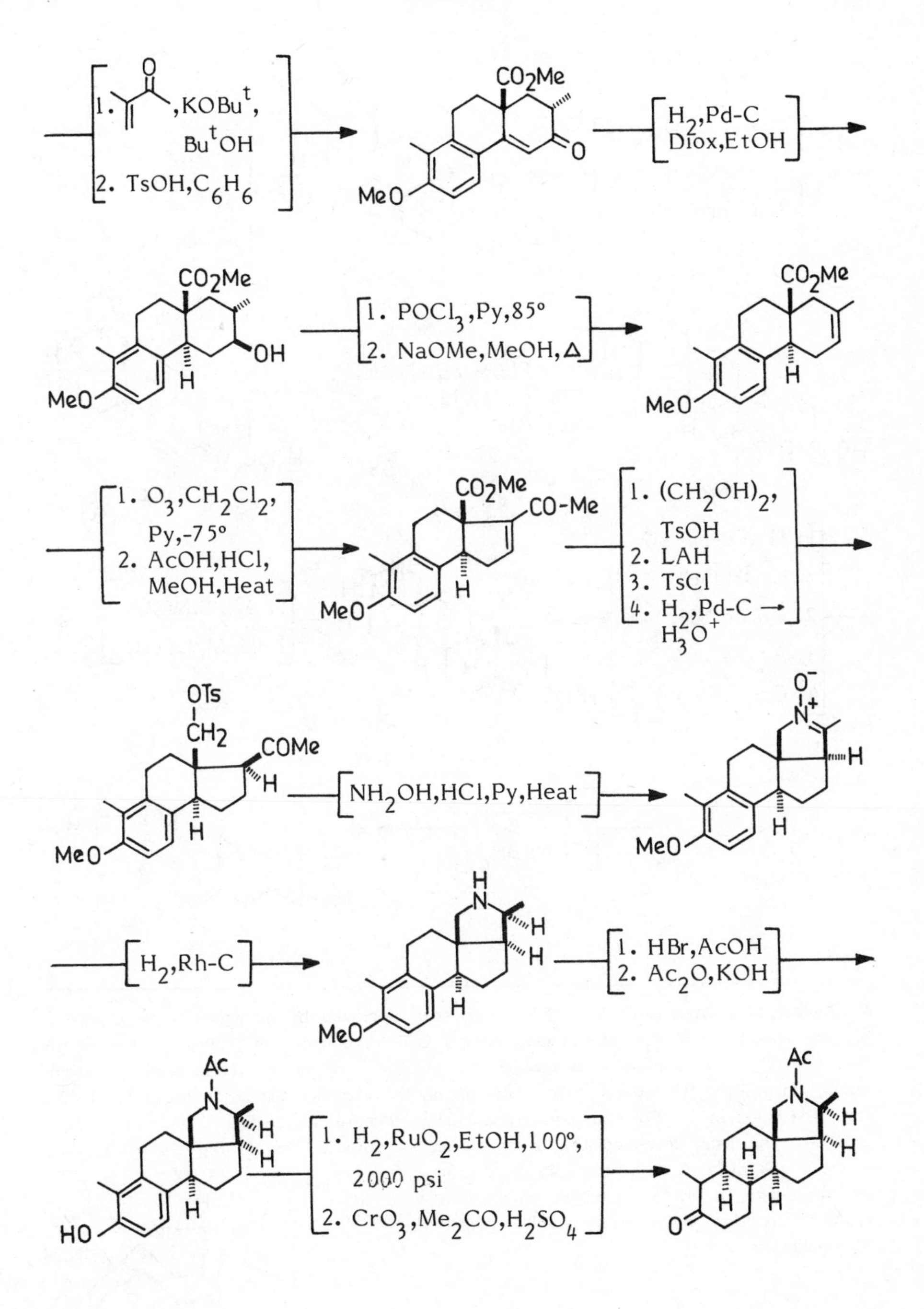
1. ,KOBu^t, Bu^{t}OH
2. TsOH,C_6H_6
CO_2Me
MeO
H_2,Pd-C
Diox,EtOH
CO_2Me
OH
H
MeO
1. $POCl_3$,Py,85°
2. NaOMe,MeOH,Δ
CO_2Me
H
MeO
1. O_3,CH_2Cl_2, Py,-75°
2. AcOH,HCl, MeOH,Heat
CO_2Me
CO-Me
H
MeO
1. $(CH_2OH)_2$, TsOH
2. LAH
3. TsCl
4. H_2,Pd-C → H_3O^+
OTs
CH_2
COMe
H
H
MeO
NH_2OH,HCl,Py,Heat
O^-
N^+
H
H
MeO
H_2,Rh-C
H
N
H
H
H
MeO
1. HBr,AcOH
2. Ac_2O,KOH
Ac
N
H
H
H
HO
1. H_2,RuO_2,EtOH,100°, 2000 psi
2. CrO_3,Me_2CO,H_2SO_4
Ac
N
H
H
H
H
H
O

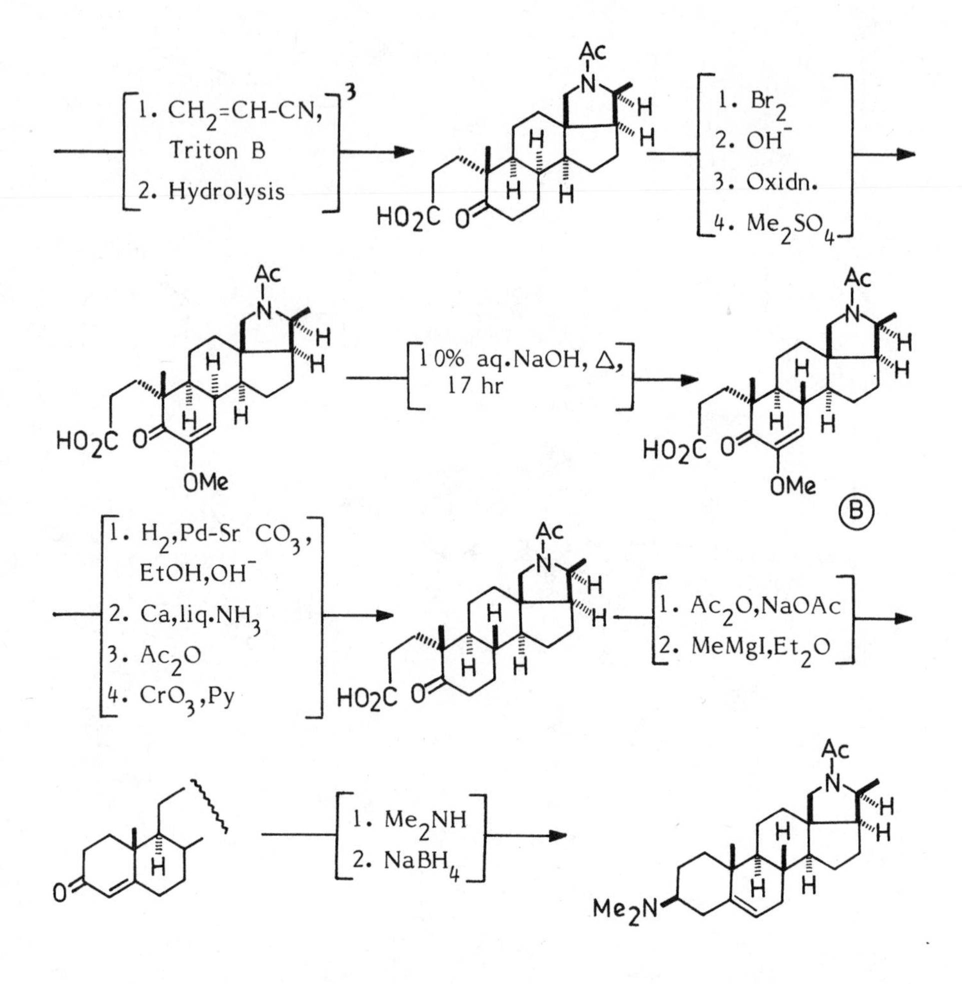

3. Control of stereochemistry at C-10 has been achieved by introducing the elements of ring A into an 8-iso intermediate having B/C rings cis. The "folded" structure of the molecule (i) now makes the position 10, already carrying the future angular methyl group, inaccessible from the α-face, thus forcing the incoming alkylating group to attack from the β-face; cf. Stork,G.; Loewenthal,H.J.E.; Mukharji,P.C. J. Am. Chem. Soc., 1956, 78, 501. The 8-iso intermediate having served its purpose, the desired trans-B/C ring fusion in the molecule is restored in an ingenious manner by introducing a double bond in conjugation with the carbonyl function which now permits inversion at C-8 by equilibration.

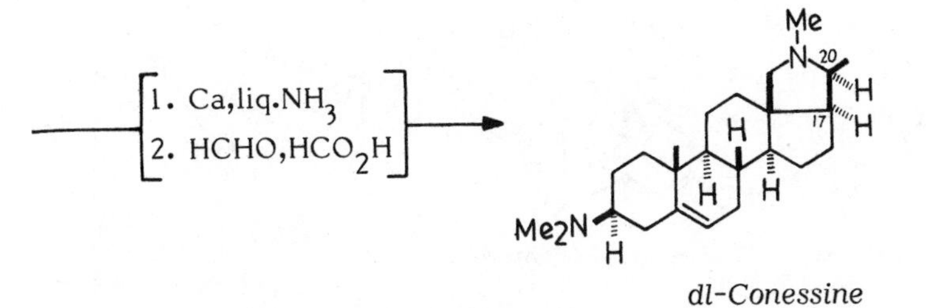

dl-Conessine

The synthesis by Johnson et al. (4) begins along lines laid down earlier in the "hydrochrysene approach" for construction of the D-ring homo-steroid (A) followed by the usual ring contraction procedure. Introduction of the C-18 group and elaboration of the heterocyclic ring E is based on the conjugate addition of cyanide ion to an α,β-unsaturated ketone system (5).

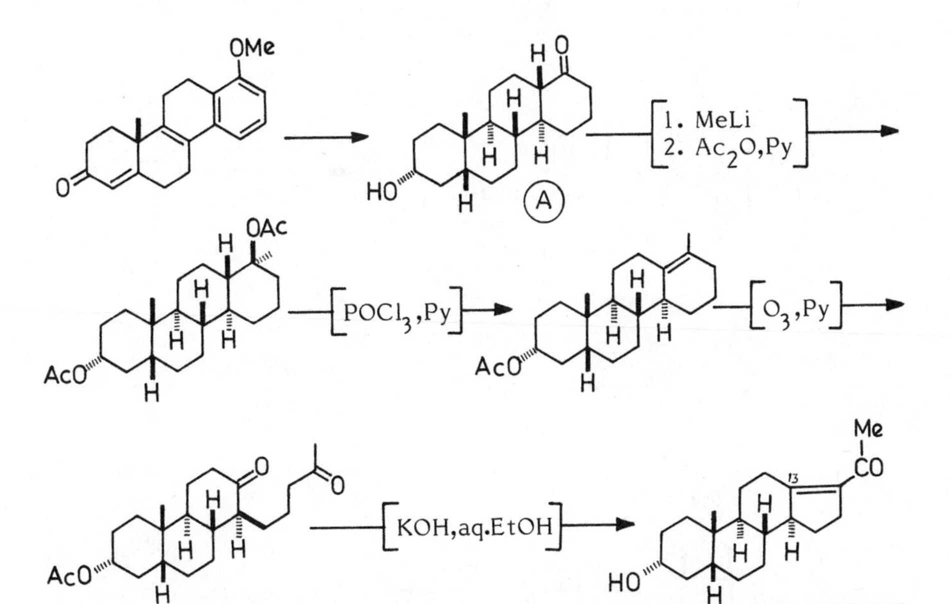

4. Marshall,J.A.; Johnson,W.S. J. Am. Chem. Soc., 1962, 84, 1485; Johnson,W.S.; Marshall,J.A.; Keana,J.F.W.; Franck,R.W.; Martin,D.G.; Bauer,V.J. Tetrahedron, Suppl.8, 1966, 22, Part II, 541.

5. This hydrocyanation procedure was developed by Nagata and also used by him in his total synthesis of conessine, Nagata,W.; Terasawa,T.; Aoki,T. Tetrahedron Lett., 1963, 869.

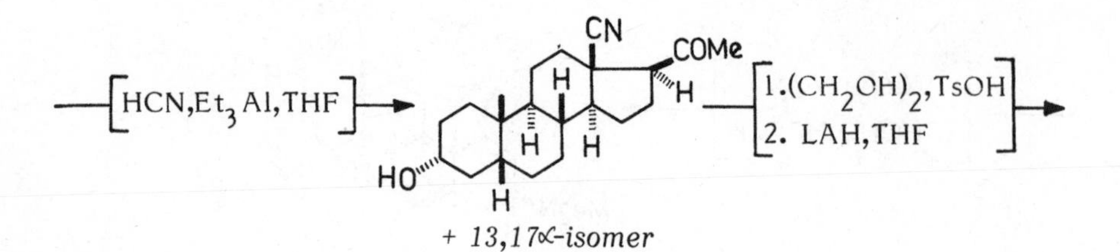

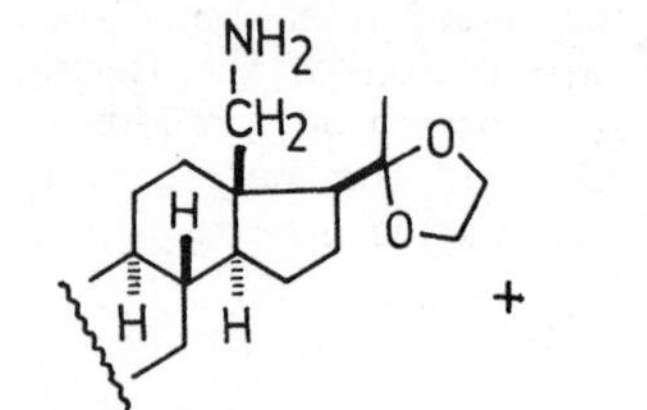

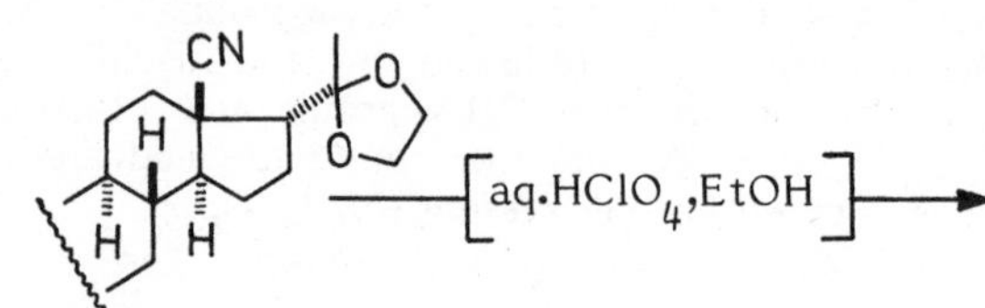

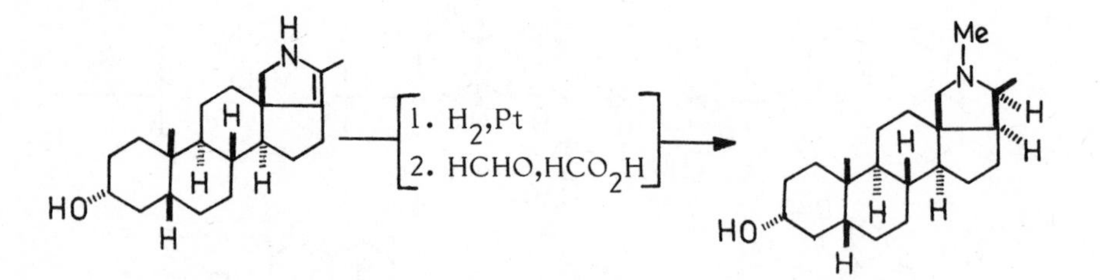

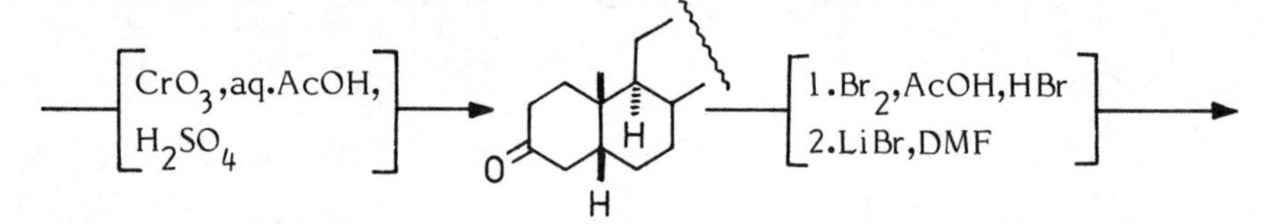

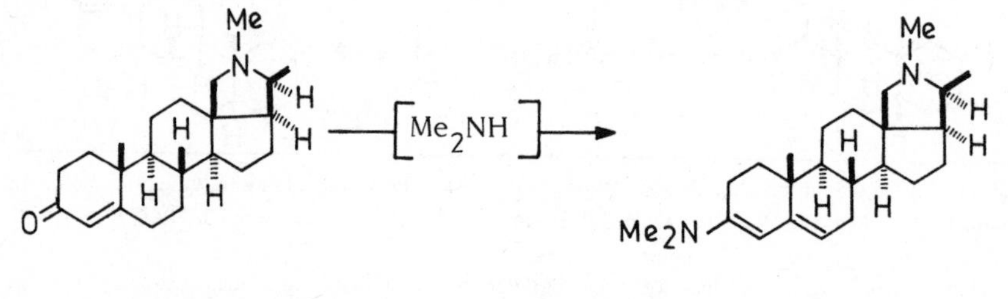

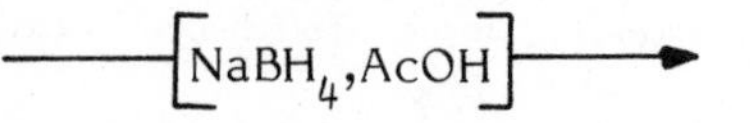

dl-Conessine

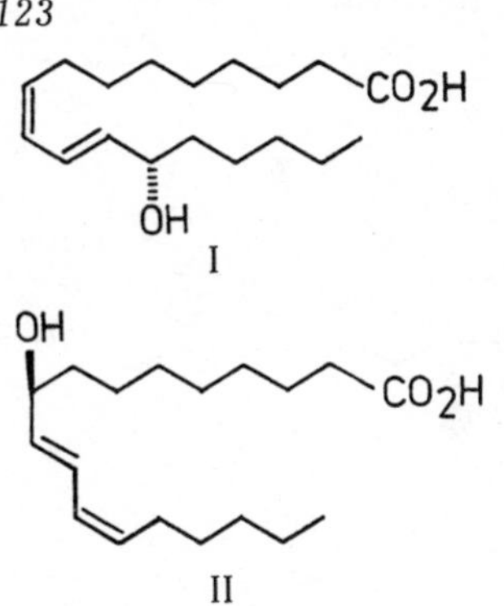

CORIOLIC ACID
DIMORPHECOLIC ACID

Unsaturated fatty acids play important roles in biological systems. Coriolic acid (I) and dimorphecolic acid (II), belonging to the family of oxyoctadecadienoates, though isolated first from bovine heart mitochondria (1), and shown to possess cation-specific ionophoric activity, have recently been isolated from rice plant Fukuyuki (2), and shown to act as self-defensive substances against rice blast disease. Rama Rao and his associates have presented a stereoselective syntheses of these acids, which have general applicability to this class of compounds (3).

Coriolic acid

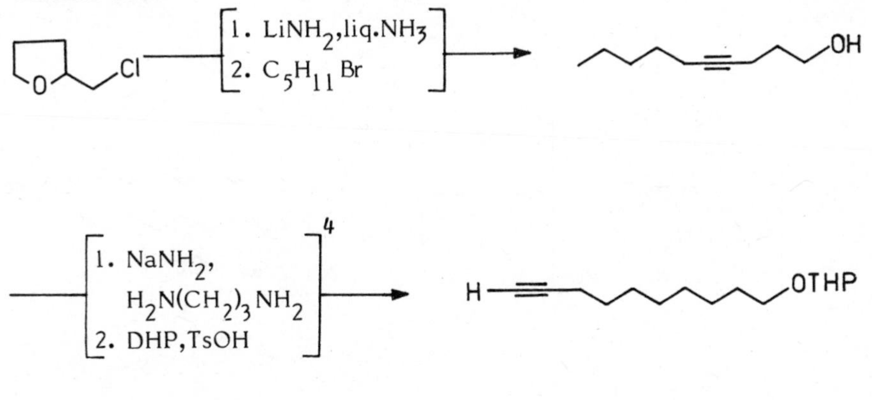

1. Blondin,G.A. Ann. N.Y. Acad. Sci., 1975, 264, 98.

2. Kato,T.; Yamaguchi,T.; Yokoyama,T.; Yyehara,T.; Namai,T.; Yamanaka,S.; Harada,N. Chem. Lett., 1984, 409.

3. Rama Rao,A.V.; Reddy,E.R.; Sharma,G.V.M.; Yadgiri,P.; Yadav,J.S. Chem. Pharm. Bull., 1985, 33, 2168; Tetrahedran, 1986, 42, 4523.

4. Brown,C.A.; Yamashita,A. J. Chem. Soc. Chem. Commun., 1976, 959.

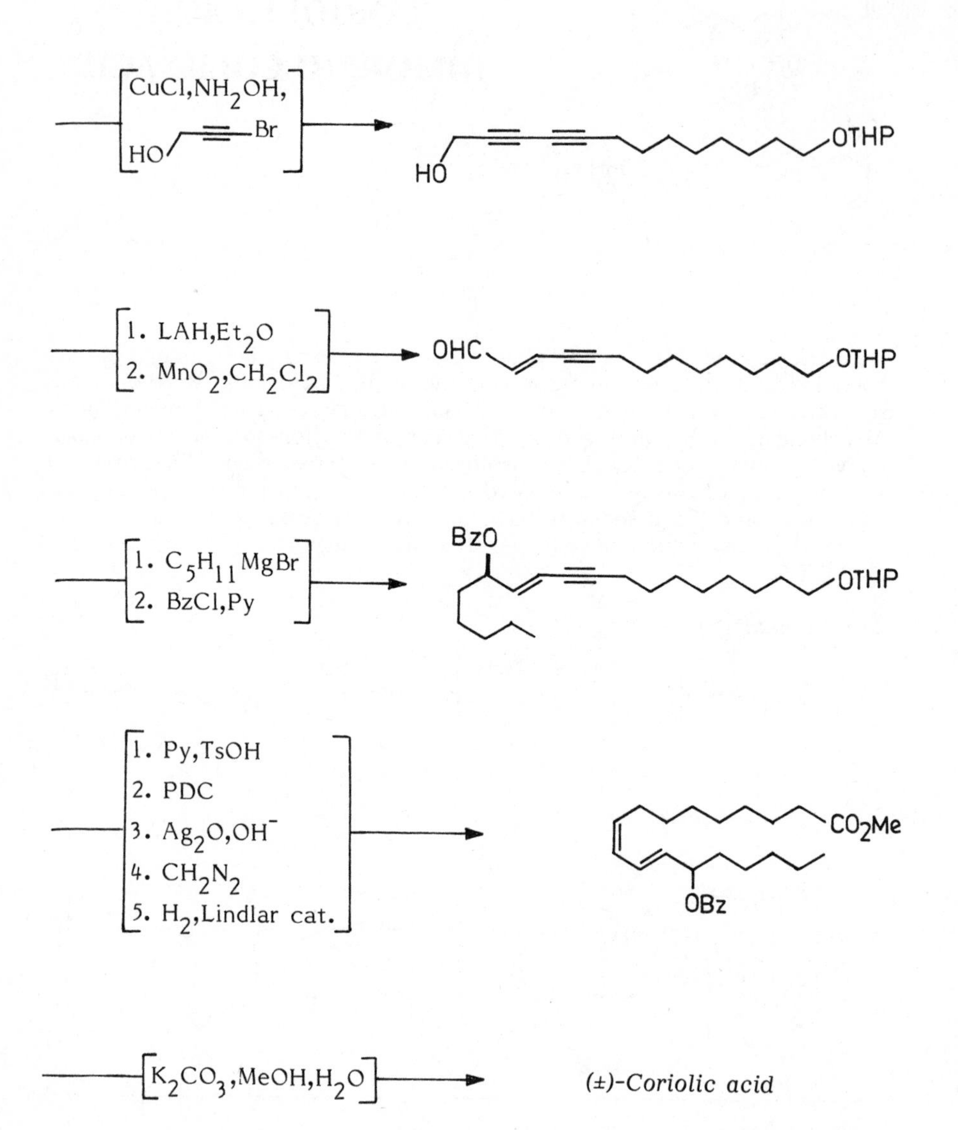
CuCl, NH2OH,
HO
Br
HO
OTHP
1. LAH, Et2O
2. MnO2, CH2Cl2
OHC
OTHP
1. C5H11MgBr
2. BzCl, Py
BzO
OTHP
1. Py, TsOH
2. PDC
3. Ag2O, OH⁻
4. CH2N2
5. H2, Lindlar cat.
CO2Me
OBz
K2CO3, MeOH, H2O
(±)-Coriolic acid

Dimorphecolic acid

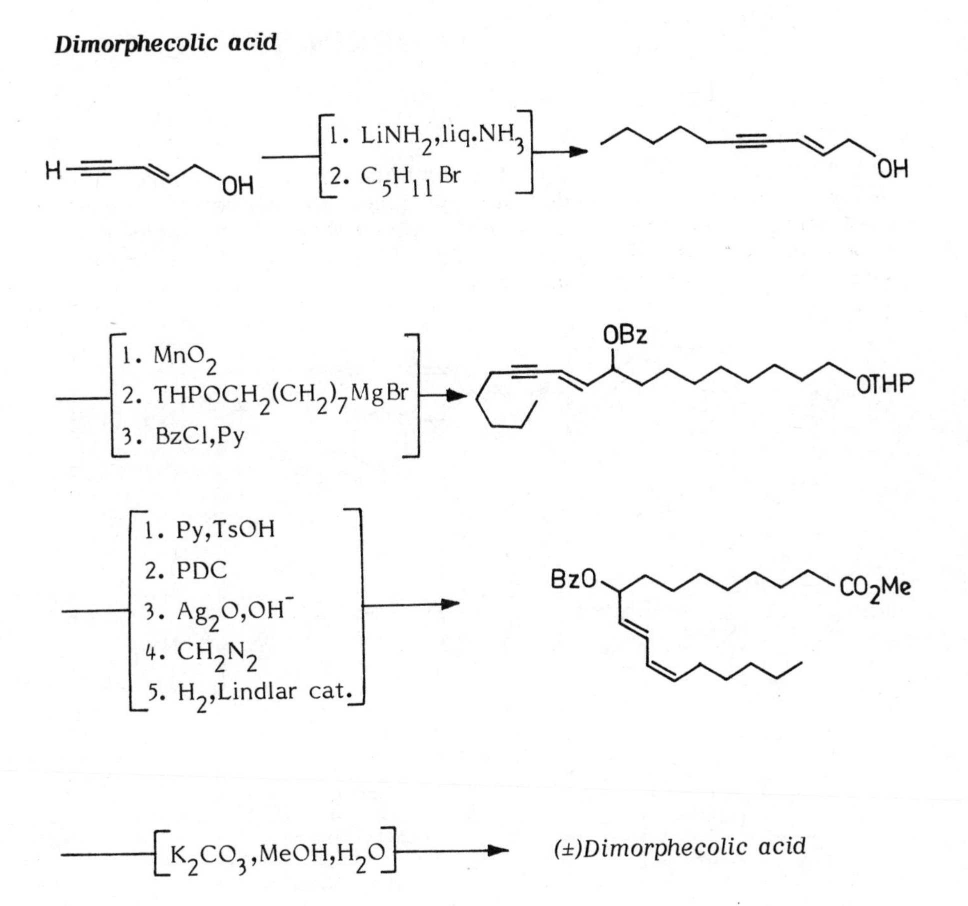

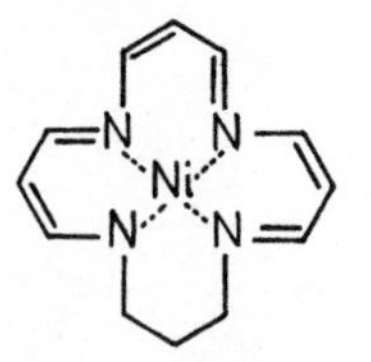

CORRIN TEMPLATE

The current interest in the function of metal complexes in biological systems has added a new facet to synthetic inorganic chemistry, namely, the creation of simple models of the biologically important systems. This aspect is best illustrated with the brilliant synthesis of the corrin template I: (1)

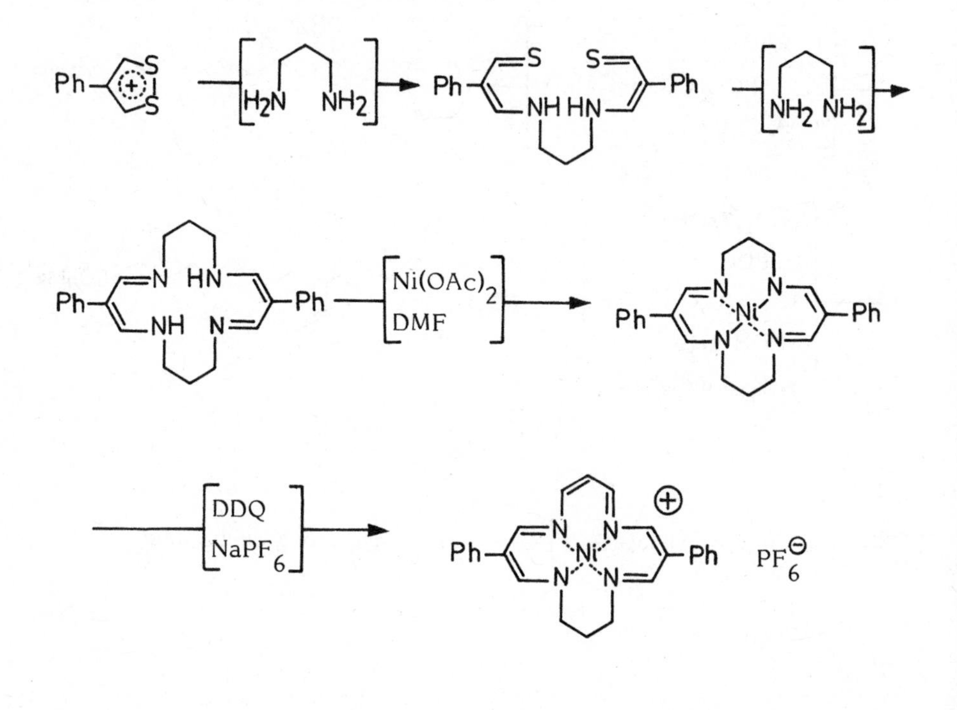

---

1. Tang,S.C., Weinstein,G.N., Holm,R.H., J. Am. Chem. Soc., 1973, 95, 613.

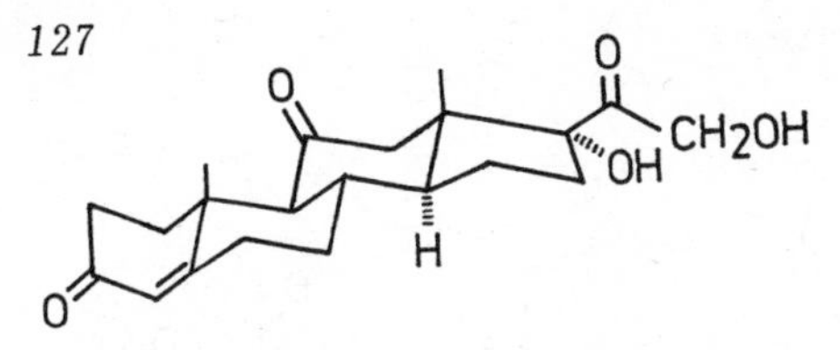

CORTISONE

The first synthesis of cortisone was reported in 1951 by Woodward and his colleagues (1). Shortly afterwards Sarett and his coworkers at Merck Laboratories (2) reported a highly stereoselective (3) route to cortisone which did not involve the use of relays (4). In this synthesis the crucial anti-trans tricyclic nucleus (D) representing rings A, B and C of cortisone has been obtained from a cis-fused bicyclic adduct (A); the cis ring junction in (A) (later equilibrated to trans) serves an important function in directing the addition of ring A from the less hindered convex α-face and in ensuring formation of the angular hydrophenanthrene derivative during the transformation (B) → (C) (7). Equilibration at C-8 during the Oppenauer oxidation furnishes the desired anti-trans ring fusion.

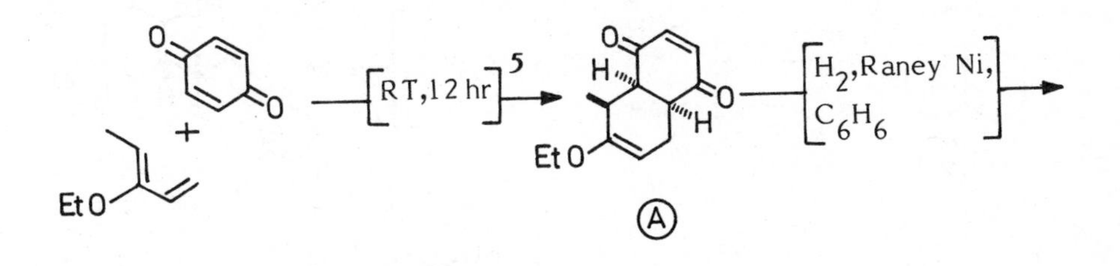

1. Woodward,R.B.; Sondheimer,F.; Taub,D.; Heusler,K.; McLamore,W.M. J. Am. Chem. Soc., 1951, 73, 4057; 1952, 74, 4223. Woodward's synthesis of $\Delta^{9(11),16}$-bis-dehydro-20-norprogesterone is described in detail under the synthesis of cholesterol, and from this point a practical route to cortisone, involving the fashioning of the cortical side-chain and addition of hypobromous acid to the 9(11)-double bond for introduction of the 11-oxygen function, is described by Barkley,L.B.; Farrar,M.W.; Knowles,W.S.; Raffelson, H. J. Am. Chem. Soc., 1954, 76, 5017.
2. Sarett,L.H.; Arth,G.E.; Lukes,R.M.; Beyler,R.E.; Poos,G.I.; Johns,W.F.; Constanin,J.M. J. Am. Chem. Soc., 1952, 74, 4974.
3. These authors use the term "stereospecific" which is taken to mean that "in each reaction producing a fixed asymmetric centre, the ratio of isomer having the same configuration as the end product to all other isomers is greater than unity" (ref.2). By this definition, ratios of 8:1 or better have been obtained at all steps in this synthesis. For use of the terms "stereospecific" and "stereoselective", see: Eliel,E.L. in "Stereochemistry of Carbon Compounds", McGraw-Hill, New York, 1962, Chapter 15.
4. For an account of the gigantic effort that went into devising partial and totally synthetic routes to cortisone, see: Fieser,L.F.; Fieser,M. in "Steroids", Reinhold, New York, 1959, 600; Djerassi,C. Vitamins & Hormones, 1953, 11, 205.
5. Sarett,L.H.; Lukes,L.M.; Poos,G.I.; Robinson,J.M.; Beyler,R.E.; Vandergrift,J.M.; Arth,G.E. J. Am. Chem. Soc., 1952, 74, 1393.

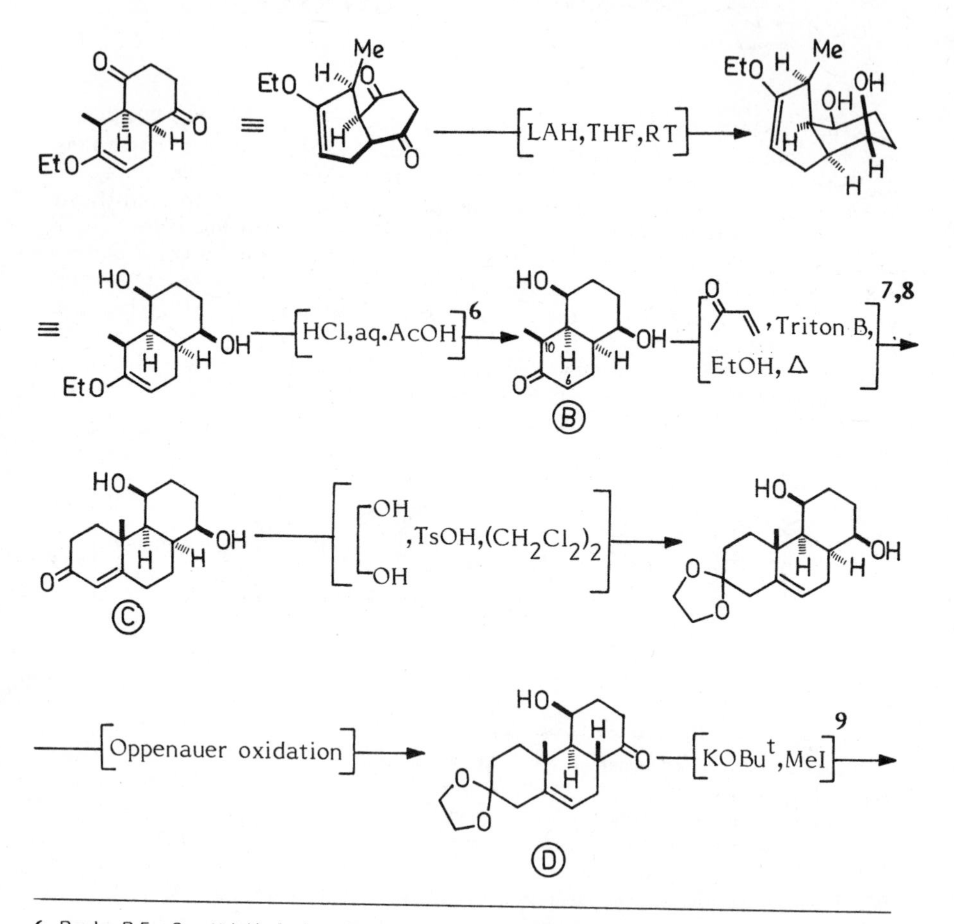

6. Beyler,R.E.; Sarett,L.H. J. Am. Chem. Soc., 1952, 74, 1406.

7. Formation of the angular hydrophenanthrene derivative instead of a linear anthracene is favoured because enolization of the carbonyl function occurs towards the ring junction in a cis-fused decalin. Addition of methyl vinyl ketone to (B) therefore occurs at C-10 rather than C-6, cf. the preferential formation of Δ^3-enols by 3-oxo-A/B cis steroids, Corey,E.J.; Sneen,R.A. J. Am. Chem. Soc., 1955, 77, 2505; Velluz,L.; Valls,J.; Nomine,G. Angew. Chem. Internat. Ed., 1965, 4, 181. The bond formation at C-10, however, occurs from both the β- and α-face, and gives a 2:1 ratio of unnatural to natural isomer.

8. Poos,G.I.; Arth,G.E.; Beyler,R.E.; Sarett,L.H. J. Am. Chem. Soc., 1953, 75, 422.

9. Lukes,R.M.; Poos,G.I.; Beyler,R.E.; Johns,W.F.; Sarett,L.H. J. Am. Chem. Soc., 1953, 75, 1707.

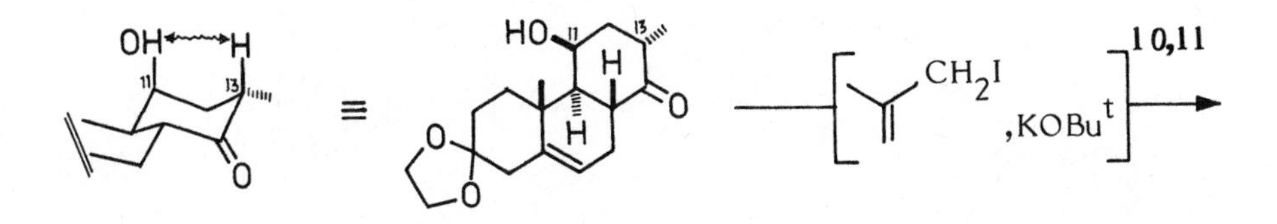

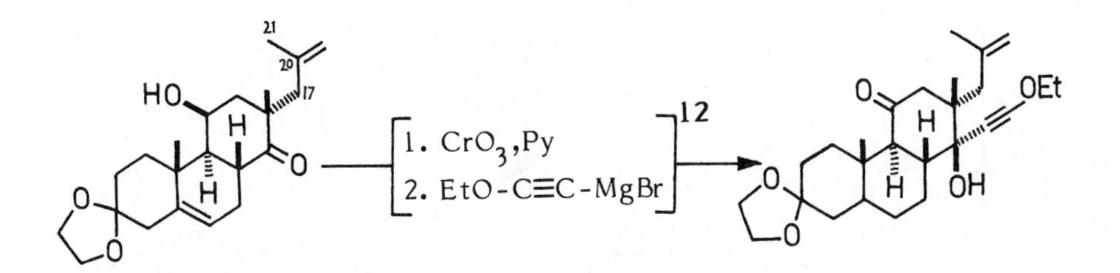

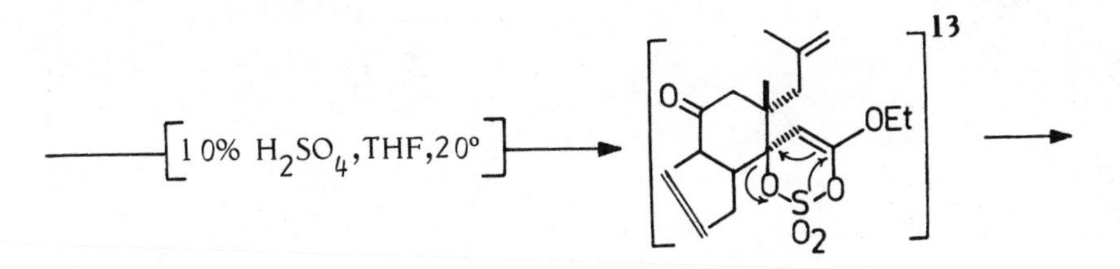

---

10. Sarett,L.H.; Johns,W.F.; Beyler,R.E.; Lukes,R.M.; Poos,G.I.; Arth,G.E. J. Am. Chem. Soc., 1953, 75, 2112.

11. Due to 1,3-interaction between the 11-hydroxyl group and the approaching electrophile (i), instead of the usual axial alkylation at position 13, the incoming isopropenyl group attaches itself from the α-face (ii). This is an example of "equatorial or twist-axial alkylation", see: Stork,G.; Darling,S.D. J. Am. Chem. Soc., 1964, 86, 1761.

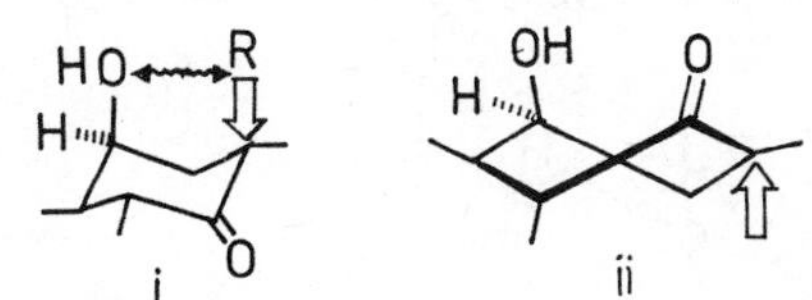

12. Arth,G.E.; Poos,G.I.; Lukes,R.M.; Robinson,F.M.; Johns,W.F.; Feurer,M.; Sarett,L.H. J. Am. Chem. Soc., 1954, 76, 1715.

13. cf. Arens,J.F. in "Advances in Organic Chemistry", Vol.II, eds. R.A. Raphael, E.C. Taylor and Wynberg,H., Interscience, New York, 1960, 159.

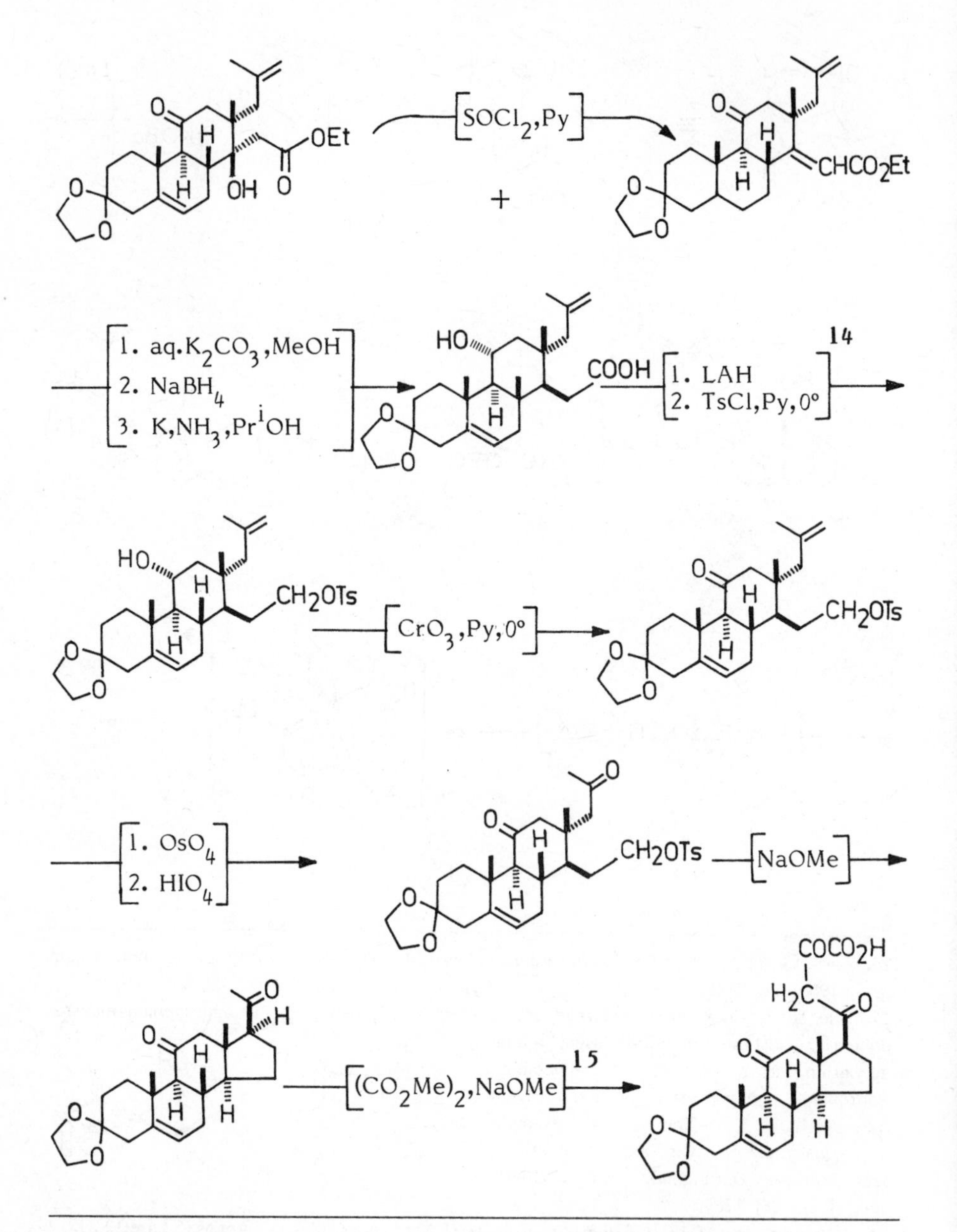

14. Johns,W.F.; Lukes,R.M.; Sarett,L.H. J. Am. Chem. Soc., 1954, 76, 5026.
15. Poos,G.I.; Lukes,R.M.; Arth,G.E.; Sarett,L.H. J. Am. Chem. Soc., 1954, 76, 5031.

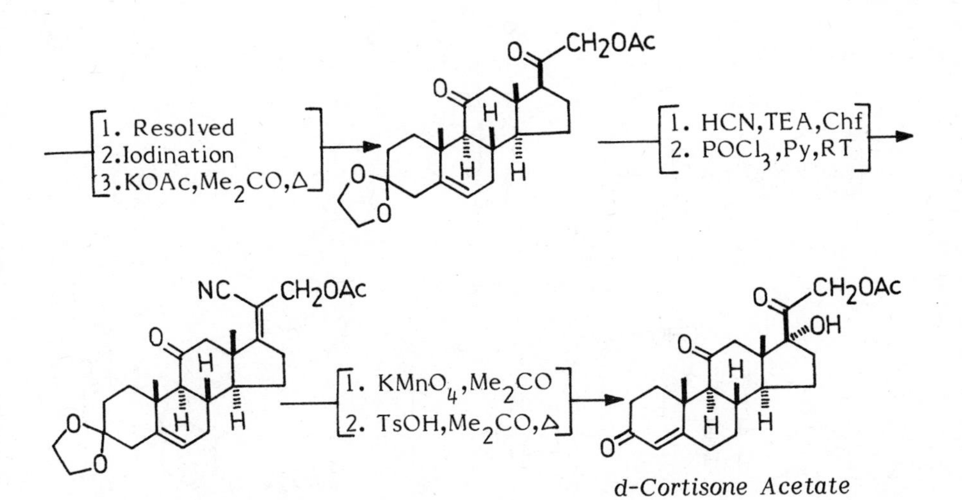

d-Cortisone Acetate

One drawback in Woodword's synthesis of cortisone (1) is the lack of stereoselectivity during construction of ring A to form the tricyclic ketone (C). The alkylating group preferentially enters from the β-face and formation of the quaternary centre at C-10 invariably results in an unfavourable 2:1 ratio of unnatural to natural isomer. The problem was resolved independently by Stork et al. (16) and a group of workers at Monsanto (17). Since it is the existing group at C-10 which finally occupies the α-configuration, these authors reversed the sequence of alkylation and introduced the methyl group last, which now entered predominantly from the β-face as desired. The Monsanto workers also found that blocking of the 6-position during introduction of ring A is unnecessary if ring D is five-membered. This route to cortisone is outlined below.

Monsanto approach

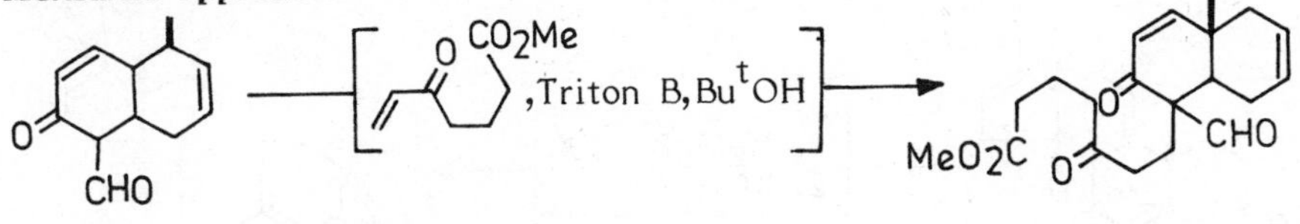

16. Stork,G.; Loewenthal,H.J.E.; Mukherji,P.C. J. Am. Chem. Soc., 1956, 78, 501.
17. Barkley,L.B.; Knowles,W.S.; Raffelson,H.; Thompson,Q.E. J. Am. Chem. Soc., 1956, 78, 4111.

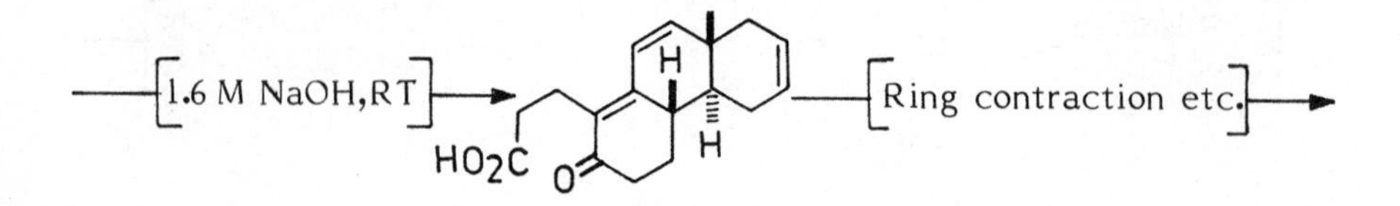
1.6 M NaOH,RT
H
H
HO2C
O
Ring contraction etc.

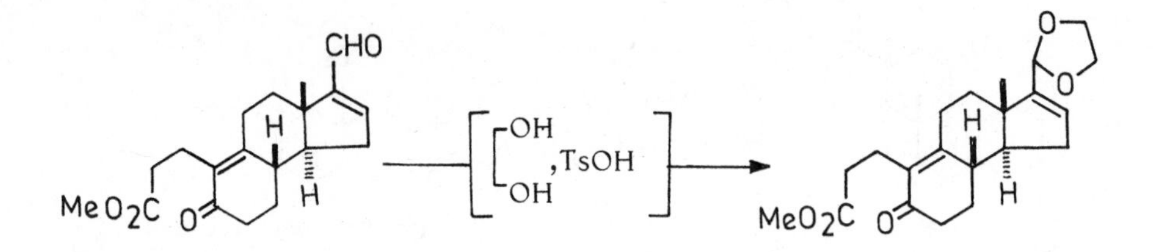
CHO
H
H
MeO2C
O
OH
,TsOH
OH
O
O
H
H
MeO2C
O

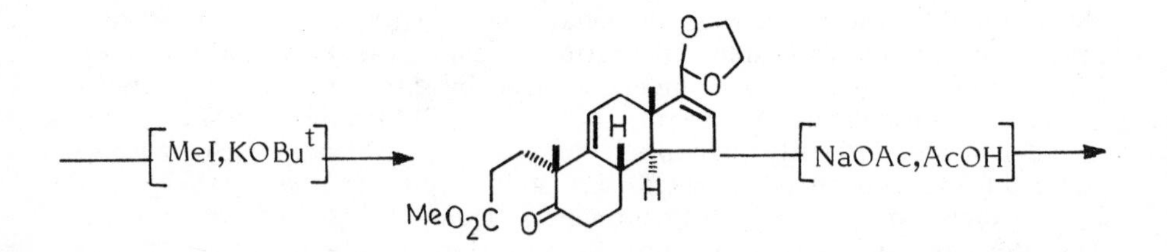
MeI,KOBut
O
O
H
H
MeO2C
O
NaOAc,AcOH

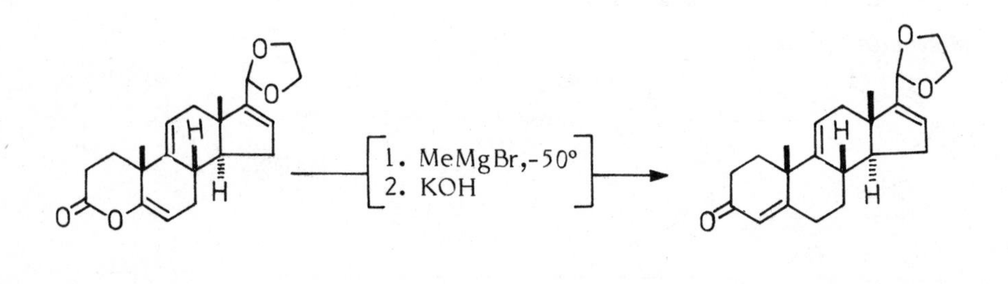
O
O
H
H
O
O
1. MeMgBr,-50°
2. KOH
O
O
H
H
O

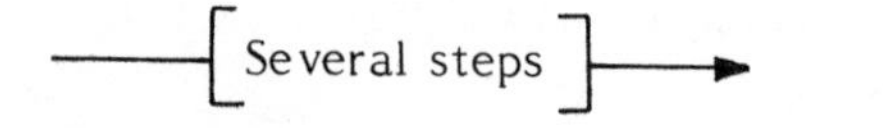
Several steps

Cortisone

Stork approach:

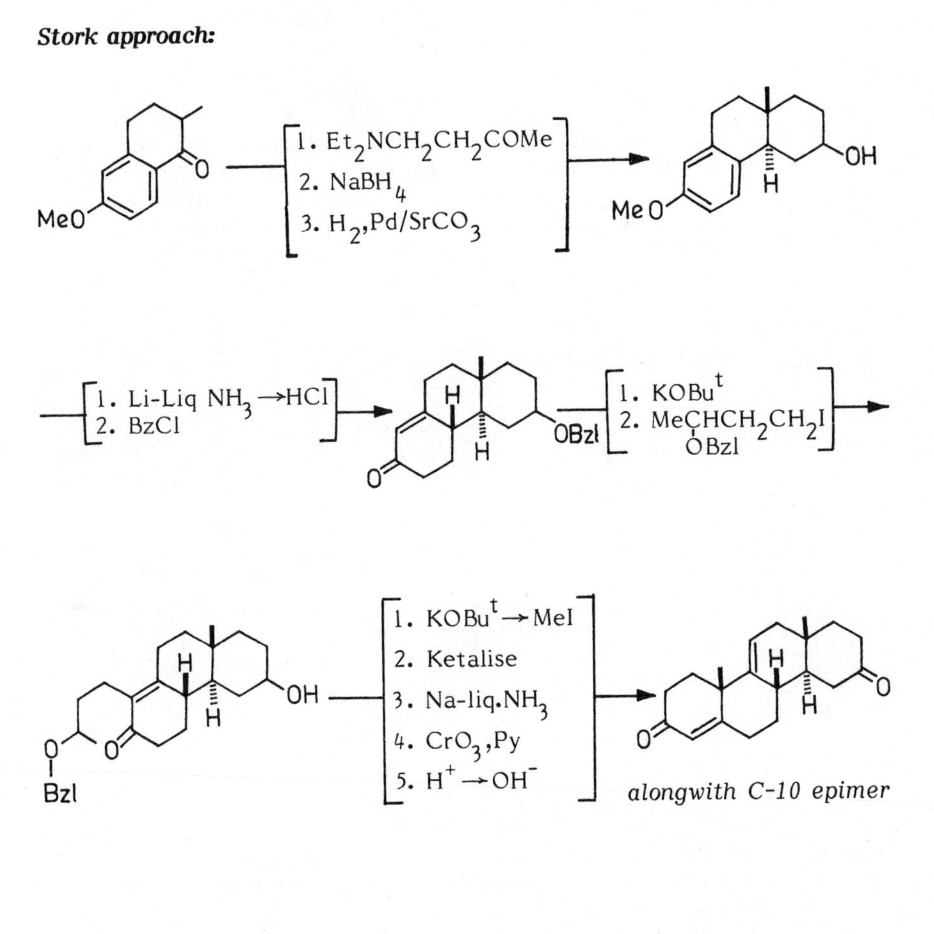

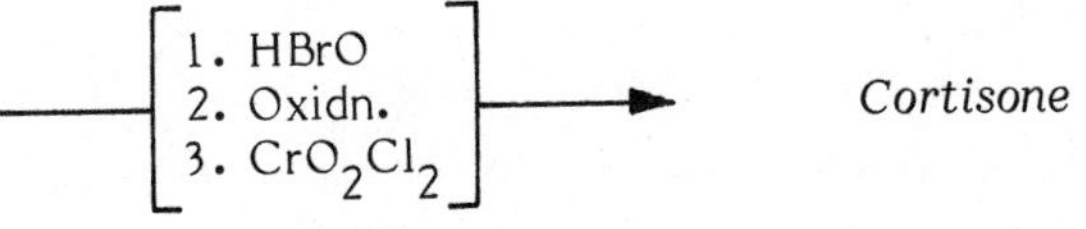

CYCLOBUTADIENE CUBANE

For an exceptionally long period cyclobutadiene was an enigma and the amount of work that has gone into this problem is enormous. The synthesis and characterization of cyclobutadiene (1), therefore, is a big event in organic chemistry. Perhaps the most interesting reaction involving cyclobutadiene is in the cubane synthesis (2).

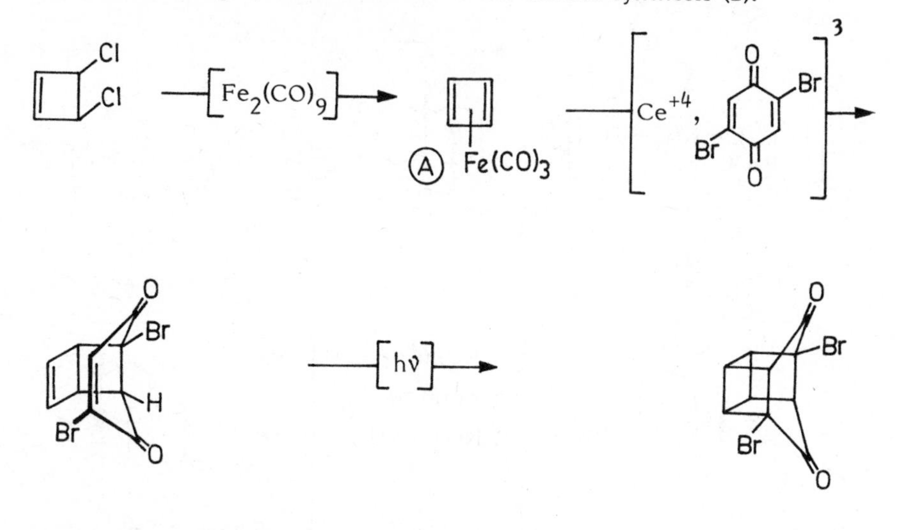

---

1. Cyclobutadiene, though per se unavailable for organic synthesis, because of its extreme instability, gets stabilized as an organometallic complex, cyclobutadiene tricarbonyl iron (A), which seems to provide a useful source of the hydrocarbon, Emerson,G.F.; Watts,L.; Pettit,R. J. Am. Chem. Soc., 1965, 87, 131.

2. Barborak,J.C.; Watts,L.; Pettit,R. J. Am. Chem. Soc., 1966, 88, 1328.

3. During the course of the oxidative decomposition of (A) with ceric ion in the presence of dienophiles a molecule of cyclobutadiene can be transferred from the iron atom to the dienophile. For example, with acetylenes its oxidative decomposition gives rise to several derivatives of Dewar benzene, of Watts,L.; Fitzpatrick,J.D.; Pettit,R. J. Am. Chem. Soc., 1965, 87, 3253; Burt,G.D; Pettit,R. Chem. Commun., 1965, 517.

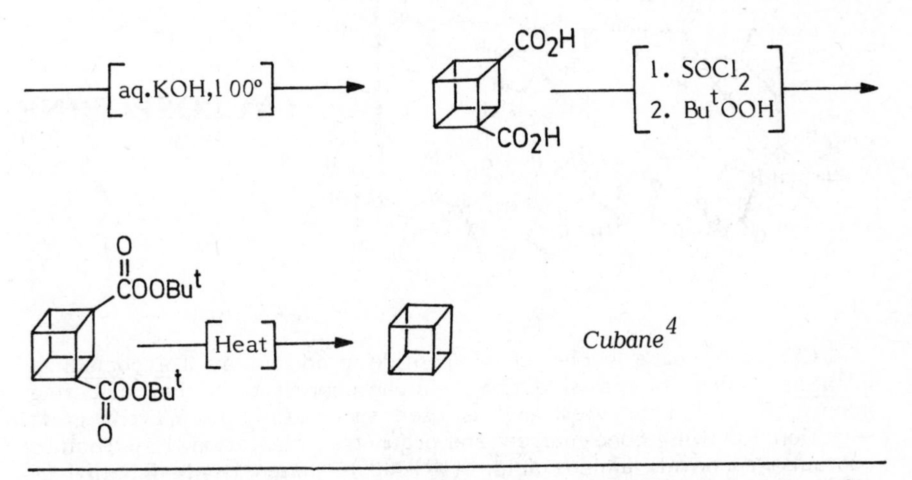

4. For an earlier synthesis of cubane, see: Eaton,P.E.; Cole,T.W.Jr. J. Am. Chem. Soc., 1964, 86, 962, 3157.

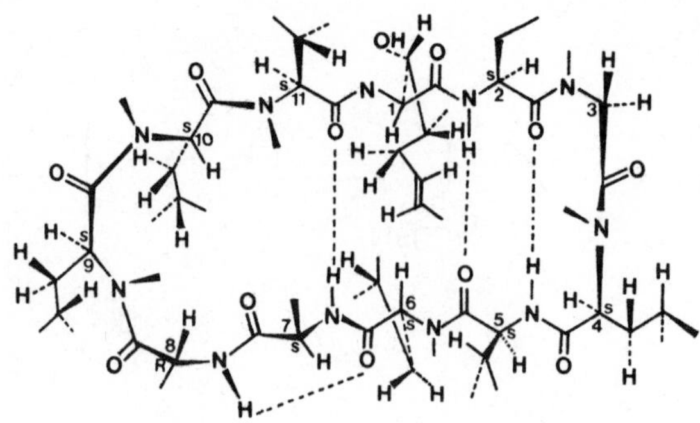

CYCLOSPORINE

Cyclosporine, a cyclic undecapeptide produced by Tolypocladium inflatum Gams, has a selective immunosuppressive action affecting mainly the T-lymphocytes, and is used successfully to prevent graft rejection following bone-marrow and organ transplantation. Its structure contains a novel amino acid, (4R)-4[(E)-2-butenyl]-4,N-dimethyl-L-threonine (MeBmt) and several known N-methylated amino acids (1); this unusual new amino acid could play a significant role in determining the pharmacological activity of cyclosporin. The new amino acid was synthesised (2) using (R,R)-(+) tartaric acid as the basic chiral building block. The strategy followed for the synthesis of cyclosporin by the Sandoz scientists (3) was to synthesise the linear peptide with 7-Ala and 8-D-Ala, the only consecutive non-N-methylated amino acids, as the carboxy and N-terminus respectively, as the bond formation between non-N-methylated amino acids would be more facile, and that the intramolecular hydrogen bonds between the amino groups of this linear peptide would operate so as to stabilise the open chain in a folded conformation approximating the cyclic structure of cyclosporin and thus assist in cyclisation. For the synthesis of the linear undecapeptide fragment-condensation technique was employed using mixed pivalic anhydride activation method as adapted for N-methylated amino acids; MeBmt was introduced at the end to reduce to the minimum the number of steps after introducing this acid. Cyclosporin was obtained in a yield of 27.5% with respect of MeBmt.

---

1. Ruegger,A.; Kuhn,M; Lichti,H.; Loosli,H.R.; Huguenin,R.; Quiquerez,C.; von Warburg,A. Helv. Chim. Acta, 1976, 59, 1072; Petcher,T.J.; Webber,H.P.; Ruegger,A. ibid, 1480.

2. Wenger,R.M. Helv. Chim. Acta, 1983, 66, 2308.

3. Wenger,R.M. Ang. Chem. (Int. Edn.),1985, 24, 77.

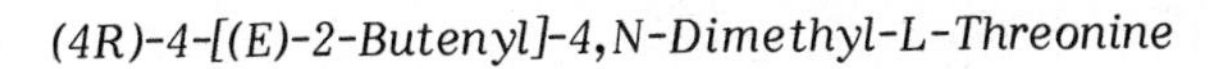

(4R)-4-[(E)-2-Butenyl]-4,N-Dimethyl-L-Threonine

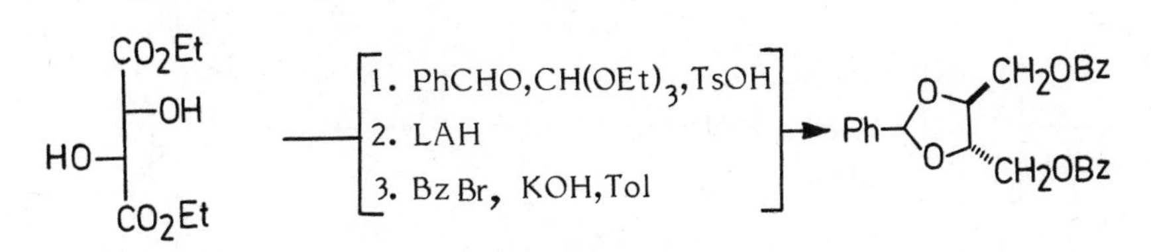

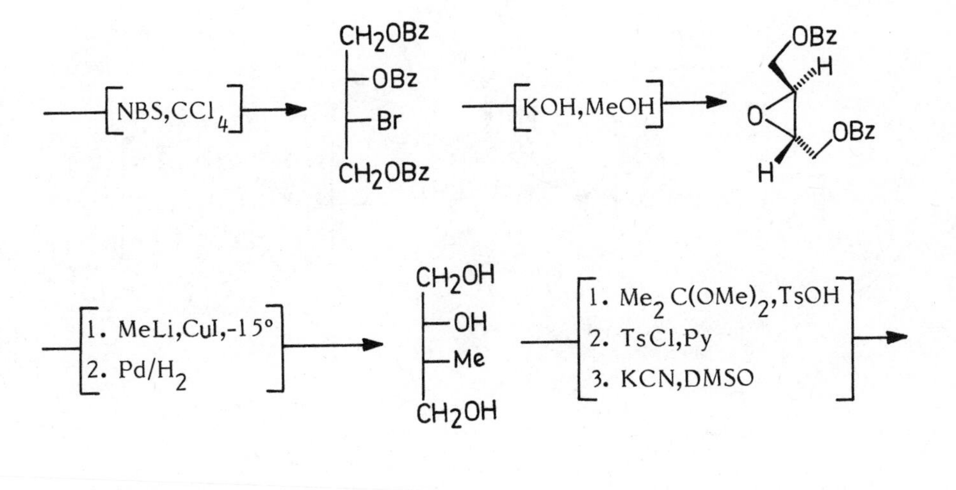

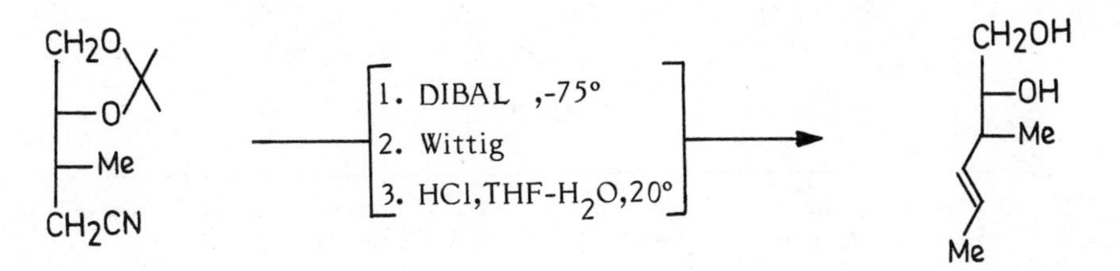

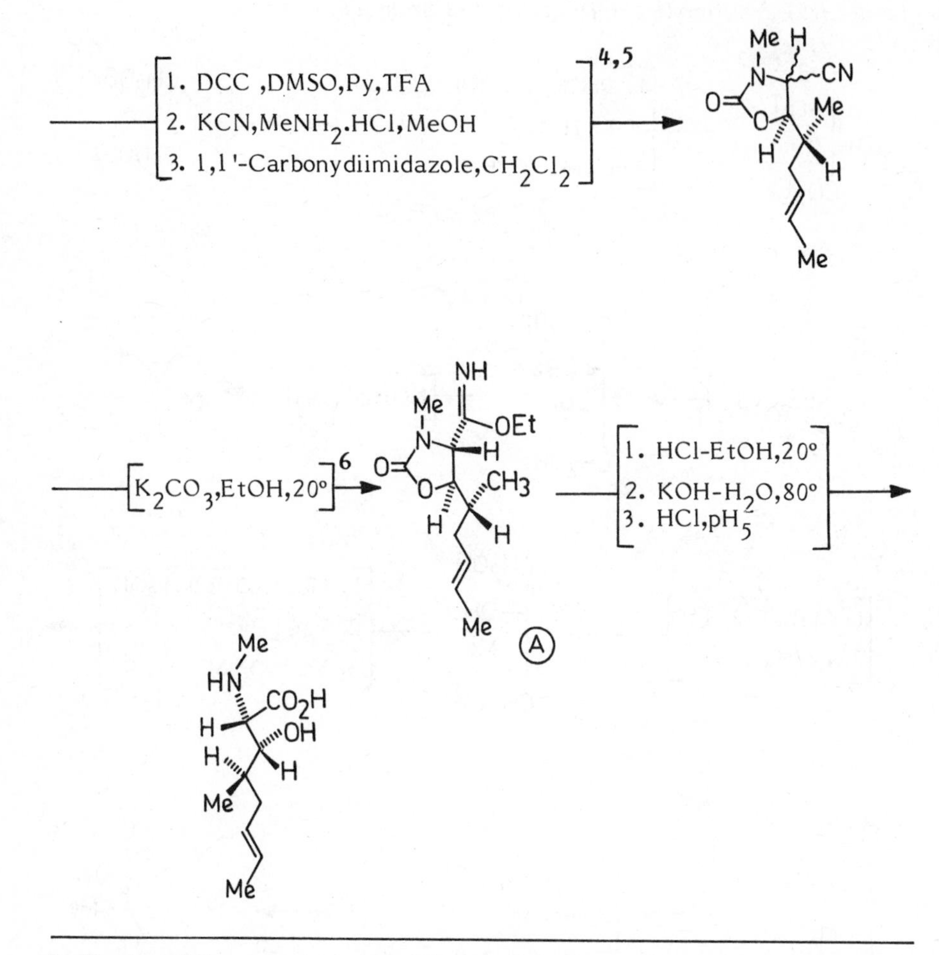

4. Direct oxidation gave low yield, but protection of secondary hydroxyl, followed by oxidation and deprotection gave the aldehyde in much higher yield.

5. The cyanhydrin was obtained as a mixture of diastereomers, and the 2-oxazolidinone as 6:1 mixture of cis: trans isomers.

6. The intermediate iminomethylene derivative reacts stereo-specifically to give the thermodynamically more stable trans-imidate (A), which on hydrolysis furnished the enantiomerically pure N-methylamino acid with the O- and N-functional groups in the desired threo- configuration.

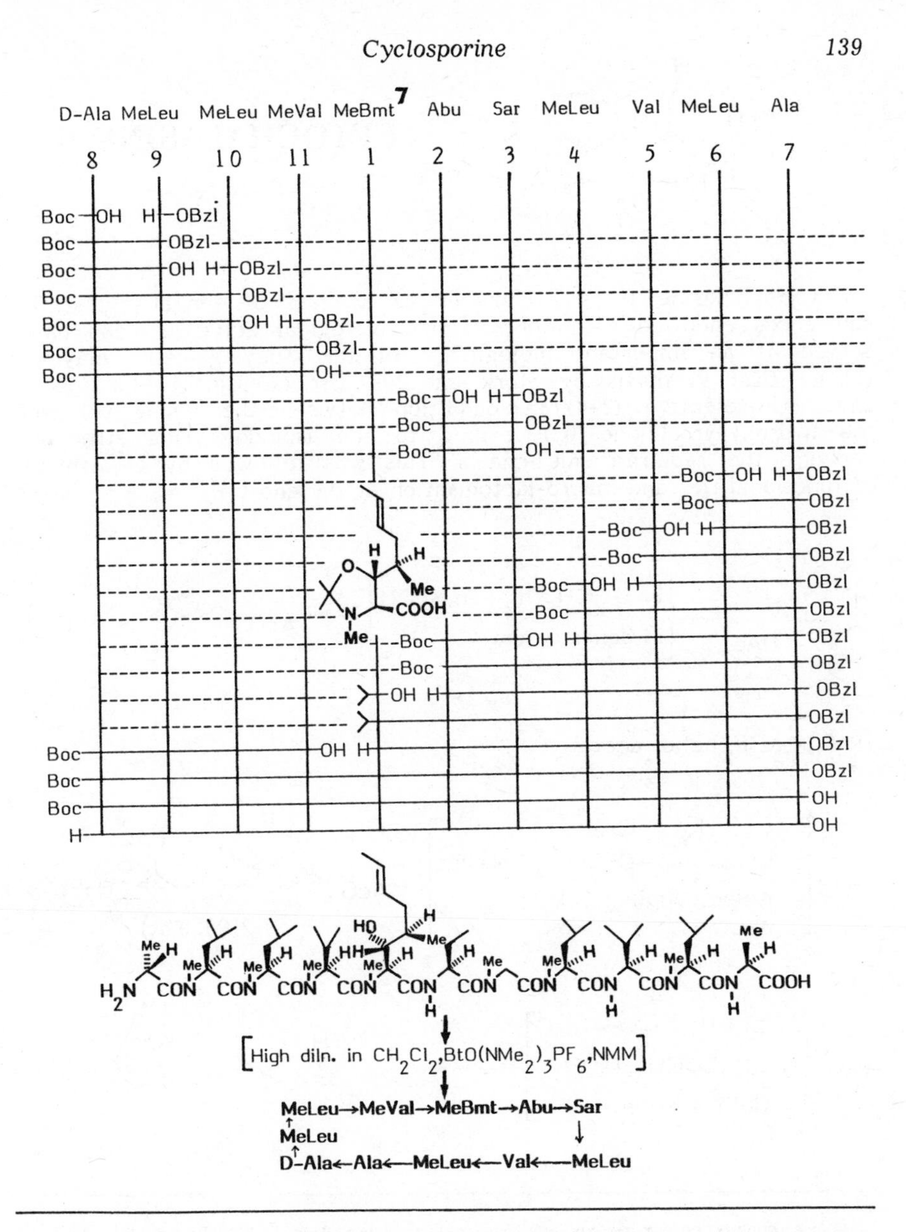

7. During incorporation of MeBmt the hydroxy and $NHCH_3$ were protected as isopropylidene derivative to avoid epimerisation and the peptide bond formed using DCC and N-hydroxybenzotriazole. The final amide bond between the tetra and hepta-peptides was made using $BtOP(NMe_2)_3^+ PF_6^-$, NMM (8).

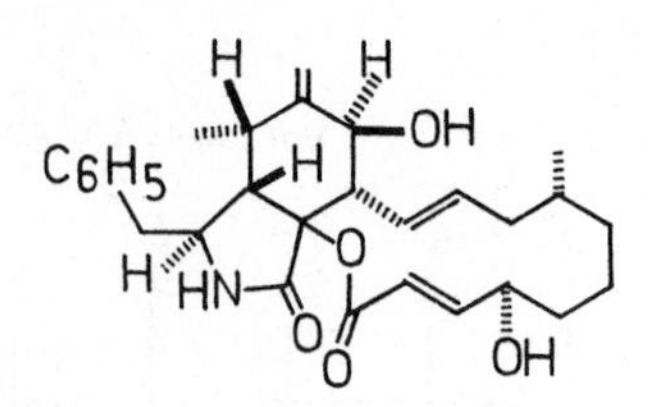

CYTOCHALASINE B

Cytochalasine B is a member of a class of fungal cytostatic substances endowed with very unusual biological activities, and have served as an important biological tool for studying cell behaviour (1). Its first synthesis by Stork and his associates involved a stereo- and regio-selective (2+4) cycloaddition between the triene (B) and the hydroxypyrrolinone (C) leading to the isoindole ring structure carrying the required side chains. This was followed by adjustment of functionalities and macro-lactonisation at the end (2).

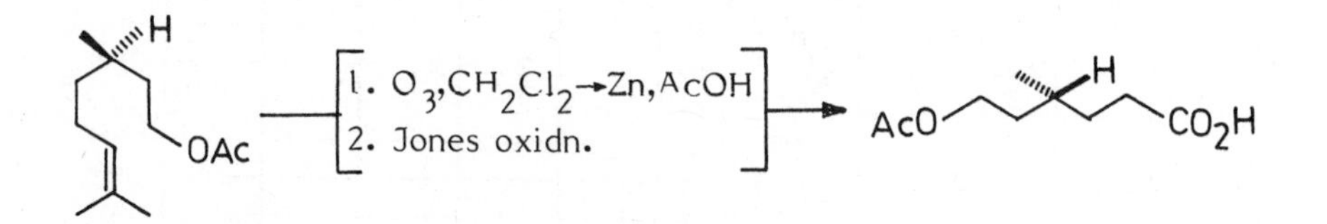

R-(+)-β-Citronellol acetate

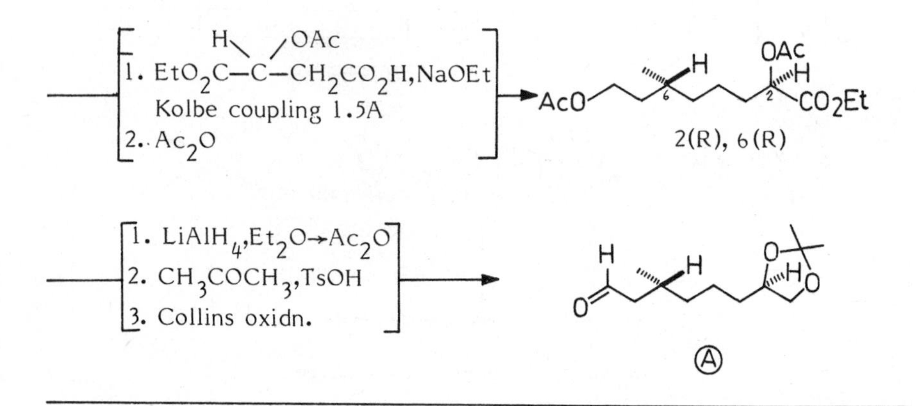

---

1. For a comprehensive review on Cytochalasins see: Binder, M., Tamm, Ch. Angew. Chem., Int. Ed. Engl., 1973, 12, 370.

2. Stork,G.; Gilbert, Nakahara, Yoshiaki, Nakahara, Yuko, Greenlee, W.J. J. Am. Chem. Soc., 1978, 100, 7775.

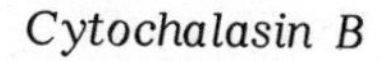

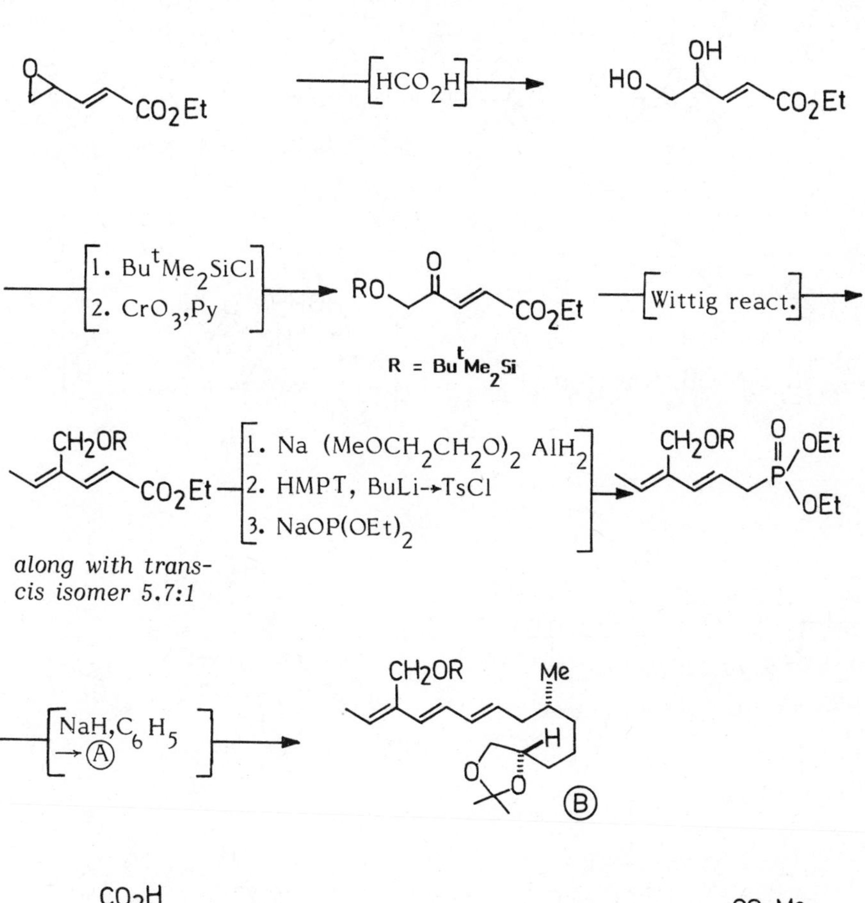
CO2Et
HCO2H
HO
OH
CO2Et
1. ButMe2SiCl
2. CrO3,Py
RO
O
CO2Et
R = ButMe2Si
Wittig react.
CH2OR
CO2Et
1. Na (MeOCH2CH2O)2 AlH2
2. HMPT, BuLi→TsCl
3. NaOP(OEt)2
CH2OR
O
OEt
P
OEt
along with trans-
cis isomer 5.7:1
NaH,C6H5
→A
CH2OR
Me
H
O
O
B

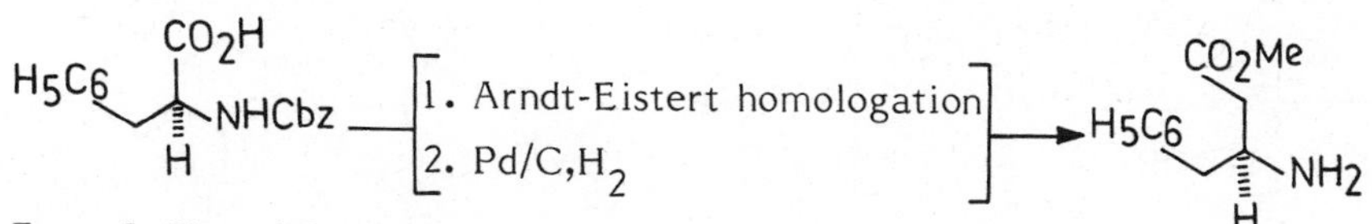
CO2H
H5C6
NHCbz
H
1. Arndt-Eistert homologation
2. Pd/C,H2
CO2Me
H5C6
NH2
H
From L-Phenylalanine

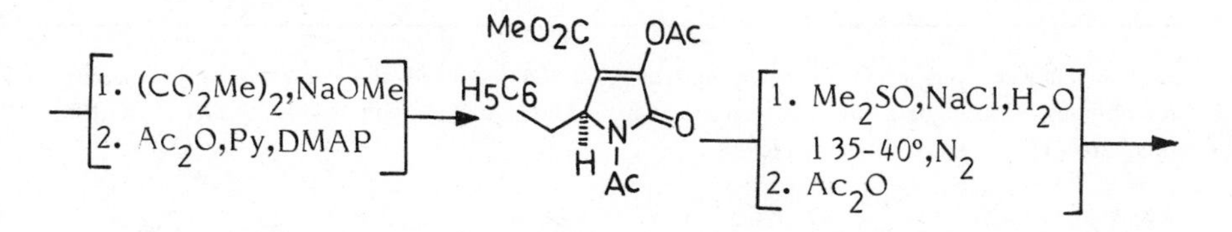
1. (CO2Me)2,NaOMe
2. Ac2O,Py,DMAP
MeO2C
OAc
H5C6
O
N
H
Ac
1. Me2SO,NaCl,H2O
135-40°,N2
2. Ac2O

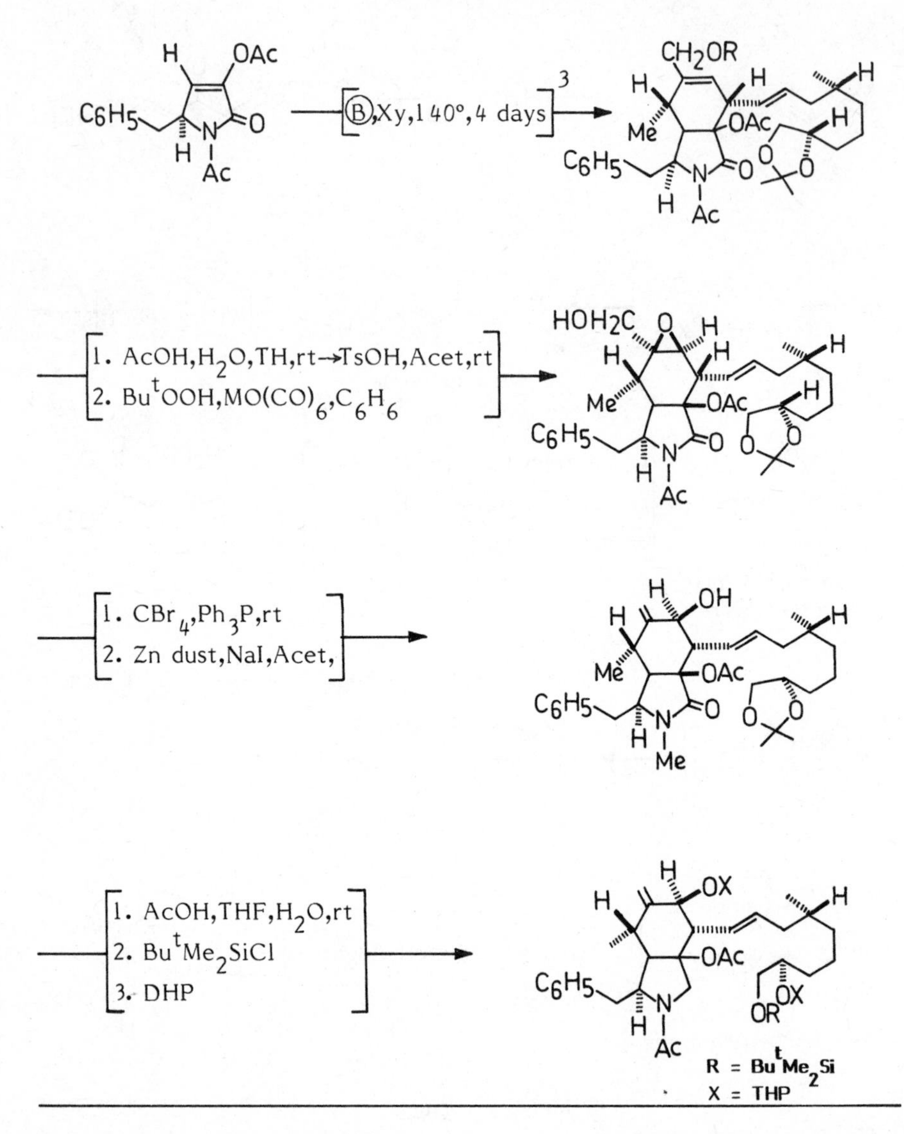

3. This addition was regio-selective and gave a 96:4 ratio of the required to the wrong regioisomer (total yield of the adduct was 40%), along with some quantity of the recovered triene and the pyrrolinone.

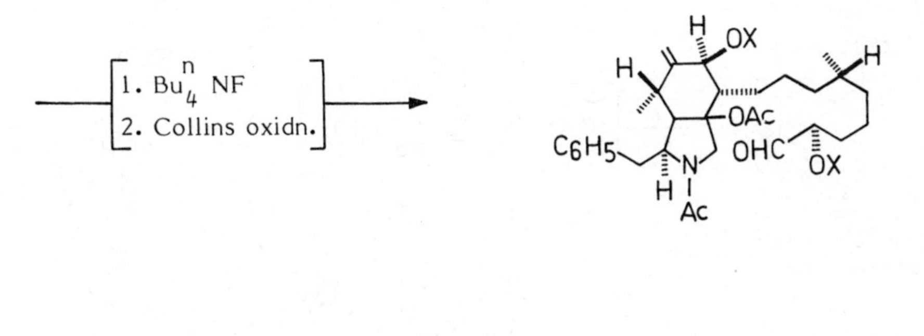

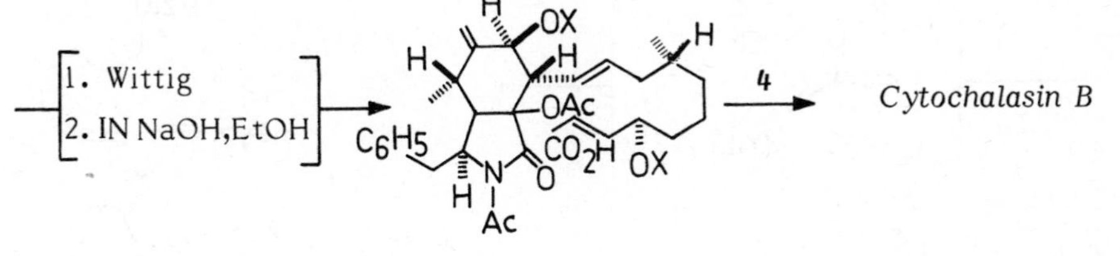

Stork & Nakamura in a refinement of the above synthesis constructed the macrocyclic and the hexane rings on the pyrrolone by an intramolecular Diels-Alder reaction on the tetraene (B) (5), which avoided the inefficient lactonisation at the end of the earlier synthesis (2).

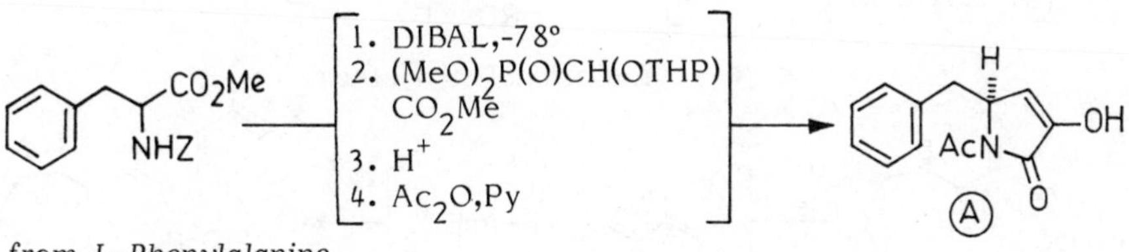

from L-Phenylalanine

---

4. The cyclisation of the di-tetrahydropyranyl unsaturated acid to cytochalasin B has been accomplished previously; Massamune,S.; Hayase,Y.; Schilling,W.; Chan,W.K.; Bates,G.S. J. Am. Chem. Soc., 1977, 99, 6756.

5. Stork,G.; Nakamura,E. J. Am. Chem. Soc., 1983, 105, 5510.

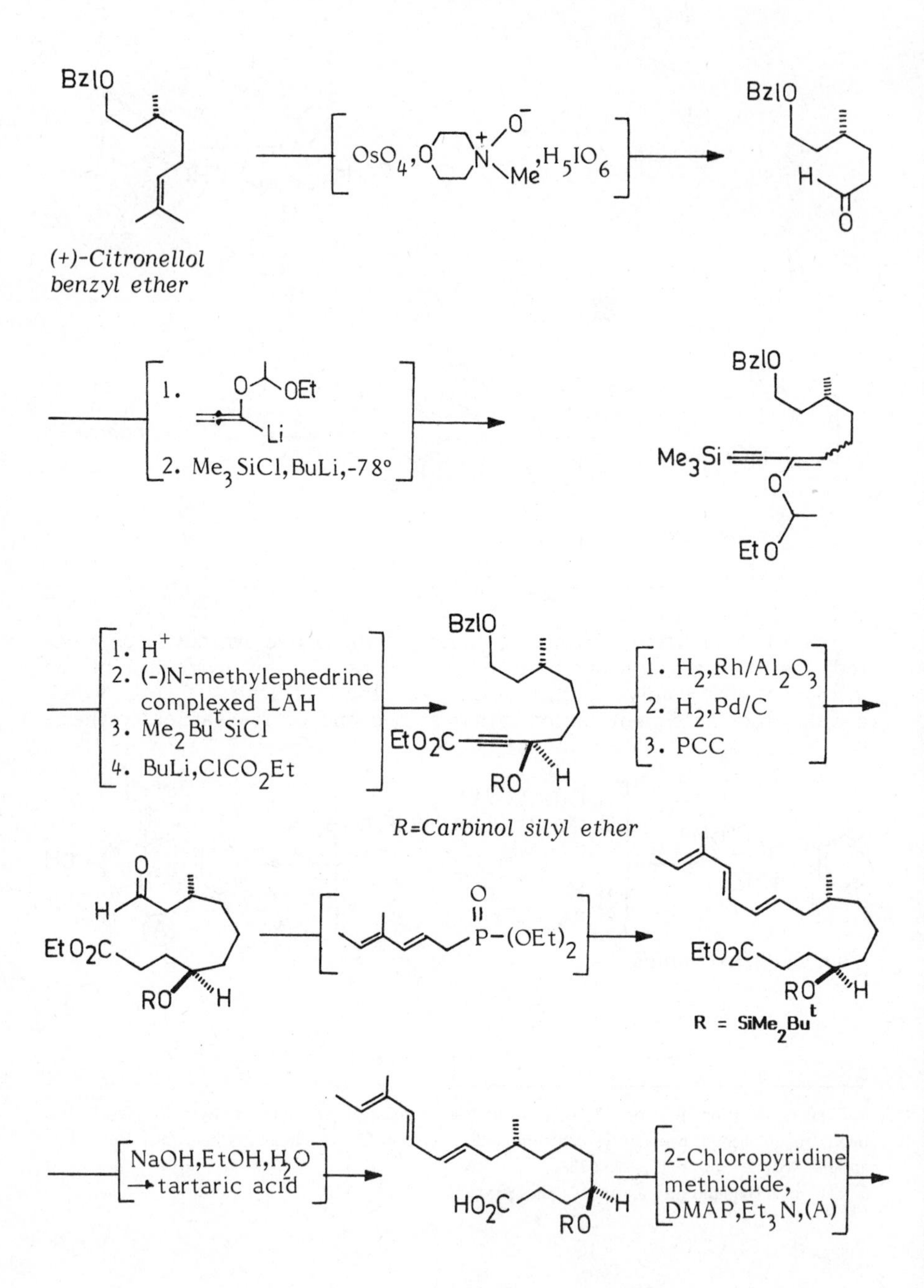
BzlO
(+)-Citronellol
benzyl ether
OsO4, O N Me O⁻ , H5IO6
BzlO
H
O
1. O OEt Li
2. Me3SiCl, BuLi, -78°
BzlO
Me3Si
O
Et O
1. H+
2. (-)N-methylephedrine
complexed LAH
3. Me2ButSiCl
4. BuLi, ClCO2Et
BzlO
EtO2C
RO H
R=Carbinol silyl ether
1. H2, Rh/Al2O3
2. H2, Pd/C
3. PCC
O
H
Et O2C
RO H
O
P-(OEt)2
EtO2C
RO H
R = SiMe2But
NaOH, EtOH, H2O
→ tartaric acid
HO2C
RO
H
2-Chloropyridine
methiodide,
DMAP, Et3N, (A)

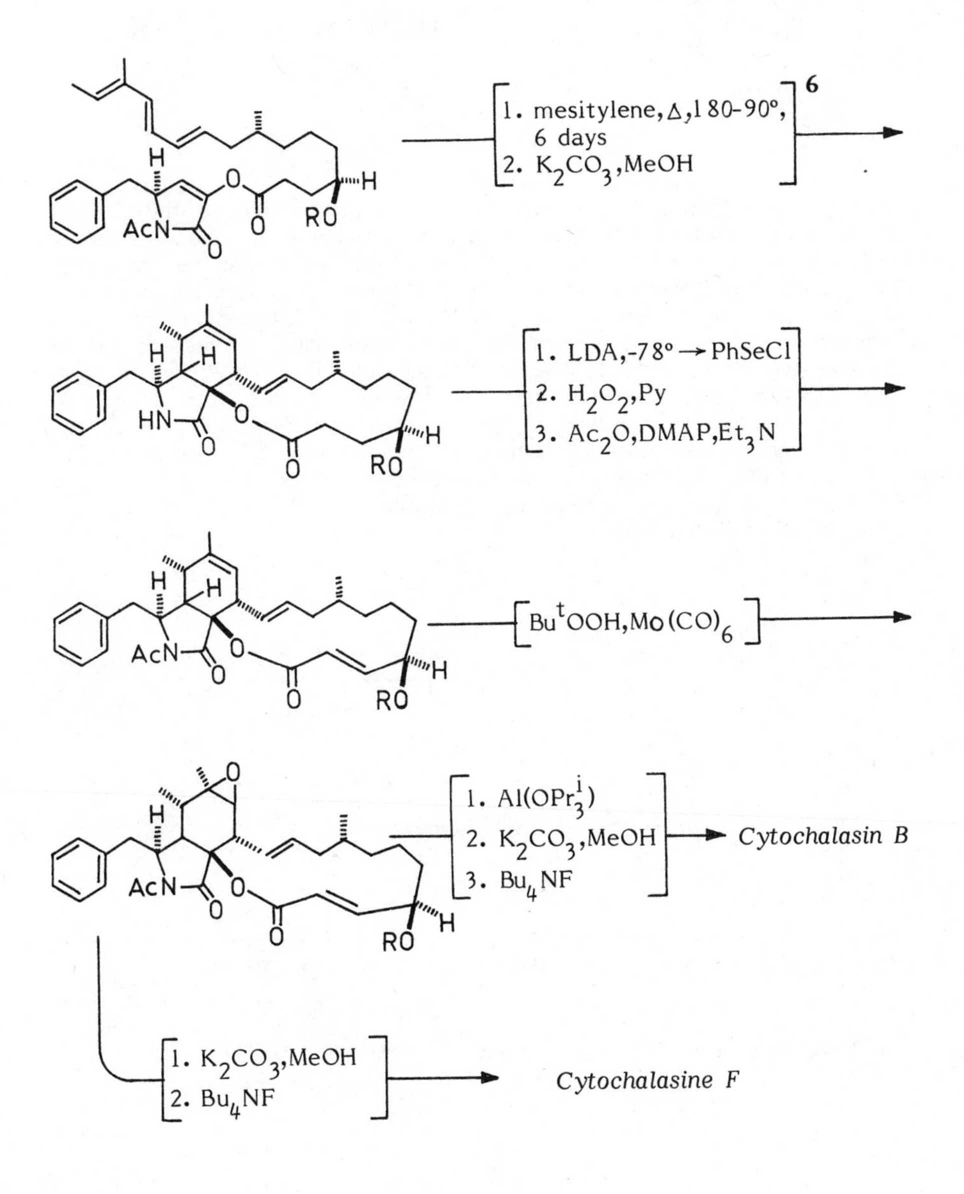

---

6. The required cycloaddition product was formed along with its diastereomer in a ratio of 4:1 separable on silica after deacetylation.

DEWAR BENZENE

The view is generally held that James Dewar in 1867 had proposed a para bond structure for benzene as an alternative to the Kekulé structure of 1865. This representation which visualizes a bond connecting 1-4 positions of a planar hexagon should of course be incapable of existence. However, if the para bonded structure is regarded as the classical non-planar bicyclo [2.2.0] hexa-2,5-diene, a modified version of the original 'Dewar benzene', then such a molecule should be capable of rational synthesis. Taking advantage of steric factors, van Tamelen and Pappas(1) were the first to prepare such a bicyclic system. The parent member of this class, 'Dewar benzene' was prepared by the same group(2) using an intramolecular photolytic 2+2 cycloaddition, followed by oxidative decarboxylation. This highly strained hydrocarbon is remarkably stable having a half life of two days at room temperature.

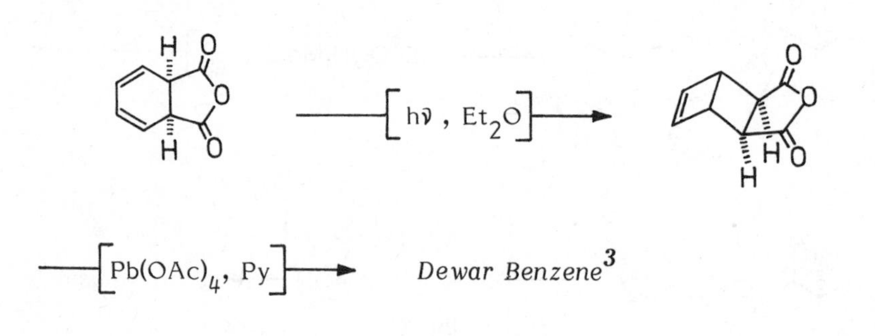

1. van Tamelen, E.E., Pappas, S.P. J. Am. Chem. Soc., 1962, 84, 3789.

2. van Tamelen, E.E., Pappas, S.P. J. Am. Chem. Soc., 1963, 85, 3297.

3. Review: Schafer, W., Hellemann, H. Angew. Chem. Internat. Ed., 1967, 6, 518.

E,E-1,4-DIACETOXY-1,3-BUTADIENE

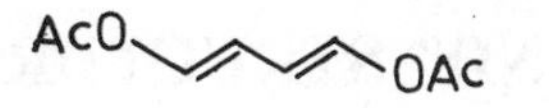

The Woodward-Hoffmann rules (1) have created a new dimension in the art of organic synthesis and this facet is best illustrated with the practical synthesis of E,E-1,4-diacetoxy-1,3-butadiene [I] from cyclooctatetraene (2). This transformation involves the removal of the Z,Z-1,3-butadiene unit from the intermediate III arising from a disrotatory cyclization of II. The required removal of a 1,3-butadiene has been carried out by employing a 4+2 addition followed by a 4+2 reversal sequence to give a trans-diacetoxycyclobutene, which then undergoes the expected conrotatory opening to give [I].

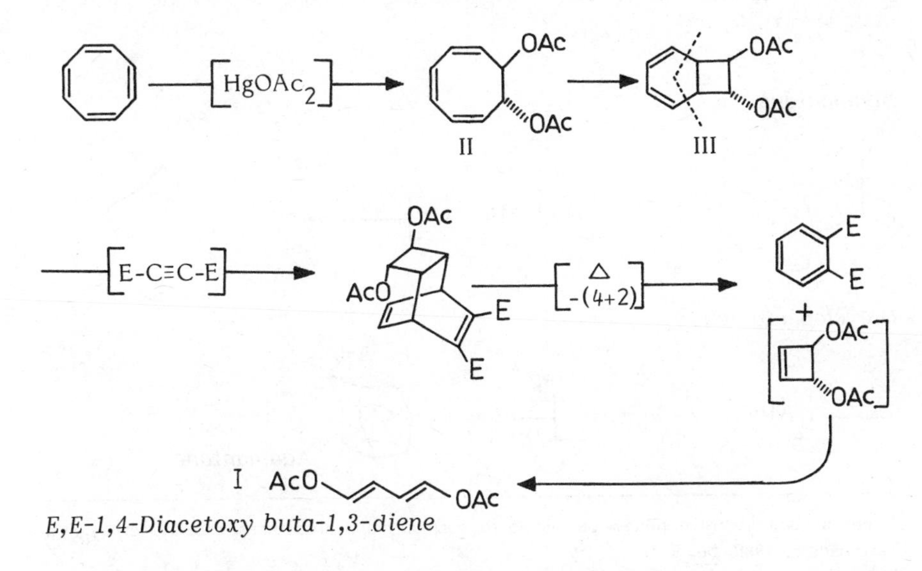

E,E-1,4-Diacetoxy buta-1,3-diene

1. Woodward, R.B., Hoffmann, R. "The Conservation of Orbital Symmetry", Academic Press Inc., New York, 1970.

2. Carlson, R.M., Hill, R.Y., Org. Syn., 1970, 50, 24.

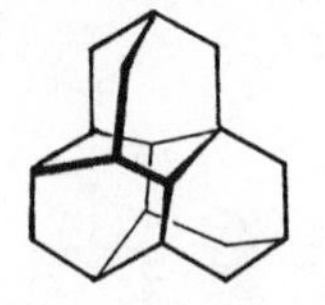

"DIAMOND" STRUCTURES

Covalent linking of carbon atoms in space and in a tetrahedral fashion gives rise to diamond. Adamantane has a "diamond" type structure in the sense that in principle its polymerization involving replacement of the C-H bonds with C-C bonds would lead to diamond (1). Adamantane occurs in coal-tar and thus gave rise to the speculation that hydrocarbons when placed under conditions where the possibility of C-H bond-forming and bond-breaking exist, could ultimately lead to a strain free "diamond" type structure. The assumption was proved correct with the transformation of a C_{10} hydrocarbon to adamantane (2). Similar methods have given rise to an "admantalogous series" The series has progressed upto triamantane, where one of the carbons is linked covalently to four neighbouring carbons as in the case of diamond (3-5).

Adamantane

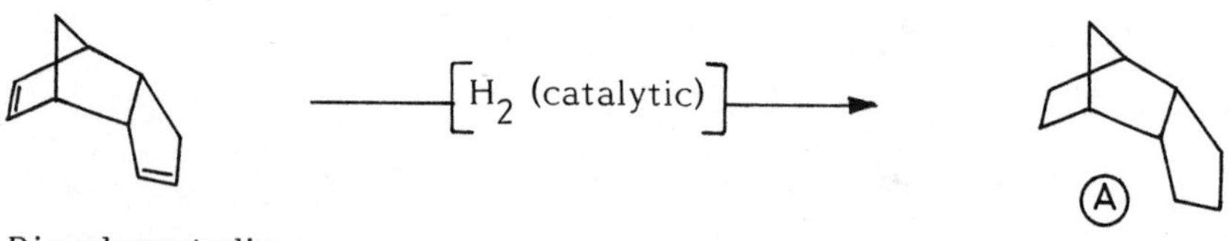

Dicyclopentadiene

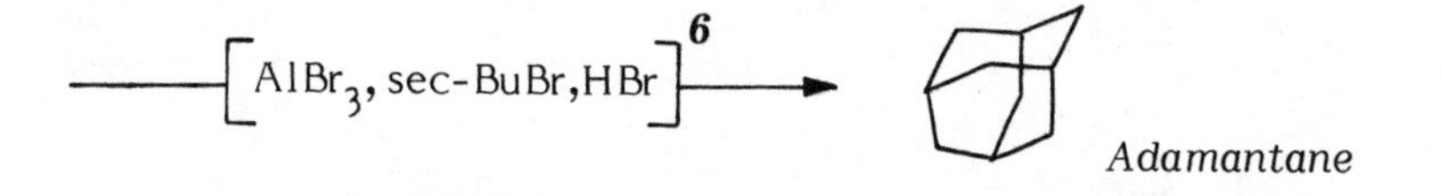

---

1. For a comprehensive review pertaining to adamantoid hydrocarbons, see: McKervey,M.A. Tetrahedron, 1980, 36, 971.

2. Schleyer,P.von R.; Donaldson,M.M. J. Am. Chem. Soc., 1960, 82, 4645.

3. Cupas,C.; Schleyer,P.von R.; Trecker,D.J. J. Am. Chem. Soc., 1965, 87, 917; Karle, I.L.; Karle,J. J. Am. Chem. Soc., 1965, 87, 918.

4. Williams Jr.von Z.; Schleyer,P.von R.; Gleicher,G.J.; Rodewald,C.B. J. Am. Chem. Soc., 1966, 88, 3862.

5. An attempt to synthesize tetramantane led to "Bastardane"; Schleyer,P.von R.; Osawa, E.; Drew,M.G.B. J. Am. Chem. Soc., 1968, 90, 5034.

Diamantane (Congressane)

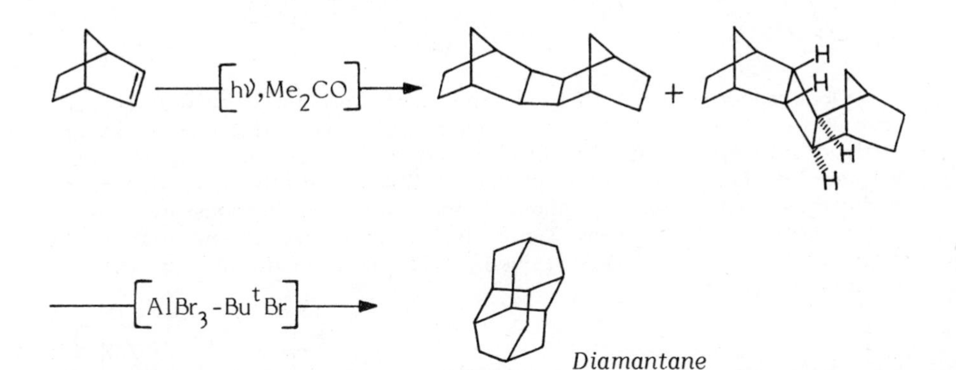

Congressane can now be prepared in quantities from the 4+4 dimer of bicycloheptadiene (7).

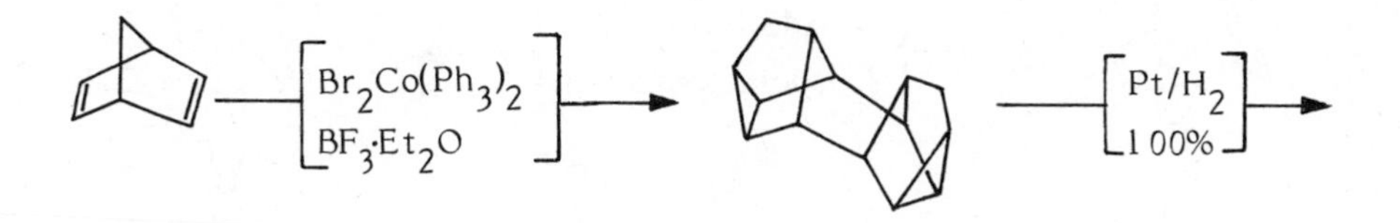

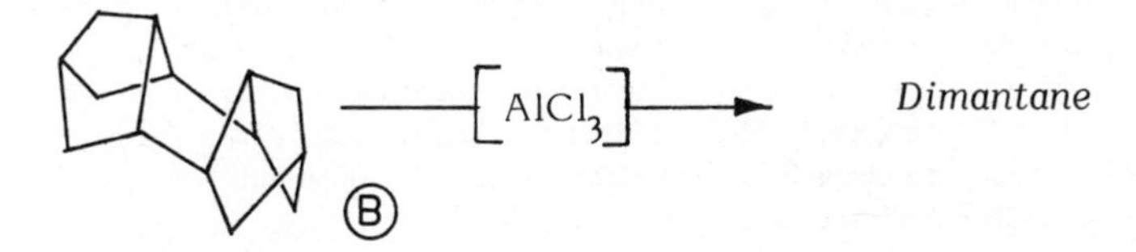

---

6. This profound rearrangement is pictured as proceeding by a multistage ionic process:

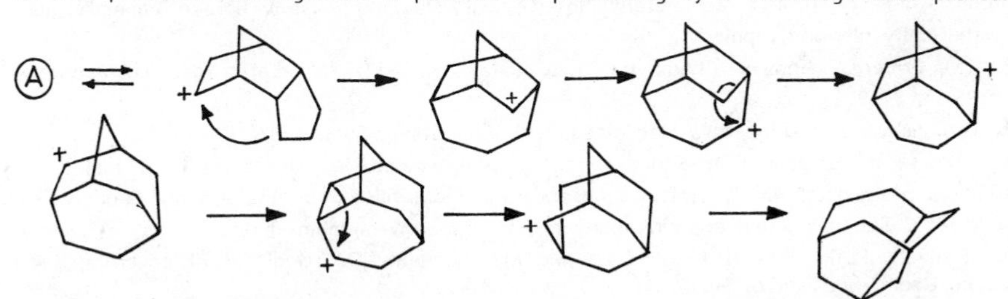

7. McKervey,M.A. J. Chem. Soc. Perkin I, 1972, 2691.

DODECAHEDRANE

Dodecahedrane (1) with I_h symmetry (icosahedral group), characterised by a very high torsional strain and a very small angle strain, is structurally the most complex, symmetric and aesthetically appealing member of the CnHn convex polyhedra (2). The synthesis of dodecahedrane has been intensely pursued for over two decades by several group (2b) and its recent synthesis (3,6) described below marking a culmination of years of sustained effort is a land mark in the art of organic synthesis.

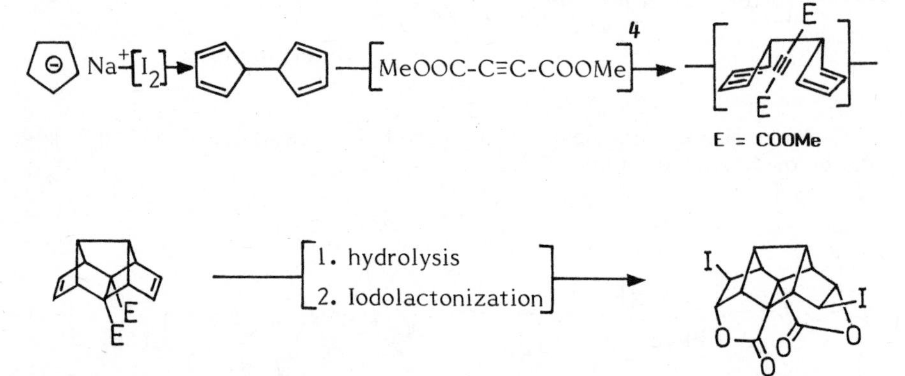

---

1. Plato with commendable insight limited to five the number of covex polyhedrons whose faces are congruent regular polygons forming equal dihedral angles at each edge. If 'm' number of regular 'n' sided polygons meet at each vertex they must satisfy the equation: $1/m + 1/n = 1/2 + 1/E$ where E represents the total number of edges of the polyhedron. The five fundamental solids that fulfil this requirement, called Platonic solids, are tetrahedron, cube, dodecahedron, octahedron and eicosahedron. Of these the valencies of carbon would permit the construction of only the first three and amongst these dodecahedrane holds a pre-eminent position because of its relatively strain-free array of 12 polyfused cyclopentane rings, which can generate the highest known point group symmetry (I_h), a completely closed cavity lacking solvent uptake capacity and aesthetically pleasing topology.

2. (a) Ermer,O. Angew. Chem. Int. Ed., 1977, 16, 411; (b) Eaton,P.E. Tetrahedron, 1979, 35, 2189.

3. Ternansky,R.J.; Balogh,D.W.; Paquette,L.A. J. Am. Chem. Soc., 1982, 104, 4503.

4. The term "domino Diels-Alder reaction" was coined for this inter/intra molecular addition; Paquette,L.A.; Wyvratt,M.J.; Berk,H.C.; Moerch,R.E. J. Am. Chem. Soc., 1978, 100, 5845. The compound contains many cis, syn fused 5-membered rings.

5. Paquette,L.A.; Wyvratt,M.J.; Schal ner,O.; Schneider,D.F.; Begley,W.J. Blankenship,R.M. J. Am. Chem. Soc., 1976, 98, 6744.

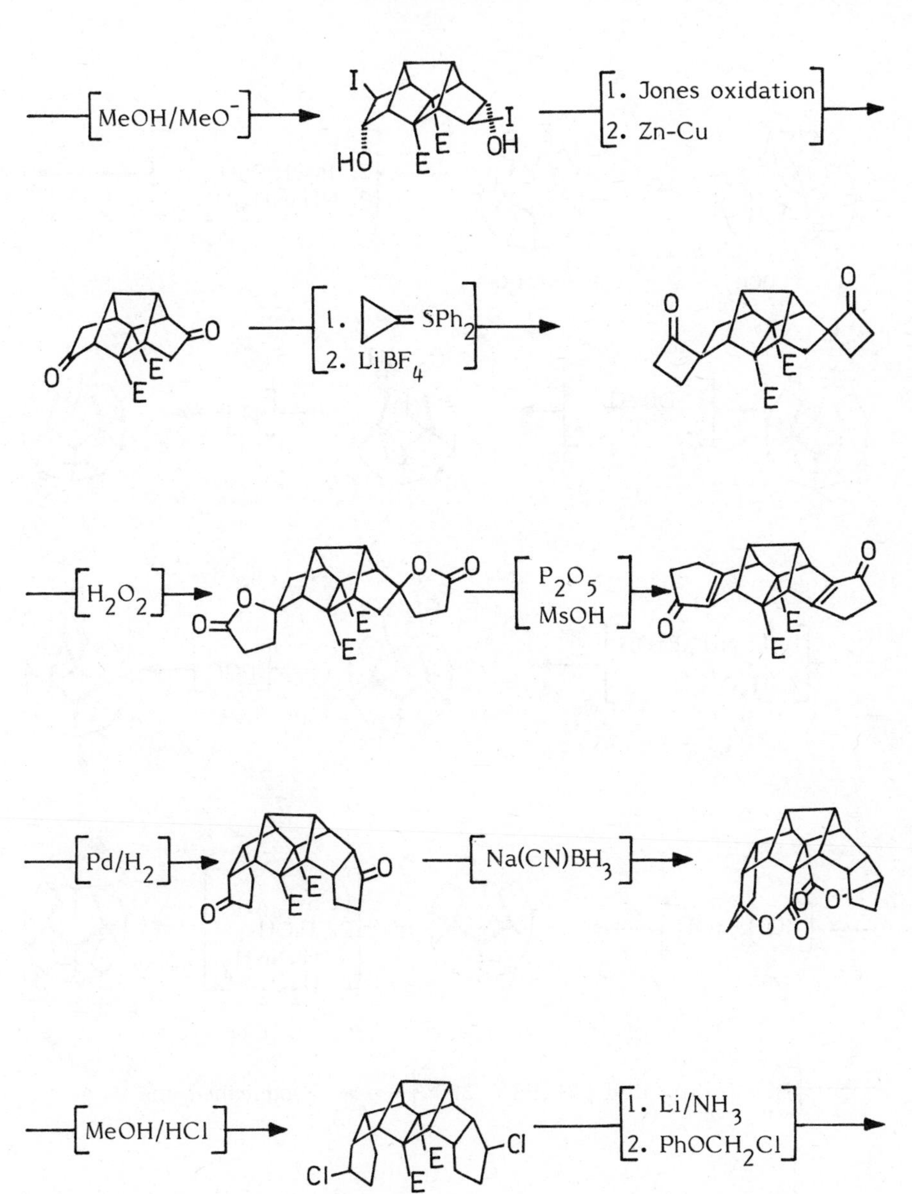

6. (a) Gallucci,J.B.; Doecke,C.W.; Paquette,L.A. J. Am. Chem. Soc., 1986, 108, 1343; (b) Paquette,L.A.; Miyahara,Y.; Doecke,C.W. ibid, 1986, 108, 1716.

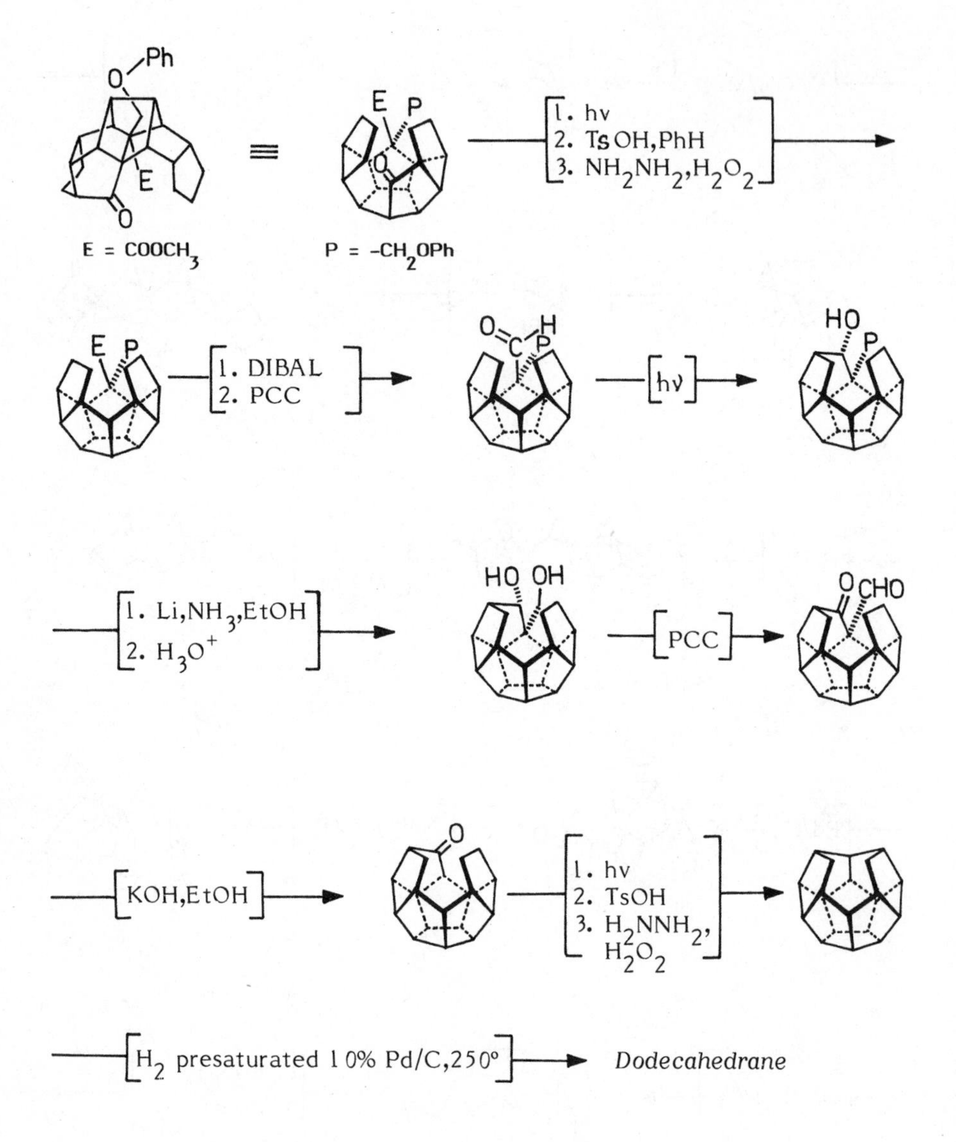

—[H_2 presaturated 10% Pd/C, 250°]→ *Dodecahedrane*

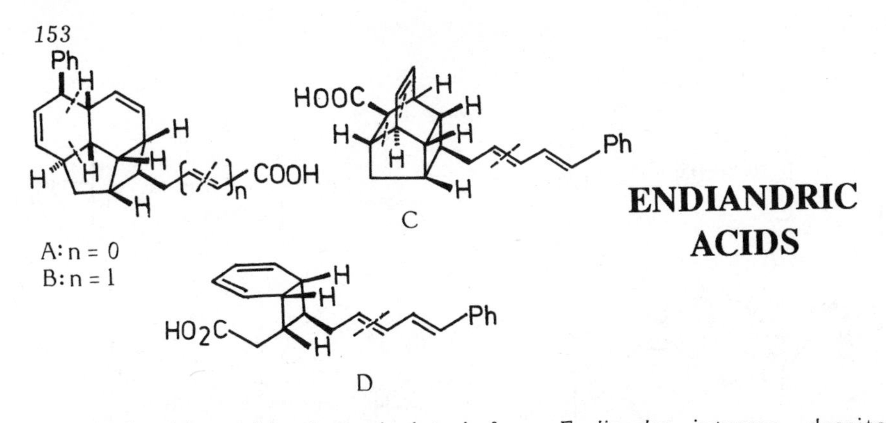

ENDIANDRIC ACIDS

Endiandric acids A-D, isolated from Endiandra introrsa, despite the presence of eight asymmetric centres, occur in nature in racemic form (1). In view of the presence of all the possible structural types of endiandric acid cascade i.e. A,B,C and D in the same plant and in a racemic form, Black et al. (2) were led to propose that endiandric acids A-D are biosynthesised from achiral precursors by a series of electrocyclisations thermally allowed by Woodward-Hoffmann rules, namely an 8 π e conrotatory electrocyclisation, followed by a 6 π e disrotatory cyclisation, followed by an 4 π S + 2 π S intramolecular cycloaddition reaction. Nicolaou and his associates have reported both a step-wise stereocontrolled total synthesis of endiandric acids as also a biomimetic synthesis consisting of the synthesis of the polyunsaturated carboxylic acids from suitable precursors followed by one step conversion to various members of the endiandric acid cascade (4), which provide support to this hypothesis (5).

1. Bandaranayake,W.M.; Banfield,J.E.; Black,D.St.C.; Fallon,G.D.; Gatehonse,B.M. J. Chem. Soc., Chem. Commun., 1980, 162.

2. Bandaranayake,W.M.; Banfield,J.E.; Black,D.St.C. J. Chem. Soc., Chem. Commun., 1980, 902.

3. Nicolaou,K.C.; Petasis,N.S.; Zipkin,R.E.; Uenishi,J. J. Am. Chem. Soc., 1982, 104, 5555; Nicolaou,K.C.; Petasis,N.S.; Uenishi,J.; Zipkin,R.E. ibid, 1982, 104, 5557.

4. Nicolaou,K.C.; Zipkin,R.E.; Petasis,N.A. J. Am. Chem. Soc., 1982, 104, 5558; Nicolaou,K.C.; Petasis,N.A.; Zipkin,R.E. ibid, 1982, 104, 5560.

5. The syntheses were based on a retro-synthetic analysis of the complex polycyclic frameworks, which provides a clear picture of the endiandric acid cascade and the causal relationship of different structural types in the likely sequence of their formation in the plant.

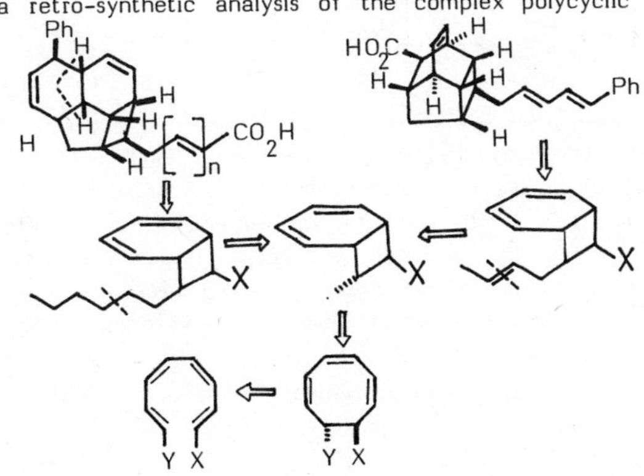

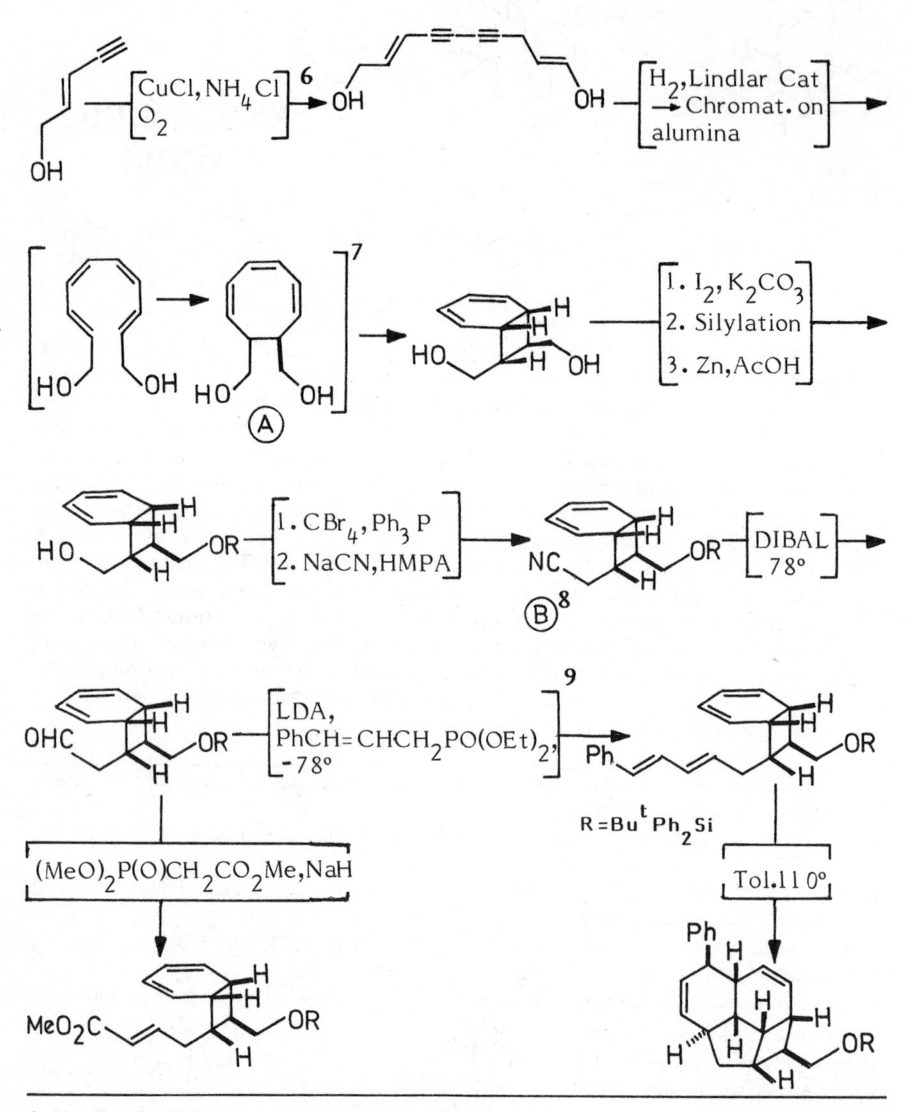

6. Haynes,L.J.; Heilbron,I.; Jones,E.R.H.; Sondheimer,F. J. Chem. Soc., 1947, 1583; Heilbron,I.; Jones,E.R.H.; Sondheimer,F. ibid, 1947, 1586.

7. The bicyclic-diol (A) was obtained directly after column chromatography and the presumed intermediates were not detected under these conditions.

8. The cyanide (B) served as a key intermediate for the synthesis of all the endiandric acids, some of which have not even been isolated from natural sources or isolated after their synthesis.

9. This Witting reagent yielded E:Z ratio of 20:1.

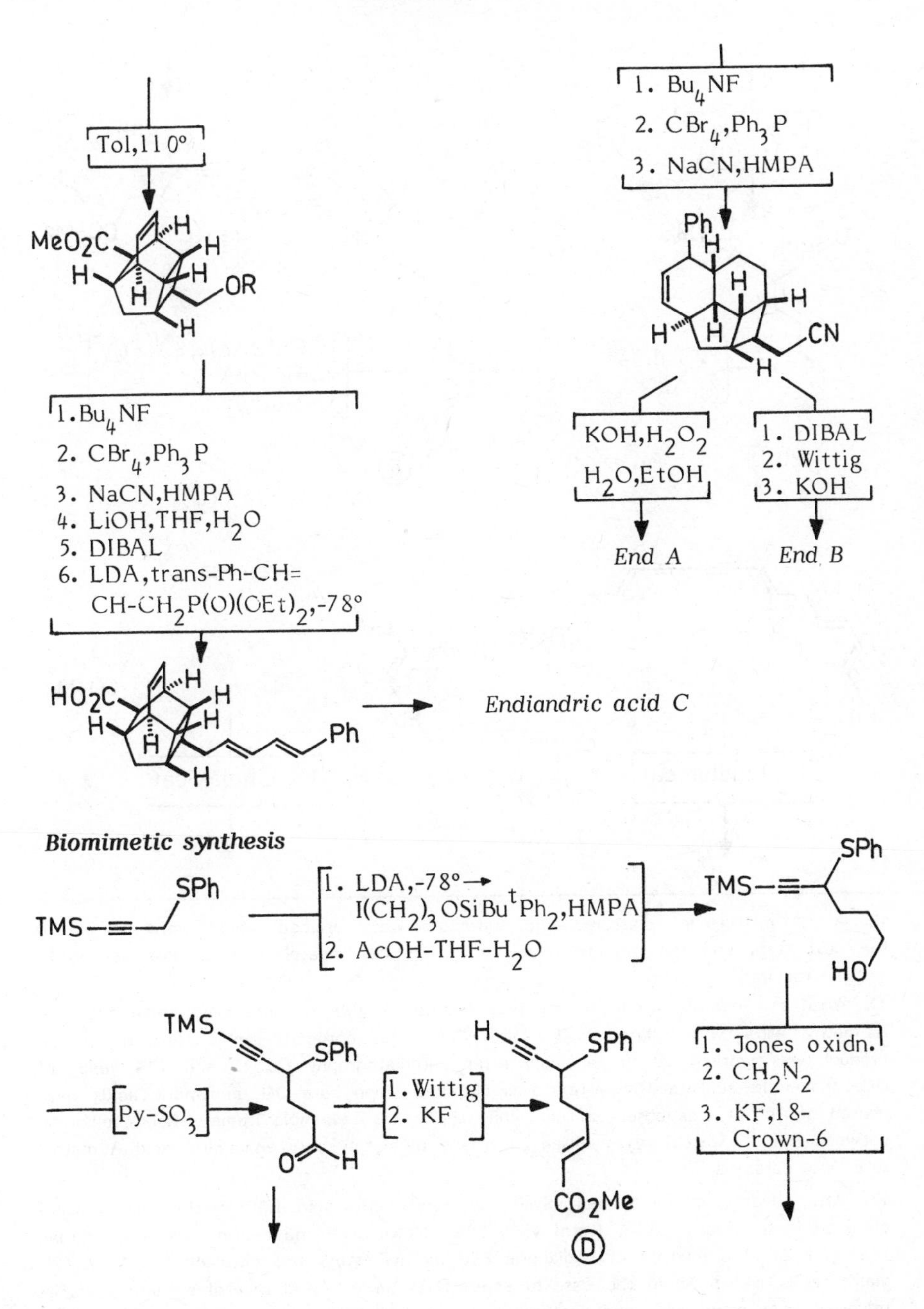

Biomimetic synthesis

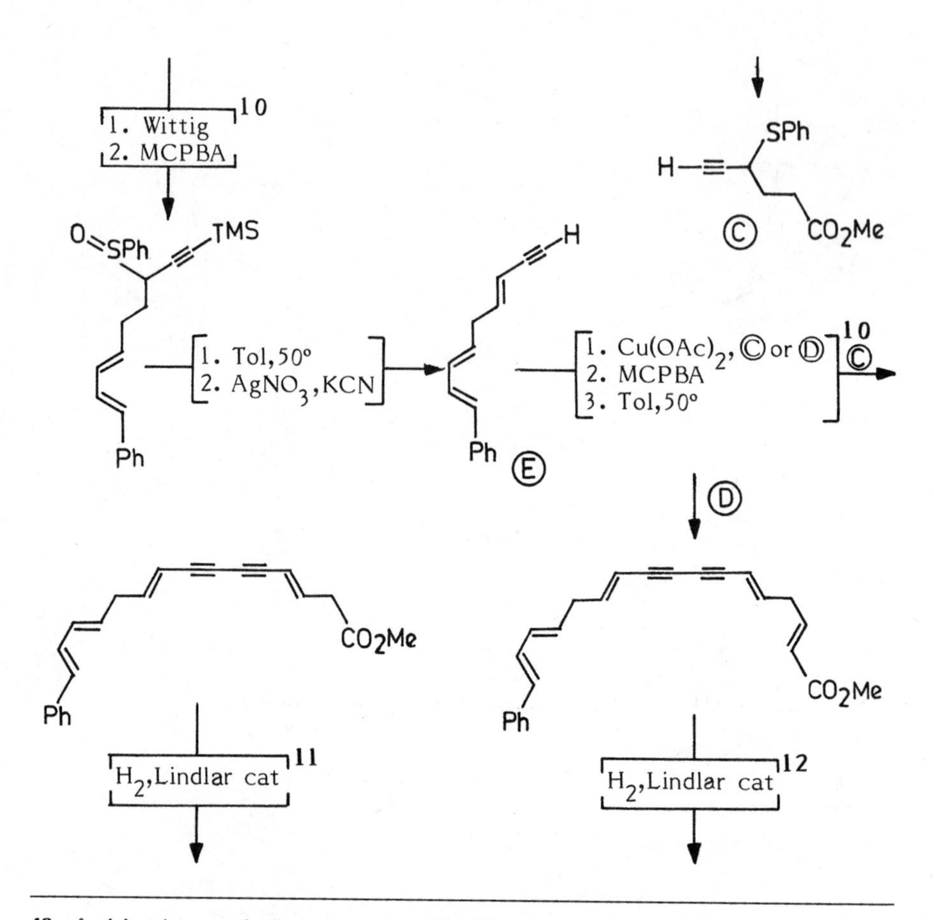

10. A 1:1 mixture of diastereomeric sulfoxides was obtained which were carried to the next step, and the mixture of E and Z olefines separated on a flash column to get the required E isomer.

11. When the hydrogenated product was heated at 100°m tolune only endiandric acid A methyl ester was isolated in Ca 30% yield after chromatography. However, if the product was worked up in the cold after chromatography Ca 12 and 10% yields of D&E endiandric acids methyl esters was obtained when pure D&E endiandric acids were heated at 70°, it was observed that while there was reversible isomerisation equilibrium between D&E, F was slowly cyclised to A and ultimately only endiandric acid A methyl ester was obtained.

12. After hydrogenation & thermolysis only endiandric acid B&C methyl esters were obtained in a ratio of 4.5:1 (total yield 28%). However, if no heating was done during workup (25°C) a mixture of endiandric F&G methyl esters was obtained (Ca 15 & 12% yields respectively). As in the case of esters D&E, there was a reversible thermal equilibrium between esters F&G, and final conversion to B&C methyl esters.

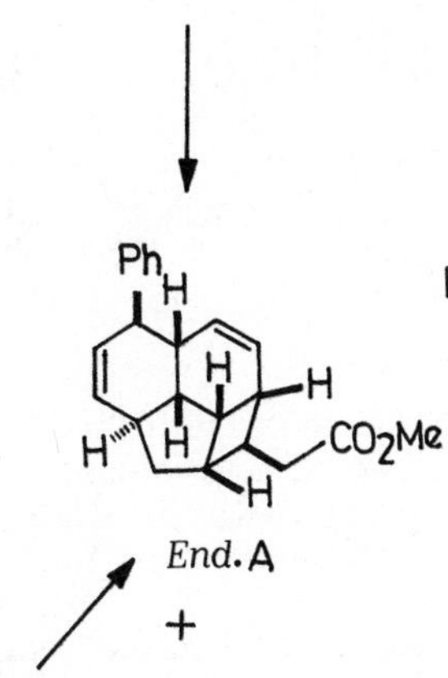

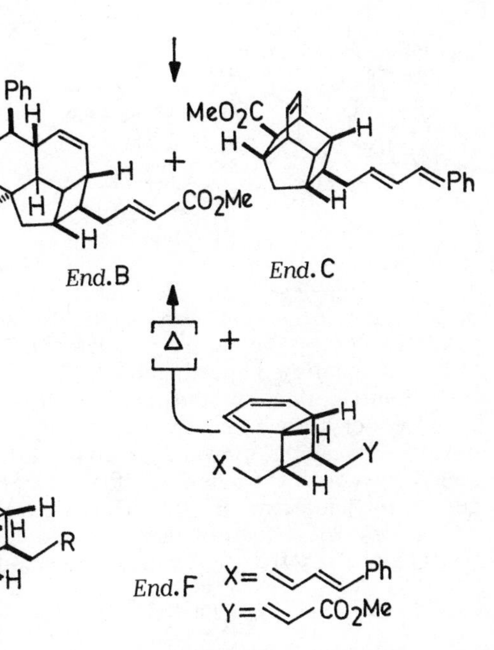

+

End. E

*End.*D

R= ⟋⟍⟋⟍Ph

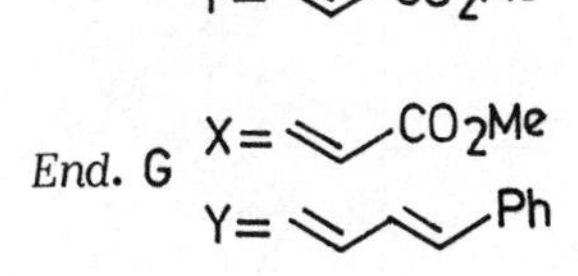

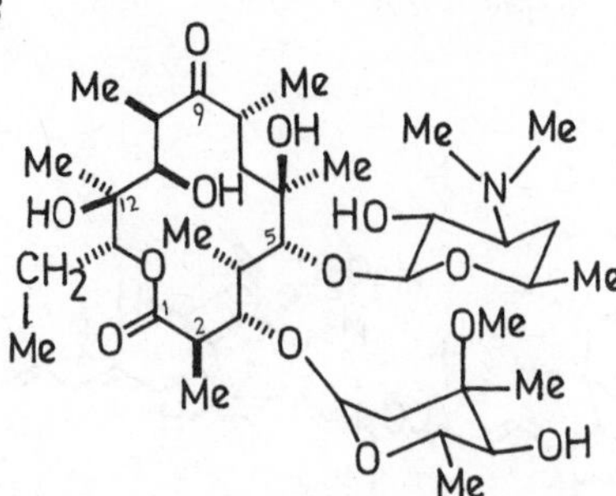

ERYTHROMYCIN

The "hopelessly complex" erythromycin molecule with "its plethora of asymmetric centers" was originally identified as a formidable synthetic challenge by Woodward in 1956 (1). Its synthesis was finally completed and published posthumously in 1981 (2), becoming Woodward's final contribution to the art of organic synthesis (3). Synthesis of the key erythronolide Ⓐ derivative proceeds via macrolactonization of the seco acid Ⓐ, which is constructed by aldol coupling of enantiomerically correct fragments Ⓑ (C3-C8) and Ⓒ (C9-C13), with subsequent introduction of the C1-C2 fragment using a thiopropionate derivative. The two fragments Ⓑ and Ⓒ, which share a hidden symmetry, are crafted from a common intermediate Ⓓ.

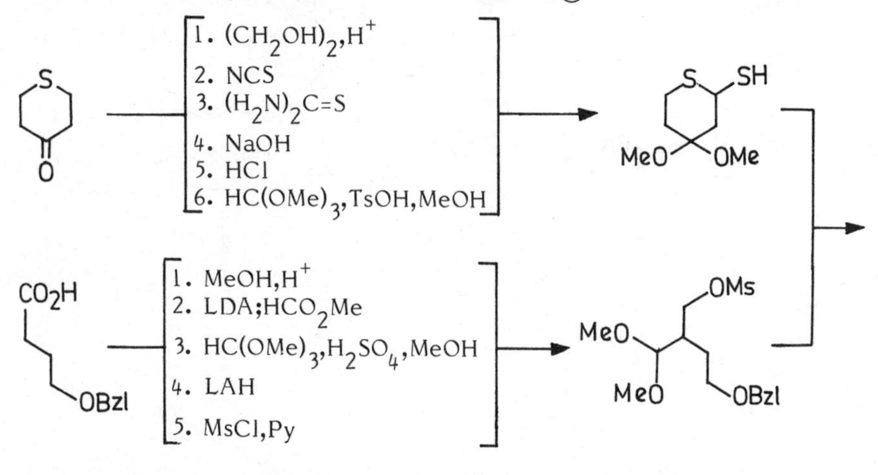

1. Woodward,R.B.; in "Perspective in Organic Chemistry, ed., A. Todd, 155, Wiley (Interscience), New York, 1956.

2. Woodward,R.B.; Logusch,E.; Nambiar,K.P.; Sakan,K.; Ward,D.E.; Au-Yeung,B.W.; Balaram,P. Browne,L.J.; Card,P.J.; Chen,C.H.; Chenevert,R.B.; Fliri,A.; Frobel,K.; Gais,H.J.; Garratt,D.G.; Hayakawa,K.; Heggie,W.; Hesson,D.P.; Hoppe,D.; Hoppe,I.; Hyatt,J.A.; Ikeda,D.; Jacobi,P.A.; Kim,K.S.; Kobuke,Y.; Kojima,K.; Krowicki,K.; Lee,V.J.; Leutert,T.; Malchenko,S.; Martens,J.; Matthews,R.S.; Ong,B.S.; Press,J.B.; Rajan Babu,T.V.; Rousseau,G.; Sauter,H.M.; Suzuki,M.; Tatsuta,K.; Tolbert,L.M.; Truesdale,E.A.; Uchida,I.; Ueda,Y.; Uyehara,T.; Vasella,A.T.; Vladuchick,W.C.; Wade,P.A.; Williams,R.M.; Wong,H.N.C. J.Am.Chem.Soc., 1981, 103, 3210,3213,3215.

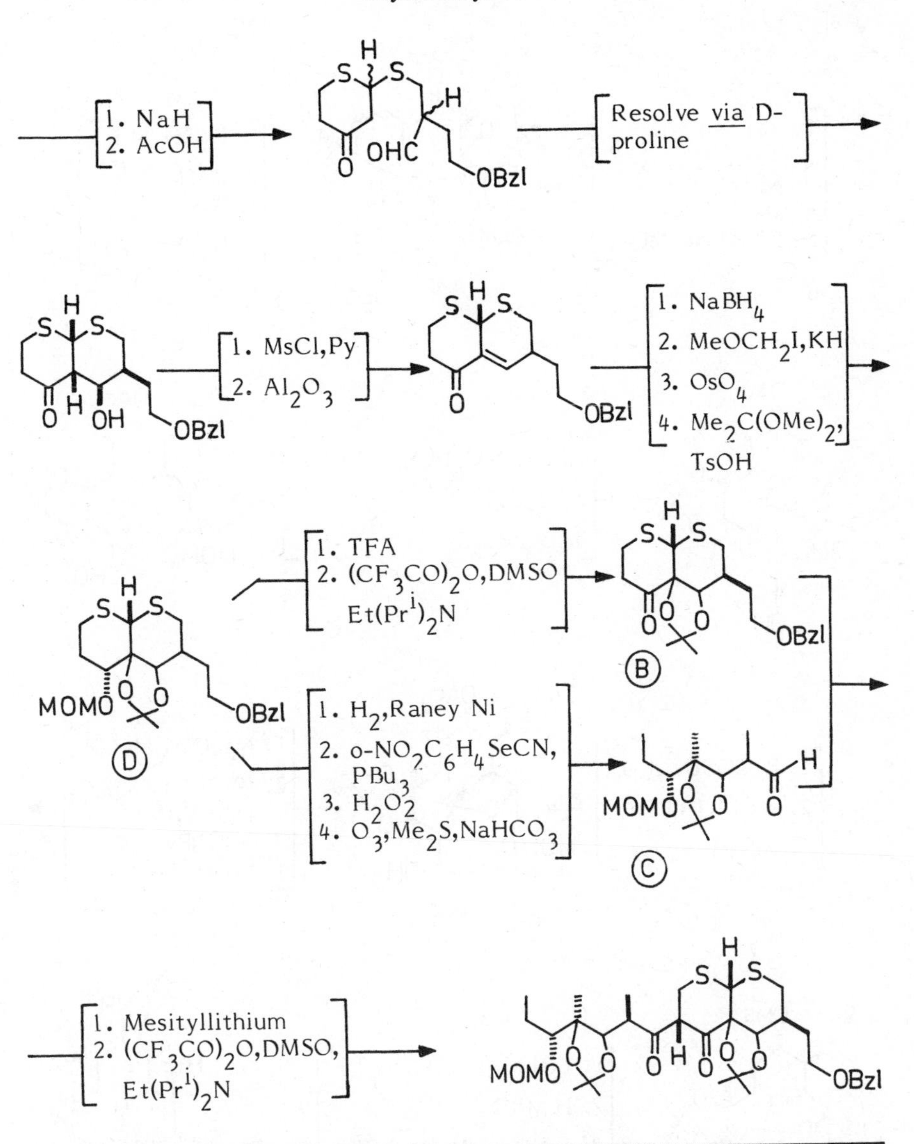

3. For reviews on the strategy of macrolide synthesis, including the erythronolides, see: Masamune,S.; Bates,G.S.; Corcoran,J.W. Angew. Chem. Int. Ed. Engl., 1977, 16, 586; Nicolaou,K.C. Tetrahedron, 1977, 33, 683; Back,T.G. Tetrahedron, 1977, 33, 3041; Masamune,S; McCarthy,P.A. Macrolide Antibiotics, ed. Omura,S. p.127, Academic Press, New York, 1984; Paterson,I.; Mansuri,M.M. Tetrahedron, 1985, 41, 3569.

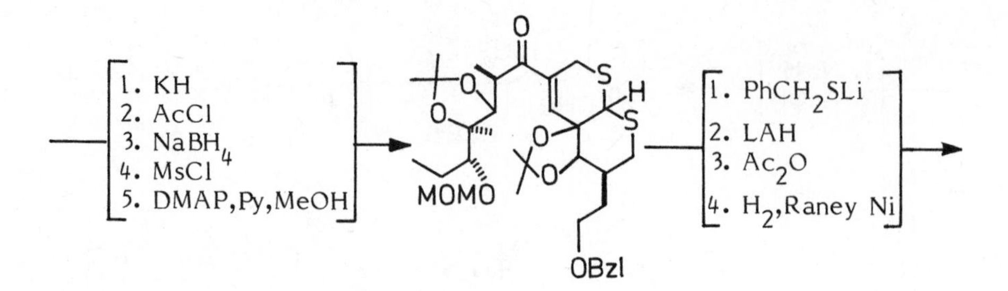
1. KH
2. AcCl
3. NaBH4
4. MsCl
5. DMAP,Py,MeOH
MOMO
OBzl
1. PhCH2SLi
2. LAH
3. Ac2O
4. H2,Raney Ni

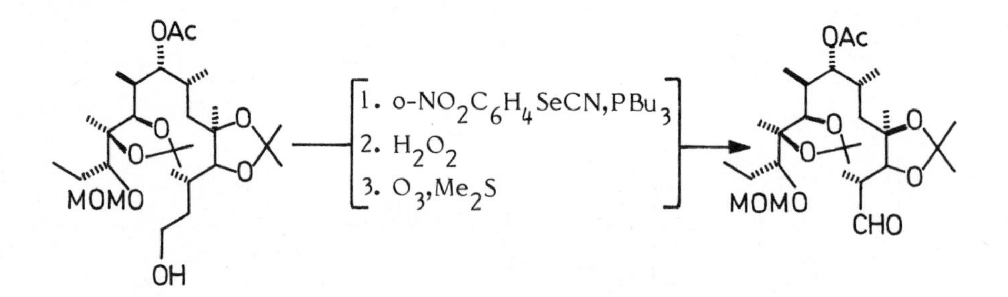
OAc
MOMO
OH
1. o-NO2C6H4SeCN,PBu3
2. H2O2
3. O3,Me2S
OAc
MOMO
CHO

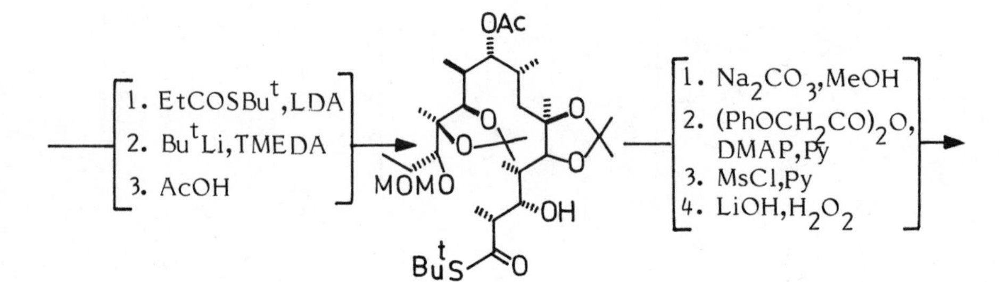
1. EtCOSBut,LDA
2. ButLi,TMEDA
3. AcOH
OAc
MOMO
OH
ButS
1. Na2CO3,MeOH
2. (PhOCH2CO)2O, DMAP,Py
3. MsCl,Py
4. LiOH,H2O2

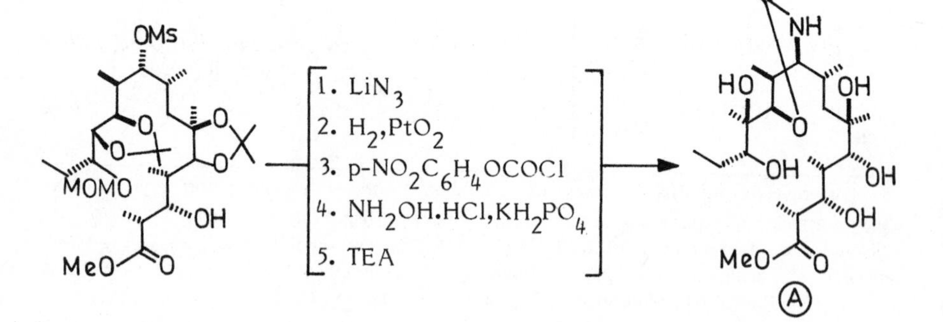
OMs
MOMO
OH
MeO
1. LiN3
2. H2,PtO2
3. p-NO2C6H4OCOCl
4. NH2OH.HCl,KH2PO4
5. TEA
NH
HO
OH
OH
OH
OH
MeO
A

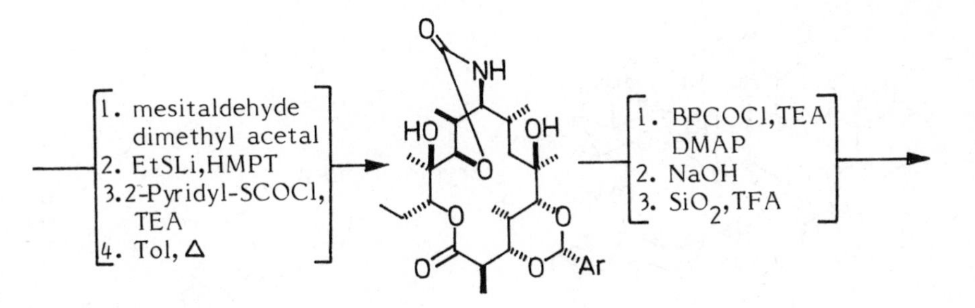

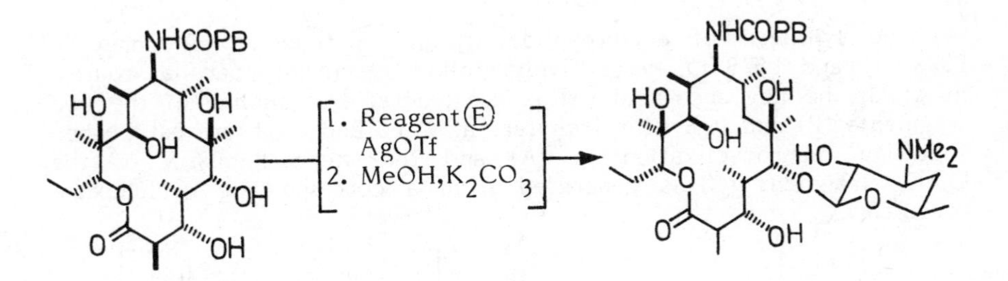

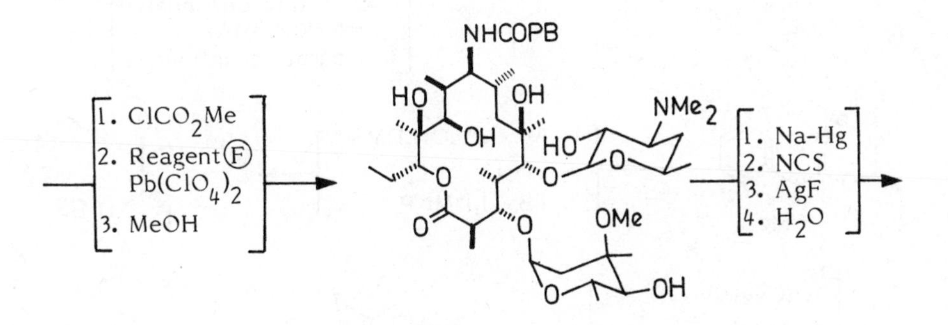

PBCO = p-phenylbenzoyl

Ar = 2,4,6-trimethylphenyl

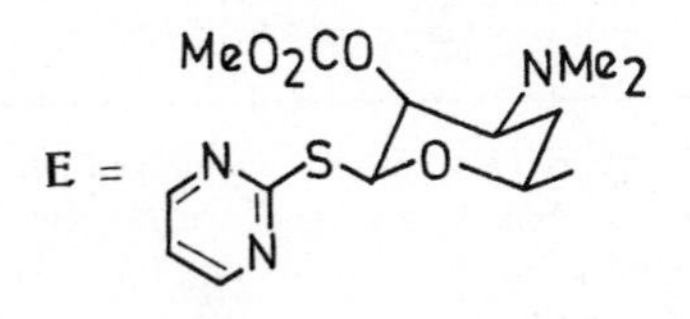

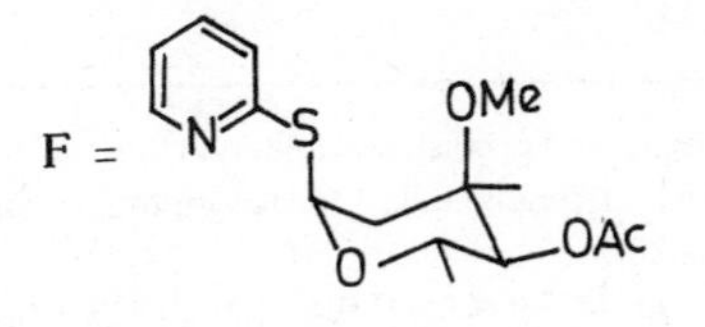

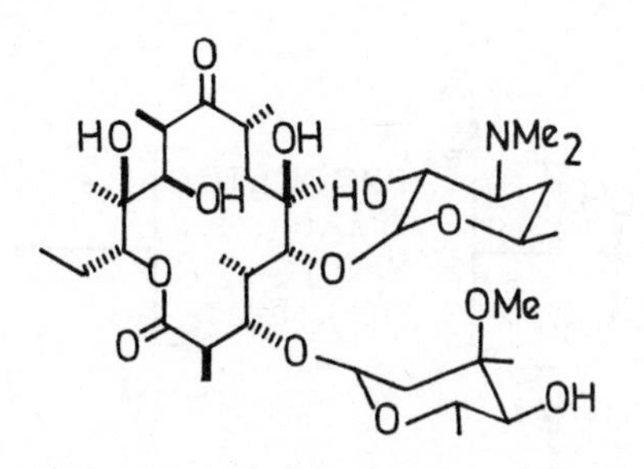

Erythromycin A

Corey synthesis of Erythronolide A

The synthesis of erythronolides Ⓑ and Ⓐ reported by Corey in 1978 (4) and 1979 (5) respectively, follow essentially parallel routes, in which the key seco-acid (A) is constructed by coupling of the two fragments Ⓑ and Ⓒ. Chiral centers at C-10 and C-11 are established following macrocyclization of Ⓐ and the stereochemistry of the C_1-C_9 fragment Ⓒ is generated from a corresponding cyclohexane precursor.

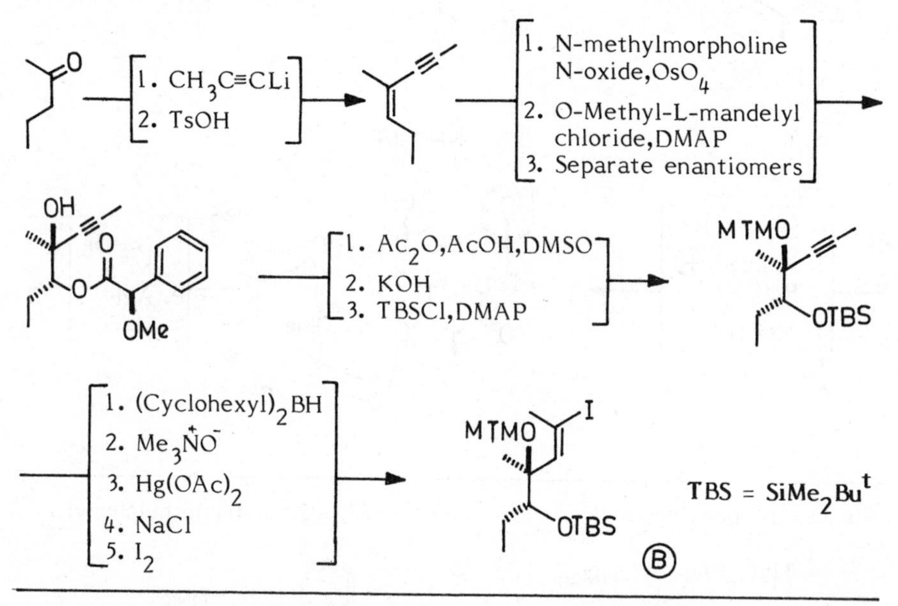

4. Corey, E.J.; Trybulski, E.J.; Melvin, L.S.Jr.; Nicolaou, K.C.; Secrist, J.A.; Lett, R.; Sheldrake, P.W.; Falck, J.R.; Brunelle, D.J.; Haslanger, M.F.; Kim, S.; Yoo, S. J. Am. Chem. Soc., 1978, 100, 4618, 4620.

5. Corey, E.J.; Hopkins, P.B.; Kim, S.; Yoo, Y.; Nambiar, K.P.; Falck, J.R. J. Am. Chem. Soc., 1979, 101, 7131.

Fragment C:

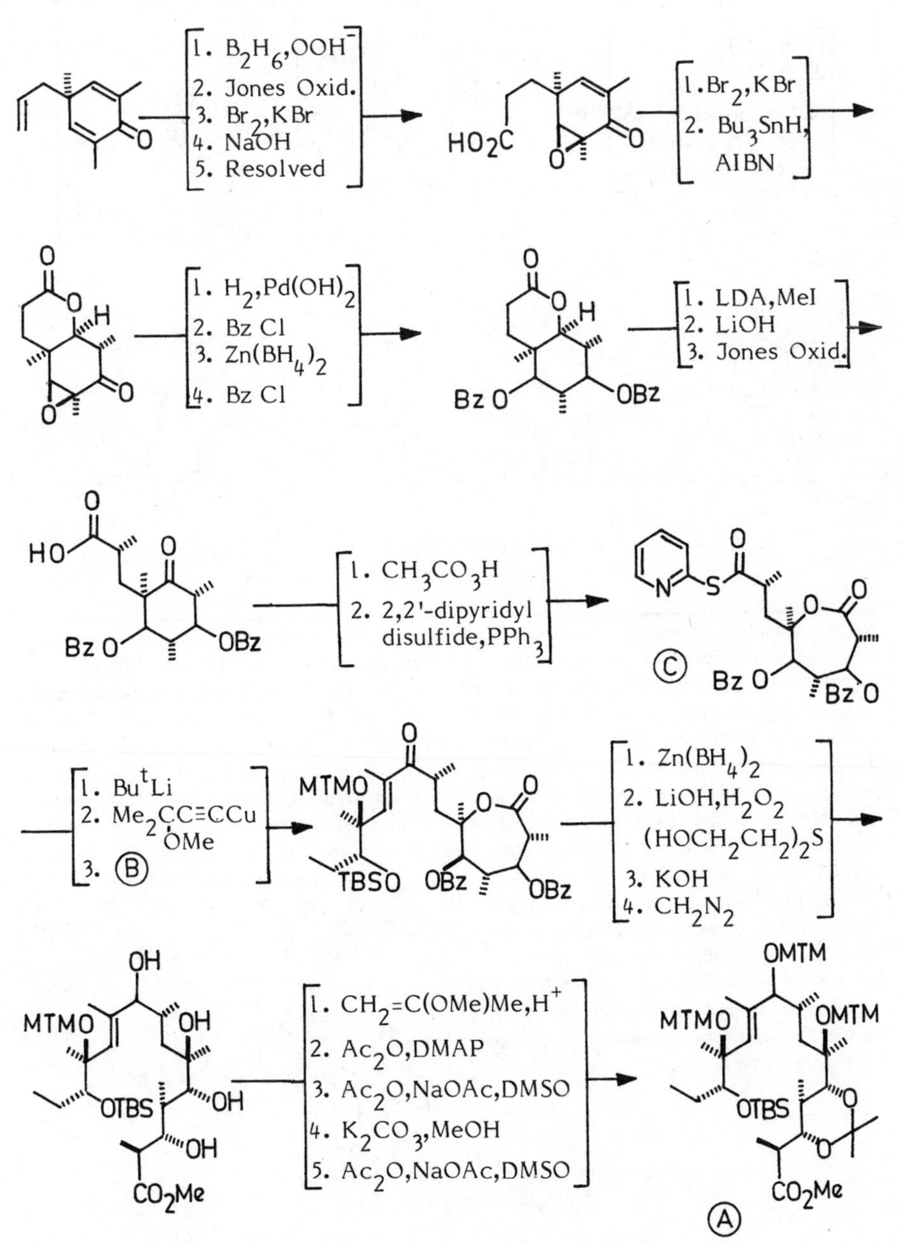

1. B_2H_6, OOH^-
2. Jones Oxid.
3. Br_2, KBr
4. NaOH
5. Resolved
HO_2C
1. Br_2, KBr
2. Bu_3SnH, AIBN
1. H_2, $Pd(OH)_2$
2. Bz Cl
3. $Zn(BH_4)_2$
4. Bz Cl
BzO
OBz
1. LDA, MeI
2. LiOH
3. Jones Oxid.
HO
1. CH_3CO_3H
2. 2,2'-dipyridyl disulfide, PPh_3
C
1. Bu^tLi
2. $Me_2C(OMe)C\equiv CCu$
3. B
MTMO
TBSO
1. $Zn(BH_4)_2$
2. LiOH, H_2O_2 $(HOCH_2CH_2)_2S$
3. KOH
4. CH_2N_2
OH
OTBS
CO_2Me
1. $CH_2=C(OMe)Me, H^+$
2. Ac_2O, DMAP
3. Ac_2O, NaOAc, DMSO
4. K_2CO_3, MeOH
5. Ac_2O, NaOAc, DMSO
OMTM
A

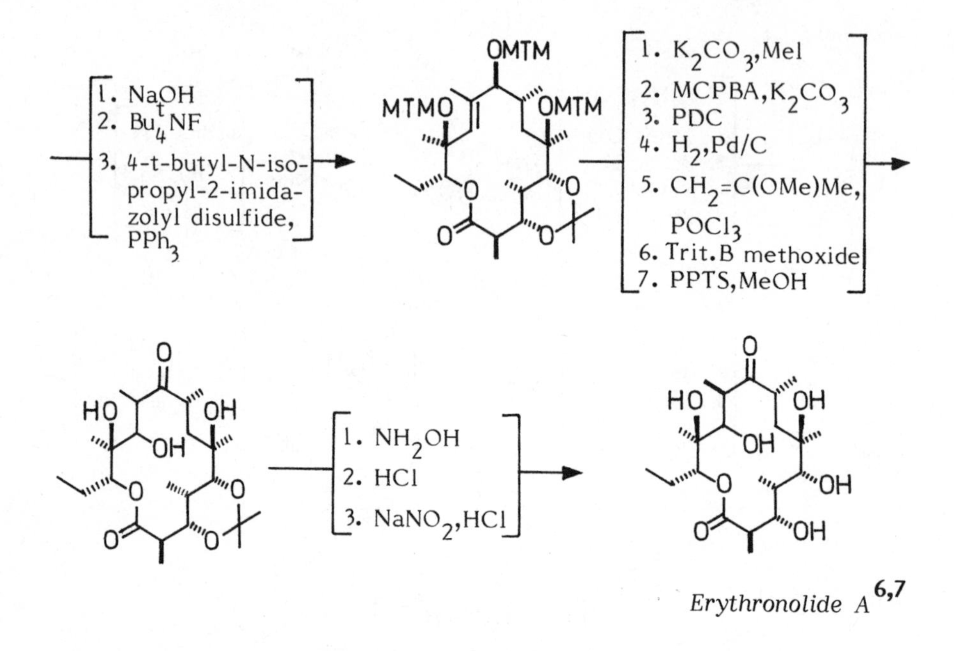

Erythronolide A[6,7]

6. Masamune's synthesis of 6-deoxyerythronolide B, the biogenetic precursor of all the erythromycins, portrays the application of stereocontrolled aldol reactions using chiral enolates, for construction of the macrolide seco-ring, see: Masamune,S.; Hirama,M.; Mori,S.; Ali,S.A.; Garvey,D.S. J. Am. Chem. Soc., 1981, 103, 1568; Masamune,S.; Choy,W.; Kerdesky,F.A.J.; Imperiali,B. J. Am. Chem. Soc., 1981, 103, 1566.

7. Viewing the erythronolide A seco-acid as an appropriately substituted 1,9-dihydroxy-ketone, which can be converted to and obtained from a 1,7-dioxaspiro[5.5]undecane, (an internal acetal of the former), Deslongchamps et al have developed a strategy for synthesis based on using 1,7-dioxaspiro[5.5]undecane as a template to introduce several chiral centers with a high degree of stereochemical control. Sauve,G.; Schwartz,D.A.; Ruest,L.; Deschlongchamps,P. Canadian J. Chem., 1984, 62, 2929. This also shows a possible structural relationship with avermectins, which contain a 1,7-dioxaspiroundecane residue as a part of their structure.

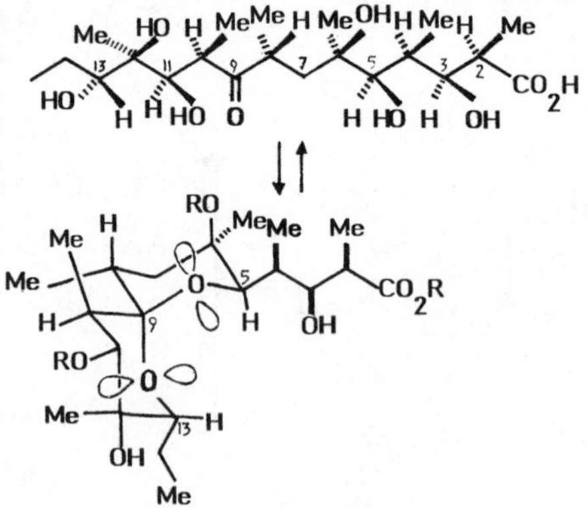

165

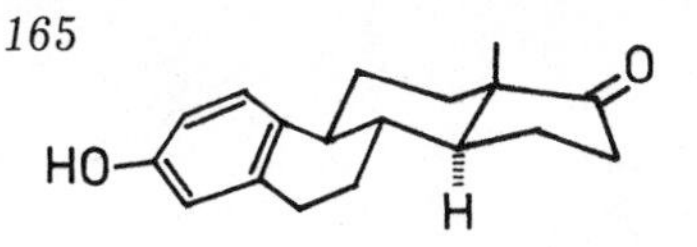

ESTRONE

The synthesis of d-estrone has been of special interest to chemists both because of its own important physiological activity and because it is a crucial intermediate in the production of 19-norsteroids in contraceptive use; it is one of the few steroids produced commercially by total synthesis (1). Its first synthesis was reported in 1948 (2), while the latest has been reported in 1986 with a number of synthesis covering varied approaches, each ingenious in its own right, appearing in between; the main thrusts of these syntheses have been stereo-selectivity, brevity of path and to get optically active product, either through the use of optically active intermediates or through asymmetric induction. The syntheses described below are illustrative of the various approaches followed.

Some noteworthy features of the Johnson-Walker synthesis (3), outlined below, were the elements of stereoselectivity achieved at every step. Introduction of ring D, accomplished by a diene addition, selective removal of a carbonyl function after equilibrating the adjacent C/D ring junction to trans, followed by the usual angular methylation-ring contraction sequence applied to the D-norhomosteroids Ⓐ. The unfavourable stereochemistry at the C/D ring junction which generally results during the angular methylation step was reversed in an ingenious manner by the introduction of a 9,11-double bond (4).

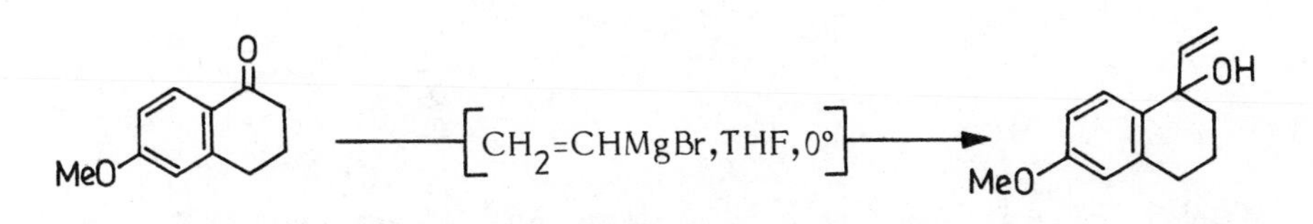

---

1. For two comprehensive accounts of the progress in steroid synthesis, see: Torgov,I.V. Pure Appl. Chem., 1963, 6, 525; Windholz,T.B.; Windholz,M. Angew. Chem. Internat. Ed., 1964, 3, 353. For a review of the early work on the synthesis of estrone, see: Morand,P.; Lyall,J. Chem. Revs., 1968, 68, 85.

2. Anner,G.; Miescher,K. Helv. Chim. Acta, 1948, 31, 2173; 1949, 32, 1957; 1950, 33, 1379; some of the other early synthesis of estrone include: Johnson,W.S.; Banerjee,D.K.; Schneider,W.P.; Gutsche,C.D.; Shelberg,W.E.; Chinn,L.J. J. Am. Chem. Soc., 1950, 72, 1426; 1951, 73, 4987; Johnson,W.S.; Christiansen,R.G. J. Am. Chem. Soc., 1951, 73, 5511; Johnson,W.S.; Christiansen,R.G.; Ireland,R.E. ibid, 1957, 79, 1995; Banerjee,D.K.; Sivanandaiah,K.M. Tetrahedron Lett., 1960, 20; J. Ind. Chem. Soc., 1961, 38, 652.

3. Cole,J.E.; Johnson,W.S.; Robins,P.A.; Walker,J. Proc. Chem. Soc., 1958, 114; J. Chem. Soc., 1962, 244.

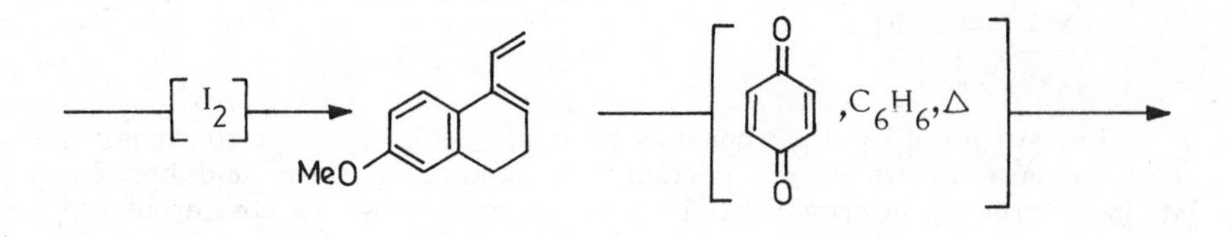
I2
MeO
O
,C6H6,Δ
O

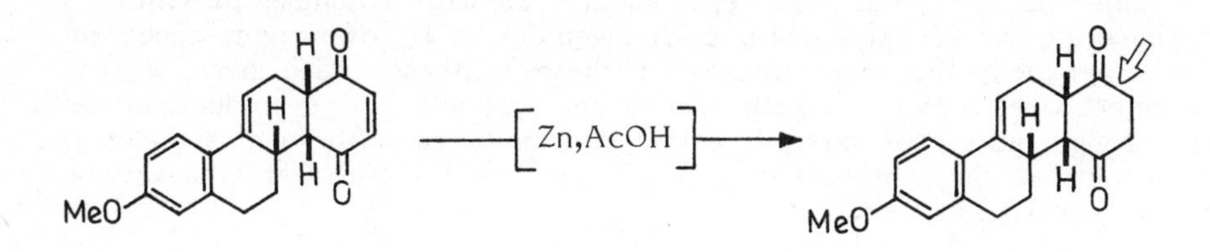
H O
H
H
O
MeO
Zn,AcOH
H O
H
H
O
MeO

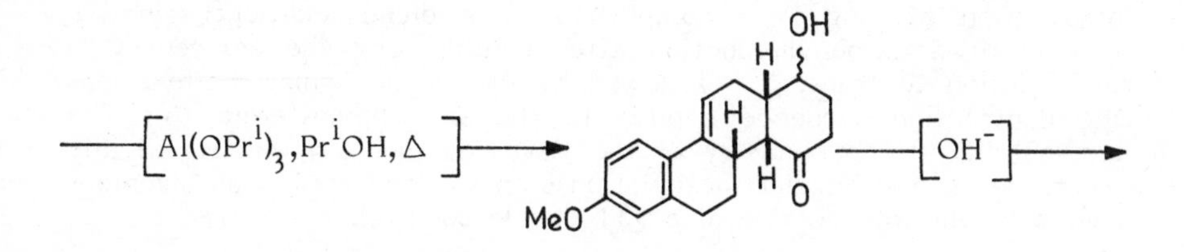
Al(OPri)3,PriOH,Δ
OH
H
H
H
O
MeO
OH−

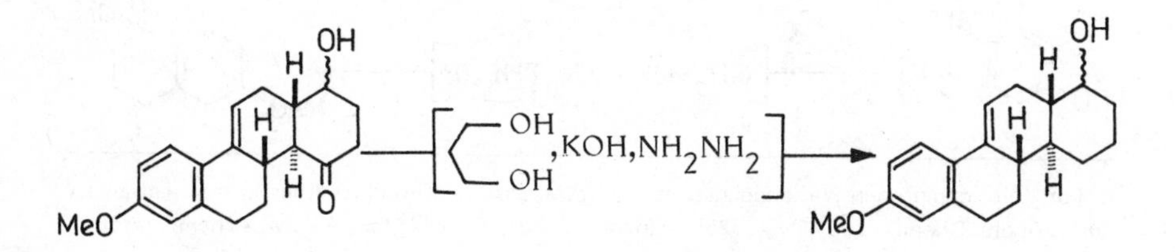
OH
H
H
H
O
MeO
OH
OH
,KOH,NH2NH2
OH
H
H
H
MeO

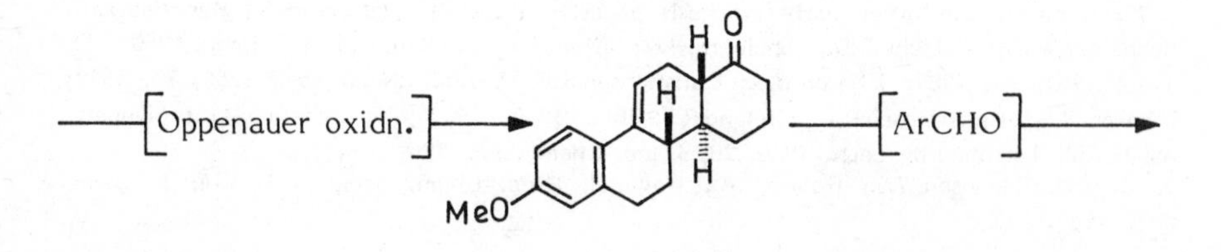
Oppenauer oxidn.
H O
H
H
MeO
ArCHO

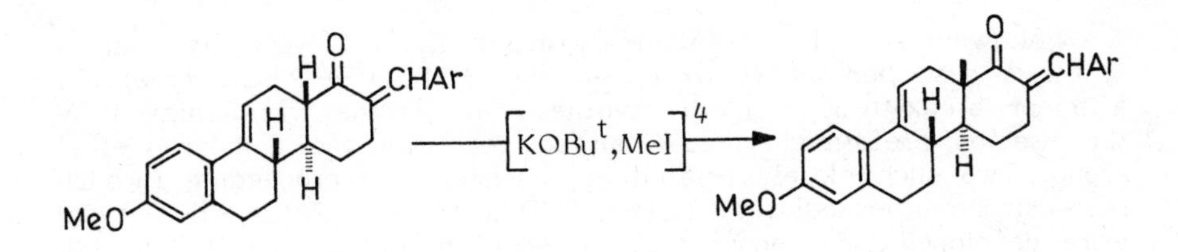

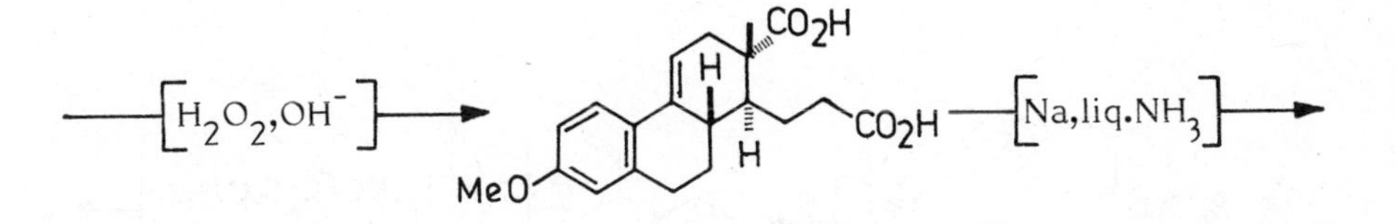

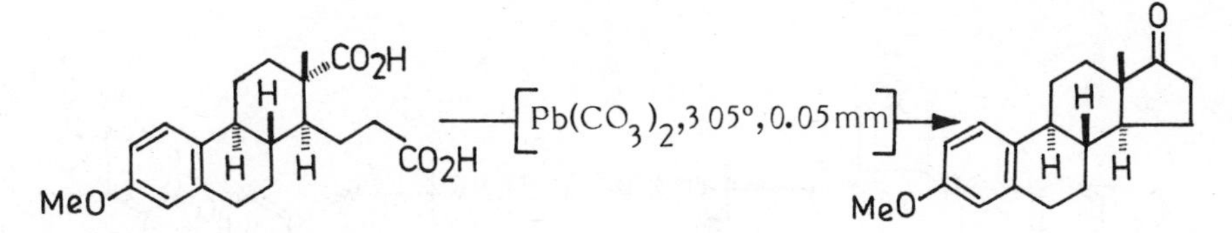

dl-Homomarrianolic acid methyl ether

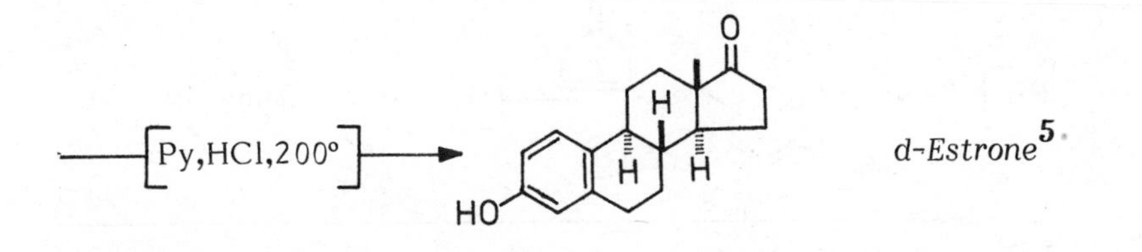

---

4. Methylation of 1-decalones is directed to the angular position by blocking the reactive 2-position using an arylmethylene group. However, during alkylation, an axial hydrogen at C-7 (decalone numbering) interferes with the trans-approach of an electrophile to the angular position (1,3-interaction) (i), resulting in a preponderance of the less desirable C/D cis-methylated isomer; Johnson,W.S.; Allen,D.S. J. Am. Chem. Soc., 1957, 79, 1261. Introduction of a trigonal C-11 (steroid numbering) clearly results in a removal of the offending hydrogen, permitting a free access to the anion from the β-face.

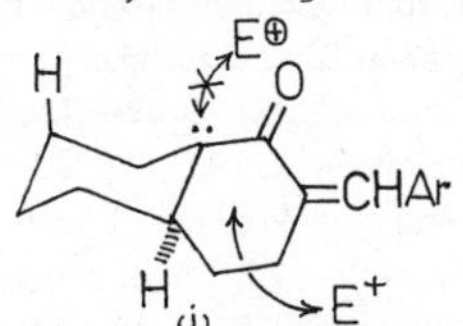

5. Resolved via 1-menthoxy acetate.

Following the improvements wrought by the versatile steroid synthesis described by Velluz et al. (BC → BCD → ABCD type) (6), a major break-through in the synthesis of estrone was achieved by the use of preformed ring D already carrying the angular methyl group. Two such closely related approaches, both proceeding through seco-estrone intermediates (ACD→ABCD and AB → ABD → ABCD type) were developed independently by Herchel Smith (7,8) and Torgov (9). The highly stereoselective synthesis of estrone by Smith et al., which involves the cyclodehydration of a trione Ⓐ to give bisdehydroestrone Ⓑ, is outlined below.

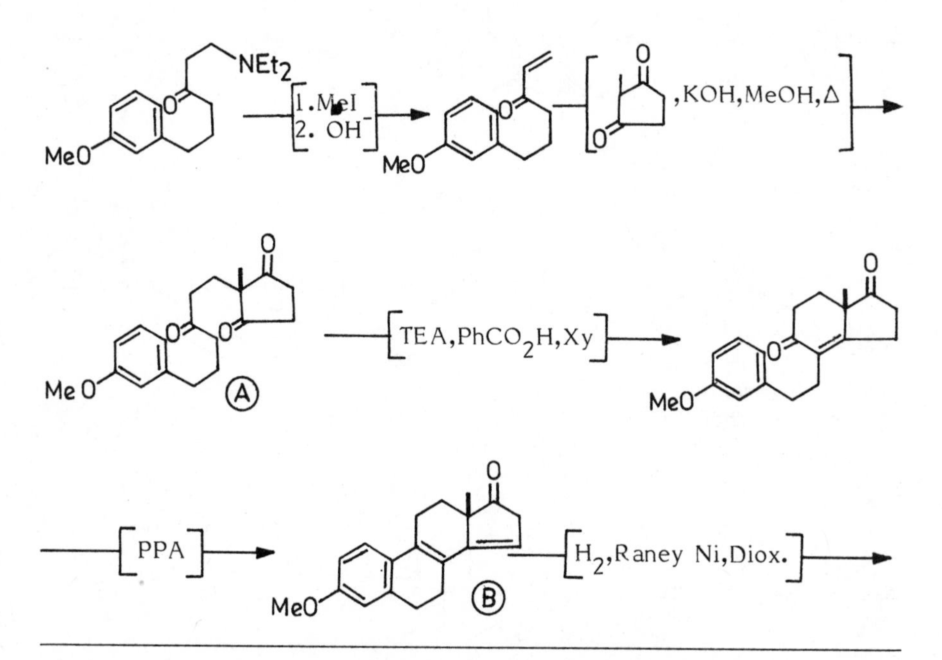

6. Velluz,L.; Nomine,G.; Mathieu,J.; Toromanoff,E.; Bertin,D.; Vignau,M.; Tessier,J. Compt. Rend., 1960, 250, 1510; Velluz,L.; Nomine,G.; Mathieu,J. Angew. Chem., 1960, 72, 725. This approach has already been illustrated by the synthesis of adrenosterone.

7. Hughes,G.A.; Smith,H. Proc. Chem. Soc., 1960, 74; Chem. Ind., 1960, 1022.

8. Douglas,G.H.; Graves,J.M.H.; Hartley,D.; Hughes,G.A.; McLoughlin,B.J.; Siddall,J.; Smith,H. J. Chem. Soc., 1963, 5072.

9. Ananchenko,S.N.; Leonov,V.N.; Platonora,A.V.; Torgov,I.V. Dokl. Acad. Nauk. SSSR, 1960, 135, 73; Ananchenko,S.N.; Limanov,V.Ye;Leonov,V.N.; Rzheznikov,V.N.; Torgov,I.V. Tetrahedron, 1962, 18, 1355; Ananchenko,S.N.; Torgov,I.V. Tetrahedron Lett., 1963, 1553.

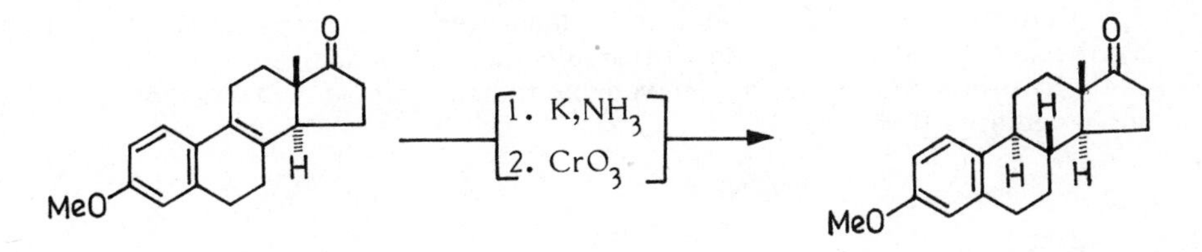

dl-Estrone Methyl Ether

Torgov's novel coupling reaction between vinyl carbinols and diketones (10) provided an unusually simple route to bisdehydro-estrone Ⓑ.

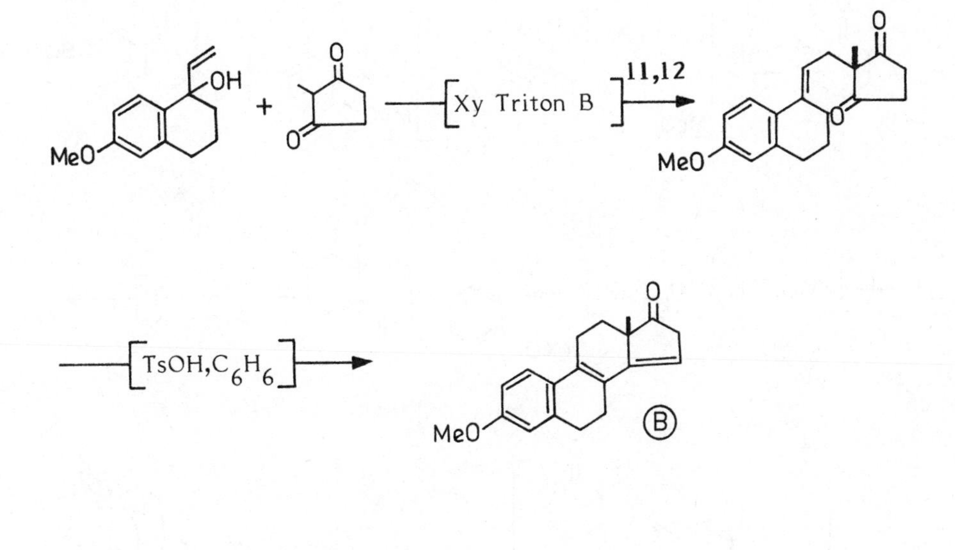

---

10. Ananchenko,S.N.; Torgov,I.V. Dokl. Acad. Nauk. SSSR, 1959, 127, 553.

11. Kuo,C.H.; Taub,D.; Wendler,N.L. Angew. Chem. Internat. Ed., 1965, 4, 1083; Chem. Ind., 1966, 1340; J. Org. Chem., 1968, 33, 3126.

12. The condensation was originally considered to be base-catalyzed but it has been shown since then by Wendler et al. that the condensation is in fact an acid-catalyzed reaction wherein the β-diketone functions autocatalytically. Significant improvements, including the use of isothiouronium salts to facilitate coupling, have resulted from a study of this reaction by the Merck group (11).

Kametani and his associates elaborated a stereoselective total synthesis of estrone by an intramolecular cycloaddition reaction of olefinic o-quinodimethanes, derived by thermolysis of benzocyclobutane intermediates (13).

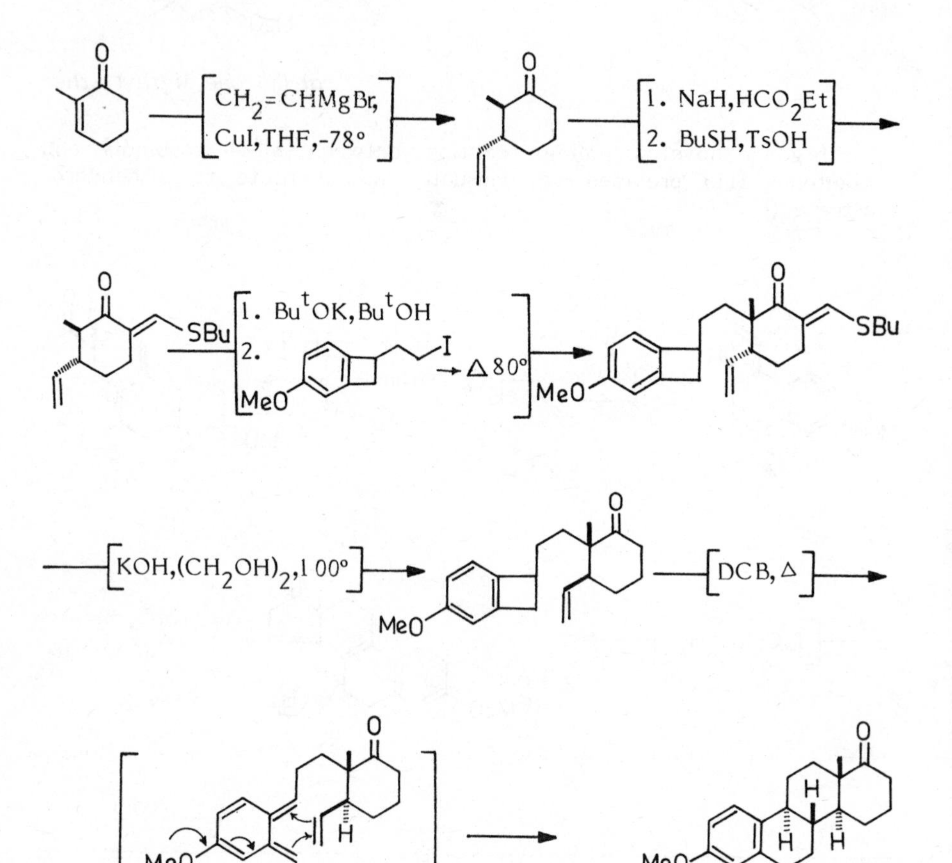

O-Methyl-D-homoestrone[14]

---

13. Kametani,T.; Nemoto,H.; Ishikawa,H.; Shiroyama,K.; Matsumoto,H.; Fukumoto,K. J. Am. Chem. Soc., 1977, 99, 3461.

14. O-Methyl-D-homoestrone has previously been converted to estrone, so this constitute a formal total synthesis of estrone.

Grieco <u>et al.</u> (15) also used the same approach of intramolecular cycloaddition via the corresponding o-quinodimethane to build a tetracyclic aromatic steroid structure but obtained directly estrone by going through bicyclo[2.2.1]heptane Ⓒ as the key intermediate.

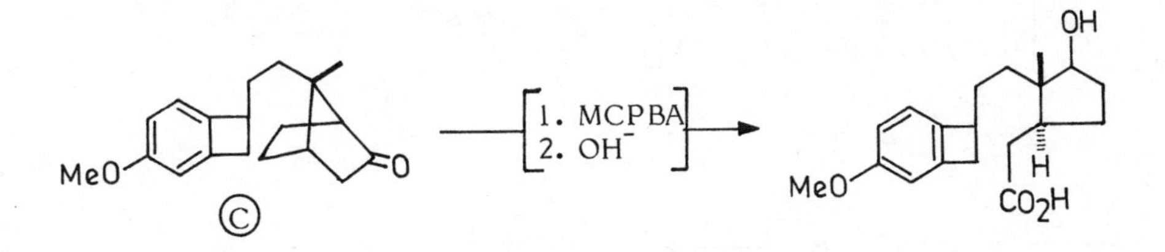

Bartlett & Johnson (16) described a fundamentally new approach to the synthesis of estrone through a cationic polyolefinic cyclisation.

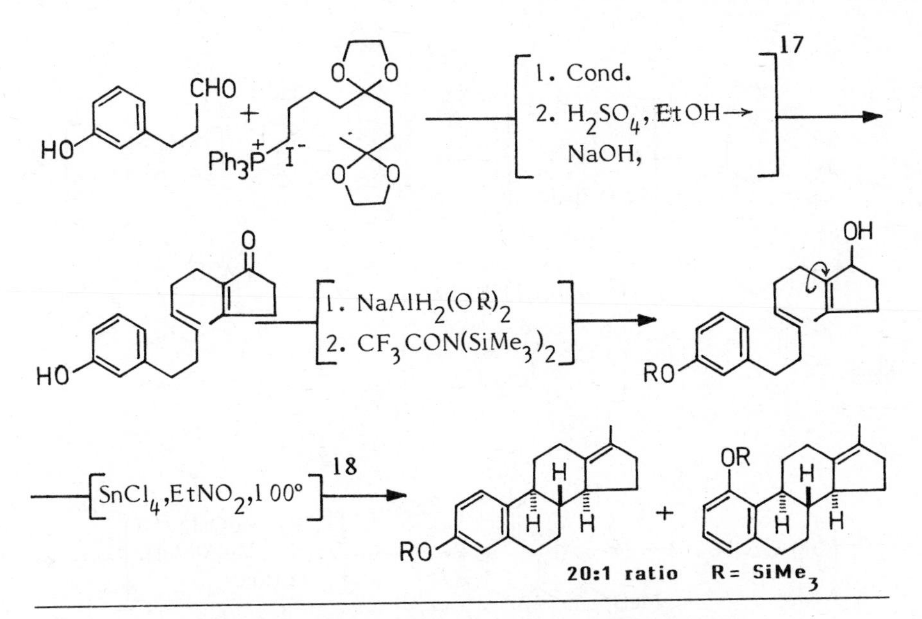

15. Grieco,P.A.; Takigawa,T.; Schillinger,W.J. J. Org. Chem., 1980, 45, 2247.
16. Bartlett,P.A.; Johnson,W.S. J. Am. Chem. Soc., 1973, 95, 7501.
17. The Wittig reaction went with greater than 98% trans-stereoselectivity.
18. With 3-methyl ether the optimum para:ortho isomers ratio obtained was 4.3:1; it was found that this ratio was strikingly dependent upon the leaving group at the substrate and suggested it to be a concerted process; Bartlett,P.A.; Brauman,J.I.; Johnson,W.S.; Volkmann,R.A. J. Am. Chem. Soc., 1973, 95, 7502.

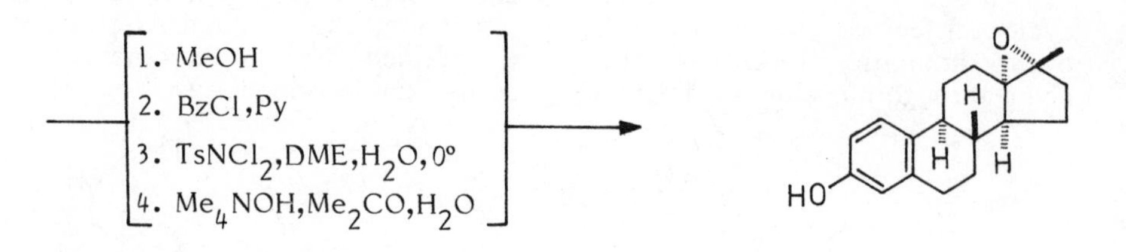

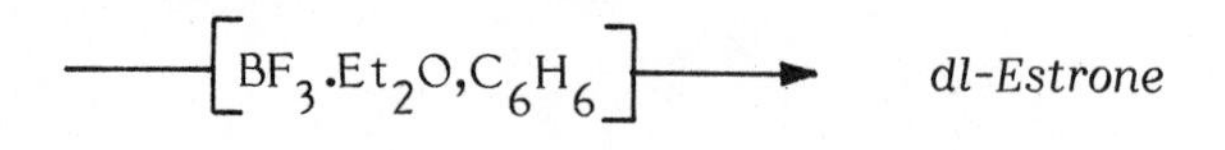

Ziegler and Wang have reported a modified polyene approach to estrone which controls the stereochemistry via a Cope arrangement (19).

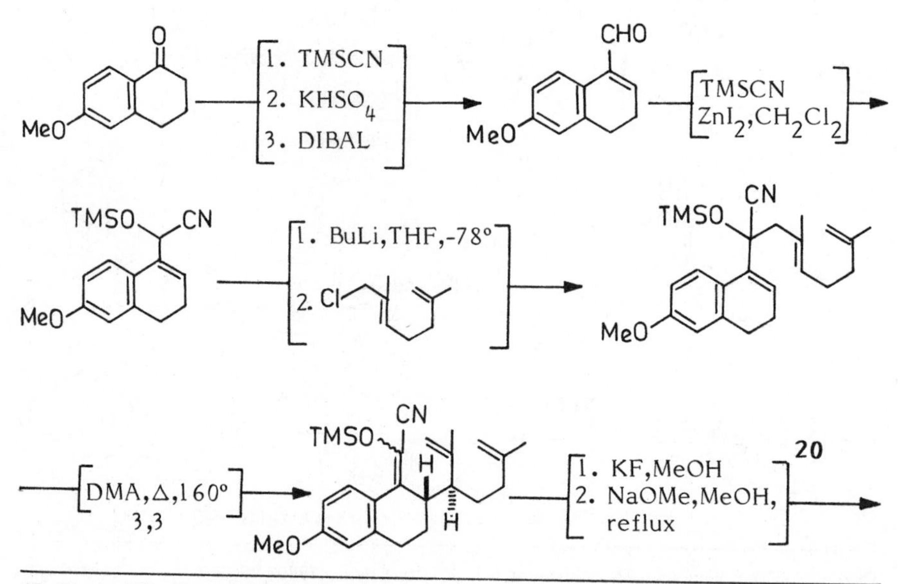

19. Ziegler,F.E.; Wang,Tein-Fu. Tetrahedron Lett., 1981, 22, 1179.

20. KF treatment gave a pair of diastereomeric esters formed in 80:20 ratio and major ester was tentatively assigned the cis stereochemistry, since NaOMe equiliberation provided the 8,9-trans:cis isomers (steroid numbering) in a ratio of 17:83; this would arise from a chair-like transition state involving pro-C_8 and pro-C_{14} atoms. The esters could be conveniently separated.

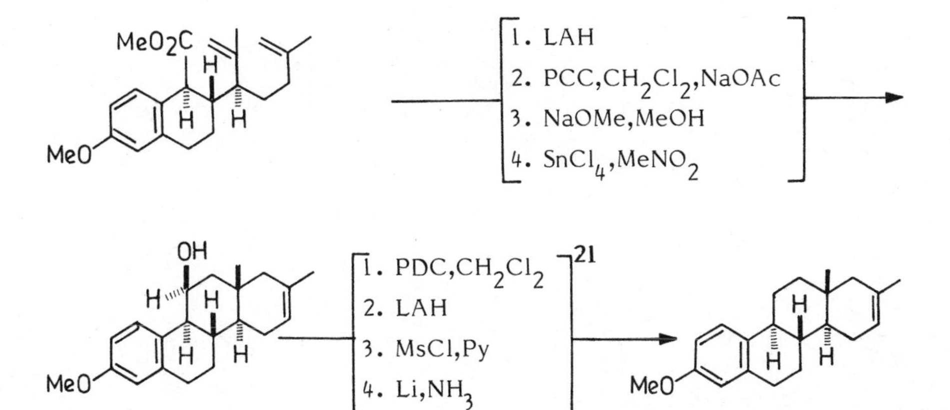

Saucy and his associates at Roche Laboratories developed a short and efficient steroid synthesis based on CD → ACD → ABCD approach and have applied it successfully to the synthesis of d-estrone (23), which involves conjugate addition of Grignard reagent destined to become ring A&B to optically active ring C-D-enone in presence of Cu^+ salts, followed by cyclisation, hydrogenation and adjustment of functionalities (23,24).

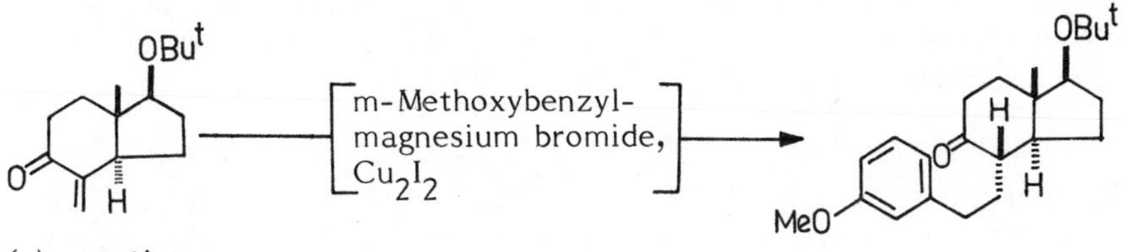

21. Attempted mesylation of the 11-axial alcohol produced a mixture of olefins arising presumably from facile trans-diaxial elimination of the expected mesylate. However, the mesylate of the equatorial alcohol as shown underwent smooth deoxygenation to provide only the tetracyclic compound. This tetracyclic compound has previously been converted to estrone by Valenta et al. (22).

22. Das,J.; Kubela,R.; MacAlpine,G.A.; Stojanac,Z.; Valenta,Z. Can. J. Chem., 1979, 57, 3308.

23. Cohen,N.; Banner,B.L.; Eichel,W.F.; Parrish,D.R.; Saucy,G.; Cassal,Jean-Marie; Meier,W.; Furst,A. J. Org. Chem., 1975, 40, 681.

24. This approach has also been used for the total synthesis of 19-norsteroids and androstanes; Scott,J.W.; Buchschache,P.; Lablar,L.; Meier,W.; Furst,A. Helv. Chim. Acta, 1974, 57, 1217; Micheli,R.A.; Hajos,Z.G.; Cohen,N.; Parrish,D.R.; Portland,L.A.; Sciammanna, W.; Scott,M.A.; Wehrli,P.A. J. Org. Chem., 1975, 40, 675.

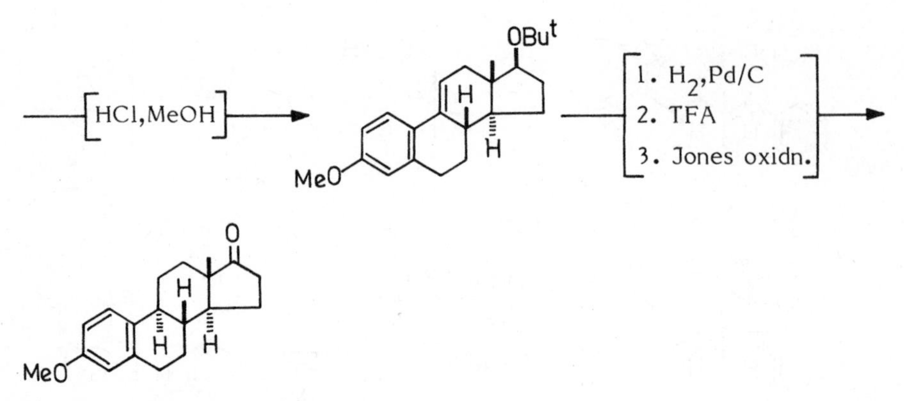

(+)-Estrone 3-methyl ether

Danishefsky and his associates in some exploratory studies on annellation reactions for steroid synthesis and using α-picolines as equivalents of cyclohexenones demonstrated that bis-annellation of Wieland-Miescher ketone with 6-vinyl-α-picoline followed by Birch reduction provided a convenient synthesis of D-homoestrone (25). This approach, dovetailed with the observed optical induction in l-amino acid promoted aldolisation in hydrindenone synthesis by Hajos et al. and Eder et al. (26) was then adopted for an elegant synthesis of d-estrone (and 19-norsteroids) by inducing asymmetry by L-amino acid promoted aldolisation of the key prochiral precursor Ⓐ, achieving high chiral selectivity in the cyclisation step (27).

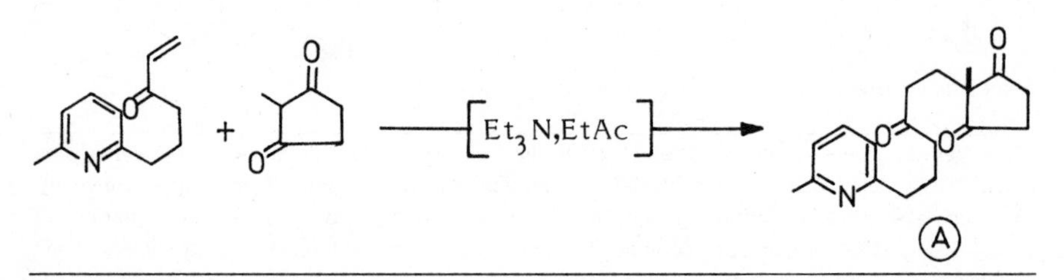

25. Danishefsky,S.; Nagel,A. J. Chem. Soc. Chem. Comm., 1972, 373; Danishefsky,S.; Cain,P.; Nagel,A. J. Am. Chem. Soc., 1975, 97, 380.

26. Hajos,Z.G.; Parrish,D.R. J. Org. Chem., 1974, 39, 1615; Eder,U.; Sauer,G.; Wiechert,R. Angew. Chem. Int. Ed., 1975, 10, 417.

27. Danishefsky,S.; Cain,P.; J Org. Chem., 1974, 39, 2925; J. Am. Chem. Soc., 1975, 97, 5282; 1976, 98, 4975.

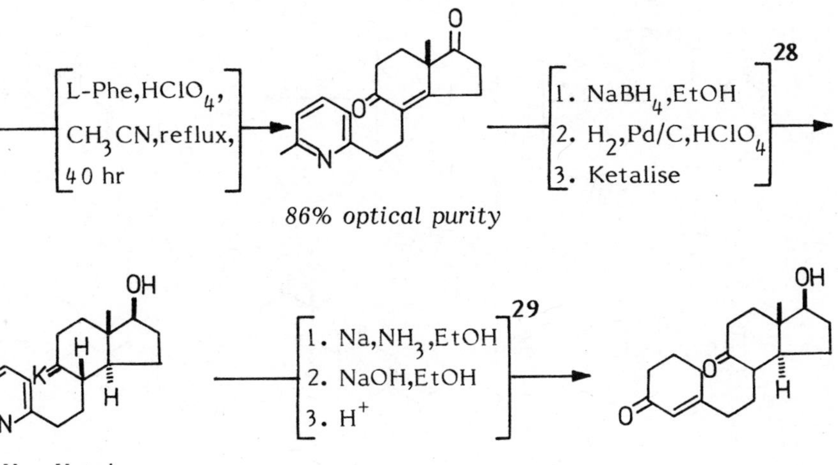

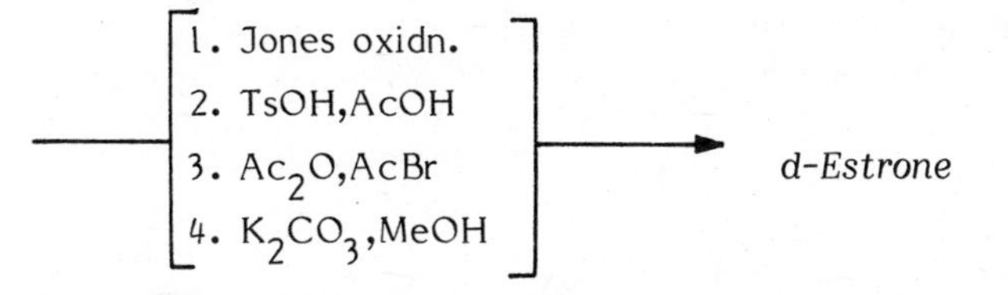

Posner and his associates carried out a highly convergent synthesis of estrone in which a tandem Michael-Michael-ring closure reaction (MIMIRC) has been developed for efficient sequential formation of three C-C bonds to provide a one-pot synthesis of (±)-9,11-dehydroestrone, which can be easily converted to estrone (30).

---

28. Along with 14 -epimer (trans:cis, 2.6:1) and hydrogenolysed product.

29. For a discussion of the conversion of 2-CH_2R-6-methylpyridines (I) to 3-CH_2R-cyclohex-2-enones (II) see: Danishefsky,S.; Cavanaugh,R. J. Am. Chem. Soc., 1968, 90, 520; Danischefsky,S.; Nagel,A.; Peterson,D. J. Chem. Soc. Chem. Comm., 1972, 374.

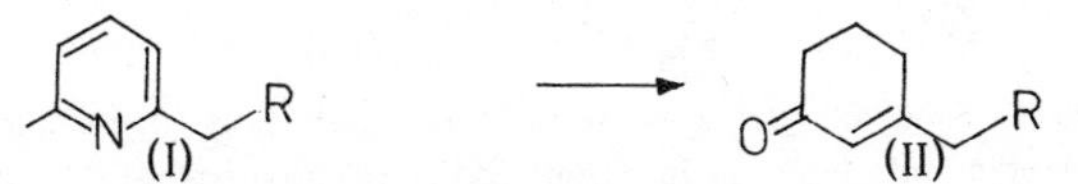

30. Posner,G.H.; Mallamo,J.P.; Black,A.Y. Tetrahedron, 1981, 37, 3921.

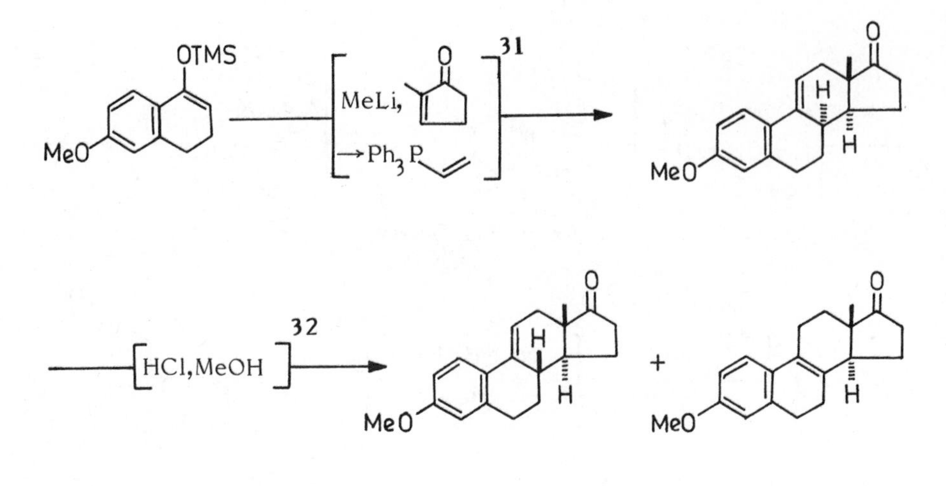

In a further refinement of this AB→ ABD→ ABCD approach to get optically active product, Posner and Switzer have employed a highly diastereoselective Michael addition of a ketone enolate ion to an enantiomerically pure 2-(arylsulfinyl)-2-cyclopentenone (33) as the key step in an effective asymmetric total synthesis giving esterone in over 97.3% enantiomeric purity and with natural absolute stereochemistry, which involves induction of asymmetry at the prochiral β-carbon of an enantiomerically pure α,β-ethylenic sulfoxide (34,35).

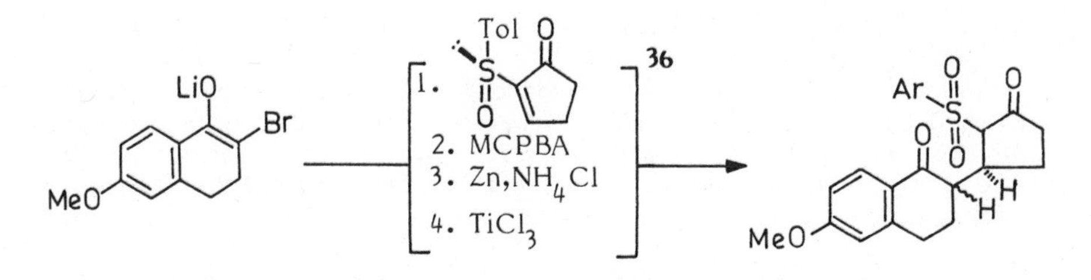

31. This reaction goes through the intermediacy of the enolate (A); (B) can be isolated.

32. Such a mixture of 9,11- and 8,9-dehydroestrones has previously been converted into estrone and into estradiol in high yield; Douglas,G.H.; Graves,J.M.H.; Hartley,D.; Hughes,G.A.; McLoughlin,J.; Siddal,J.; Smith,H. J. Chem. Soc., 1963, 5072; Quinkert,G.; Weber,W.D.; Schwartz,U.; Durner,G. Angew. Chem. Int. Ed., 1980, 19, 1027.

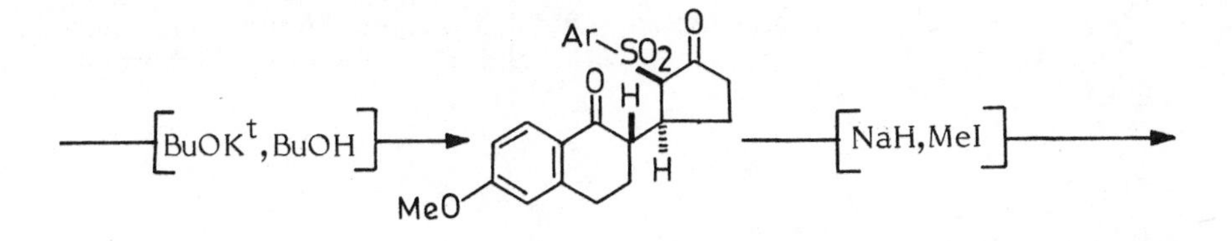

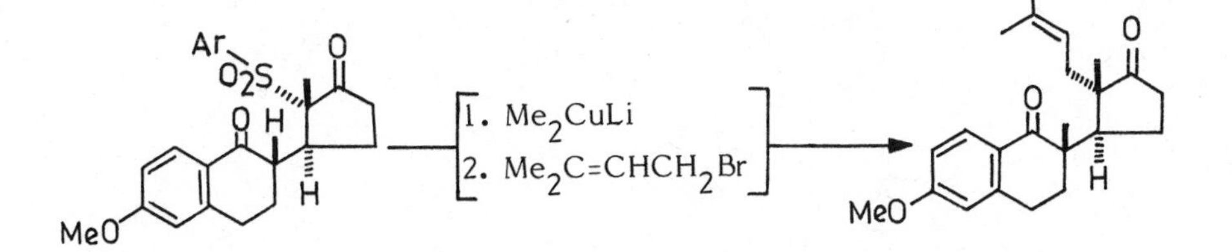

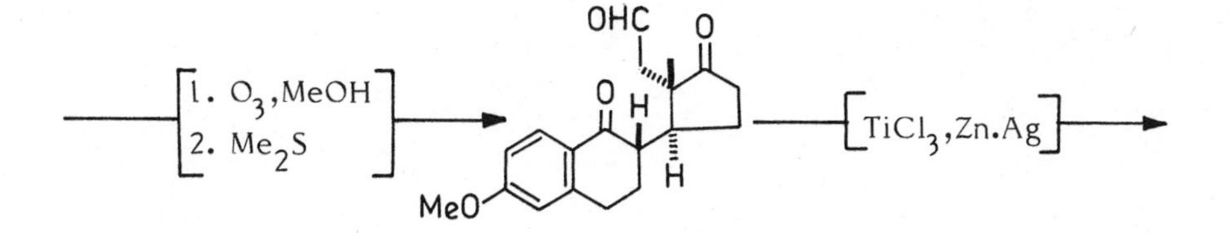

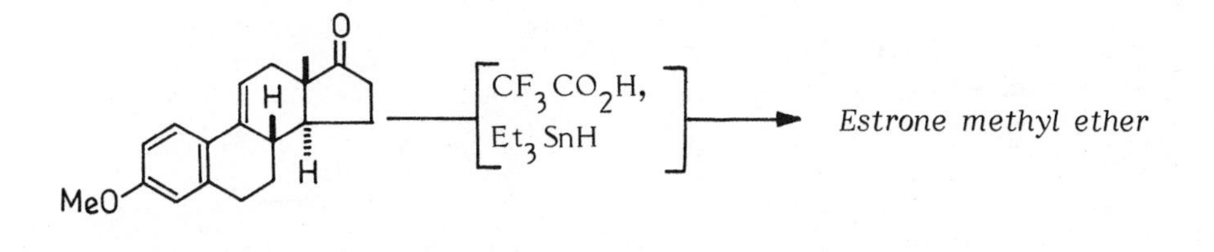

---

33. For R-(-)-2-(p-tolylsulfinyl)-2-cyclopentenone; see: Frye,L.L.; Kogan,T.P.; Mallamo,J.P.; Posner,G.H. Org. Synth., 1985, 64, 196.

34. Posner,G.H.; Switzer,C. J. Am. Chem. Soc., 1986, 108, 1239.

35. For review see: Posner,G.H. In "Asymmetric Synthesis"; Morrison,J.D. Ed. Academic Press, New York, 1983; Vol.2, Chapter 8, p.225.

36. When lithium enolate itself was used in Michael addition only 54% diastereomeric selectivity was achieved; however, the use of 2-bromoenolate gave a product with over 90% diastereomeric selectivity which was followed by removal of Br by Zn reduction.

Funk and Vollhardt have described yet another approach to the synthesis of A-ring aromatic steroids, which is based on construction of rings ABC constellation through cobalt-mediated cooligomerisation of bis (trimethylsilyl) acetylene with almost complete chemoregio and stereospecificity (37).

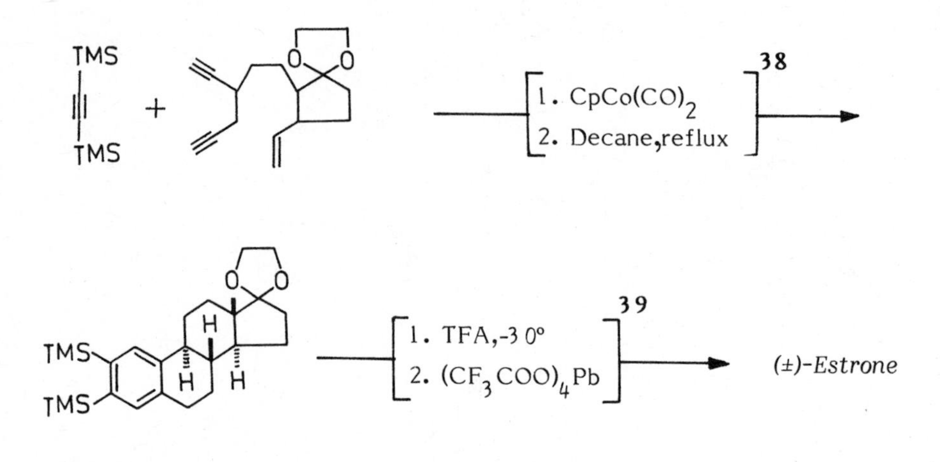

---

37. Funk,R.L.; Vollhardt,K.P.C. J. Am. Chem. Soc., 1979, 101, 215.

38. The cyclisation proceeds through the benzocyclobutene intermediates A&B which could be isolated by work up before heating in decalin. Both Ⓐ and Ⓑ on heating gave estratriene on heating, but could not be converted into each other, thus suggesting an appreciately lower barrier to intramolecular Diels-Alder reaction than to ring closure once the intermediate o-quinomethine is generated.

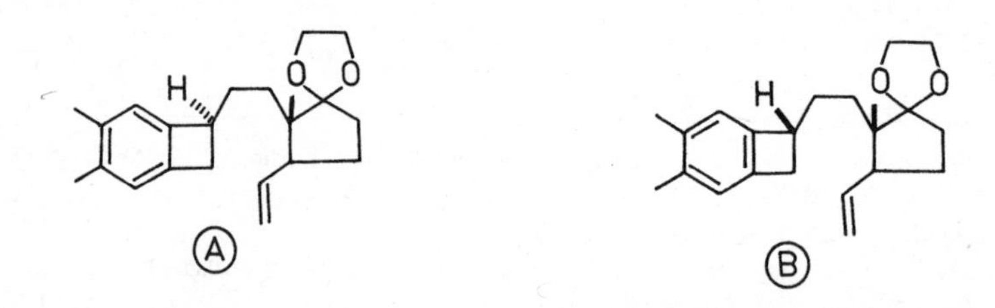

39. CF_3CO_2H treatment gave 3-trimethylsilylestra-1,3,5(10)-trien-17-one with 9:1 selectivity in 90% yield; oxidative aryl silicon cleavage occured almost quantitatively with lead tetrakis (trifluoracetate).

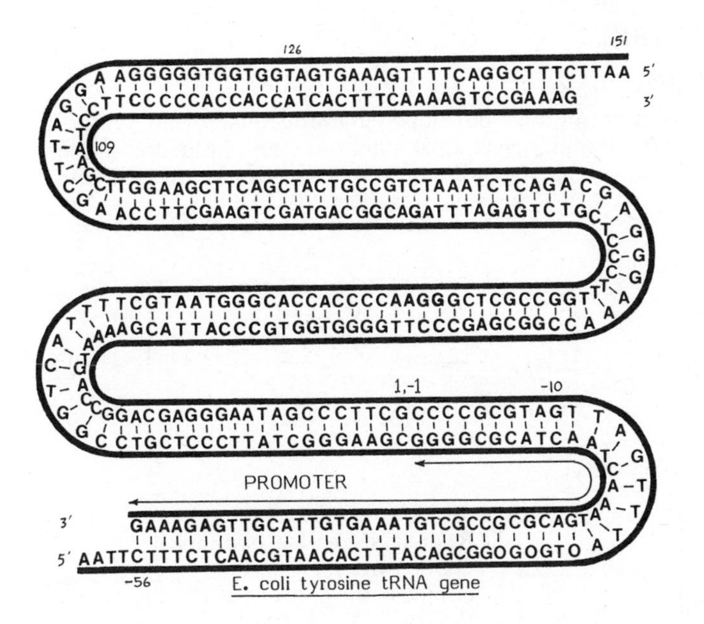

E. coli tyrosine tRNA gene

One of the great achievements of contemporary organic synthesis is the ability to construct rapidly polynucleotides of specific sequences, which has made it possible to make synthetic genes (1). This has been due primarily to dramatic improvements in the methods of polynucleotide synthesis in the last two decades (2). The relatively easy availability of DNAs of any required sequence as a result of these developments has been a big factor for the recent advances in molecular biology and biotechnology (3).

---

1. For authoritative review of the earlier developments in this field see: Khorana,H.G.; Science, 1979, 203, 614; XIII Feodor Lynen Lecture, 1981.

2. The major recent developments in the methods responsible for the outstanding successes in polynucleotide synthesis are: (i) switch over from phosphodiester approach to triester methodology, and from solution phase to solid support synthesis; (ii) use of DNA ligases to join oligodeoxyribonucleotide chains between juxtaposed 5'-phosphoryl and 3'-hydroxy groups and of DNA kinases to form 5'-phosphate on a preformed oligonucleotide; (iii) availability of instrumental methods for automated synthesis and for preparative separation. For some comprehensive reviews of methods of synthesis see: (a) Brown,E.L.; Belgaje,R.; Ryan,M.J.; Khorana,H.G. Meth. in Enzymology, 1979, 68, 109; (b) Narang,S.A. Tetrahedron, 1983, 39, 3; (c) Reese,C.B. Tetrahedron, 1978, 34, 3143; (d) Gassen,H.G.; Lang,A. Chemical and Enzymatic Synthesis of Gene Fragments: A Laboratory Manual, Verlag Chemie, Weinheim, 1982; also see ref.17.

3. For an appreciation of the impact of polynucleotide synthesis in genetic engineering see: (a) Itakura,K. Trends Bioch. Sci., 1982, 442; (b) Davies,J.E.; Gassen,H.G. Angew. Chem. Int. Ed., 1983, 22, 13; (c) Caruthers,M.H. Science, 1985, 231, 281.

METHODS OF SYNTHESIS

Phosphodiester method

This was the first method developed for oligonucleotide synthesis by Khorana and his associates and dominated the field for almost two decades and was employed for the first synthesis of a bihelical DNA corresponding to the major yeast Ala tRNA (1,12).

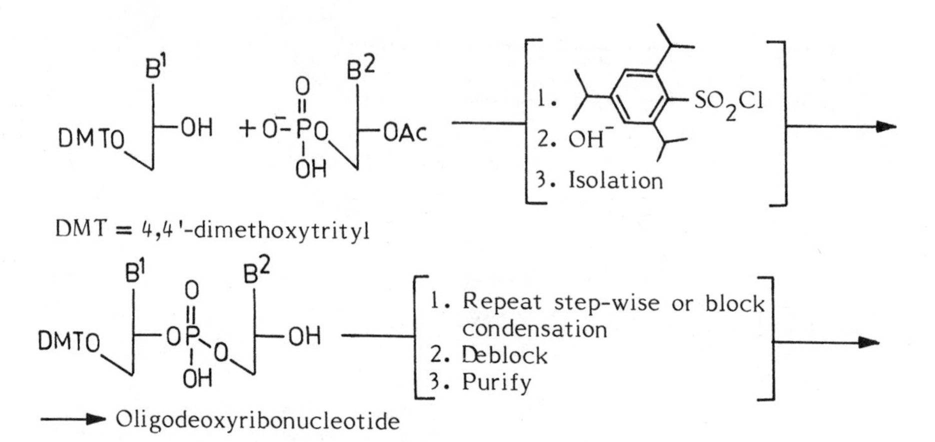

Low solubility of reactants in organic solvents, long reaction times, decreasing yields as the chain length grows and time consuming purification procedures required are the limitations of this method.

Triester methods

A solution to some of these problems was found by masking the third dissociation of the phosphate by lipophilic and easily removable blocking groups, which increased organic solvent solubility and reaction rates, reduced side product formation and simplified purification procedure (2,4). Another big advantage with triester approach is compatibility with polymer supported synthesis which is now extensively employed for oligonucleotide synthesis, and the repetitive procedures required in coupling cycles in solid phase synthesis are amenable to automation. These developments have greatly enhanced the efficiency of polynucleotide synthesis (3c,5).

---

4. (a) Itakura,K.; Bahl,C.P.; Katagiri,N.; Michniewicz,J.; Wightman,R.H.; Narang,S.A. Can. J. Chem., 1973, 51, 3469; (b) Catlin,J.C.; Cramer,F. J. Org. Chem., 1973, 38, 245; Narang, S.A.; Hsiung,H.M.; Brousseau,R. Methods in Enzym., 1979, 68, 90.

5. (a) Gait,M.J.; Matthes,H.W.D.; Singh,M.; Sproat,B.S.; Titmas,R.C. in 2(c) page 1; (b) Seliger,H.; Klein,S.; Narang,Ch.K.; Seemann-Preising,B.; Eiband,J.; Hauel,H.; in 2(c) page1.

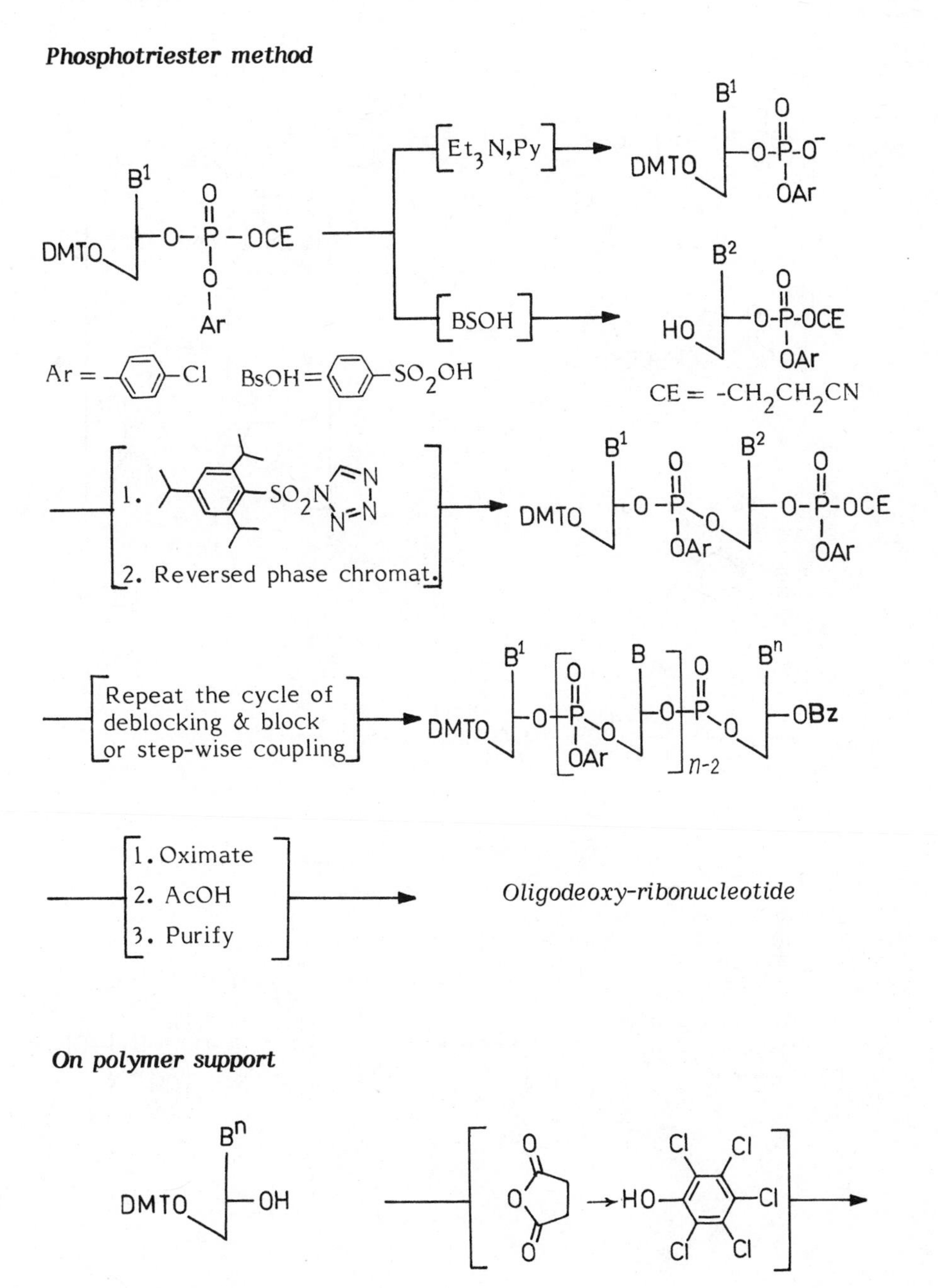
Phosphotriester method
DMTO
O-P-OCE
OAr
Et3N,Py
BSOH
DMTO
O-P-O⁻
OAr
HO
O-P-OCE
OAr
Ar = -C6H4-Cl
BsOH = C6H5-SO2OH
CE = -CH2CH2CN
1.
SO2N
2. Reversed phase chromat.
DMTO
OAr
O-P-OCE
OAr
Repeat the cycle of deblocking & block or step-wise coupling
DMTO
OAr
n-2
OBz
1. Oximate
2. AcOH
3. Purify
Oligodeoxy-ribonucleotide
On polymer support
DMTO
OH
HO
Cl
Cl
Cl
Cl
Cl

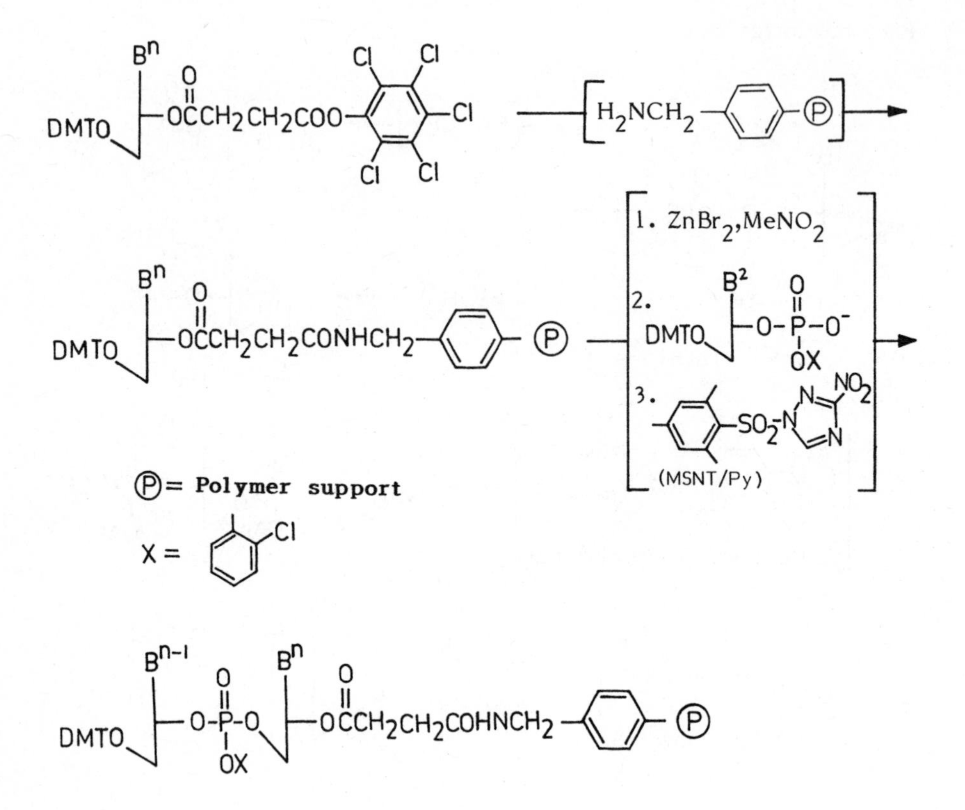

DMTO
OCCH2CH2COO
H2NCH2
OCCH2CH2CONHCH2
1. ZnBr2,MeNO2
2.
3.
(MSNT/Py)
Ⓟ= Polymer support
X =
OCCH2CH2COHNCH2

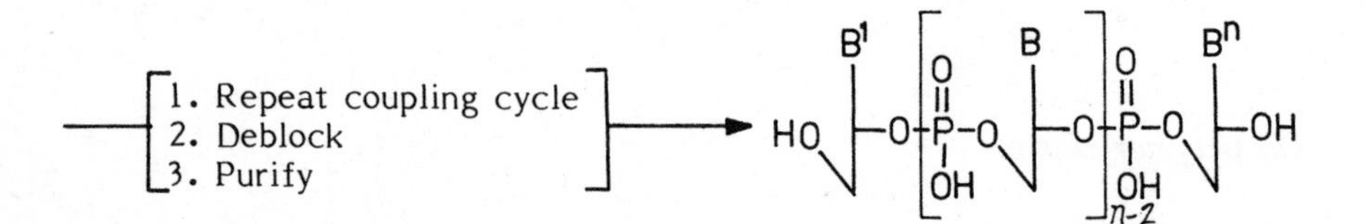

1. Repeat coupling cycle
2. Deblock
3. Purify
HO
OH
n-2

Phosphitetriester method (5)

This method overcomes the problem of negative charge on phosphate by first preparing the dinucleotide phosphite ester, which in a second step is oxidised by iodine to a phosphate bond (2,6).

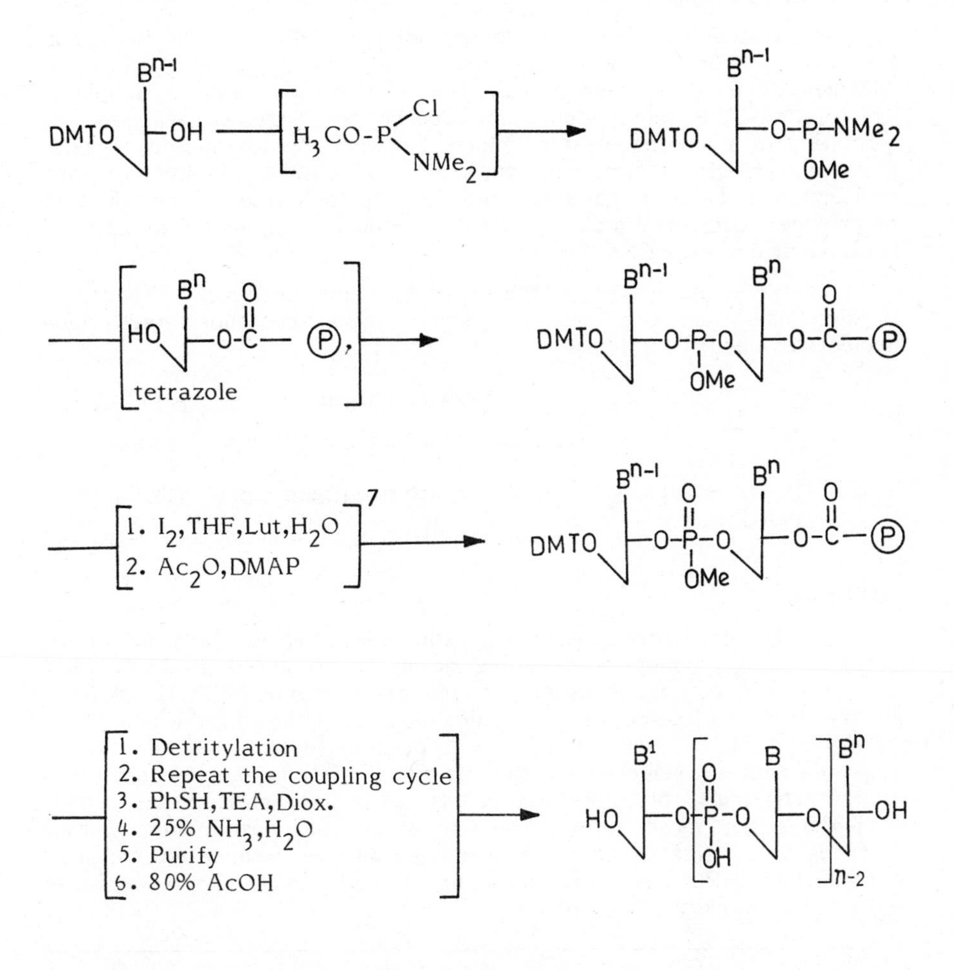

6. (a) Beancage,S.I.; Caruthers,M.H. Tetrahedron Lett., 1981, 22, 1859; (b) Matteucci,M.D.; Caruthers,M.H. J. Am. Chem. Soc., 1981, 103, 3185.

7. The product is treated with Ac_2O, Py (capping) to remove out of reaction sequence unreacted oligonucleotides attached to the polymer support.

The choice between the two methods is personal to different laboratories, but overall the phosphite–triester method does seem to offer the advantage of near quantitative yields (>98%) and much shorter reaction time (5 min at room temperature).

Enzymatic synthesis

A strategy has been developed mainly by Khorana and his group (8) whereby short synthetic oligonucleotides are joined by DNA ligase enzymes (9) in the presence of a complementary strand or template, which aligns the short segments properly; the enzyme catalyses the formation of a phosphodiester linkage between the juxtaposed 5'-phosphate and 3'-OH of two overlapping DNA chains. A minimum of four overlapping nucleotide pairs are required on each side of the junction to provide efficient template interaction for the joining of the deoxyribonucleotides with DNA ligase.

DNA kinases (10) which phosphorylate terminal 5'-hydroxyl in oligonucleotides are used to prepare the 5'-phosphorylated oligonucleotides needed for ligase catalysed coupling.

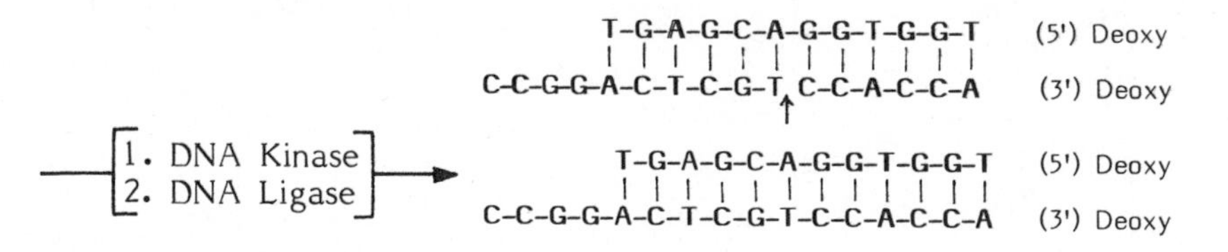

Purification

As synthesis progresses, even with 90% yield in every step, the purity of the product starts going down; the required product which comes of the column is usually 5-10% of the crude mixture relative to the first nucleoside on the solid support. Although, the materials so obtained can be used for ligation to plasmid vectors, and the correct sequence will be selected by the biological system, but for chemical characterisation, this mixture needs extensive purification which is perhaps the more time consuming step. The most commonly used methods of purification are: (1) reversed phase gel permeation chromatography; (2) HPLC, reversed phase or ion-exchange; (3) polyacrylamide gel electrophoresis.

---

8. Khorana,H.G.; Aggarwal,K.L.; Besmer,P.; Buchi,H.; Caruthers,M.H.; Cashion,P.J.; Fridkin, M.; Jay,E.; Kleppe,K.; Kleppe,R.; Kumar,A.; Loewen,P.C.; Miller,R.C.; Minamoto,K.; Panet,A.; RajBhandary,U.L.; Ramamoorthy,B.; Sekiya,T.; Takeya,T.; Van de Sande,J.H. J. Biol. Chem., 1976, 251, 565.

9. For a review of DNA joining enzymes see: Higgins,N.A.; Cozzareli,N.R. Methods in Enzym., 1979, 68, 50.

10. Richardson,C.C. Proc. Nucleic Acid Res., 1971, 2, 815.

A combination of chemical synthesis and enzymatic reactions now offers the ability to synthesis bihelical DNAs of any specific sequence. The concept which is central to the whole design of synthesis is the inherent ability of polynucleotide chains having complimentary bases to form ordered bihelical complexes of sufficient stability even in aqueous solution by virtue of base pairing, and that overlapping regions of the duplex could complex with shorter chains which could then be joined end to end with parent chain by ligases.

The strategy commonly followed for DNA synthesis consists in carefully disecting the DNA in the double strand into short single stranded segments, 10-20 bases long, with suitable overlaps in the complimentary strands; the segments are chemically synthesised by solid phase method, which are phosphorylated at terminal 5'-hydroxyl using ATP and polynucleotide kinases; a few to several neighbouring oligonucleotides are then allowed to form bihelical complexes in aqueous solution, and the latter are joined end to end by polynucleotide ligases to form covalently linked duplexes; subsequent head-to-tail joining of the short duplexes leads to the total DNA synthesis.

YEAST ALANINE tRNA GENE

The synthesis of the 77 base pair (bp) structural gene corresponding to yeast alanine tRNA by Khorana and his associates is of special historical significance. This tRNA was the first to be fully sequenced (11), and DNA corresponding to this tRNA the first DNA of specific sequence to be synthesised (12). Its synthesis was the culmination of almost two decades of orchestrated development of different elements required for the total synthesis of polynucleotides, and demonstrated that genes could be synthesised.

The plan of synthesis as shown below involved: (1) chemical synthesis of such 15 polydeoxynucleotide segments shown in brackets ranging in length from 5-20 nucleotide units, with 3'- and 5'-hydroxyl end groups free representing the entire two strands of the DNA as would give overlap of 4 to 5 nucleotides in the complimentary strands; (2) phosphorylation of the 5'-hydroxyl group of these segments using T_4 polynucleotide kinase; (3) annealing and ligase-induced head to tail joining of the appropriate segments aligned properly due to bihelical complexation.

---

11. Holley,R.W. et al., Science, 1965, 147, 1462.

12. Khorana,H.G.; Agarwal,K.L.; Buchi,H.; Caruthers,M.H.; Gupta,N.K.; Kleppe,K.; Kumar,A.; Ohtsuka,E.; RajBhandary,U.L.; van de Sande,J.H.; Sgaramella,V.; Terao,T.; Weber,H.; Yamada, T. J. Mol. Biol., 1972, 72, 209 and twelve accompanying papers in the same issue.

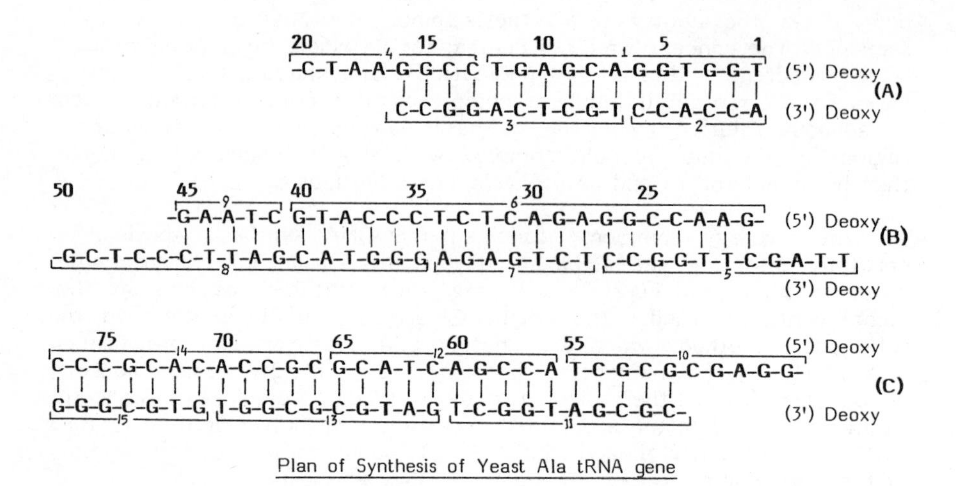

Plan of Synthesis of Yeast Ala tRNA gene

This was followed by the synthesis of a biologically functional E. coli tyrosine tRNA suppressor gene by Khorana and his associates consisting of the 126 bp sequence corresponding to the tRNA, 51 bp long promoter region and a 25 nucleotide long processing signal sequence including the endonuclease specific sequence (2a,13). Many other synthetic genes, including those for human somatostatin (14), human insulin A&B chains (15) and for human leucocyte interferon (16) have since then been constructed.

HUMAN LEUKOCYTE α_1-INTERFERON GENE

It is the synthesis of 514 bp human leucocyte α_1-interferon (166 amino acid residues) gene containing initiation and termination signals plus restriction enzyme sites for plasmid insertion, by Edge et al. (16) which has demonstrated forcefully the great jump chemical oligonucleotide synthesis has made.

It involved the synthesis of 67 oligodeoxyribonucleotide units, ranging in size from 14-21 residues (one of 10) by rapid, solid phase procedures using phosphotriester coupling method, which were separately

---

13. Belagaje,R.; Brown,E.L.; Gait,M.J.; Khorana,H.G.; Norris,K.E. J. Biol. Chem., 1979, 254, 5754; and accompanying five papers.

14. Itakura,K.; Hirose,T.; Crea,R. et al., Science, 1977, 198, 1056.

15. (a) Goeddel,D.V.; Kleid,D.G.; Bolivar,F. et al., Proc. Natl. Acad. Sci., USA, 1979, 76, 106; (b) Hsiung,H.M. et al., Nucleic Acids Res., 1979, 6, 137; (c) Narang,S.A. et al., ibid, 1980, 7, 377 and accompanying papers.

purified and 5'-phosphorylated by a kinase. This was followed by sequential enzymatic ligation of the oligonucleotides to yield a 514 base pair fragment. The synthetic gene was then ligated to a plasmid vector and cloned in Escherichia coli, and clones containing the anticipated gene sequence were obtained, and the synthetic gene charactrised by restriction enzyme digests and total sequence determination.

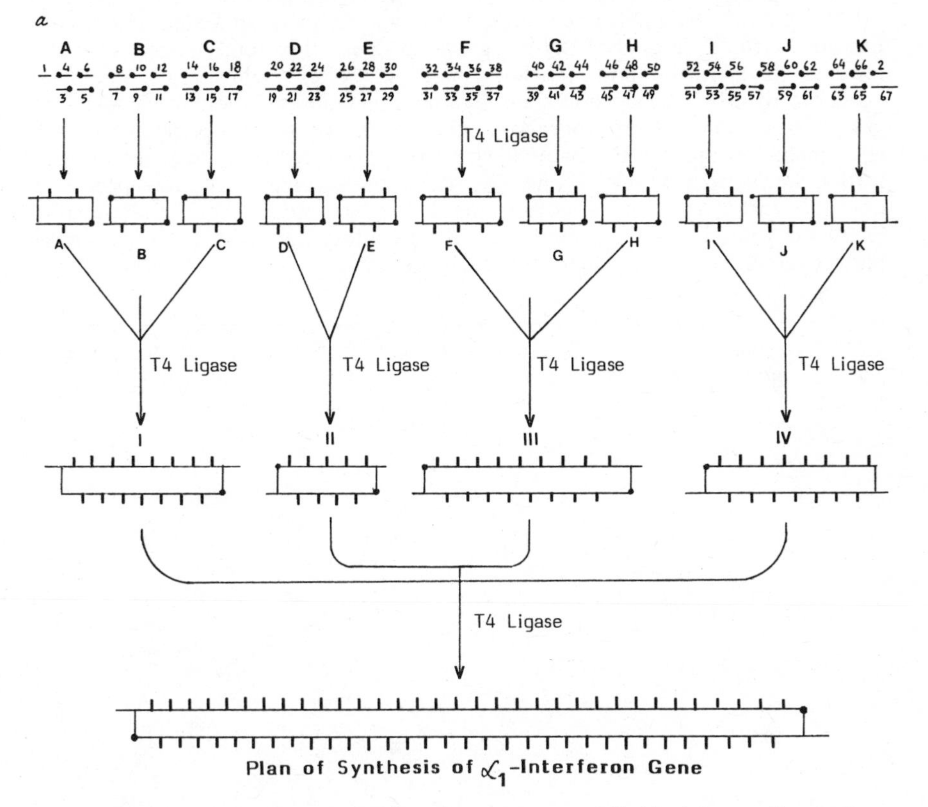

Plan of Synthesis of α_1-Interferon Gene

In planning the synthesis, the gene was divided into oligonucleotide units of ca.15 residues so as to give overlaps of at least 7 base pairs on each side of all ligation points as shown above. The phosphotriester method described above was used for synthesis of the units shown in the top portion of the diagram with arabic numerals, with

---

16. Edge,M.D.; Greene,A.R.; Heathcliffe,G.R.; Meacock,P.A.; Schuch,W.; Scanlon,D.B.; Atkinson,T.C.; Newton,C.R.; Markham,A.F. Nature, 1981, 292, 756.

small changes in experimental conditions to improve the yields; mesitylene-sulfonyl-3-nitro-1,2,4-triazole was used as condensing agent which increased yield in coupling reaction and reduced coupling time. The oligonucleotides were then ligated initially in eleven groups A-K by allowing the units to anneal followed by treatment with T_4-ligase. Ligated products of the expected size were isolated by preparative polyacrylamide gel electrophoresis in denaturing conditions, aliquots of the eleven fragments combined in four blocks, annealed and again ligated with T_4-induced DNA ligase to give the four units, and the process of isolation and ligation repeated. After isolation of the final product of the expected size by polyacrylamide electrophoresis or gel permeation chromatography, the ends were phosphorylated and the product ligated to BamHI site of pM50 plasmid and introduced into E. coli strain MRC8 and the colonies screened for the presence of the synthetic oligonucleotide insert. It was found that the plasmid DNAs obtained from a number of colonies had identical 514-base pair fragment corresponding to the Ifn gene.

---

17. Oligonucleotides Synthesis, a practical approach, Gait,M.J. Ed., IRL Press, Oxford, England, 1984.

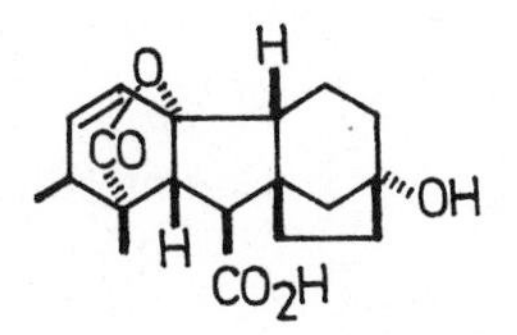

GIBBERELLIC ACID

On account of sensitivity towards reagents and high density of functionalities in some regions of the molecule, gibberellic acid remained a challenge for organic synthesis for many years and it is only relatively recently that Corey and his associates achieved its first total stereospecific synthesis (1).

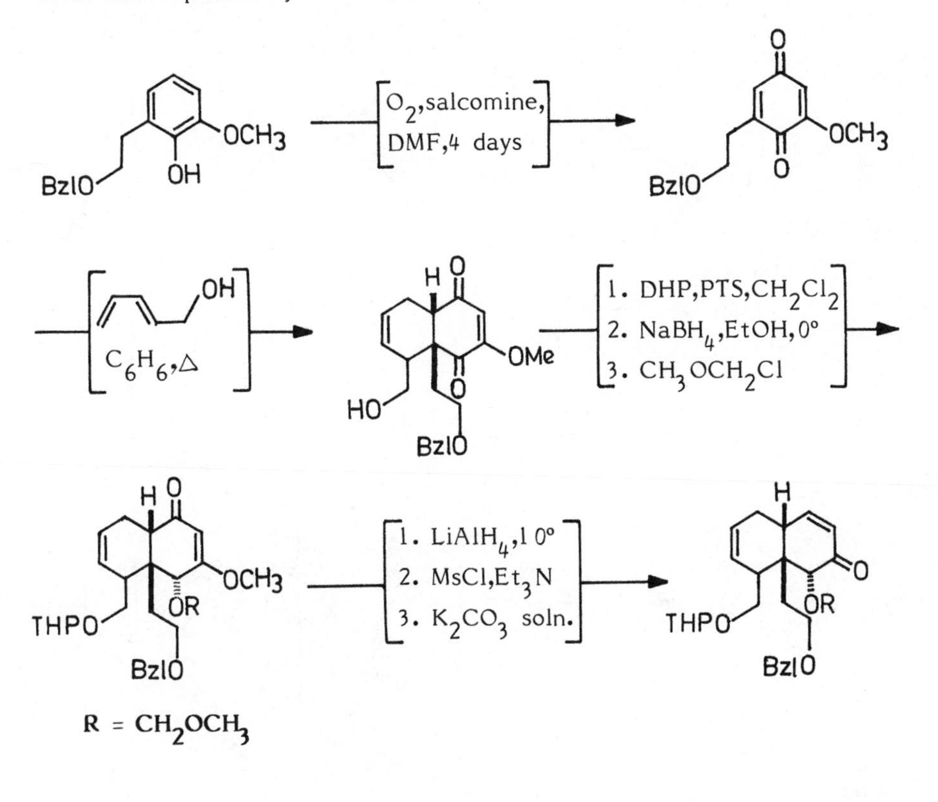

---

1. Corey,E.J.; Danheiser,R.L.; Chandrasekaran,S.; Siret,P.; Keck,G.E.; Gras,Jean-Louis, J. Am. Chem. Soc., 1978, 100, 8031;Corey,E.J.; Danheiser,R.L.; Chandra Sekaran,S.; Keck,G.E.; Gopalan,B.; Larsen,S.D.;Siret, P.; Gras, Jean-Louis, ibid, 1978, 100, 8034.

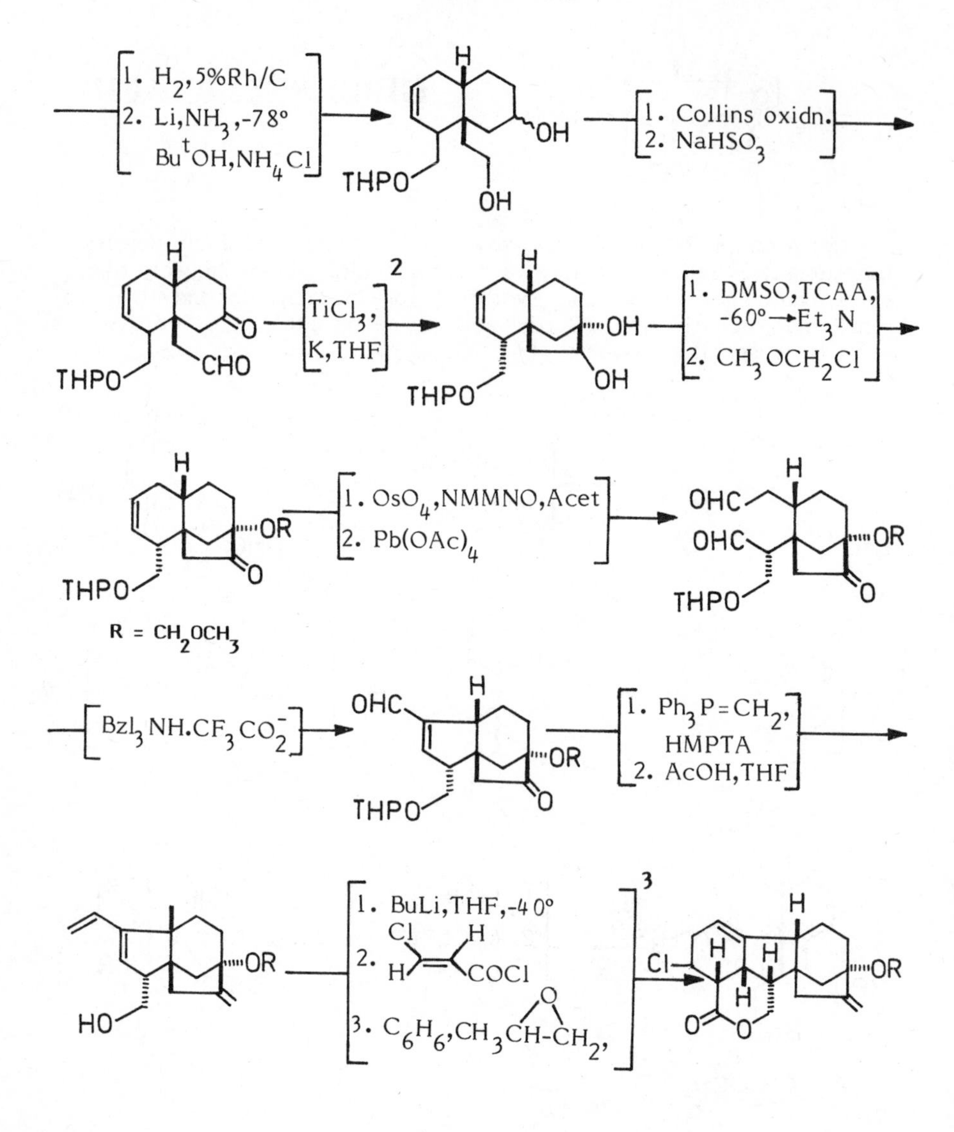

2. The reaction gave 40% of the required cis-aldol alongwith 15% of the trans isomer and 10% diol which was recycled.

3. The stereochemistry of Diels-Alder adduct is assigned on the supposition of a concerted internal α-face, 'endo' Diels-Alder addition and only one product was isolated.

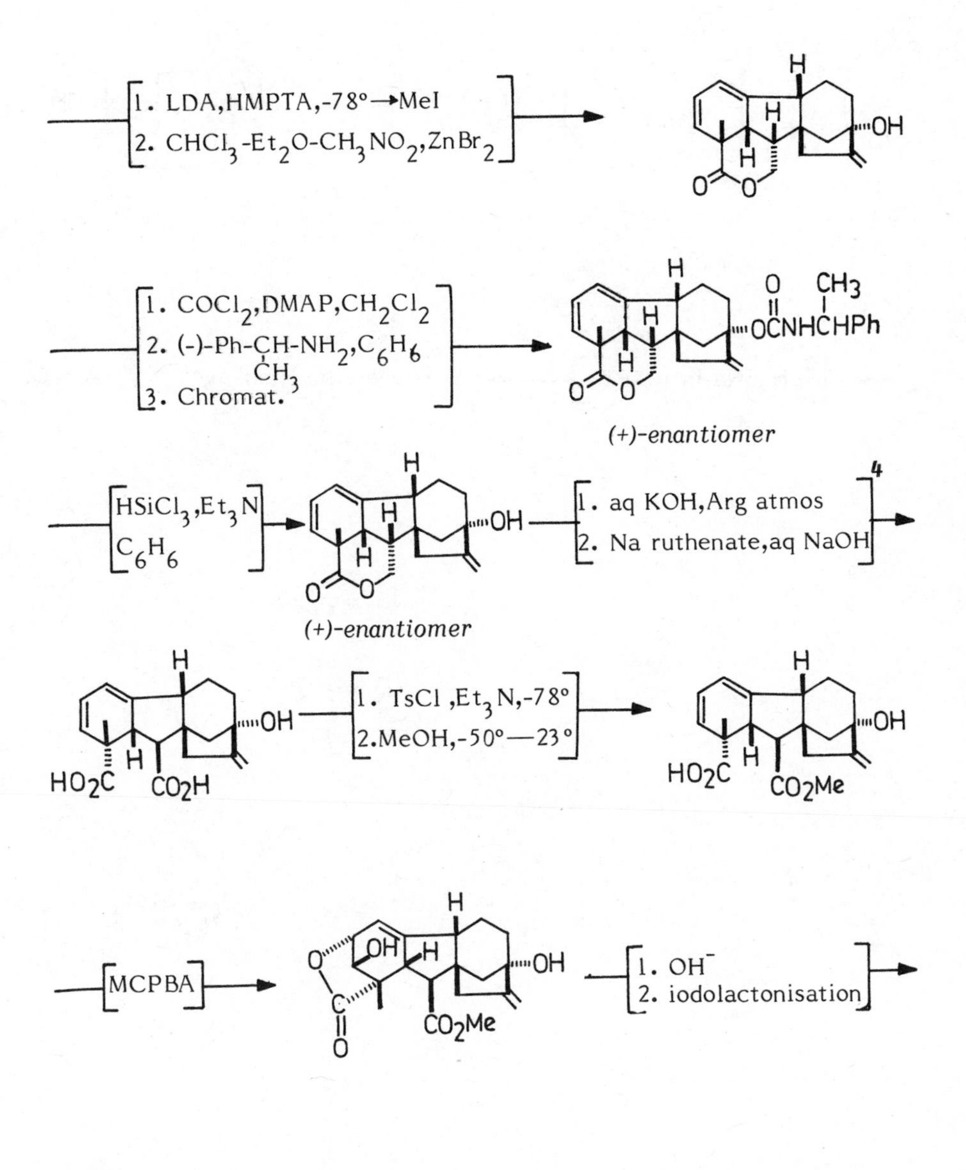

4. The formation of the di-acid very likely proceeds through an acid aldehyde intermediate which undergoes base catalysed epimerisation to the more stable 6β-formyl derivative and on further oxidation to the di-acid.

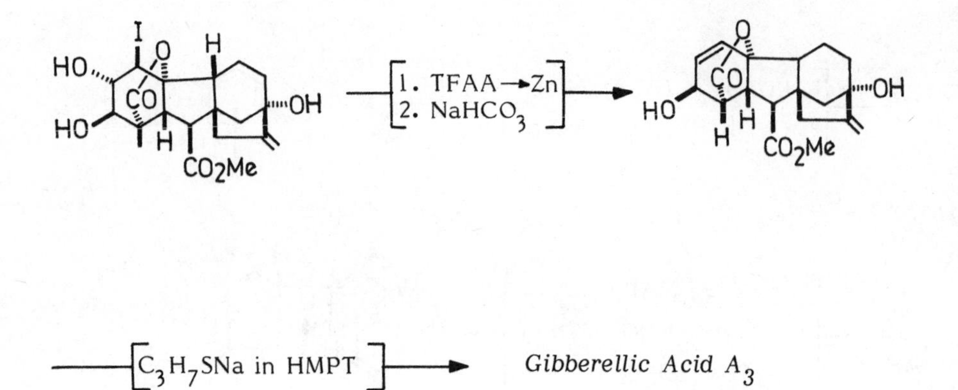
I
HO
HO
O
CO
H
H
OH
CO2Me
1. TFAA→Zn
2. NaHCO3
O
CO
HO
H
H
OH
CO2Me
C3H7SNa in HMPT
Gibberellic Acid A3

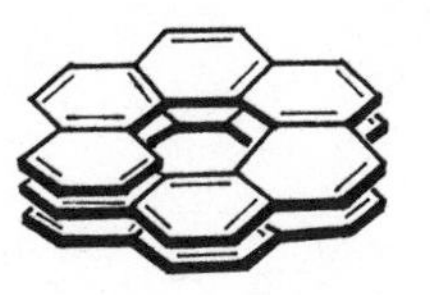

HELICENES

Successive and appropriate angular attachment of aromatic rings would lead to a helix with every sixth ring starting a new turn and the whole system constituting a chiral π orbital chromophore (1,2). As early as 1956, Newman and Lednicer (3) effected the synthesis and resolution of hexahelicene. Members of this class upto 13 rings, [13]helicene, have been synthesized by photoinduced cyclization of suitable stilbenes (4,5).

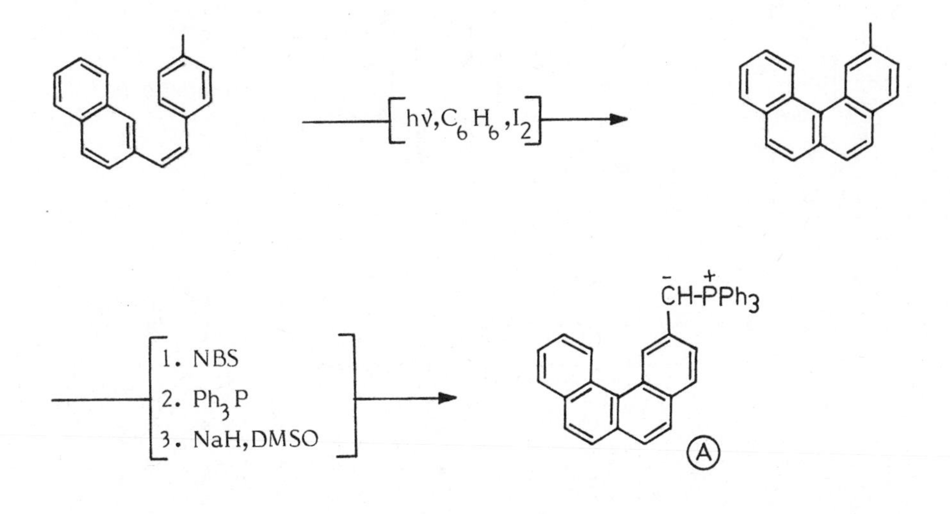

1. The NMR of these molecules show systematic changes in the ring protons as the growing end of the helix overlaps with successive rings of the previous turn; also they undergo spontaneous resolution and nonahelicene has the highest specific rotation (±15000) recorded for an organic compound.

2. There is remarkablly low barrier associated with the racemization of helicenes. For example, for nonahelicene, this barrier is a mere 43.5 kcal mol^{-1}. The low value is attributed to the fact that the necessary molecular deformations are distributed on many bonds of the molecular frame: Meurer,K.P.; Vogtle,F. "Helical Molecules in Organic Chemistry: Topics in Current Chemistry, No.127, Springer-Verlag, 1985, p.1.

3. Newman,M.S.; Lednicer,D. J. Am. Chem. Soc., 1956, 78, 4765.

4. Martin,R.H.; Flammang-Barbieux,M.; Cosyn,J.P.; Gelbcke,M. Tetrahedron Lett., 1968, 3507.

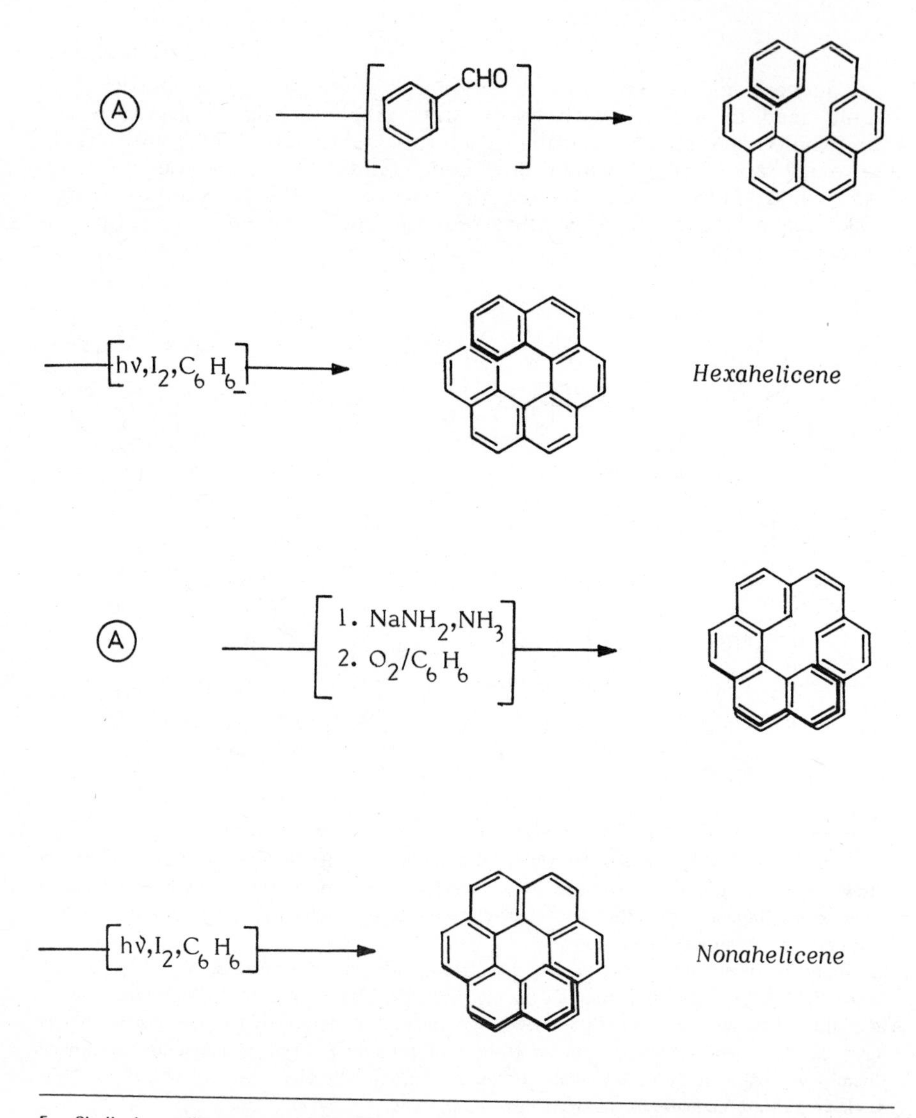

5. Similarly, using naphthalene- and anthracene-1-carboxaldehyde, hepta- and octa-helicenes have also been prepared.

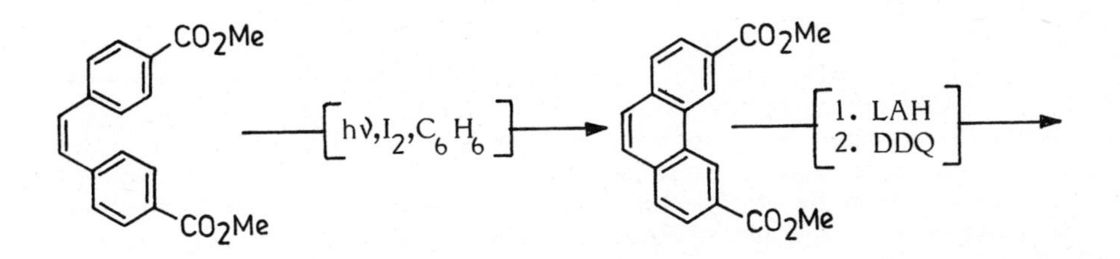

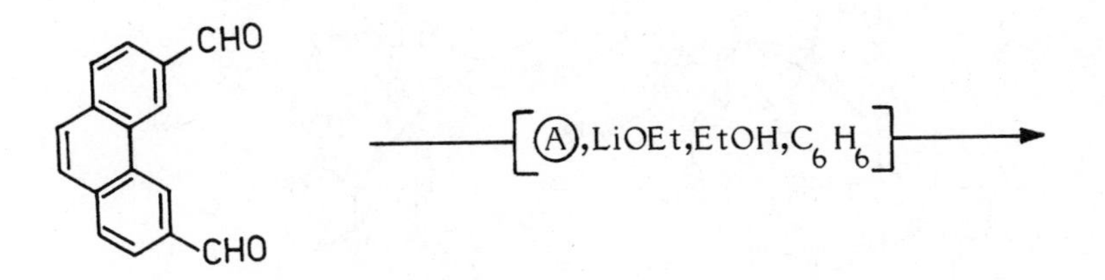

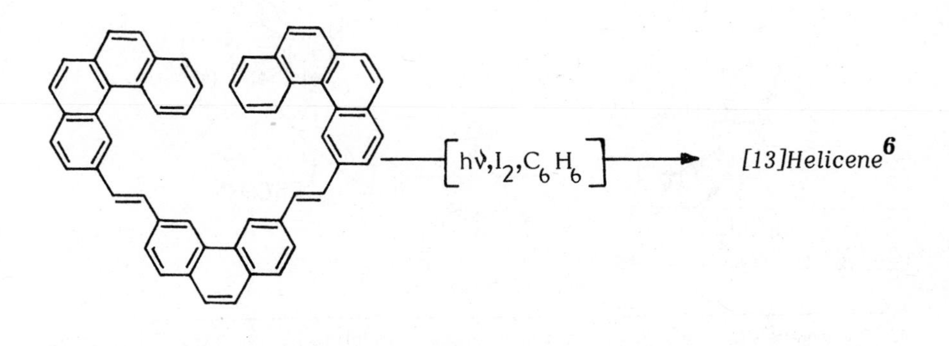

6. Martin,R.H.; Morren,G.; Schurter,J.J. Tetrahedron Lett., 1969, 3683.

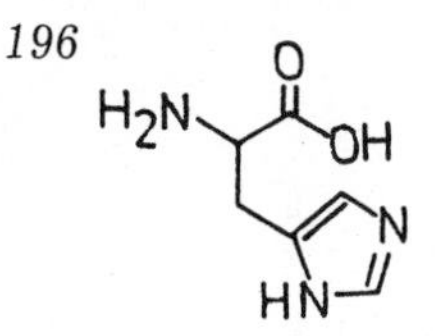

HISTIDINE

The successful chemical simulation (1) of the salient features of the ATP-Imidazole cycle which is related to the biosynthesis of histidine, represents an important addition to the art of organic synthesis in the sense that it establishes that cyclic operations, so advantageously deployed by Nature (2), could be carried out in the laboratory for regenerative synthesis. Another facet of this unique cycle is the use of an imidazole parent to generate an imidazole daughter. This endeavour has led to a practical route to 5-substituted imidazoles and offers the prospects of a heterocycle synthesising machine based on template strategy (3).

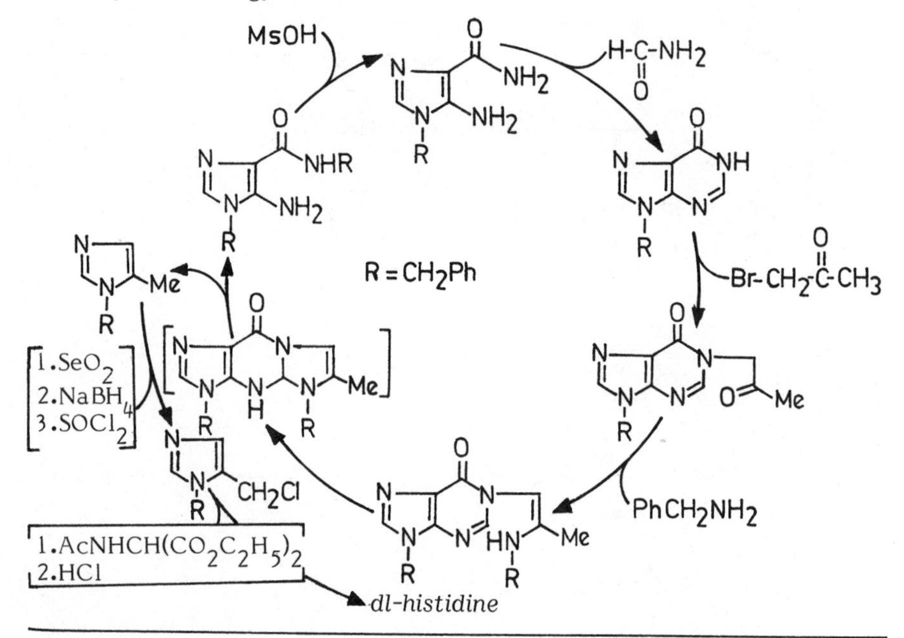

1. Ranganathan,D.; Farooqui,F.; Bhattacharyya,D.; Mehrotra,S.; Kesavan,K. Tetrahedron, 1986, 42, 4481; Ranganathan,D.; Farooqui,F. Tetrahedron Lett., 1984, 5701; Ranganathan,D; Farooqui,F.; Bhattacharyya,D. Tetrahedron Lett., 1985, 2905.

2. Metabolic pathways, vital to the sustenance of life, often proceed by cyclic pathways, as can be exemplified with the Calvin cycle, the Kreb's cycle and the Urea cycle. These pathways, undoubtedly evolved over a number of years, illustrate idyllic organic synthesis, since they represent optimization of resources with respect to yield and versatility.

3. Ranganathan,D.; Rathi,R. Tetrahedron Lett., 1986, 2491.

4. Albertson,N.F.; Archer,S. J.Am.Chem.Soc., 1945, 67, 308; Turner,R.A.; Huchner,C.F.; Scholz, C.R. J. Am. Chem. Soc., 1949, 71, 2801.

ICEANE

Iceane (tetracyclo [5.3.1.1^{2,6}O^{4,9}]dodecane) the carbon prototype of "ice" can be exhibited in any of the equivalent D_{3h} representations, as shown above. A fascinating intramolecular Diels-Alder reaction followed by a Wagner-Meerwein sequence takes tropone to iceane (1).

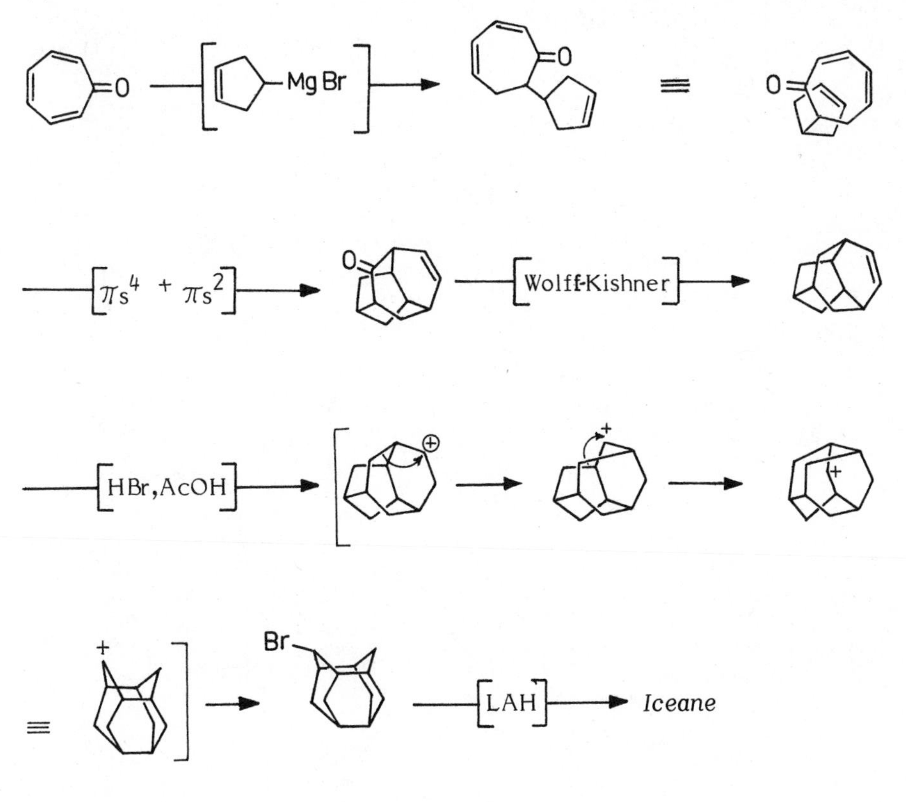

1. Cupas,C.A.; Hodakowski,L. J. Am. Chem. Soc., 1974, 96, 4668.

The transformation of adamantane to iceane requires precise regio-insertion of two carbons. This has been achieved in an ingenious manner taking advantage of the best of adamantane re-arrangements (2).

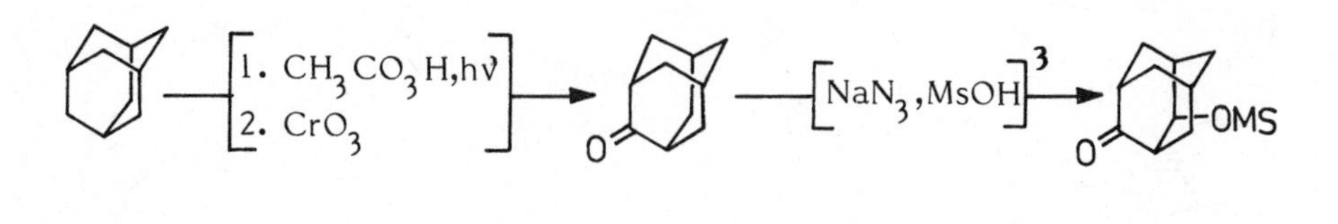

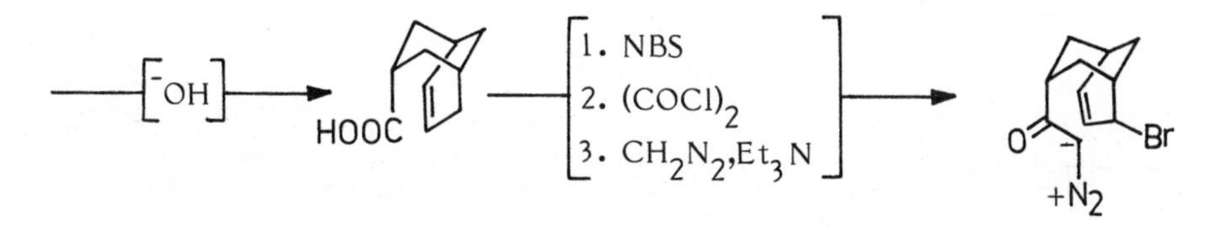

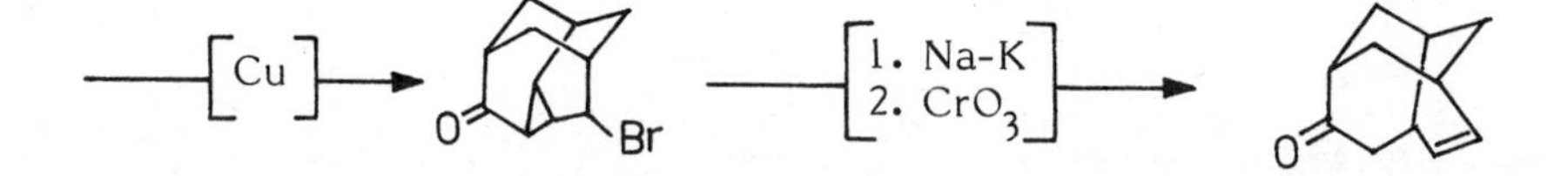

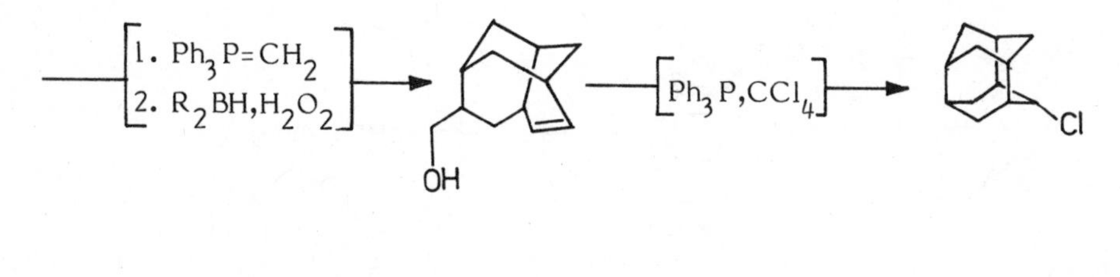

—[$(nBu)_3SnH$]→ *Iceane*

2. Hamon,D.P.G.; Taylor,G.F. Tetrahedron Lett., 1975, 155, 1623.
3. Sasaki,T.; Eguchi,S.; Toru,T. J. Am. Chem. Soc., 1969, 91, 3390.

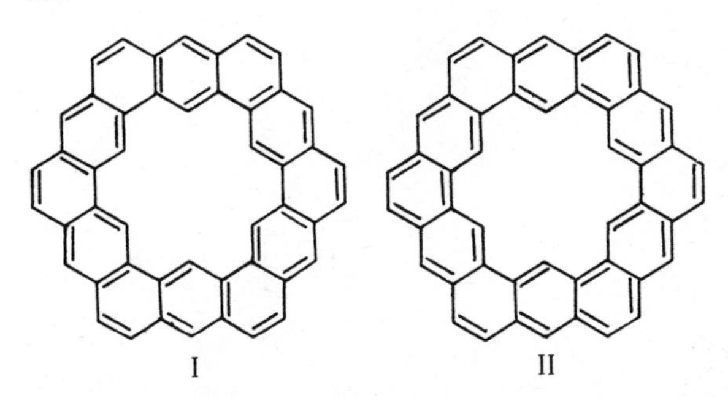

KEKULENE

Kekulene was identified as an interesting compound as early as 1965, since it can be either 12 benzene rings sitting in a circle (I) or a split, duplex aromatic system consisting of a 30 annulene at the rim and an 18 annulene at the core, sequestered by 12 spokes. The synthesis of Kekulene was only part of the adventure. The compound did not melt till at least 620°C and was extremely insoluble in all solvents. Thus, one litre of boiling (245°C) 1-methyl-naphthalene dissolved a mere 0.03 g of Kekulene. Consequently the recording of its ^{1}HNMR, needed to distinguish I and II contributions, turned out to be an awesome task. It was accomplished eventually by making 50,000 scans of a saturated (1) solution in 1,3,5-trichlorotrideuterobenzene at 215°C. The observed lack of upfield shift was a vote in favour of I. The synthesis of Kekulene incorporates as a key feature a sulfide mediated union of two symmetrical fragments (1).

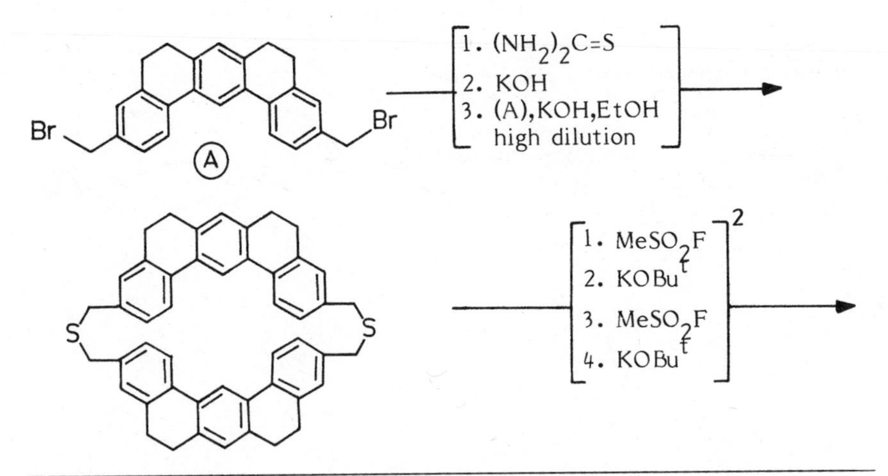

1. Diedrich,F.; Stabb,H.A. Angew. Chem. Int. Ed., 1978, 17, 372.

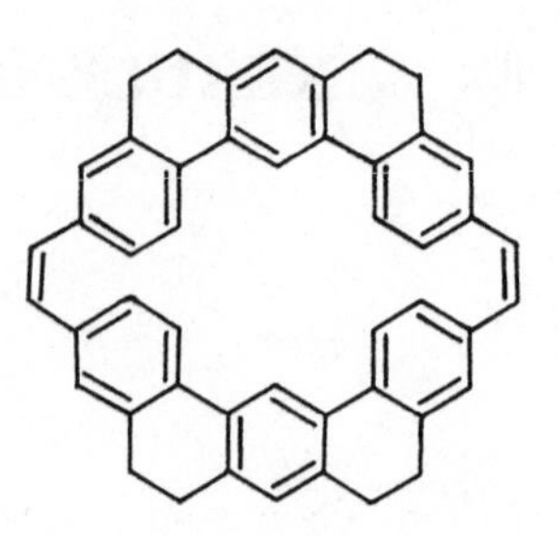

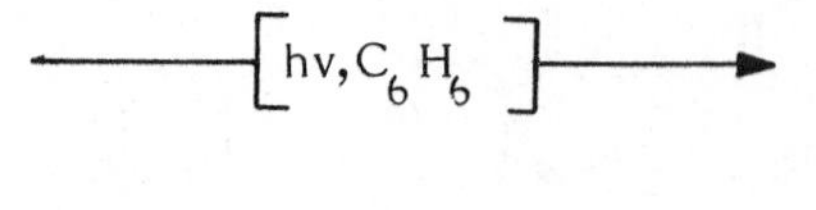

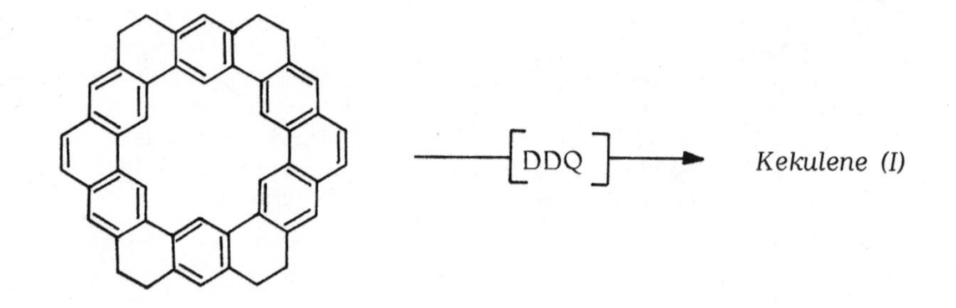

---

2. The overall transformation takes place as follows:

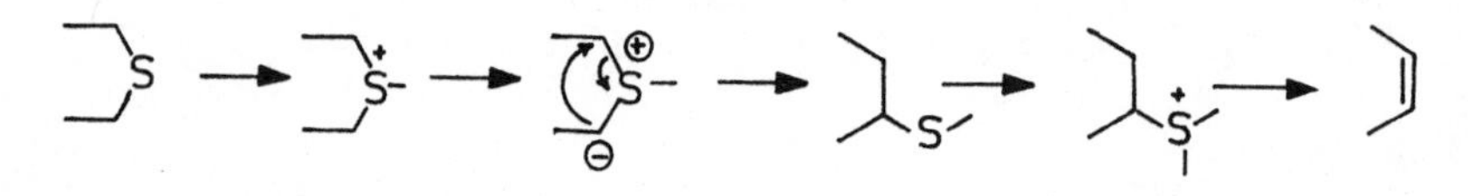

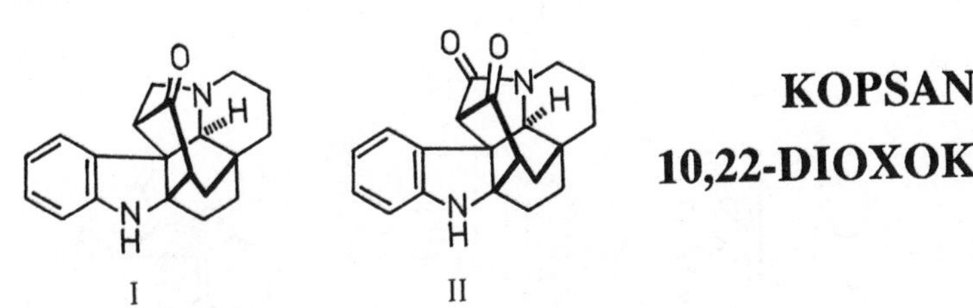

KOPSANONE
10,22-DIOXOKOPSANE

The heptacyclic indole alkaloids of the Kopsane class, though isolated as early as 1890 (1), yielded to structural elucidation only in 1960's when the spectroscopic and X-Ray crystallography methods had been greatly refined. The first and the only total synthesis of Kopsanone by Gallagher and Magnus employs two [4+2] cycloaddition reactions to create the extra-ordinarily complex cage-like structure of kopsane alkaloids (2).

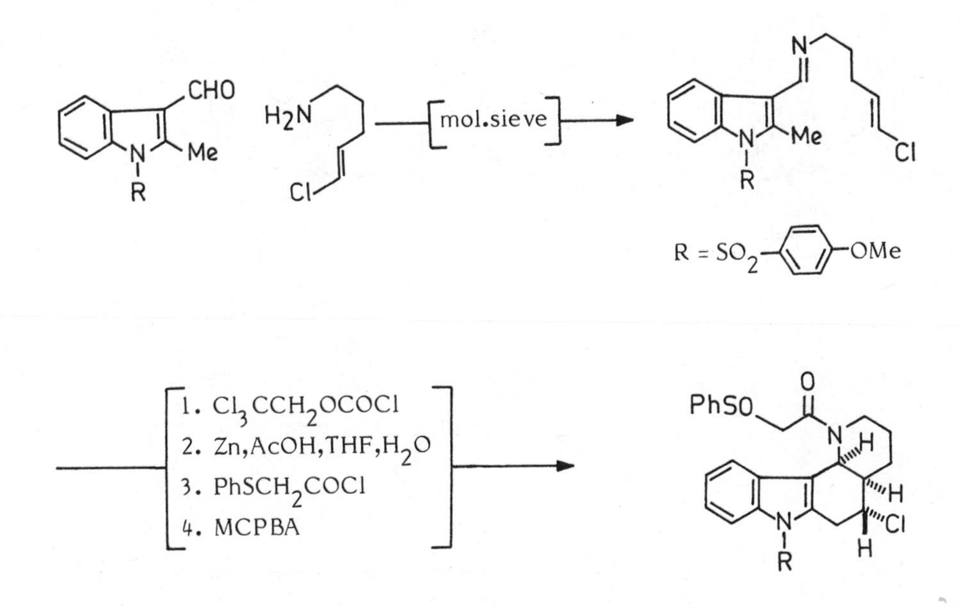

1. Greshoff,M. Ber. Dtsch. Chem. Ges., 1890, 23, 3537.

2. Gallagher,T.; Magnus,P. J. Am. Chem. Soc., 1983, 105, 2086; Magnus,P.; Gallagher,T.; Brown,P.; Huffmann,J.C. ibid, 1984, 106, 2105.

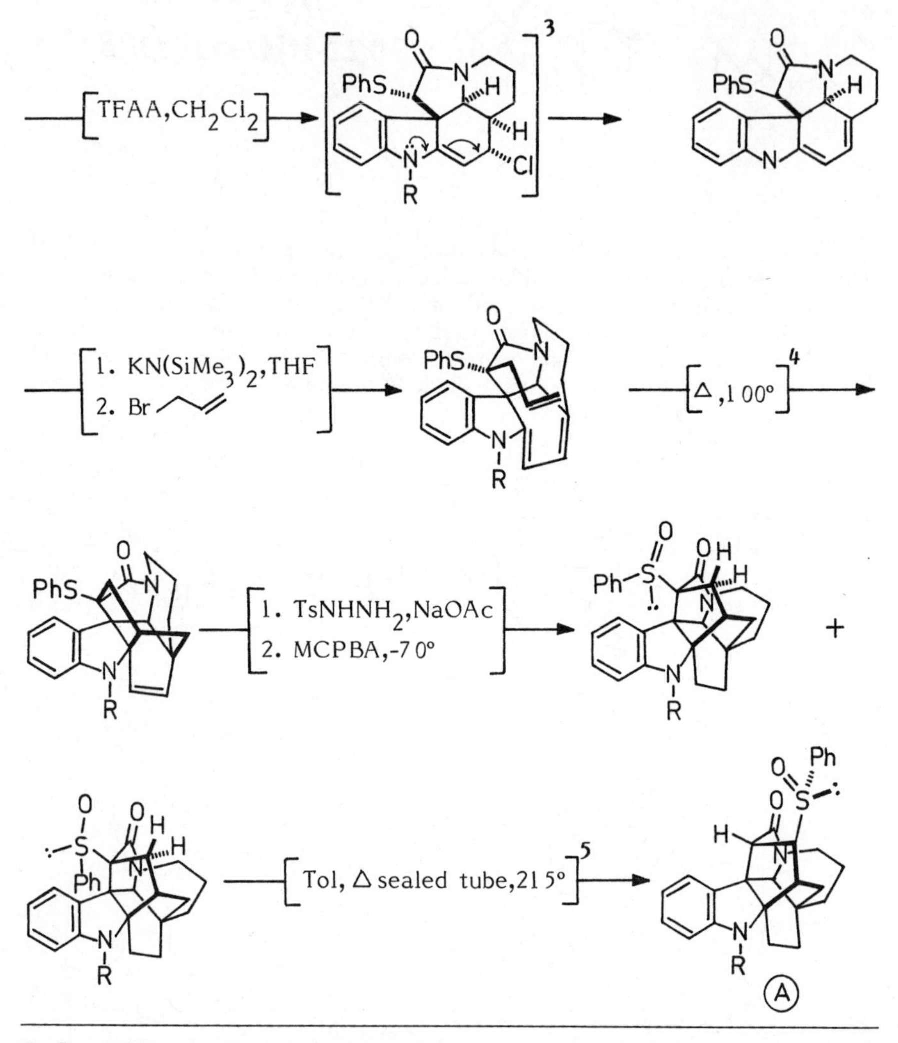

3. The indolic nitrogen is inductively deactivated, but not resonance wise; N is still able to use its lone pair of electrons. X-ray crystallographic pictures have shown that the $SO_2C_6H_4OMe$-p is not in the same plane as either of the adjacent π-systems.

4. The cycloaddition is regiospecific and none of the isomeric fruticosane structure was detected.

5. Only the sulfoxide (A) could undergo the syn elimination required for the Pummerer rearrangement; unreacted (B) was recovered which could be recycled by reduction and reoxidation to mixture of (A) & (B).

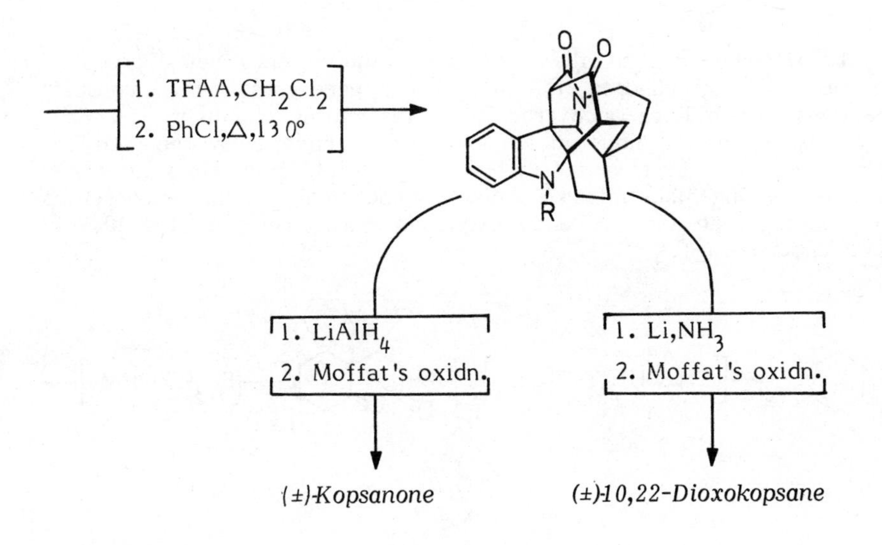
1. TFAA,CH_2Cl_2
2. PhCl,Δ,130°
O
O
N
N
R
1. $LiAlH_4$
2. Moffat's oxidn.
1. Li,NH_3
2. Moffat's oxidn.
(±)-Kopsanone
(±)-10,22-Dioxokopsane

LONGIFOLENE

Longifolene, one of the very few sesquiterpenes being produced commercially in hundred ton quantities, undergoes many "unusual" reactions, which have uncovered much fascinating chemistry (1). Four syntheses, each ingenious in its own right, have been reported for longifolene. The first synthesis by Corey et al (2) involves as a key step a most ingenious intramolecular Michael addition (3), which serves to link together the six and seven membered rings of the bicyclo [5.4.0]undecane Ⓐ.

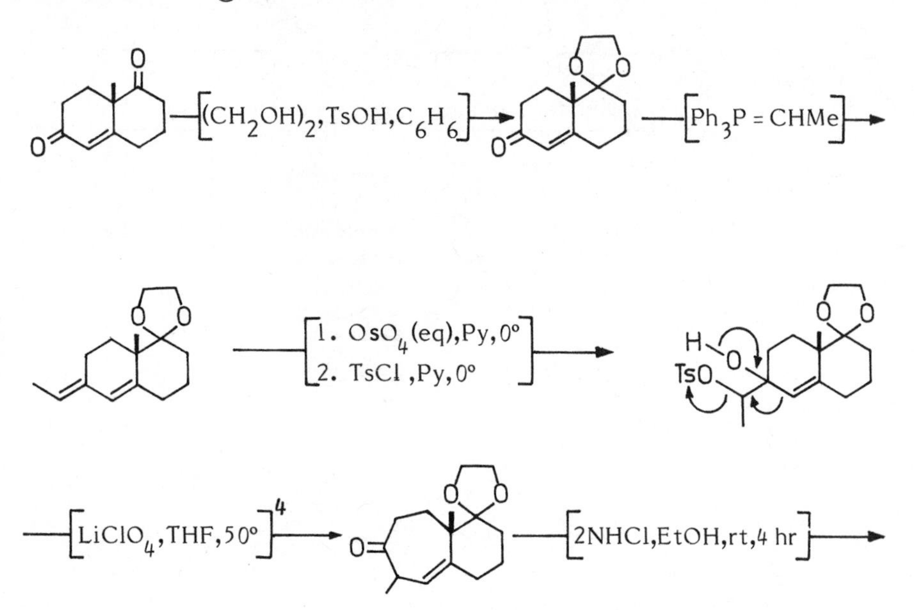

---

1. Dev,S. Progress in the Chemistry of Organic Natural Products, 1981, 40, 49.

2. Corey,E.J.; Ohno,M.;Vatakencherry,P.A.; Mitra,R.B. J. Am. Chem. Soc., 1961, 83, 1251; 1964, 86, 478.

3. This novel cyclization step perhaps finds analogy in an alkali-induced santonin-santonic acid transformation, cf. Woodward,R.B.; Brutschy,F.J; Baer,H. J. Am. Chem. Soc., 1948, 70, 4216.

4. Pinacol rearrangement of the mono-p-toluenesulfonate can occur in two directions. The observed direction of ring expansion, involving C-1 as the migrating group (π-electron participation), may be anticipated to take precedence over an alternative ring expansion involving C-3 as the migrating group (less favourable alkyl rearrangement).

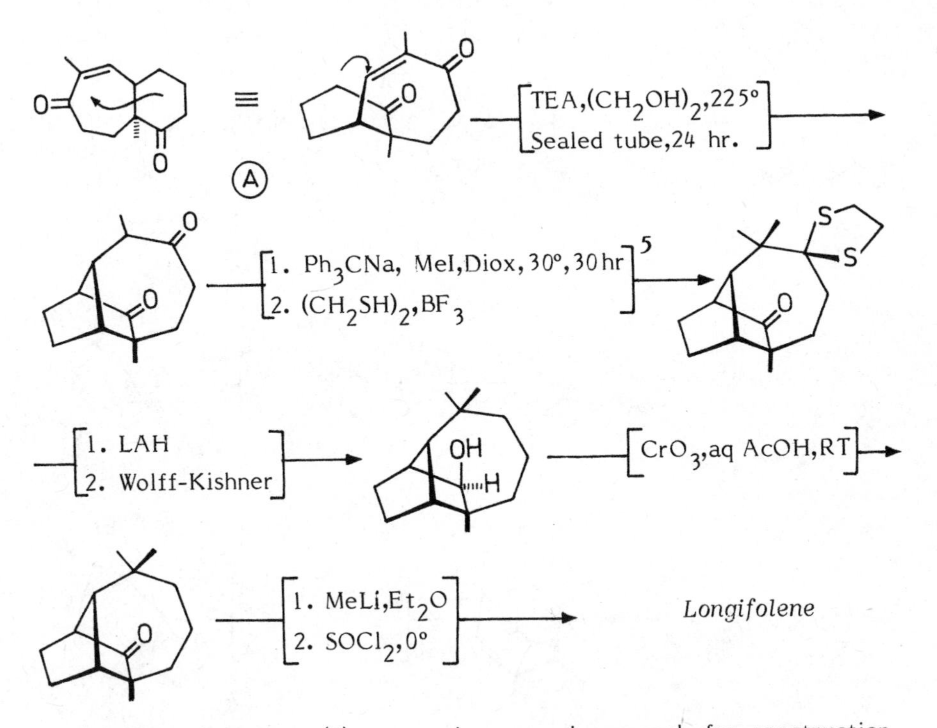

McMurry and Isser (6) reported a second approach for construction of the intricate carbon-network of longifolene based on the intramolecular alkylation of a bicyclic keto epoxide Ⓐ to the tricyclic compound Ⓑ. The olefin derived by dehydration of tertiary alcohol function in Ⓑ, after a dibromocarbene addition, undergoes silver ion assisted solvolytic ring enlargement to the allylic alcohol Ⓒ. Taking advantage of the potential enone system in Ⓒ, the remaining gem-methyl group has been introduced by a formal conjugate addition involving an unusual reductive process (→D) and fragmentation to generate the dimethyl-cycloheptane ring of the natural product.

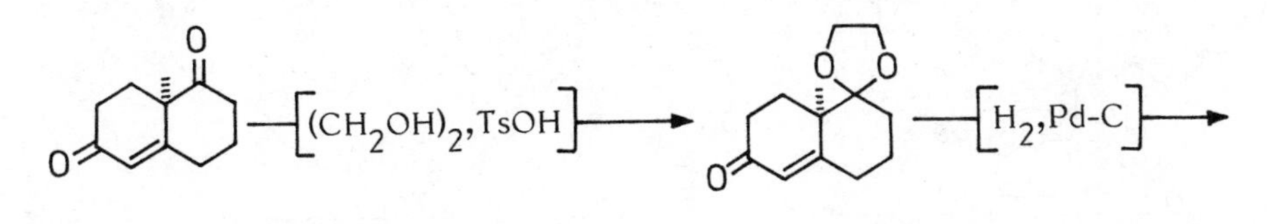

---

5. Resolution may be accomplished at this stage using L(+)-2,3-butane dithiol.
6. McMurry,J.E.; Isser,S.J. J. Am. Chem. Soc., 1972, 94, 7132.

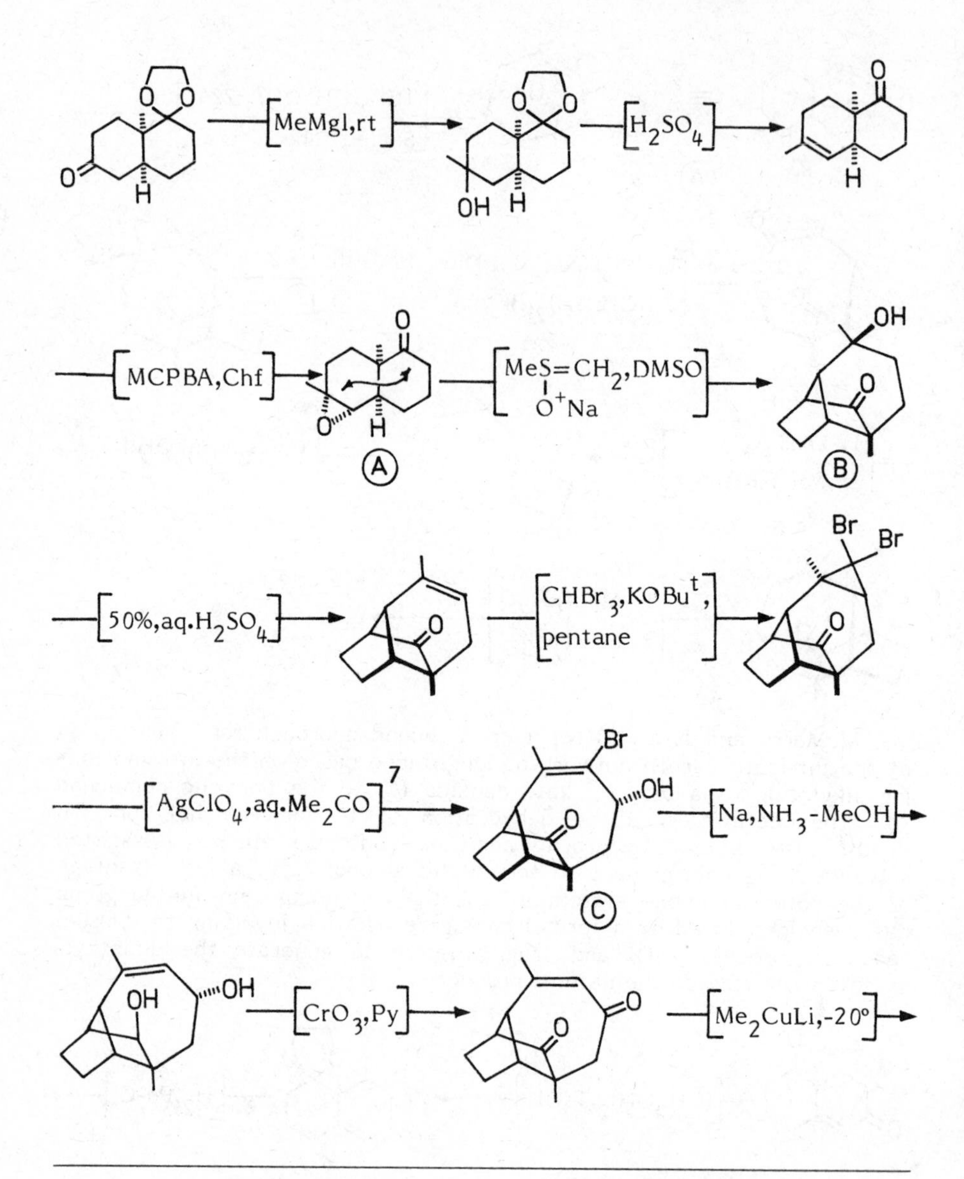

7. Other methods of ring enlargement proved inadequate. However, the sequence involving addition of a dihalocarbene to an olefin followed by silver ion assisted rearrangement and oxidation proceeded smoothly.

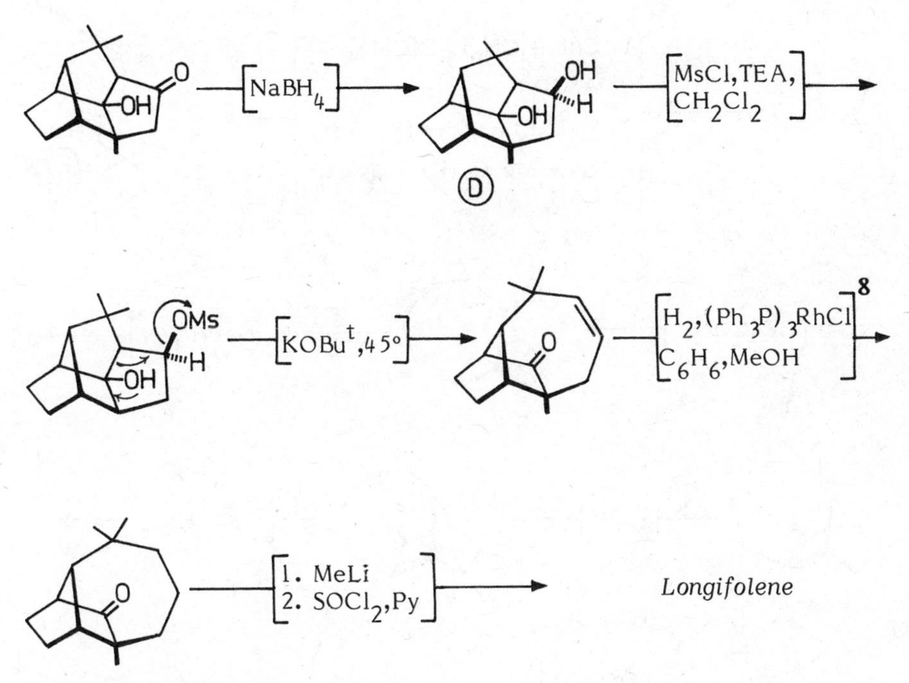

Longicamphenilone

The synthesis by Volkmann, Andrews and Johnson (9) emerged as a result of their studies on Lewis acid catalysed polyolefin cyclisation for steroids. It was observed that in stannic chloride catalysed cyclisation of heptynylmethylcyclopentenol Ⓔ in addition to the expected product Ⓕ, the tricyclic product Ⓖ was also formed (10). The structural resemblance to longifolene was obvious, and the reaction refined and exploited for a novel synthesis of longifolene as described below.

---

8. Wilkinson catalyst: Young,J.F.; Osborn,J.A.; Jardine,F.H.; Wilkinson,G. Chem. Commun. 1965, 131.

9. Volkmann,R.A.; Andrews,G.C.; Johnson,W.S. J.Am. Chem. Soc., 1975, 97, 4777.

10.

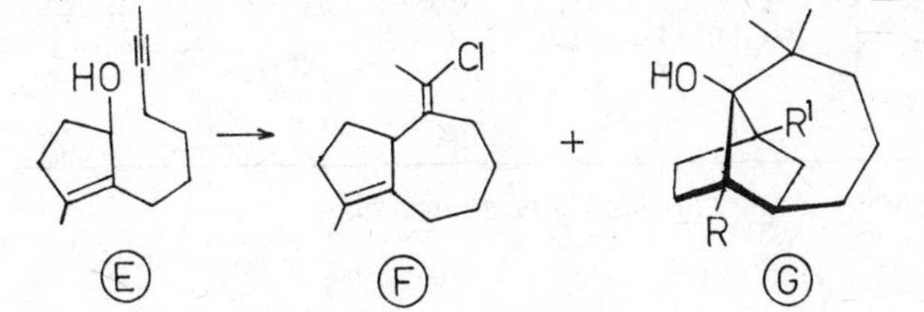

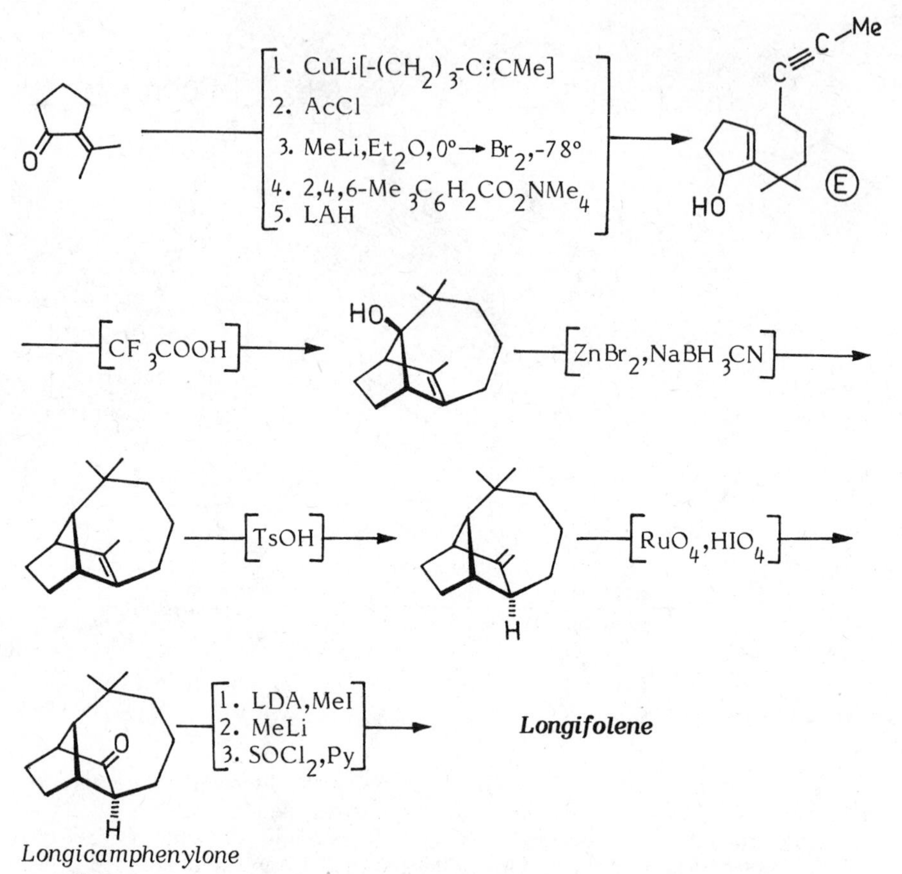

Longicamphenylone

The latest synthesis charted below (11) utilizes an intramolecular photoaddition-retroaldol reaction (H → I) as the key-step in the construction of the tricarbocyclic frame-work. The resulting diketone Ⓗ could be easily transformed into the known ketone Ⓐ which had been earlier converted into longifolene.

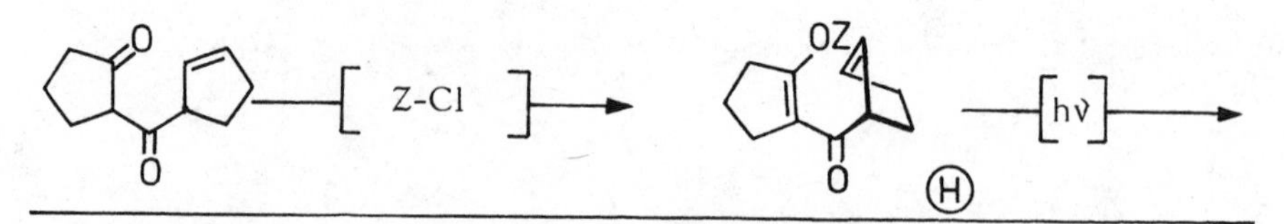

11. Oppolzer,W.; Godel,T. J. Am. Chem. Soc., 1978, 100, 2583.

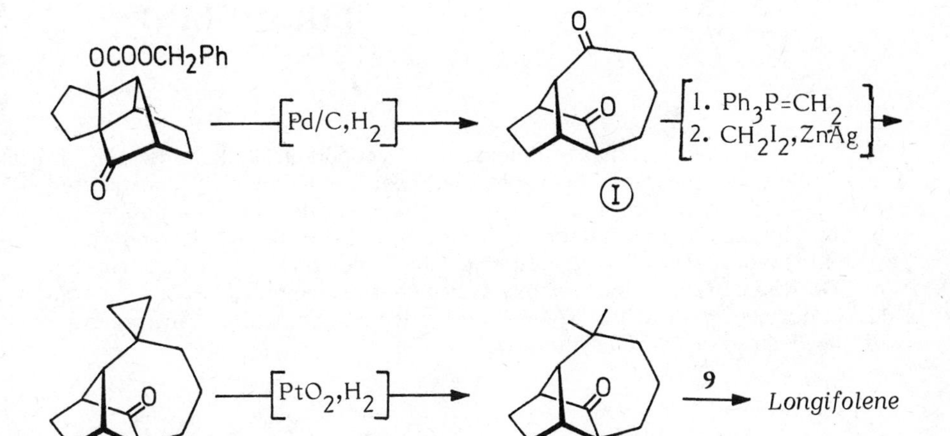
OCOOCH2Ph
O
Pd/C,H2
O
O
I
1. Ph3P=CH2
2. CH2I2,ZnAg
O
PtO2,H2
O
J
9
Longifolene

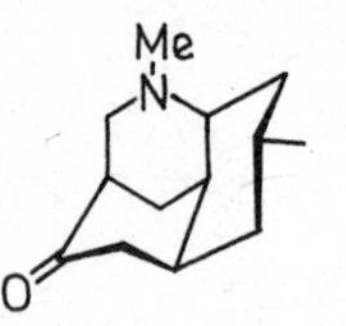

LUCIDULINE

Luciduline is the simplest of Lycopodium alkaloids (1). In an elegant synthesis of this molecule, Scott and Evans (2) constructed the desired tricyclic skeleton from the key decalin ketoamine (D) by an intramolecular Mannich reaction, C-11 being derived from formaldehyde. The cis-decalin-2,6-dione suitable for construction of (D) was obtained by the oxy-Cope rearrangement (3) of a bicyclo [2.2.2]octene derivative (A) which allowed controlled introduction of the desired sites of asymmetry.

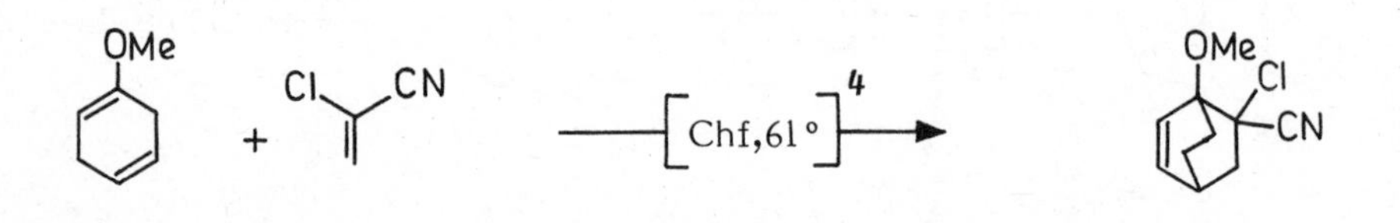

---

1. Reviews: Blumenkopf, T.A., Heathcock, C.H. in "Chemical and Biological Perspectives, vol.3", Ed., Pelletier, S.W., Wiley, New York, 1985, p.185; Inubushi, Y., Harayama, T. Heterocycles, 1981, 15, 611; Stevens in "The Total Synthesis of Natural Products, vol.3" Ed., ApSimon, J., Wiley, New York, 1977, p.489; Wiesner, K., Fortschr. Chem. Org. Naturstoffe, 1962, 20, 271.

2. Scott, W.L., Evans, D.A. J. Amer. Chem. Soc., 1972, 94, 4779.

3. The oxy-Cope rearrangement involves thermal isomerization of 3-hydroxy-1,5-hexadienes (i) to enols (ii), leading to the synthesis of carbonyl compounds [Berson, J.A., Jones, M.Jr. Amer. Chem. Soc., 1964, 86, 5019; Berson, J.A., Walsh, E.J. 1968, 90, 4729]. For extension of this method to preparation of oxygenated cis-decalin carbocycles see, Evans, D.A., Scott, W.L., Truesdale, L.K. Tetrahedron Lett., 1972, 137.

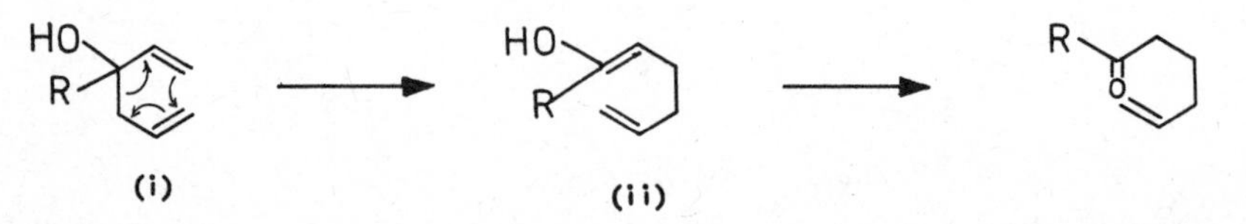

4. Chloroacrylonitrile serves as a useful ketene equivalent in the Diels-Alder reaction, Freeman, P.K., Balls, D.M., Brown, D.J. J. Org. Chem., 1968, 33, 2211; Evans, D.A., Scott, W.L., Truesdale, L.K. Tetrahedron Lett., 1972, 121. Ranganathan,S.D.; Ranganathan,D.; Mehrotra,A.K. Synthesis, 1977, 289.

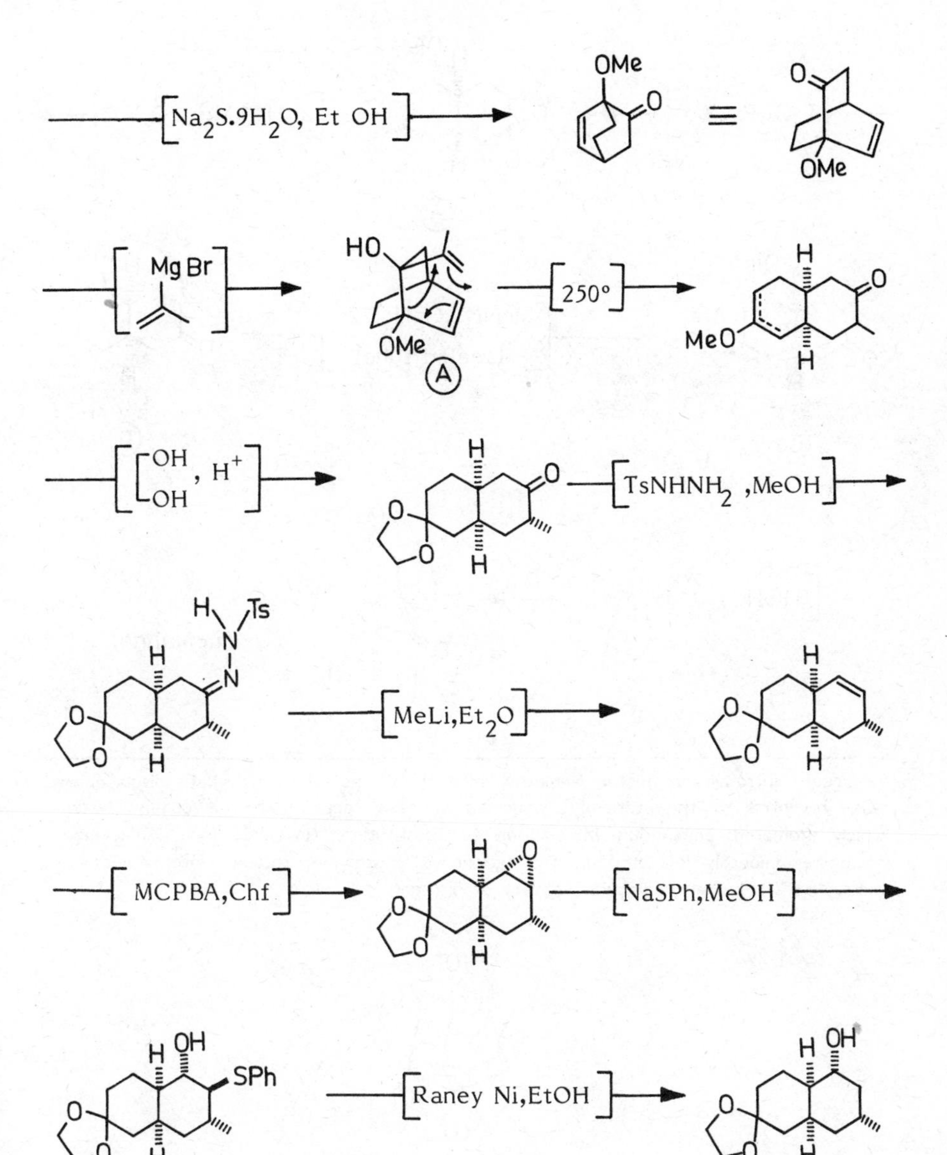

$Na_2S.9H_2O$, Et OH
OMe
O
OMe
Mg Br
HO
OMe
A
250°
H
O
MeO
H
OH
OH
, H^+
H
O
O
O
H
$TsNHNH_2$,MeOH
H
Ts
N
N
H
O
O
H
MeLi,Et_2O
H
O
O
H
MCPBA,Chf
H
O
O
O
H
NaSPh,MeOH
H
OH
SPh
O
O
H
Raney Ni,EtOH
H
OH
O
O
H

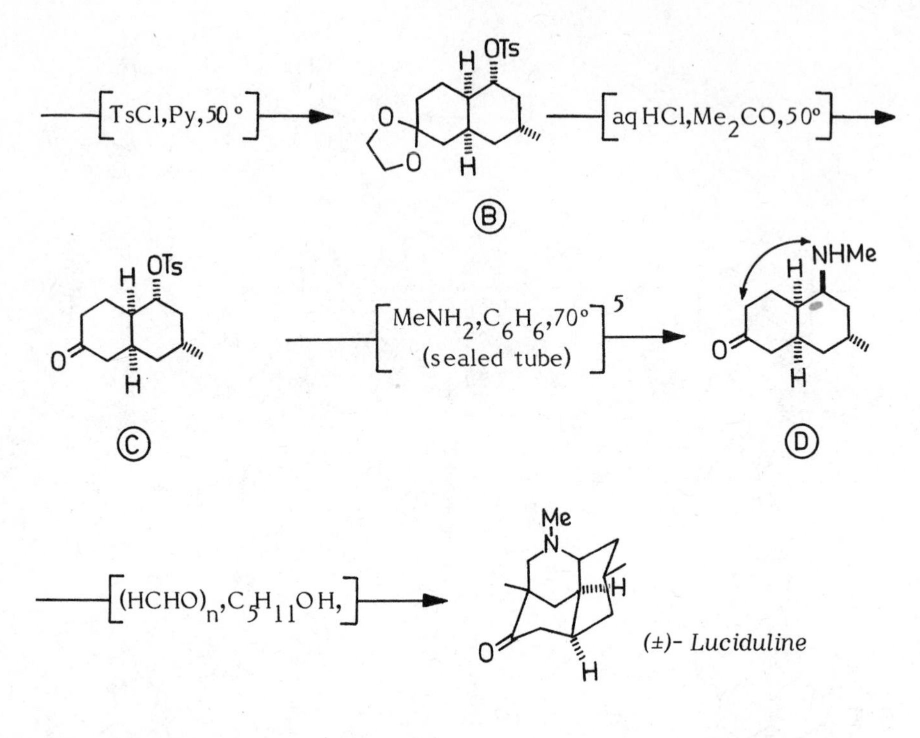

5. Smooth introduction of the N-methylamine moiety at C-6 by tosylate displacement was prevented by the sterically congested concave face of the cis decalyl system which promoted elimination in addition to substitution. However, this displacement could be smoothly effected in the ketone (C), probably through intermediacy of the aminal (i), by intramolecular delivery of nitrogen.

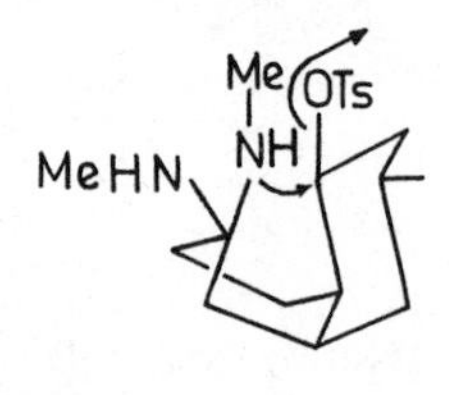

(i)

In the second synthesis of luciduline Oppolzer and Petrzilka (6) utilize an intramolecular 1,3-dipolar cycloaddition as the key step to assemble the heterocyclic ring and to introduce the oxygen functionality into the molecule.

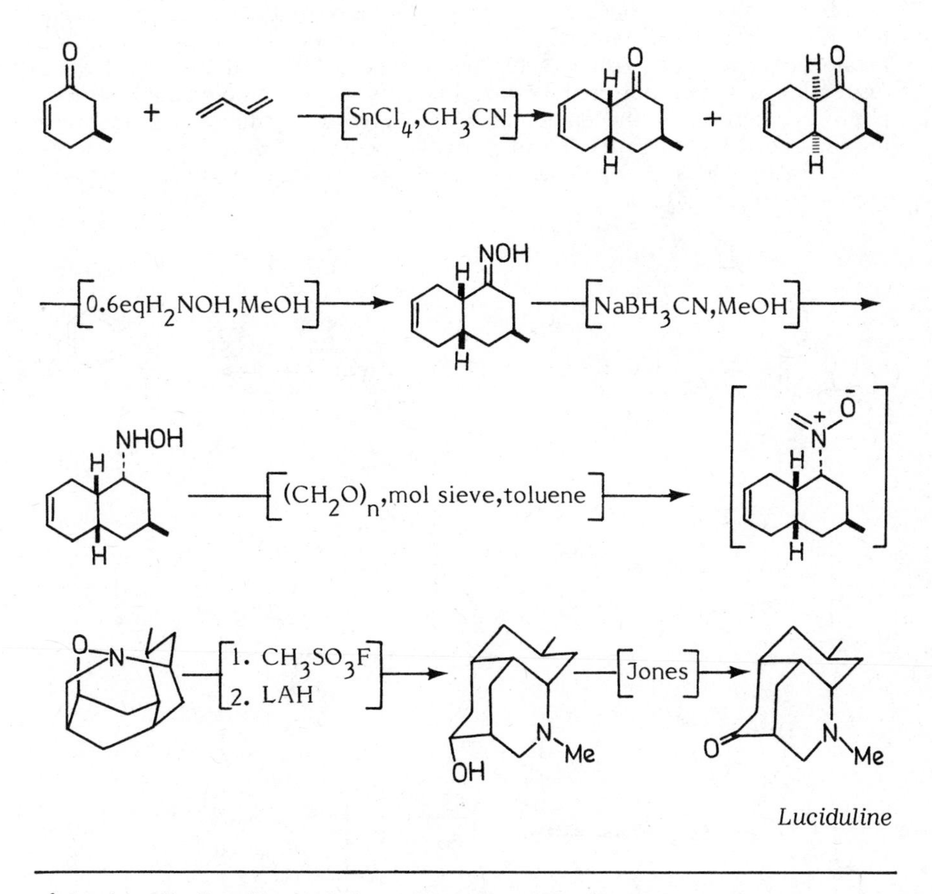

Luciduline

---

6. Oppolzer, W., Petrzilka, M. J. Amer. Chem. Soc., 1976, 98, 6722.

7. For a synthesis of luciduline patterned after the biosynthesis of Lycopodium alkaloids see, Szychowski, J., Maclean, D.B. Canad. J. Chem., 1979, 57, 1631.

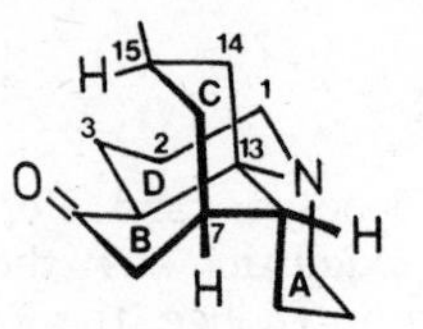

LYCOPODINE

Lycopodine is characterized by the hexahydro-julolidine system (1). A number of ingenious approaches to the synthesis of lycopodine alkaloids have been reported(2).The synthesis devised by Stork and his colleagues (3), outlined below, involves an intramolecular electrophilic cyclization of the quinolone Ⓐ to give a lycodine type structure Ⓑ incorporating rings A, B and C of lycopodine. Material for construction of ring Ⓓ of lycopodine has been obtained by dismembering the aromatic ring in Ⓑ.

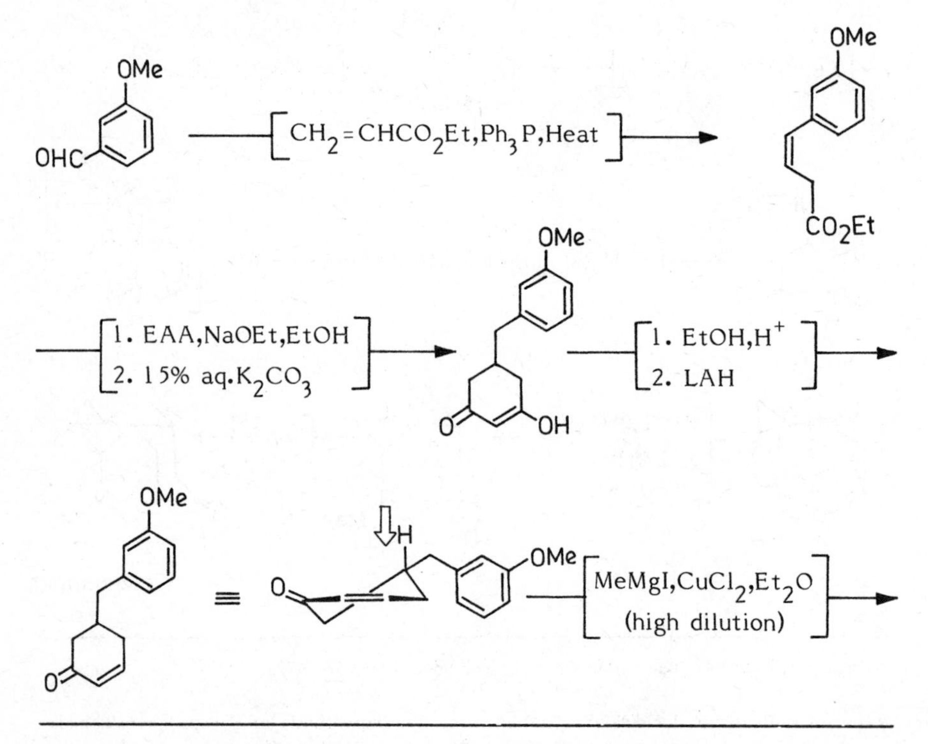

---

1. Review: Wiesner,K. Fortschr. Chem. Org. Naturstoffe, 1962, 20, 271.
2. For one of the earlier synthesis see: Ayer,W.A.; Rowman,W.R.; Joseph,T.C.; Smith,P. J. Am. Chem. Soc., 1968, 90, 1650.
3. Stork,G.; Kretchmer,R.A.; Schlessinger,R.H. J. Am. Chem. Soc., 1968, 90, 1647; Stork,G. Pure Appl. Chem., 1968, 17, 383.

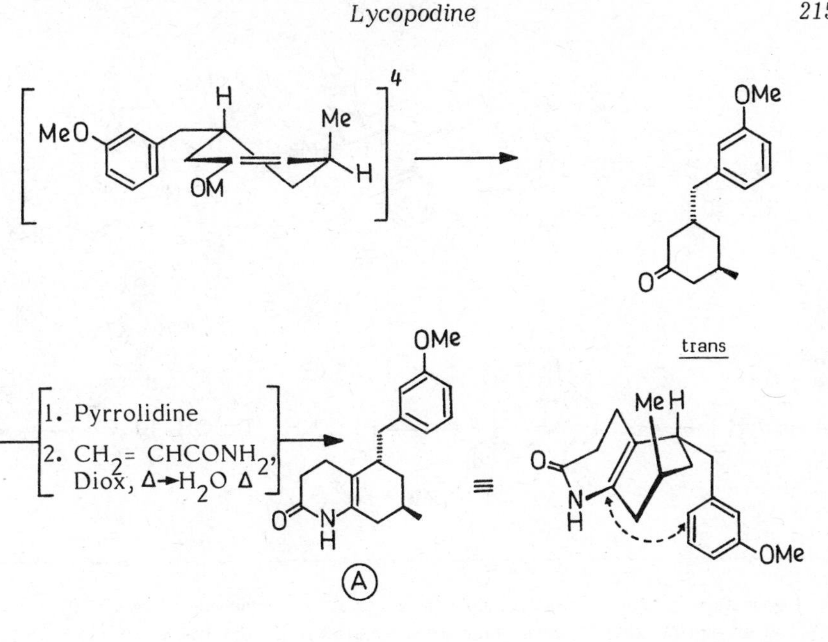

---

4. The 1,4-type Michael addition to the α,β-unsaturated ketone occurs in a transition state in which the incoming nucleophile approaches C-3 orthogonal to the plane of the double bond. There are thus four possible transition states which can lead to two different products, cis and trans, and the stereochemical outcome of the reaction is dictated by the relative energies of these transition states. The trans product results from a half-chair transition state in which the nucleophile approaches C-3 avoiding steric interference with the existing substituent on C-5. For a detailed analysis of a kinetically controlled Michael addition in an analogous case, see Allinger, N.L., Riew, C.K. Tetrahedron Letters, 1966, 1269.

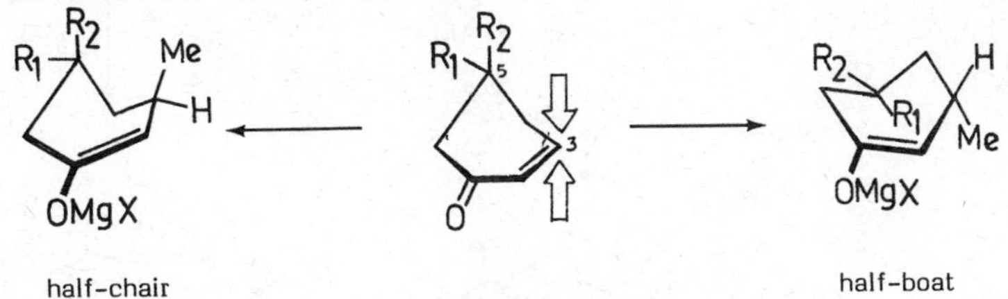

trans .. .. $R_1 = ArCH_2$, $R_2 = H$ .. .. cis

cis .. .. $R_1 = H$, $R_2 = ArCH_2$ .. .. trans

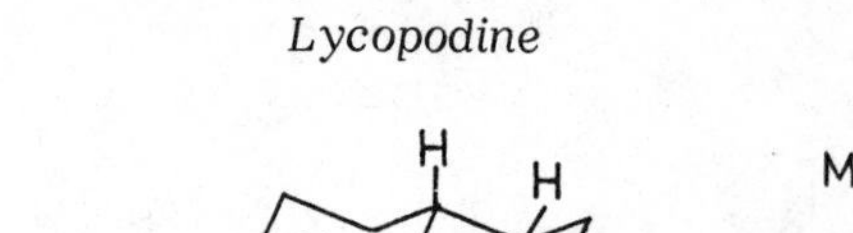

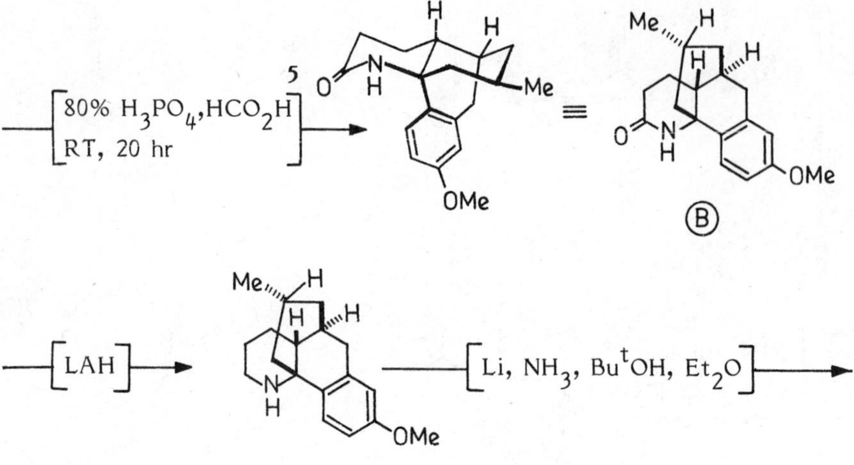

---

5. Protonation of the enamide double bond gives an acylimonium cation (i) which is involved in the crucial cyclization step. Generation of the desired stereochemistry at C-12 is due to the fact that of the two possible imonium ions (ii) and (iii), which could lead to the tetracyclic system, (iii) is not favoured because it involves ring B in a strained, boat-like configuration.

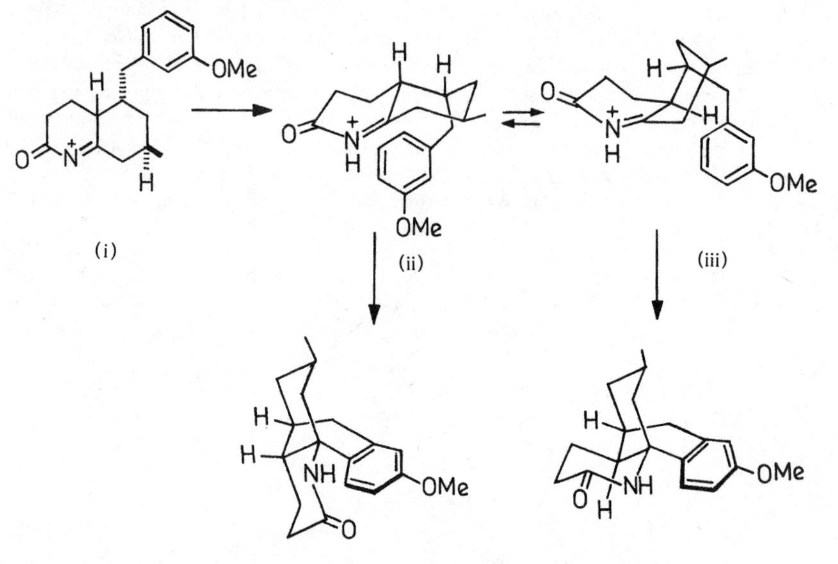

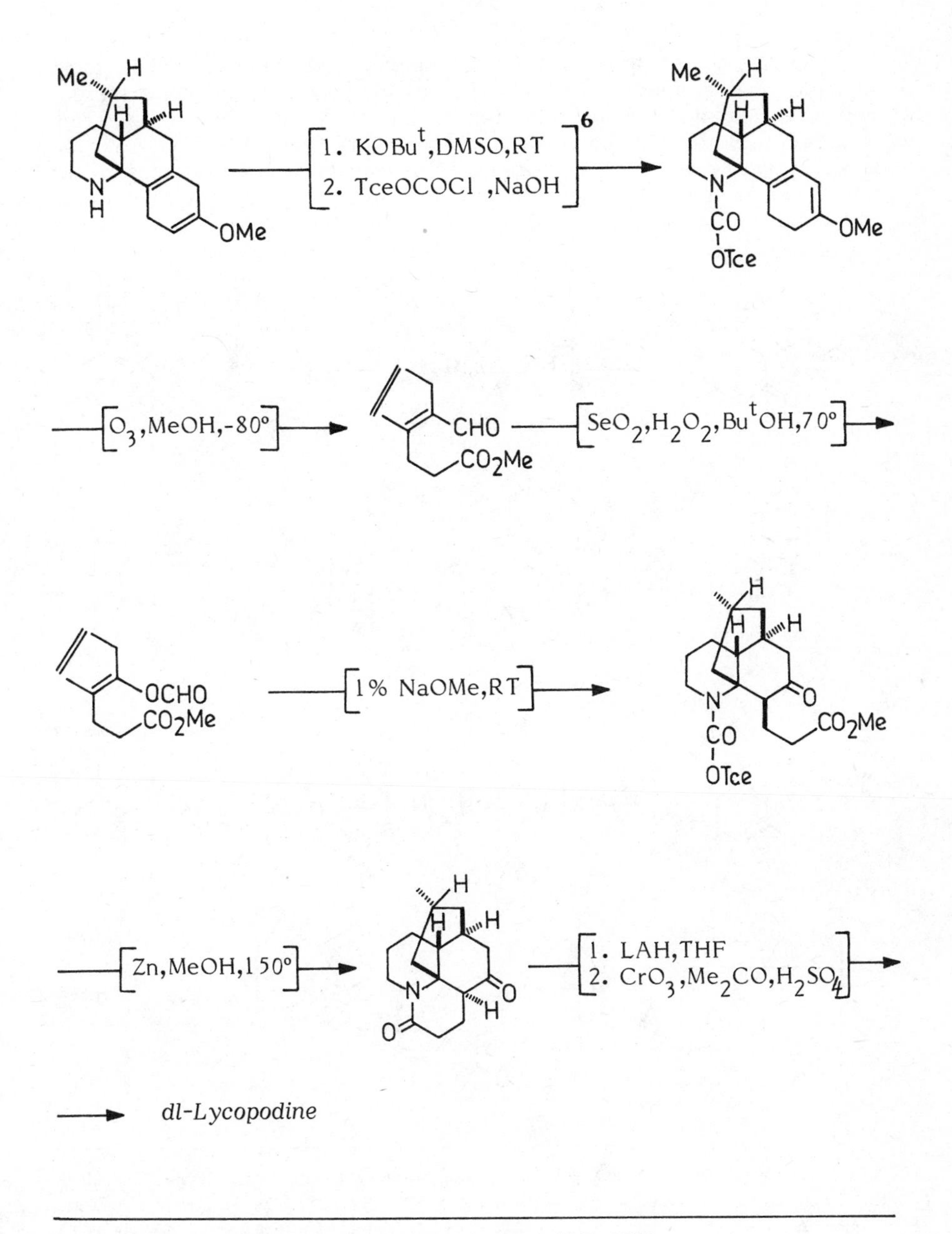

---

6. cf.Windholz,T.B.; Johnston,D.B.R. Tetrahedron Lett., 1967, 2555.

The synthesis of tetracyclic Lycopodium alkaloids by Heathcock, Kleinman and Binkley (7a) at **Berkeley** employs an intramolecular stereoselective Mannich cyclisation for construction of the A-B-C ring framework with the correct stereochemistry. Two versions of the **Berkeley** synthesis were subsequently developed, involving closure of ring-D bonds onto (a) the nitrogen, and (b) the C5-carbonyl by intramolecular alkylation (7b-d).

(a) The First Berkeley Synthesis:

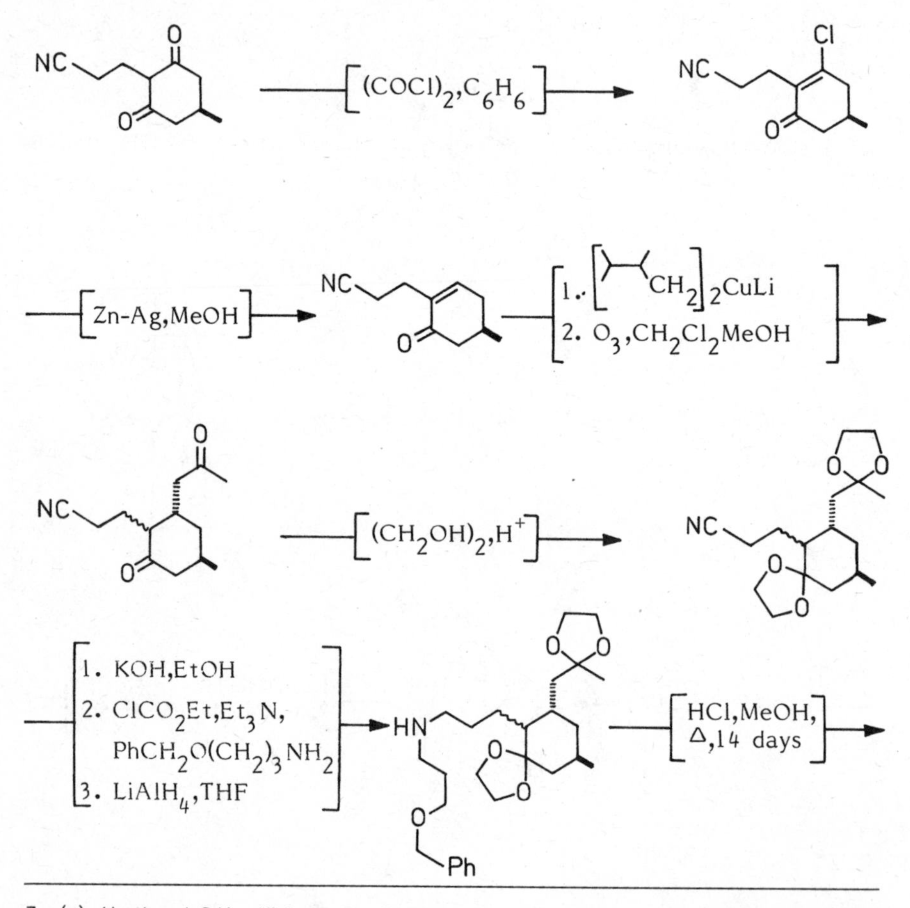

7. (a) Heathcock,C.H.; Kleinman,E.; Binkley,E.S. J. Am. Chem. Soc., 1978, 100, 8036; (b) Kleinman,E.; Heathcock,C.H. Tetrahedron Lett., 1979, 4125; (c) Heathcock,C.H.; Kleinman,E.F. J. Am. Chem. Soc., 1981, 103, 222; (d) Heathcock,C.H.; Kleinman,E.F.; Binkley,E.S. J. Am. Chem. Soc., 1982, 104, 1054.

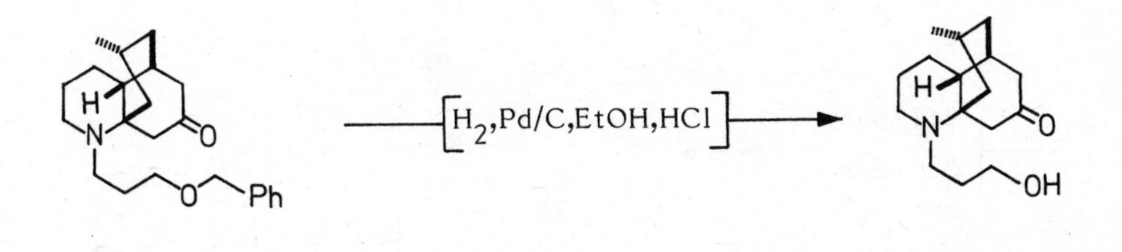

$Bu^tOK, Ph_2C=O, C_6H_6$

H

N

O

$H_2, PtO_2, EtOH$

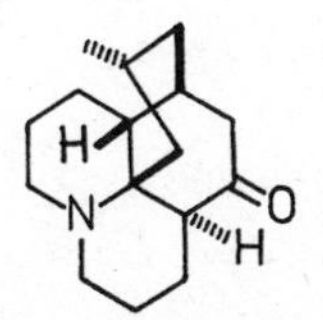

dl-Lycopodine

(b) The Second Berkeley Synthesis

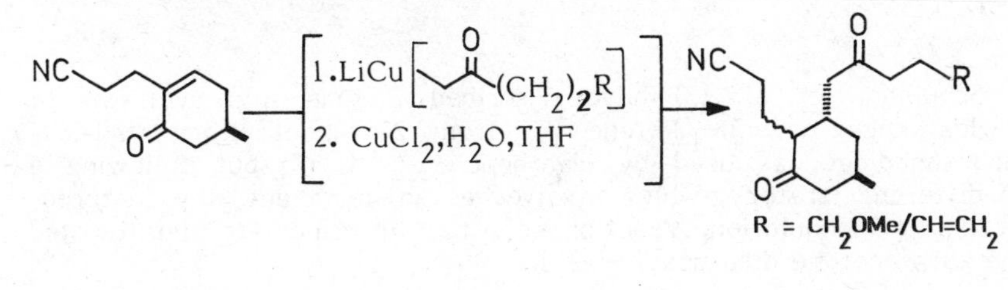

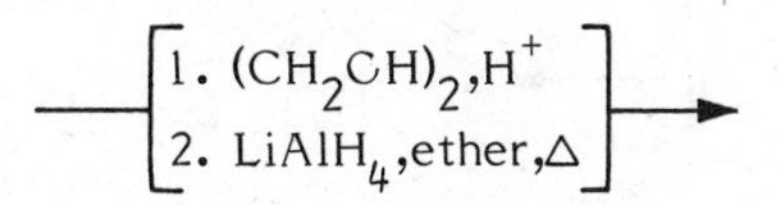

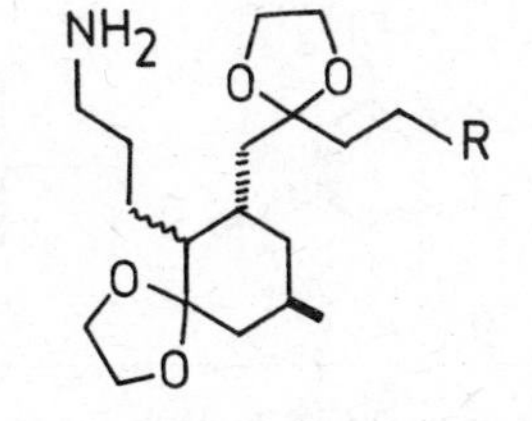

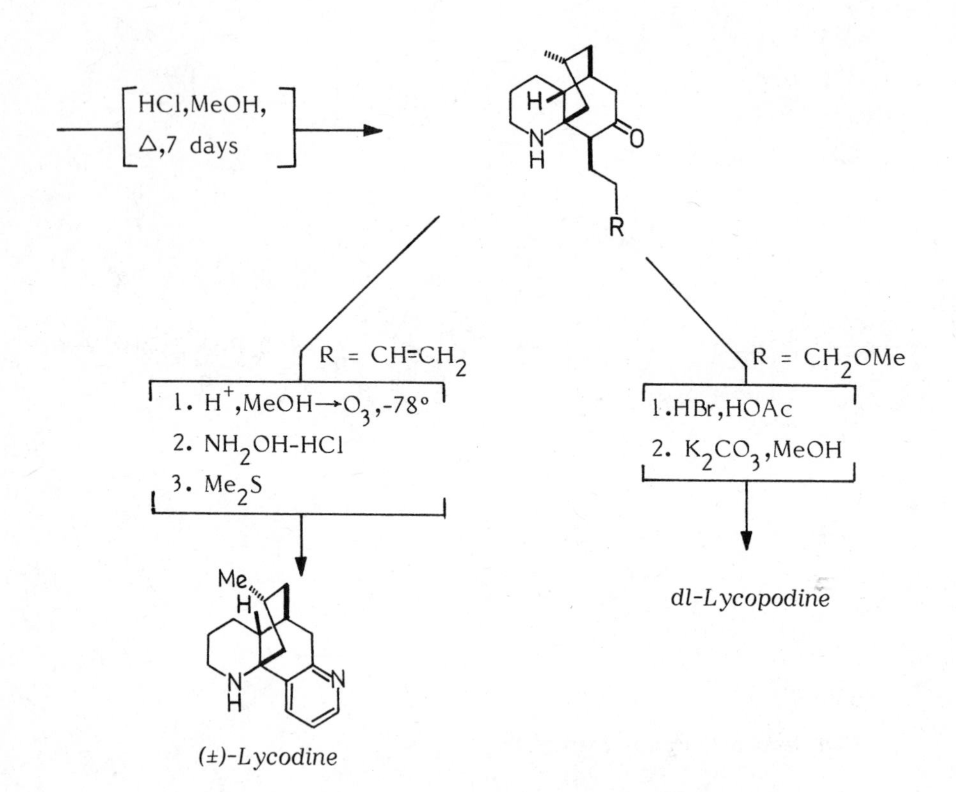

(±)-Lycodine

Schumann et al. (8) have described a four step synthesis of the Heathcock tricyclic ketone (C) from 2-cyanoethyl-5-methyl-1,3-cyclohexanedione, as used by Heathcock et al. (7) but following a very different strategy, which involves as an important step a stereoselective 1,3-annulation reaction of the intermediate unsaturated imine with acetone dicarboxylic acid.

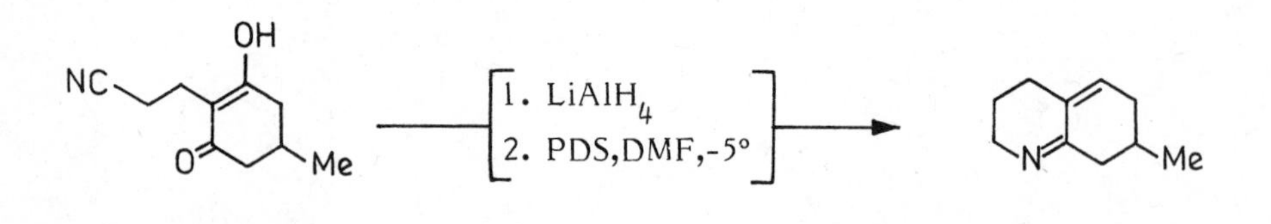

8. Schumann,D.; Muller,Hans-Jurgen; Naumann,A. Liebig Ann. Chem., 1982, 1700.

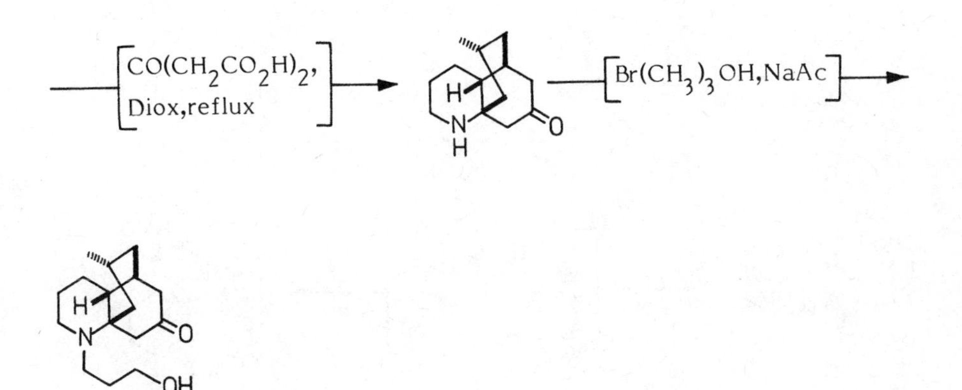

Wenkert, following his general approach to alkaloid synthesis based on partial hydrogenation of N-alkyl salts of 5-acylpyridines and acid-catalysed cyclisation of the resultant 1-alkyl-3-acyl-2-piperidines (9), has developed a new approach to the synthesis of hydrojulolidine based alkaloids including lycopodine (10); the key julolidine ring skeleton was assembled in a few steps from dimethylquinolinate. (11,12).

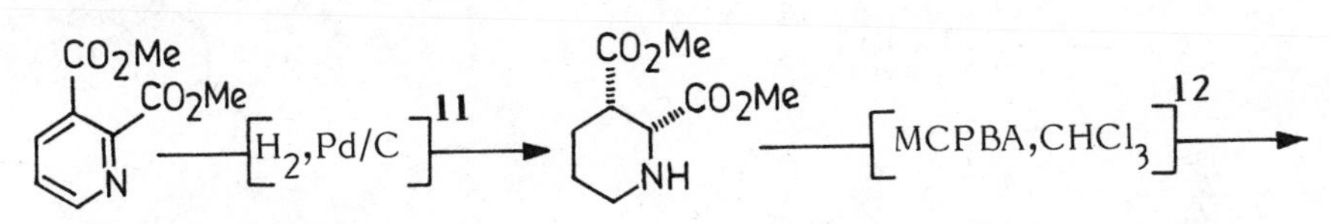

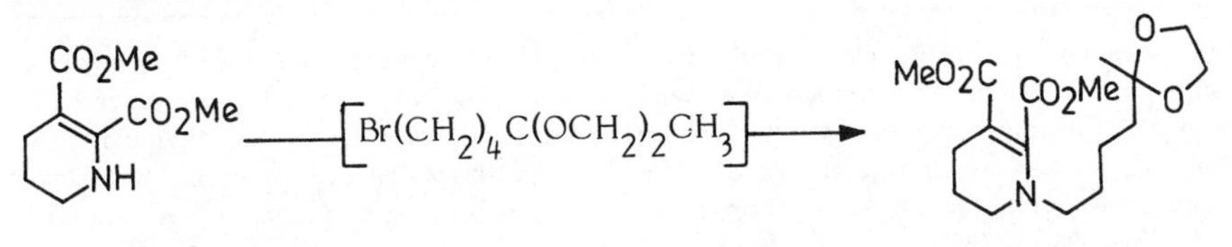

9. Wenkert,E. Accounts Chem. Res., 1968, 1, 78.
10. Wenkert,E.; Broka,C.A. J. Chem. Soc. Chem. Commun., 1984, 714.
11. Wenkert,E.; Reynolds,G.D. Aust. J. Chem., 1969, 22, 1325.
12. Wenkert,E.; Chauncy,B.; Wentland,S.H. Synth. Comm., 1973, 3, 73.

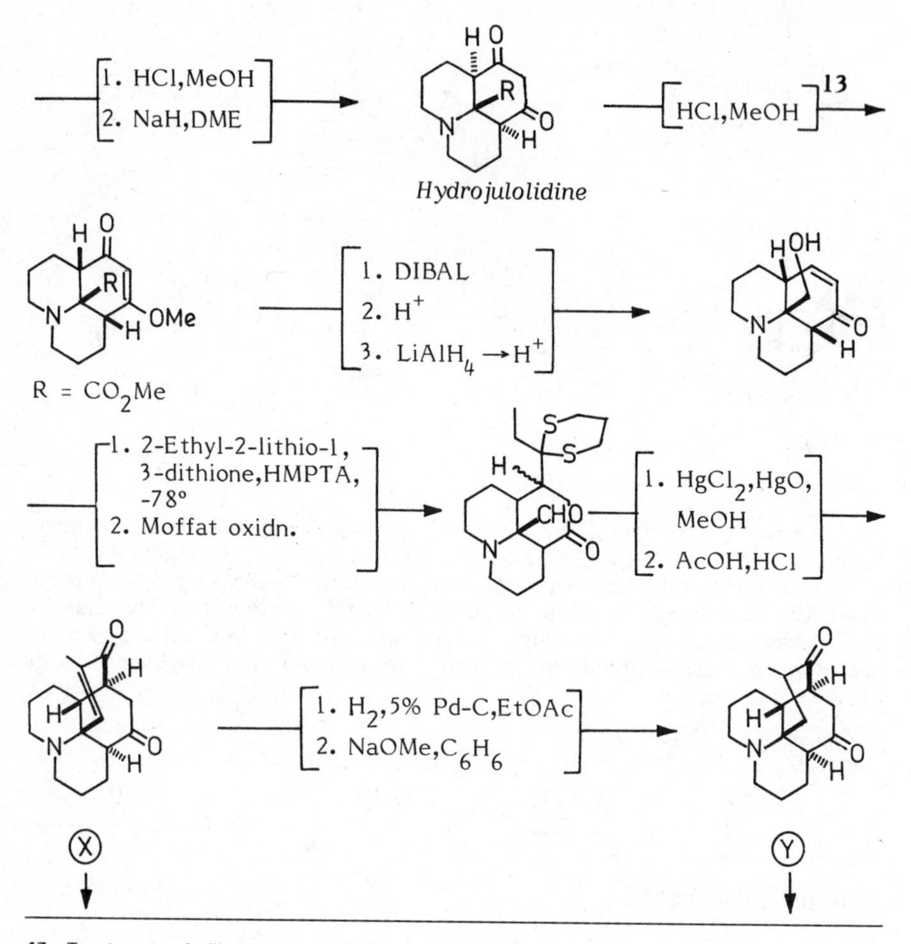

13. Treatment of diketo ester with CH_2N_2 yielded the trans enol ether (A) while exposure to methanolic HCl produced a mixture of isomers (A), (B) & (C) in a ratio of a 1:13:5, which were separated by column chromatography; this was the thermodynamic product mixture since treatment of isomer (E) with MeOH-HCl produced a mixture with same isomer ratio; the structures were assigned on the basis of ^{13}C-nmr spectroscopy. Wenkert,E. et al., J. Am. Chem. Soc., 1973, 95, 8427.

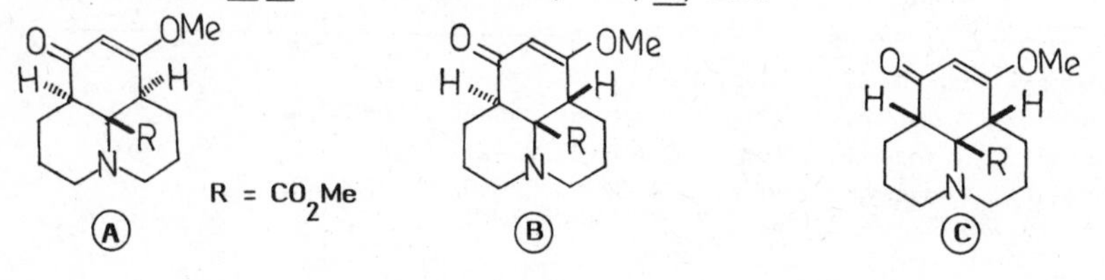

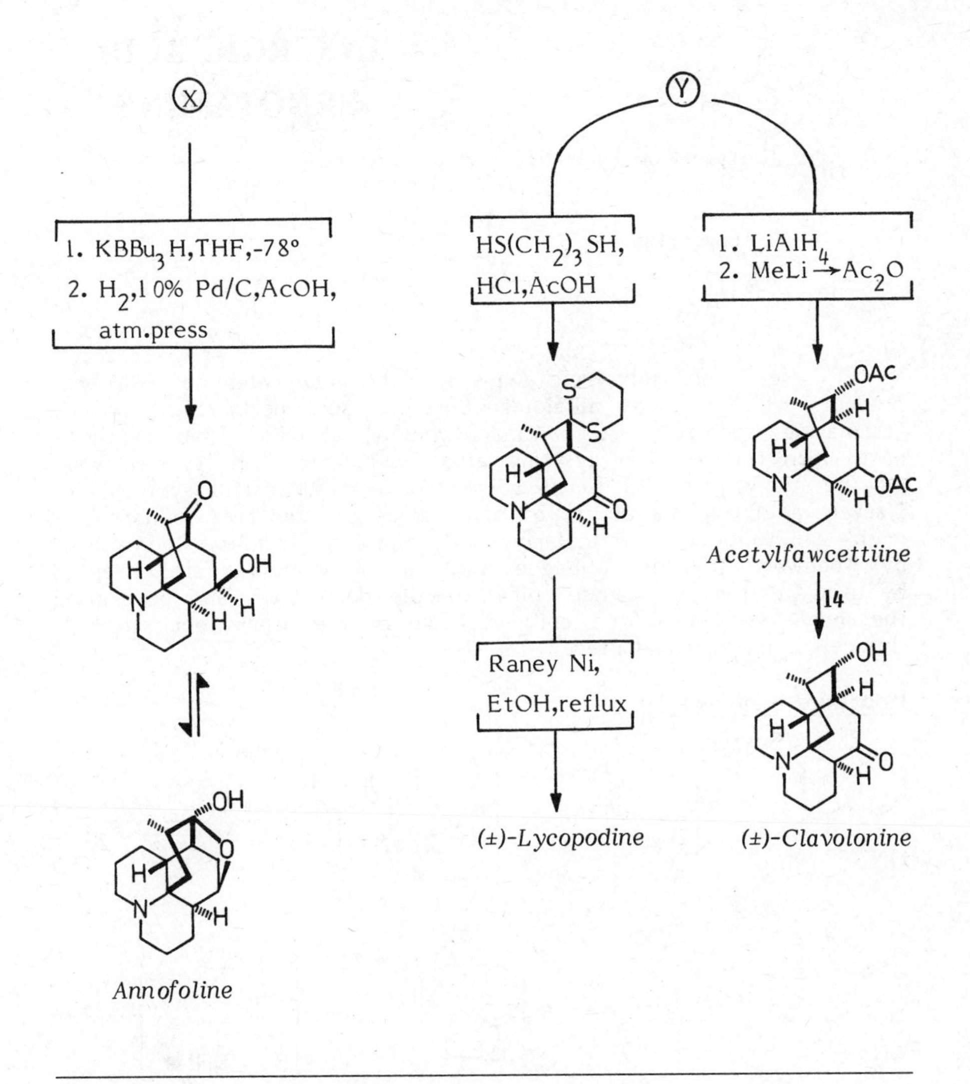

14. Nakashima,T.T.; Singer,P.P.; Brown,L.M.; Ayer,W.A. Can. J. Chem., 1975, 53, 1936.

LYSERGIC ACID
ERGOTAMINE

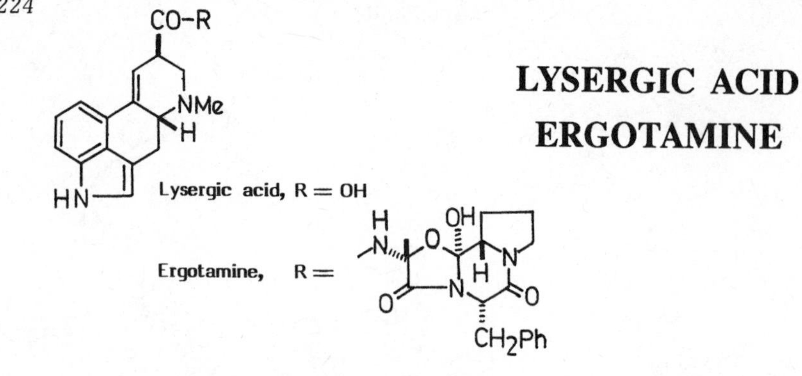

Lysergic and isolysergic acids are the main alkaline hydrolysis products of the ergot alkaloids. Lysergic acid derivatives show a fascinating spectrum of pharmacodynamic actions. Five synthesis have been reported for lysergic acid, each unique in its own way. Of the many problems to be solved for any successful synthesis of lysergic acid (1), the ability of benzindoles to isomerize to naphthalenoid structures was particularly challenging. In the classical synthesis by Woodward and his colleagues (2), this problem was circumvented by using at the very outset dihydroindole derivatives and generating the indole system at the end; in three of the other four synthesis this detour has been adopted.

Woodward Synthesis (2)

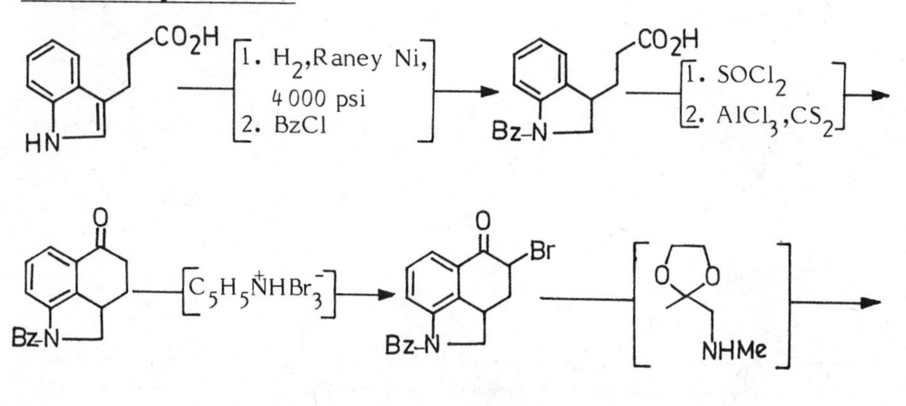

---

1. Review of earlier work: Stoll,A.; Hofmann,A. in "The Alkaloids", Vol.VIII, ed. R.H.F. Manske, Academic Press, New York, 1965, p.725.
2. Kornfeld,E.C.; Fornefeld,E.J.; Kline,G.B.; Mann,M.J.; Morrison,D.E.; Jones,R.G.; Woodward, R.B. J. Am. Chem. Soc., 1956, 78, 3087.

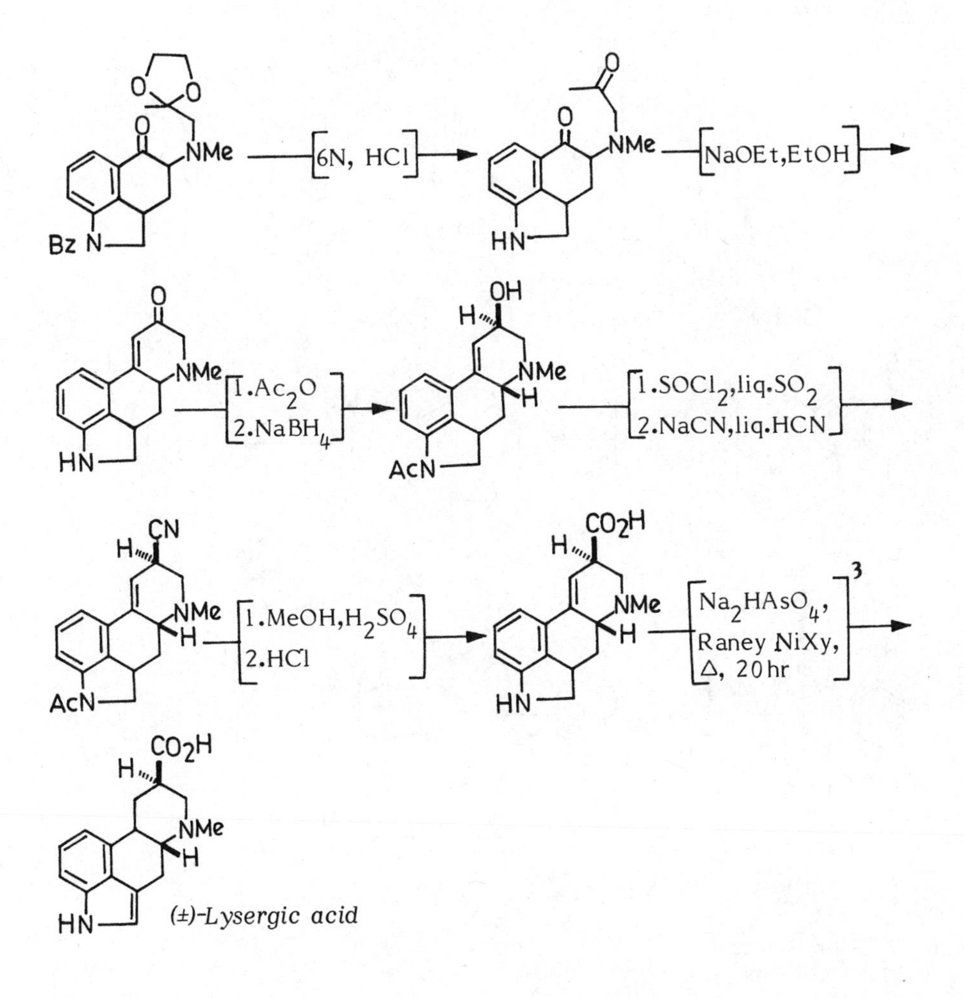

(±)-*Lysergic acid*

---

3. Recently it has been found that active MnO_2 is a more suitable reagent for dehydrogenation of indolines to indoles: Jansen,A.B.A.; Johnson,J.M.; Surtees,J.R. J. Chem. Soc., 1964, 5573.

Julia Synthesis (4)

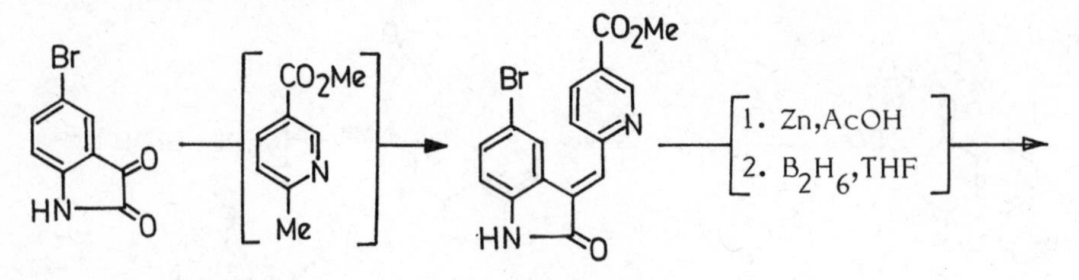

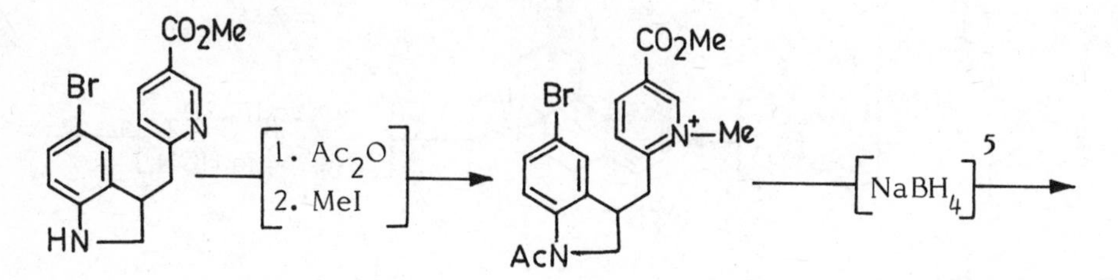

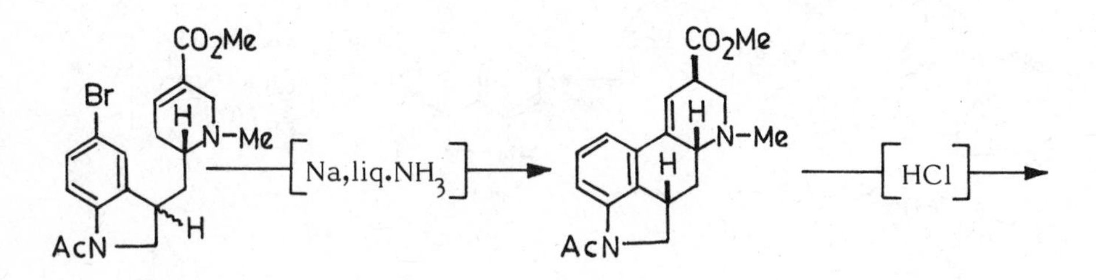

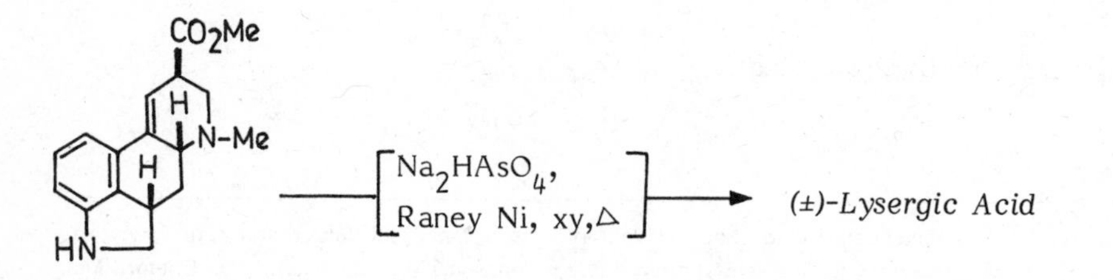

---

4. Julia,M.; LeGoffic,F.; Igolen,J.; Baillarge,M. Tetrahedron Lett., 1969, 20, 1569.

5. A mixture of both the isomers was formed from which the required isomer could be separated by chromatography after cyclisation.

Ramage Synthesis

The synthesis by Ramage and his associates (6) described below uses an intra-molecular Michael addition to construct the lysergic acid skeleton, which supports Woodwards hypothesis (2) that 2,3-dehydro derivatives of Ⓐ would be the intermediate in the epimerisation and racemisation of lysergic and isolysergic acids.

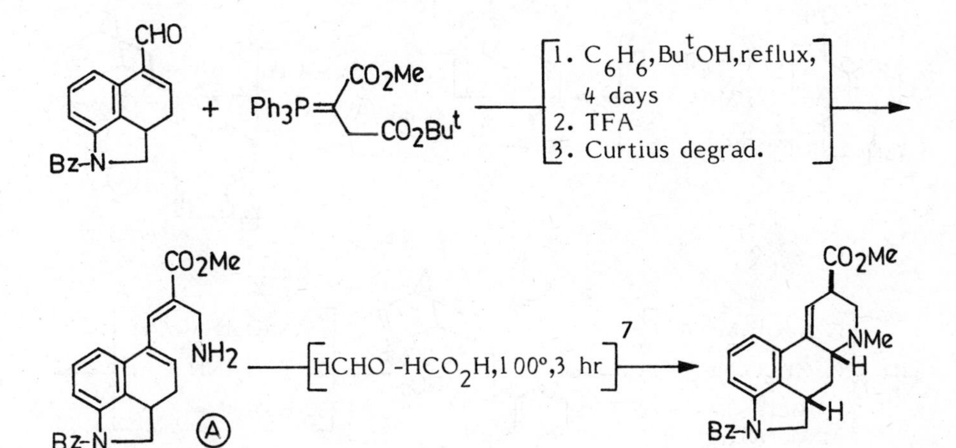

Oppolzer Synthesis (8)

The crucial step of constructing ring Ⓒ-Ⓓ was carried out by an intramolecular imino-Diels-Alder reaction at a stationary low concentration keeping the indole nucleus intact through out the reaction sequence. The labile diene unit was kept in a protected bicyclo structure and generated insitu during the thermal cycloaddition reaction.

---

6. Armstrong,V.W.; Coulton,S.; Ramage,R. Tetrahedron Lett., 1976, 47, 4311. Ramage,R.; Armstrong,V.W.; Coulton,S. Tetrahedron, 1981, 32, Suppl., 157.

7. Clark-Eschweiler reaction did not proceed to the tertiary amine but instead the secondary amine cyclised to give the product with lysergic acid structure and stereochemistry along with isolysergic acid and 10-ene products (9:3:2); these could be separated by fractional crystallisation and tlc. Methanolysis of lysergic acid or isolysergic acid structures or the mixture gave the same mixture of debenzoylated compounds, an epimeric mixture of lysergic and isolysergic acid series formed in a ratio of 3:1.

8. Oppolzer,W.; Francotte,E.; Battig,K. Helv. Chim. Acta, 1981, 64, 478.

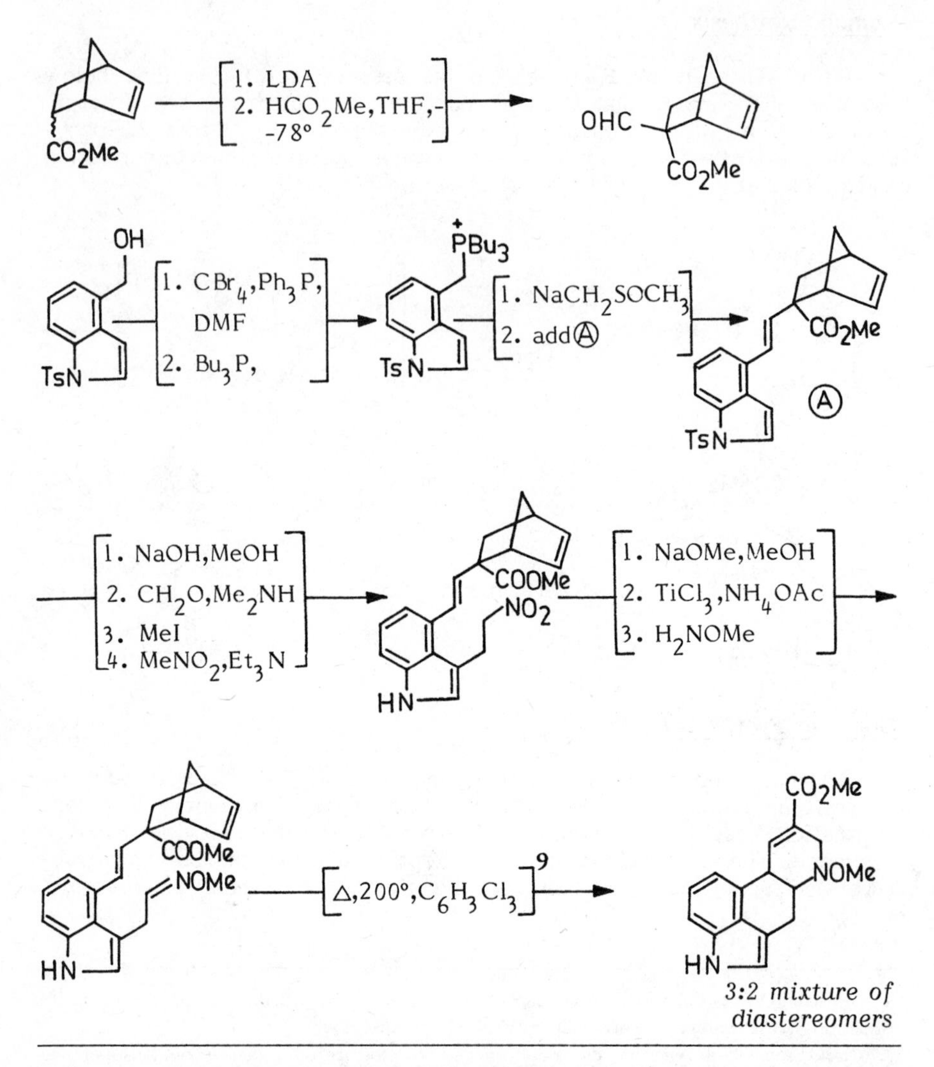

9. The reaction was forced to go intramolecularly by heating the substrate to a high temperature while keeping a stationary concentration of the transient diene as low as possible; a 1% solution of the substrate in 1,2,4-trichlorobenzene was added over 5 hr by means of a syringe drive into preheated solvent kept under argon.

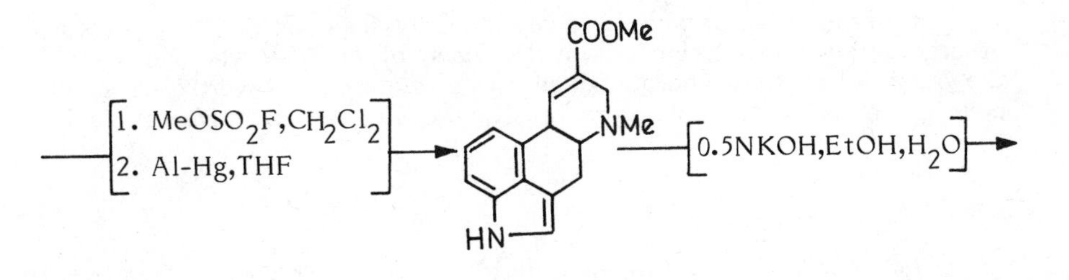

⟶ *(±)-Lysergic acid*

Rebek-Tai Synthesis (10)

This synthesis starts from tryptophan and can thus provide optically active lysergic acid directly.

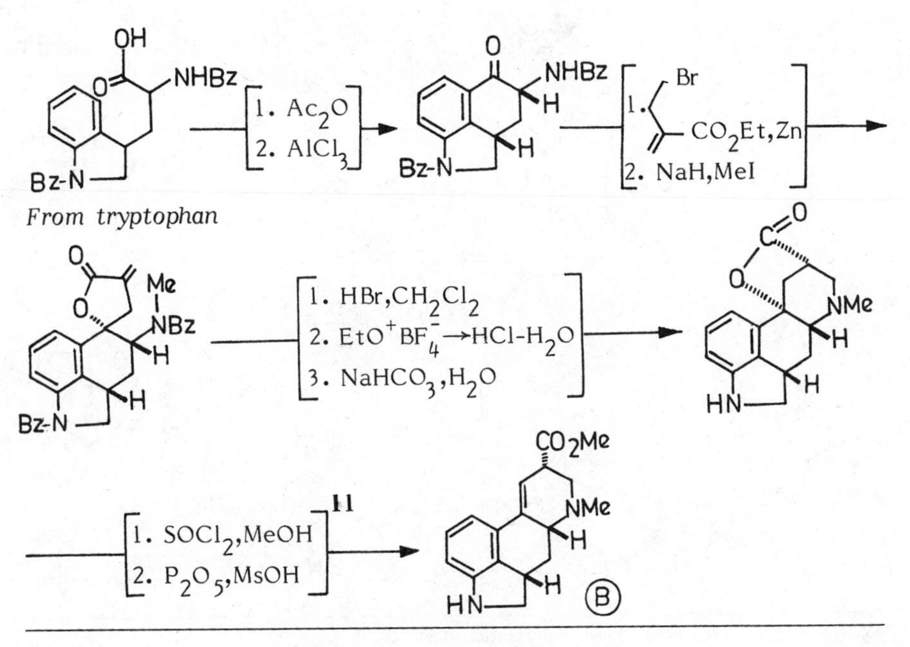

10. Rebek,Jr.J.; Tai,D.F. Tetrahedron Lett., 1983, 24, 859.

11. Substance (B) with isolysergic acid stereochemistry is known to be equilibrated to a mixture of lysergic and isolysergic acids stereochemistry by warming in MeOH (6).

Ergotamine

Ergotamine is a peptide alkaloid of lysergic acid. The synthesis of ergotamine was carried out by Hofmann et al. (12), who synthesised the peptide portion characterized by a sensitive α-hydroxy-α-amino acid grouping, and coupled the resulting cyclol Ⓒ with lysergic acid.

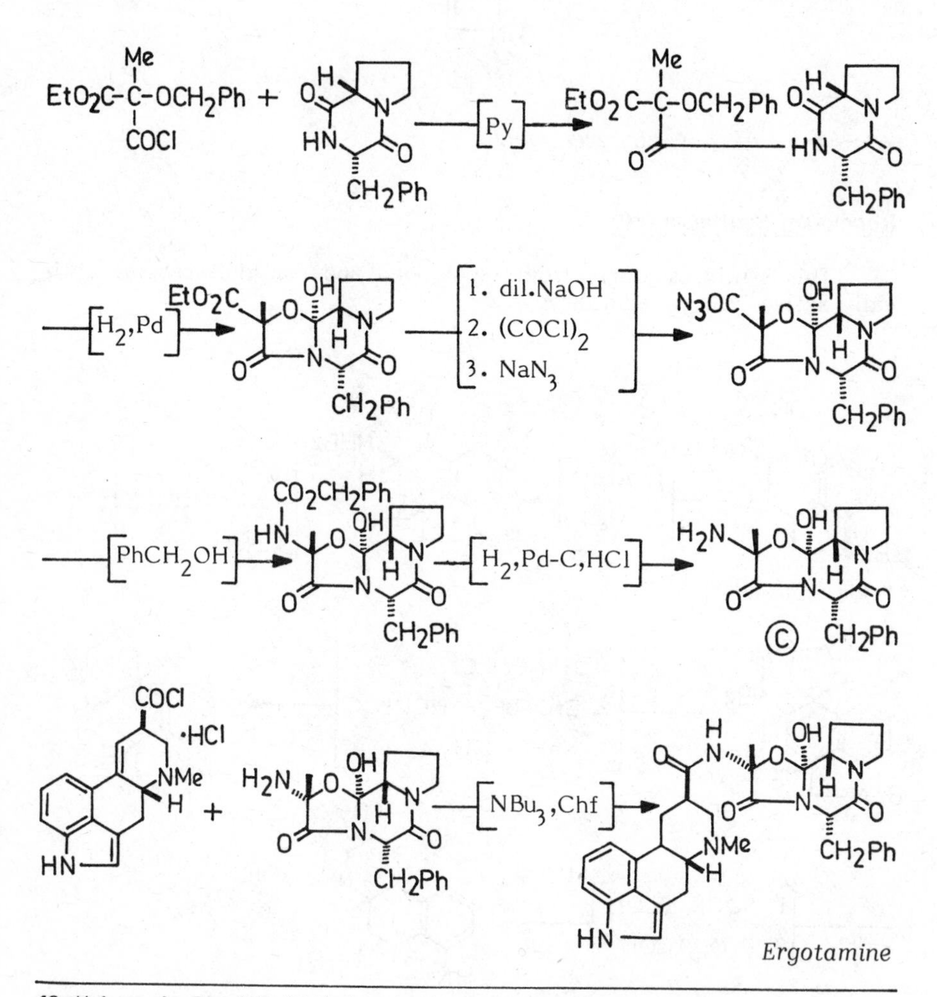

Ergotamine

12. Hofmann,A.; Frey,A.J.; Ott,H. Experientia, 1961, 17, 206.

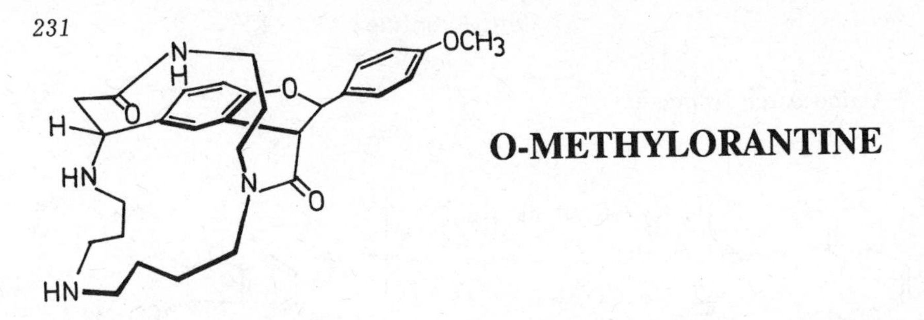

O-METHYLORANTINE

Orantine (ephedradine A), obtained from the roots of ephedra plants, belongs to the family of polyamine bicyclo-macrocyclic alkaloids. It is characterised by the presence of a dihydrobenzofuran nucleus bridging a 17 membered lactam ring containing a spermine nucleus (1). O-Methylorantine is obtained from species of Aphelandra (2). A number of these alkaloids have been synthesised by Wasserman et al. (3) by a convergent pathway by constructing two fragments separately and coupling them, making ingenious use of lactim ether of a 13 membered lactam intermediate both to link the two fragments andto construct the required 17-membered lactam by ring cleavage, which is followed by spanning of the bridge across the lactam by intramolecular lactam formation. The synthesis of O-methylorantine described below (4) is typical of the approach followed.

Lactim ether fragment (A) (3)

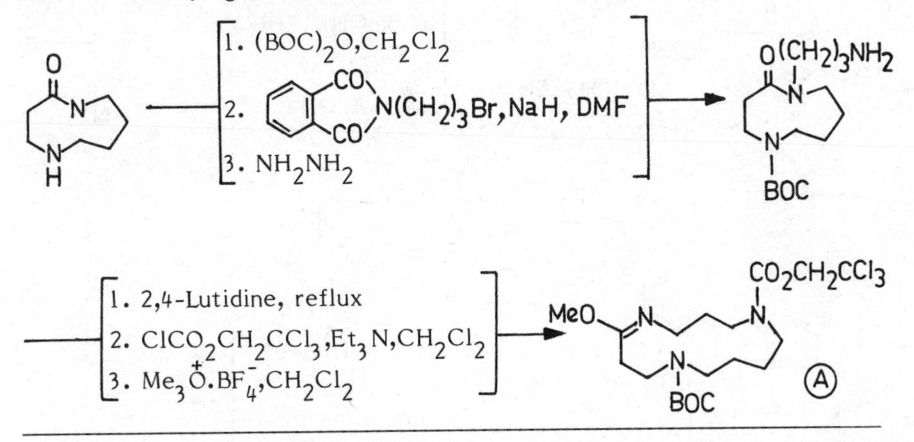

1. Hikkino,H.; Ogata,M.; Konno ,C. Heterocycles, 1982, 17, 155 and earlier references cited therein.

2. Datwyler,P.; Bosshardt,H.; Johne,S.; Hesse,M. Helv. Chim. Acta, 1979, 62, 2712; Bosshardt H.; Guggisberg,S.; John,S.; Veith,H.J.; Hesse,M.; Schmid,H. Pharm. Act. Helv., 1976, 51, 371.

3. Wasserman,H.; Robinson,R. Tetrahedron Lett., 1983, 3669; Wasserman,H.; Robinson,R.; Carter,C. J. Am. Chem. Soc., 1983, 105, 1697.

4. Wasserman,H.H.; Brunner,R.K.; Buynak,J.D.; Carter,C.G.; Oku,T.; Robinson,R.P. J. Am. Chem Soc., 1985, 107, 519.

***β*-Amino ester fragment**

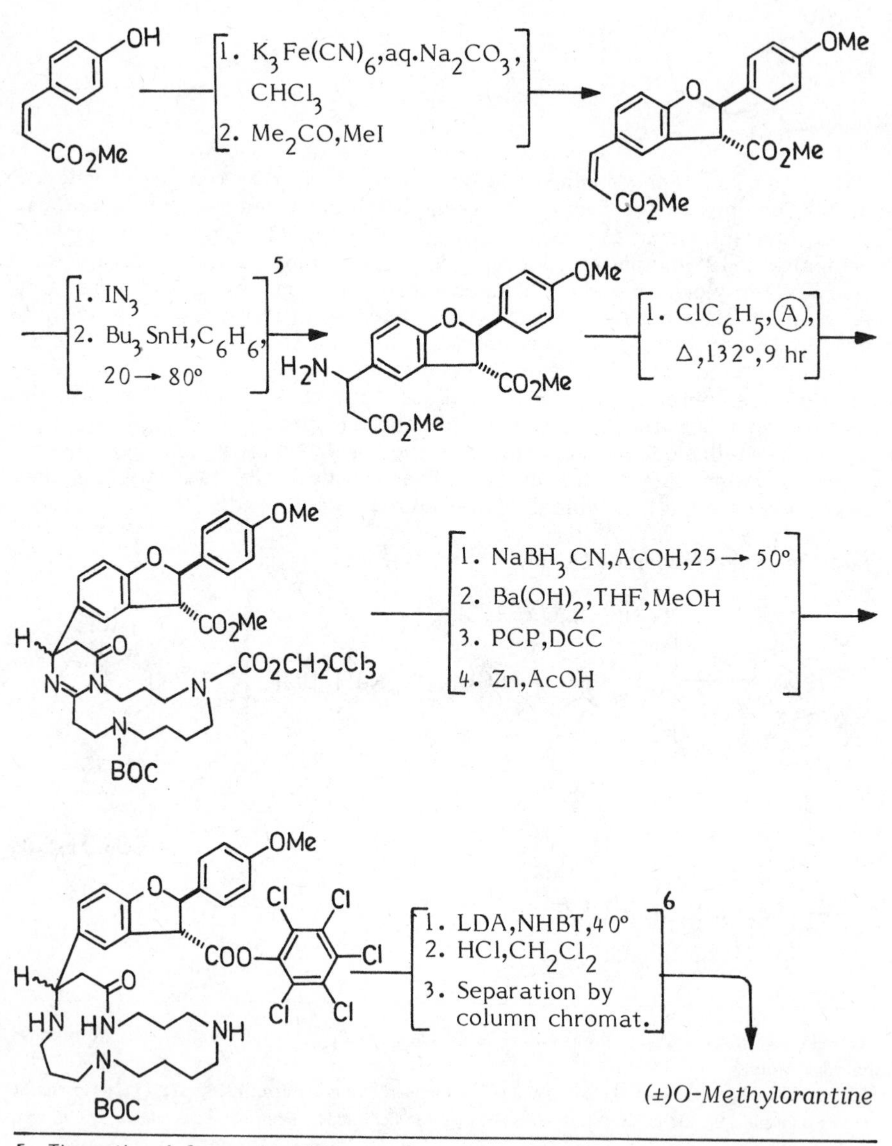

5. The ratio of β : α azidoesters was favoured 5:1, and these were separated at the amino stage by flash chromatography.

6. The cyclised product was a 1:1 mixture of the two diastereomers, which were separated by column chromatography after deblocking.

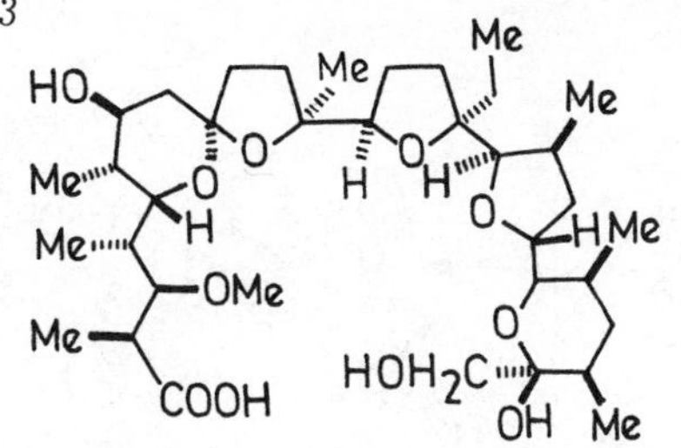

MONENSIN

Monensin, produced by a strain of Streptomyces cinnamonensis (1), is the most historic of the growing class of ionophoric polyether antibiotics, since it was the first polyether antibiotics whose structure was determined (2). Monensin presents a formidable challenge for synthesis with 17 asymmetric centers on the backbone of 26 carbon atoms. Two synthesis of monensin have been reported. Kishi and his associates reported the first total synthesis (2) and constructed the molecule from two fragments, each having the required absolute stereochemistry.

Kishi Synthesis

Left Segment (2a)

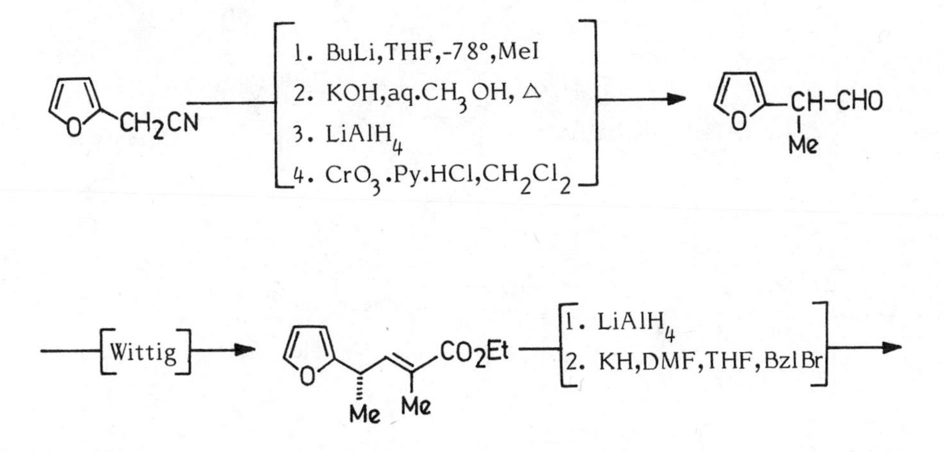

---

1. Agtarap,A.; Chamberlain,J.W.; Pinkerton,M.; Steinrauf,L. J. Am. Chem. Soc., 1967, 89, 5737; for reviews on polyether antibiotics see: Westley,J.W. Ann. Rept. Med. Chem., 1975, 10, 246; Adv. Applied Microbiology, 1977, 22, 177; Pressman,B.C. Ann. Rev. Biochem. 1976, 45, 501.

2. Schmid,G.; Fukuyama,T.; Akasaka,K.; Kishi,Y. J. Am. Chem. Soc., 1979, 101, 259 ; (b) Fukuyama,T.; Wang,C.L.J.; Kishi,Y. ibid, 262; (c) Fukuyama,T.; Akasaka,K.; Karanewsky, D.S.; Wang,C.L.J.; Schmid,G.; Kishi,Y. ibid, 263.

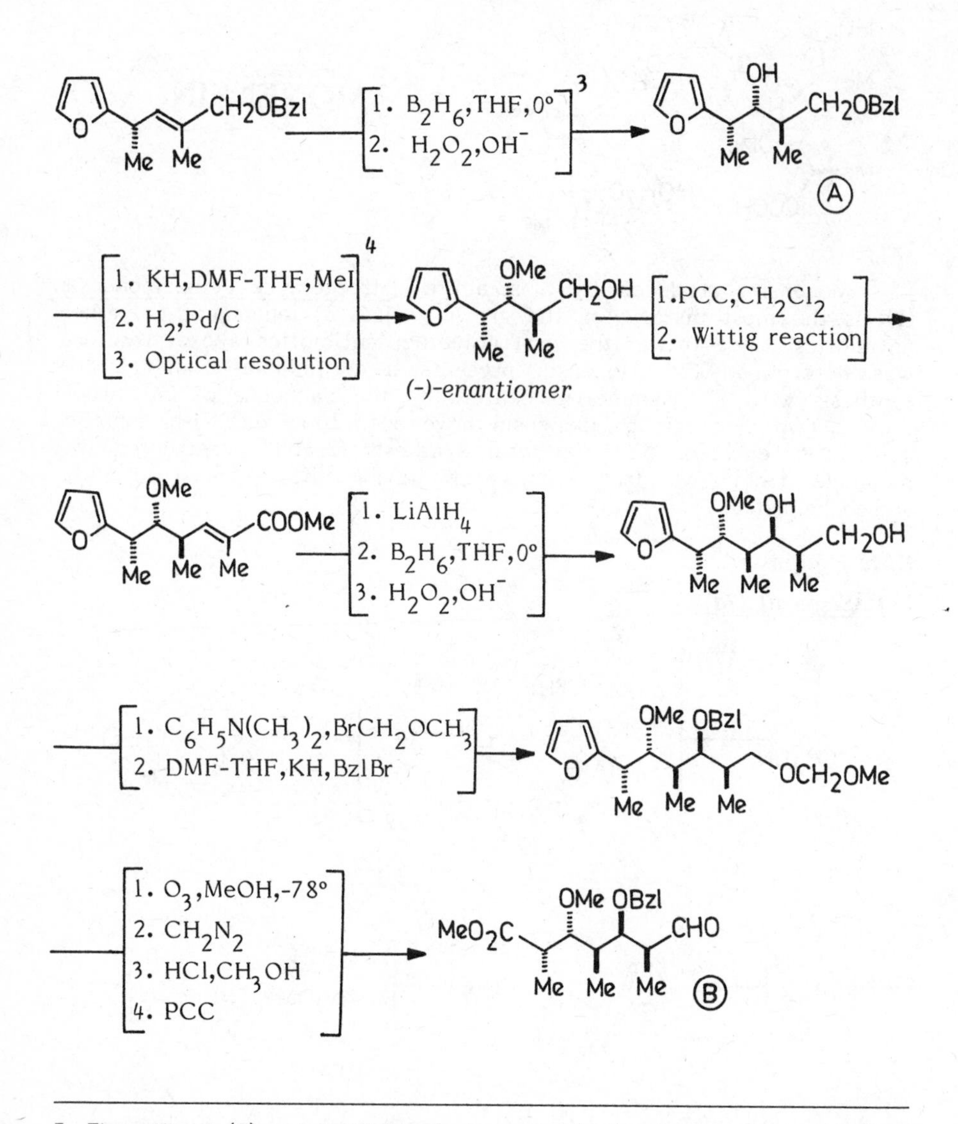

---

3. The ratio of (A) and its diastereoisomer was 8:1; the structure was assigned on the assumption that the preferred conformation of the olefin would be as shown and therefore the hydroboration would take place from the less sterically hindered α-face.

4. Optical resolution was achieved by reacting with (-)-$C_6H_5CH(CH_3)NCO$, separation of the diastereomeric urethanes by HPLC followed by $LiAlH_4$ reduction.

Right Segment (2b)

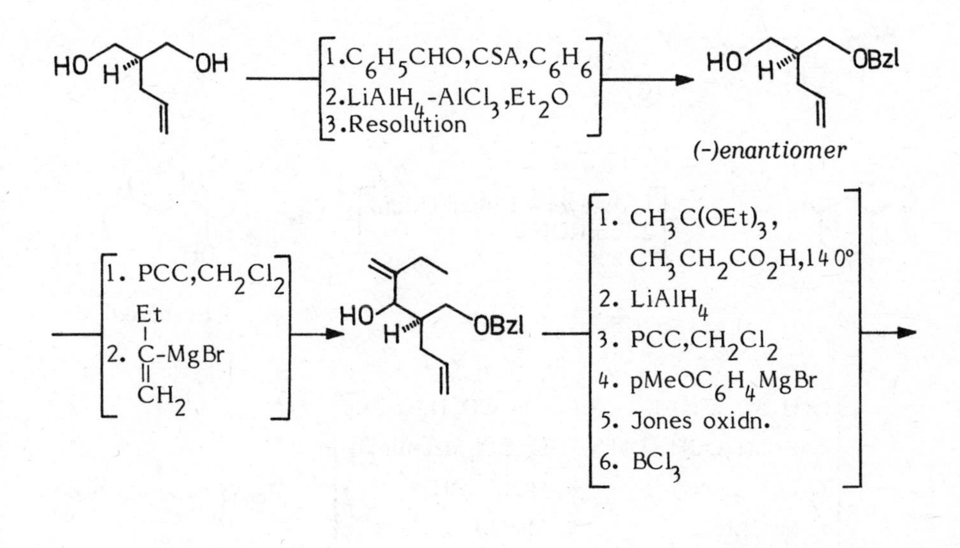

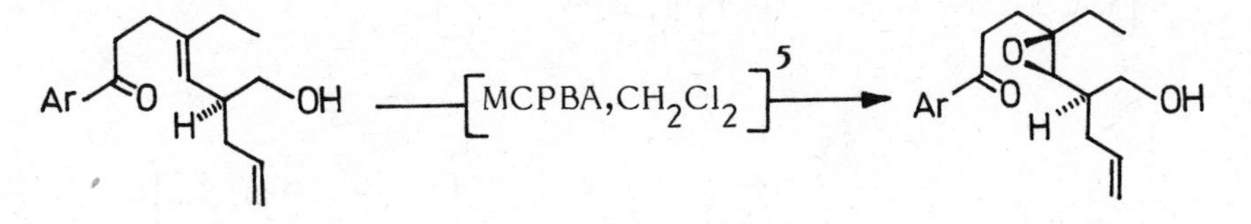

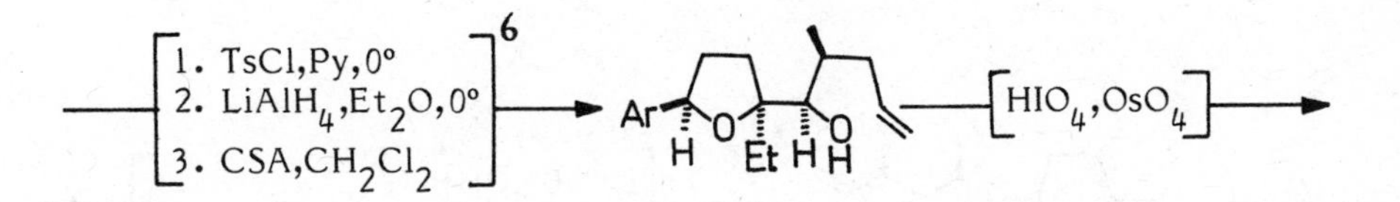

---

5. Only one product was obtained in this epoxidation and the structure was assigned on the consideration that the transition state shown would be preferred on account of lesser steric crowding with hydroxyl group complexed oxidant.

O
Ar Et
CH_2OH
H H
C_3H_5

6. The best ratio of the product shown and its diastereomer was 7:2.

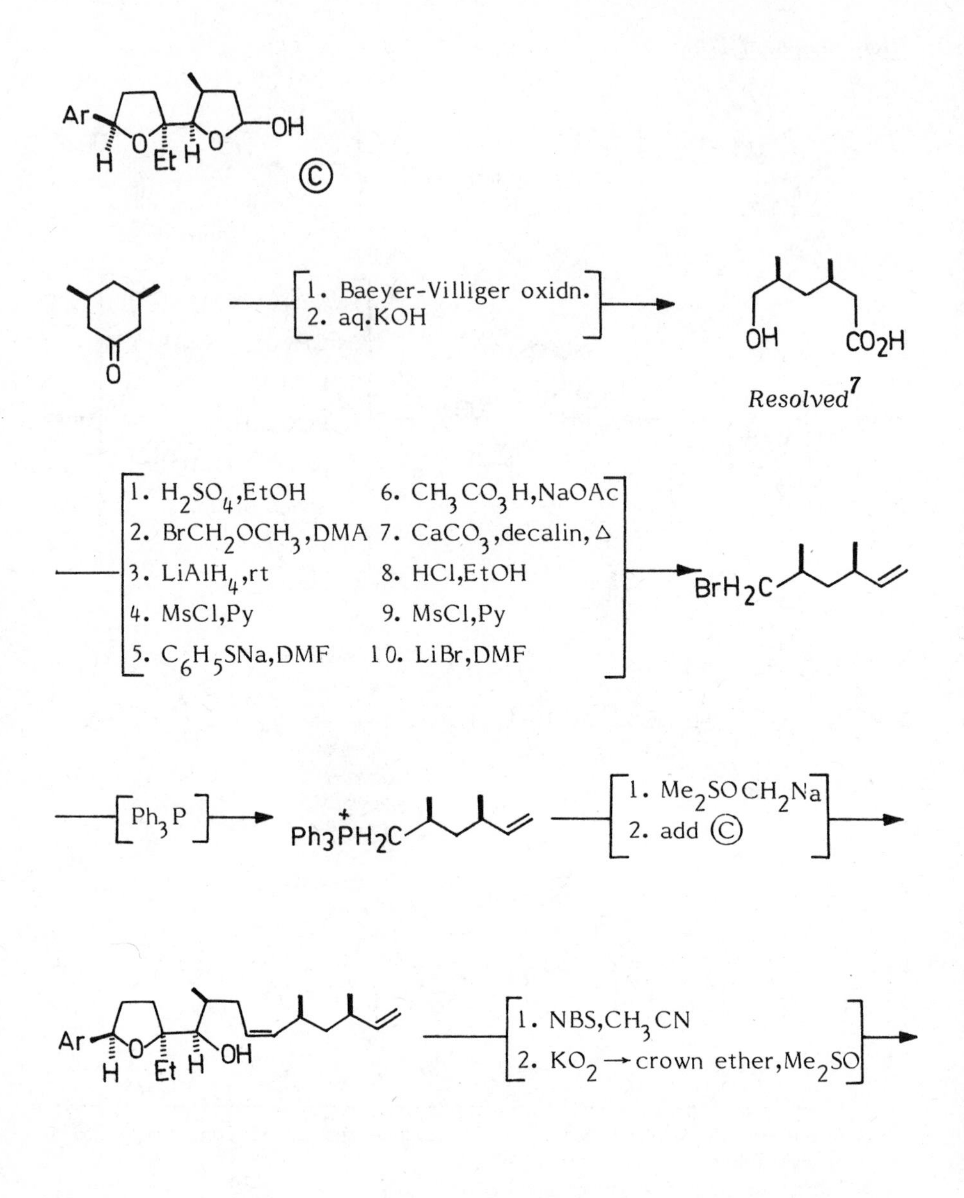

7. The optical resolution of the hydroxy acid was achieved by fractional crystallisation of the (+)- -methylbenzylamine salt, and 3R configuration assigned to the dextrorotatory isomer.

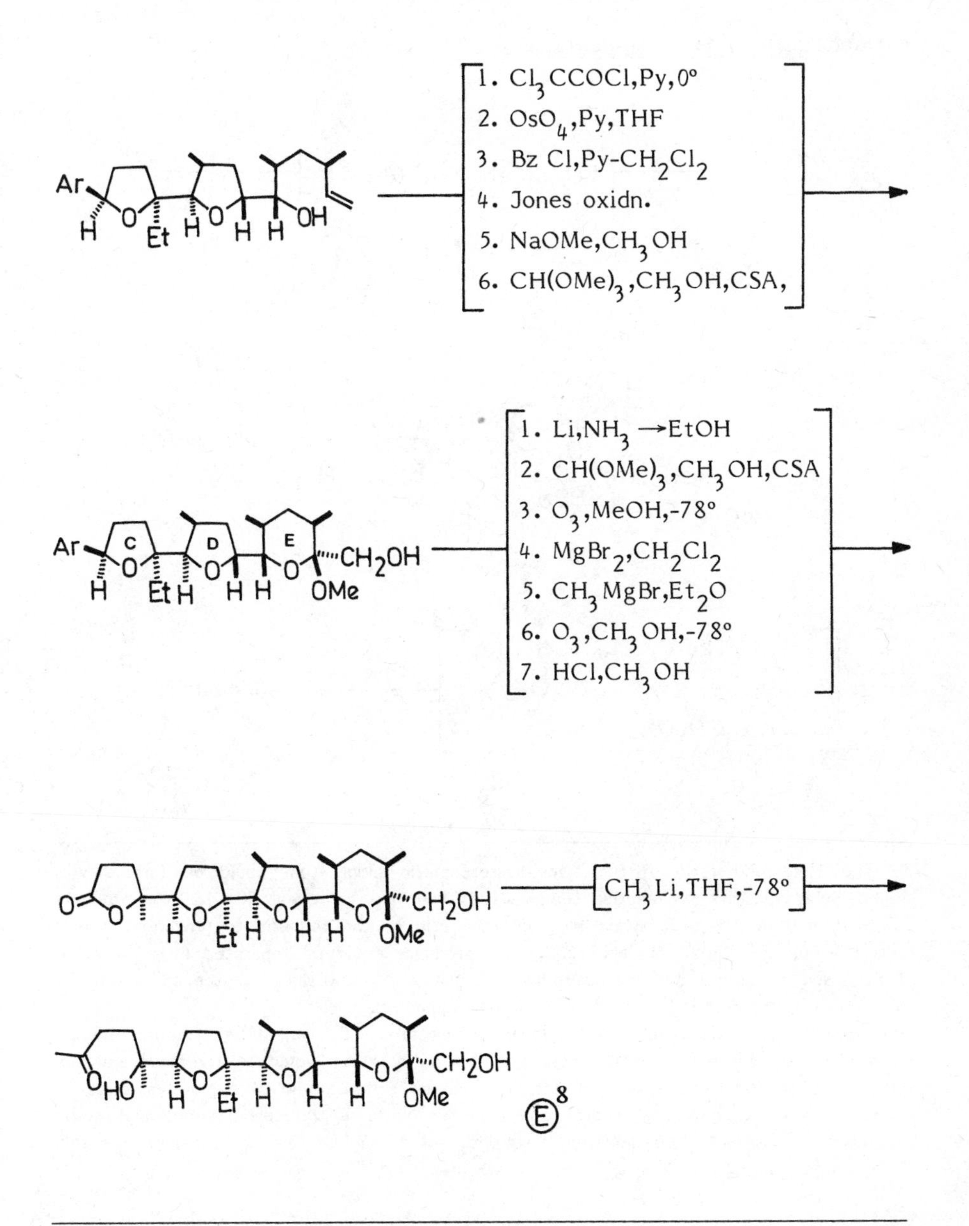

---

8. The structure of the product was confirmed by converting it to a ring E δ-lactone, which is obtained from monensin.

Condensation of two Segments (2c)

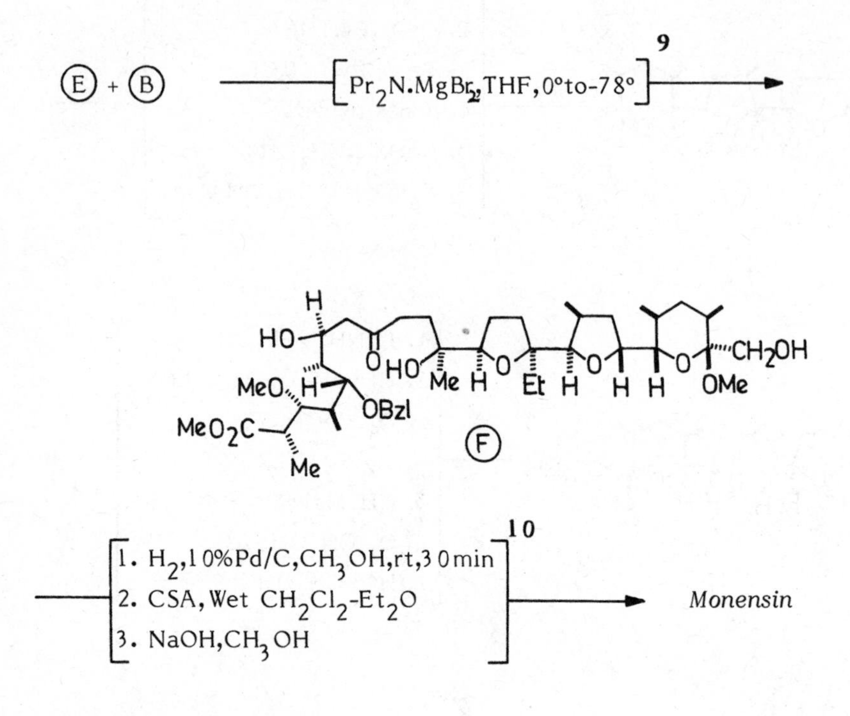

9. The yield and ratio of the two diastereomeric aldols that could be formed was found highly sensitive to reaction temperature; at 0° the yield was 70% but the products were formed in 1:1 ratio while at -78° the ratio of the required to unwanted isomer was 8:1 but the yield was only 20%. The products could be separated by preparative TLC. Based on Cram's rule the desired stereochemistry was assigned to the major product (F), which was confirmed by transformation to monensin.

10. It was anticipated that the asymmetric center at C_9 would be stereospecifically created by intramolecular ketalisation, as this configuration would be thermodynamically more stable owing to the anomeric effect. Camphor sulfonic acid treatment was required to equilibrate the spiroketal center and also to hydrolyse the tertiary methoxyl group at the C_{25} position. Pure monensin was obtained as sodium salt after preparative thin layer chromatography.

Still Synthesis

Soon after the publication of Kishi's synthesis, Still's group reported their highly convergent synthesis of monensin (11) from small fragments having the proper absolute stereochemistry (12) and the additional centers with the required stereochemistry created during coupling with control by the existing asymmetric centers. A special feature exploited for steric control is the chelation-controlled nucleophilic addition, when the stereochemistry produced may be opposite to the usual Cram's rule. The synthesis starts with (-)-malic acid and (+)- -hydroxyisobutyric acid.

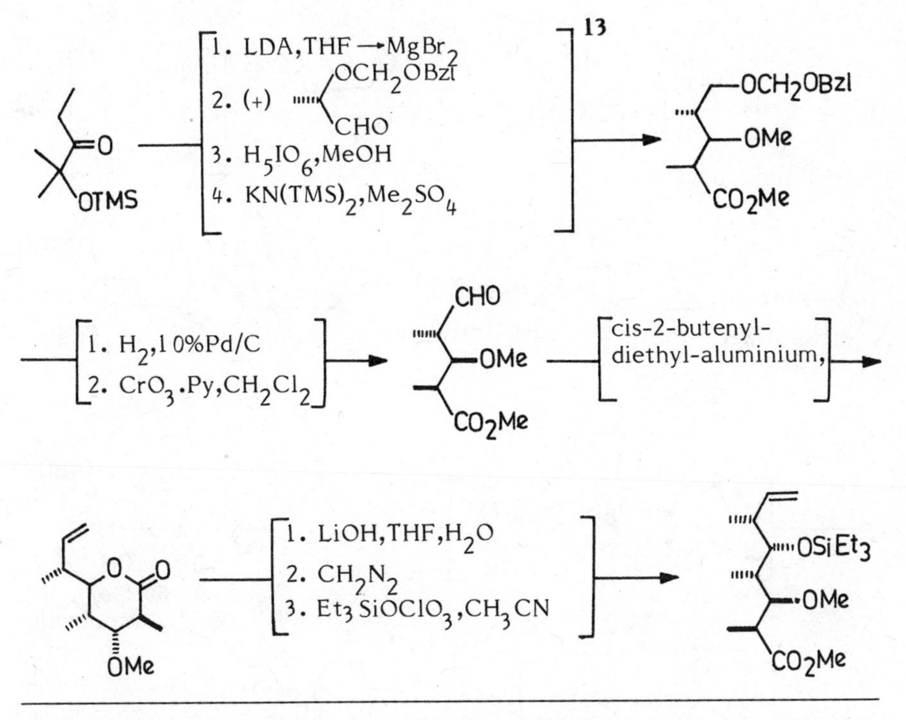

---

11. (a) Collum,D.B.; McDonald ,J.H.; Still,W.C. J. Am. Chem. Soc., 1980, 102, 2117; (b) idem, ibid, 1980, 102, 2118; (c) idem, ibid, 1980, 102, 2120.

12. The stereochemical assignments of the fragments were confirmed by comparison with the authentic products obtained by degradation of natural monensin (11a).

13. The aldol condensation gave a 5:1 mixture of diastereomeric aldols, in which the major product was that predicted by chelation-controlled (antiCram) α-induction.

14. This condensation followed the normal Cram's rule and gave 3:1 mixture of diastereomeric aldols which were separated by flash chromatography.

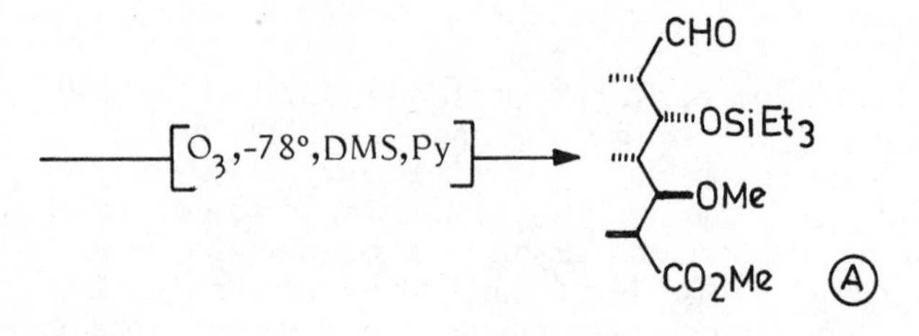

Central fragment

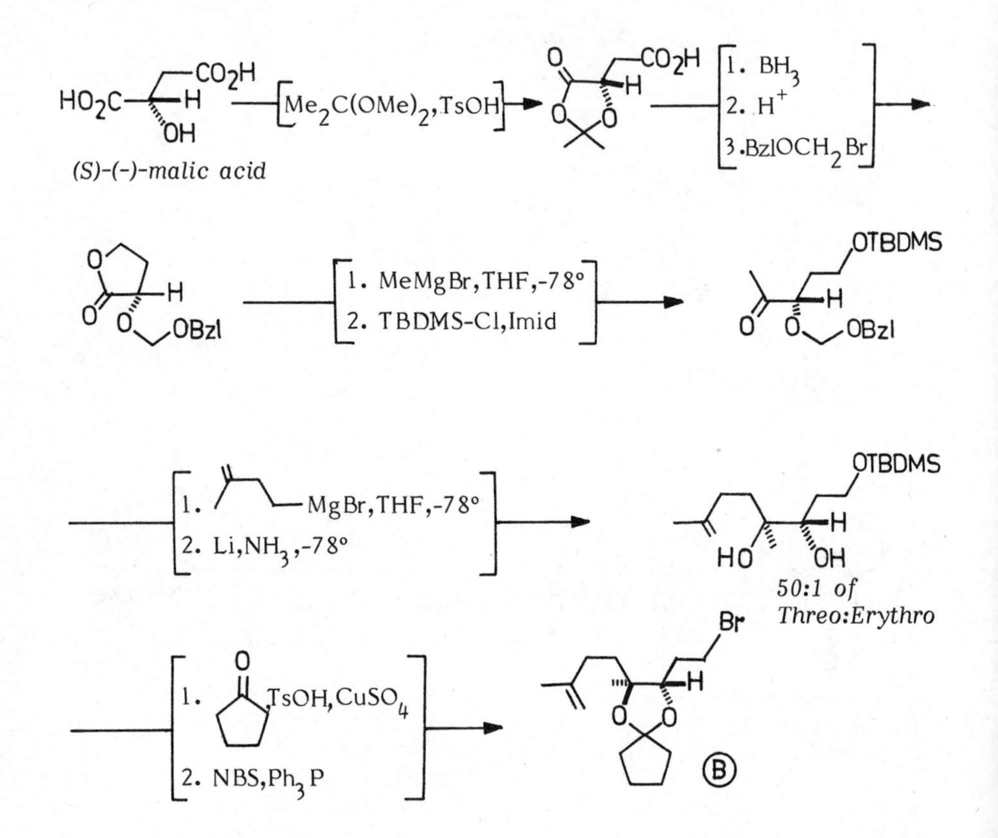

Right hand fragment

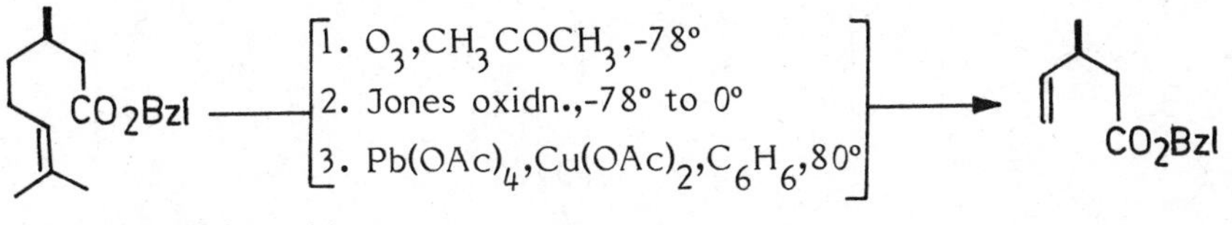

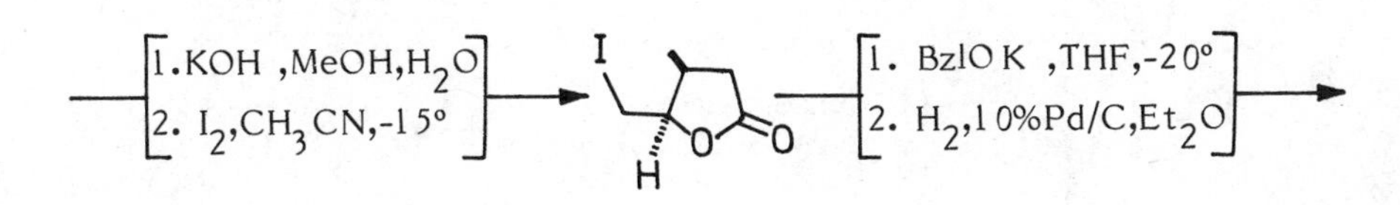

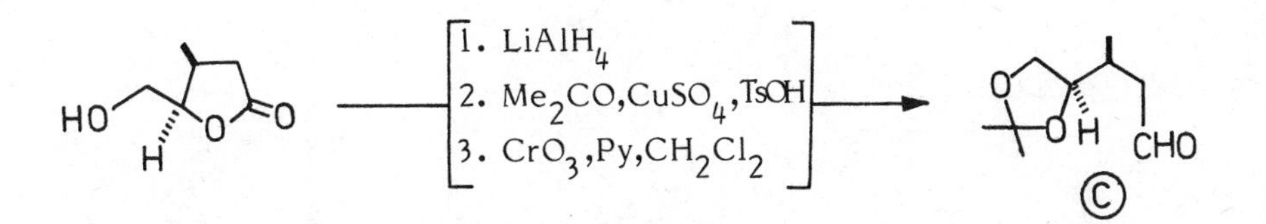

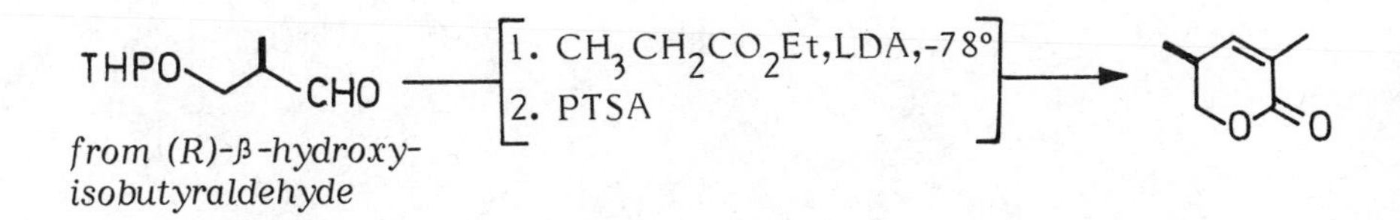

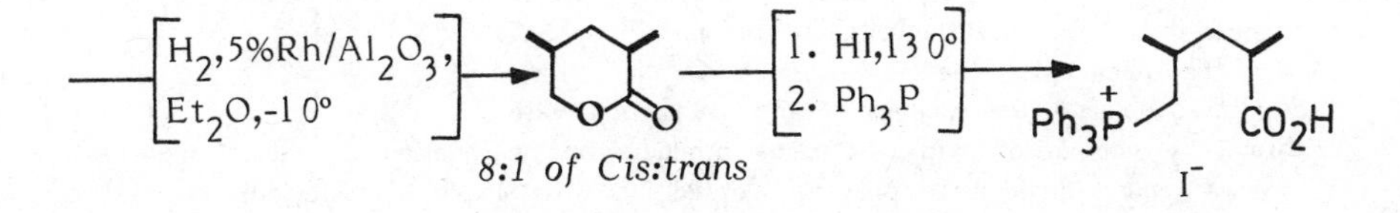

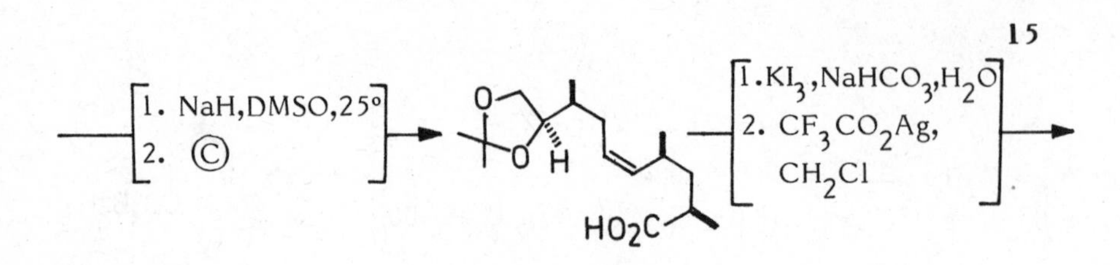

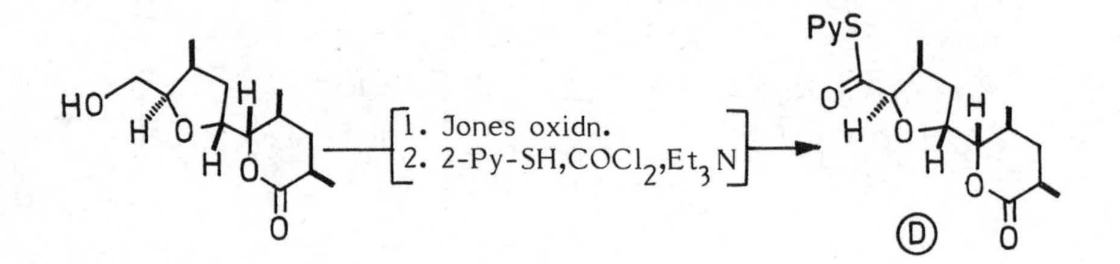

Coupling of fragments

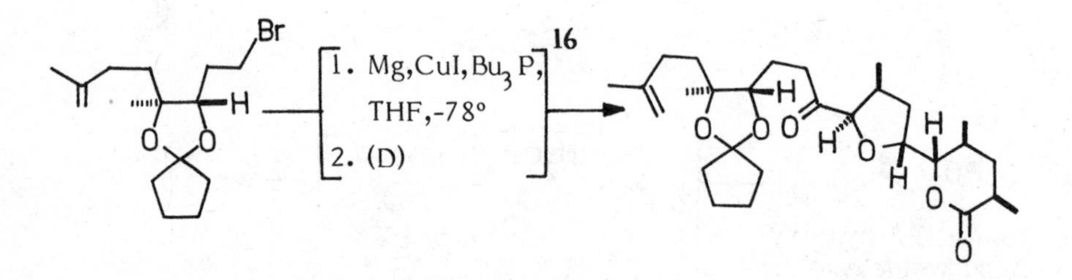

---

15. In iodolactonisation the required asymmetry at C_{20} and C_{21} was anticipated from steric considerations of lactonisation in which the cis-olefin and the adjacent asymmetric centre C_{22} would be expected to constrain the carboxylate-bearing appendage to the space below the olefine plane.

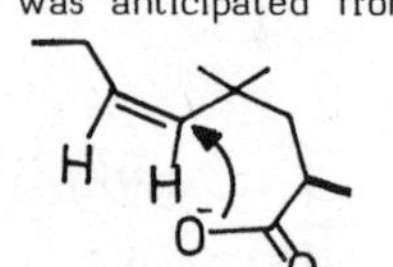

*The stereochemical assignments of the fragments were confirmed by comparison with the same products by degradation of natural monensin.

16. It was found difficult to prevent overaddition with simple Mg salt, but use of CuI with the Grignard reagent resulted in clean formation of ketolactone.

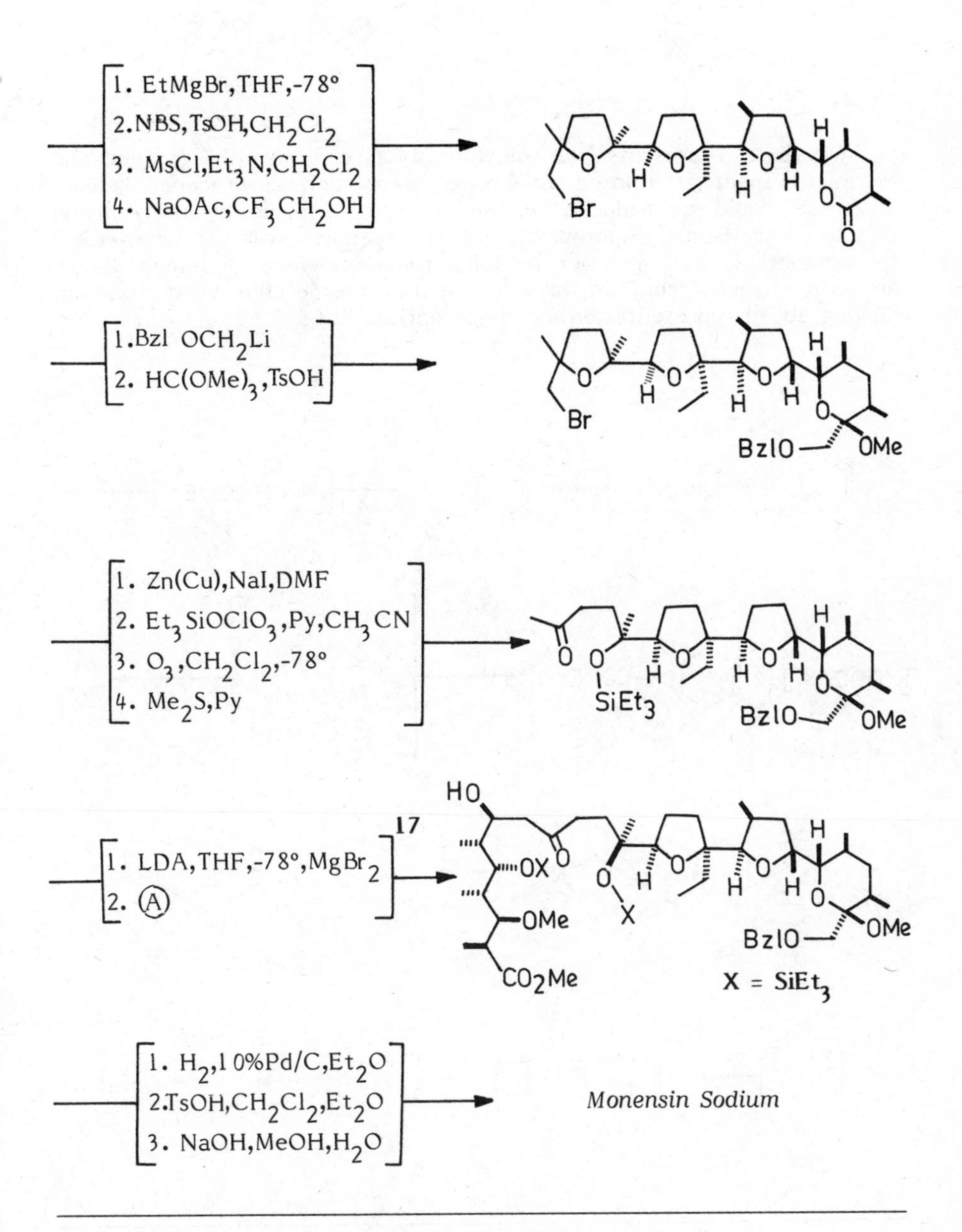

17. 3:1 mixture of the required to the unwanted aldols was produced.

OCTALENE

Octalene represents the bicyclic 14π e^- aromatic system and in this respect is related to phenanthrene and anthracene. Indeed, all these could be reduced to the 14π e^- pericycle by reorganising the bridging bonds. Appropriately, the synthesis of octalene starts from naphthalene, the lower homolog (1). Octalene is a lemon yellow aromatic hydrocarbon, as expected with considerable bond fixation, making above representation more appropriate.

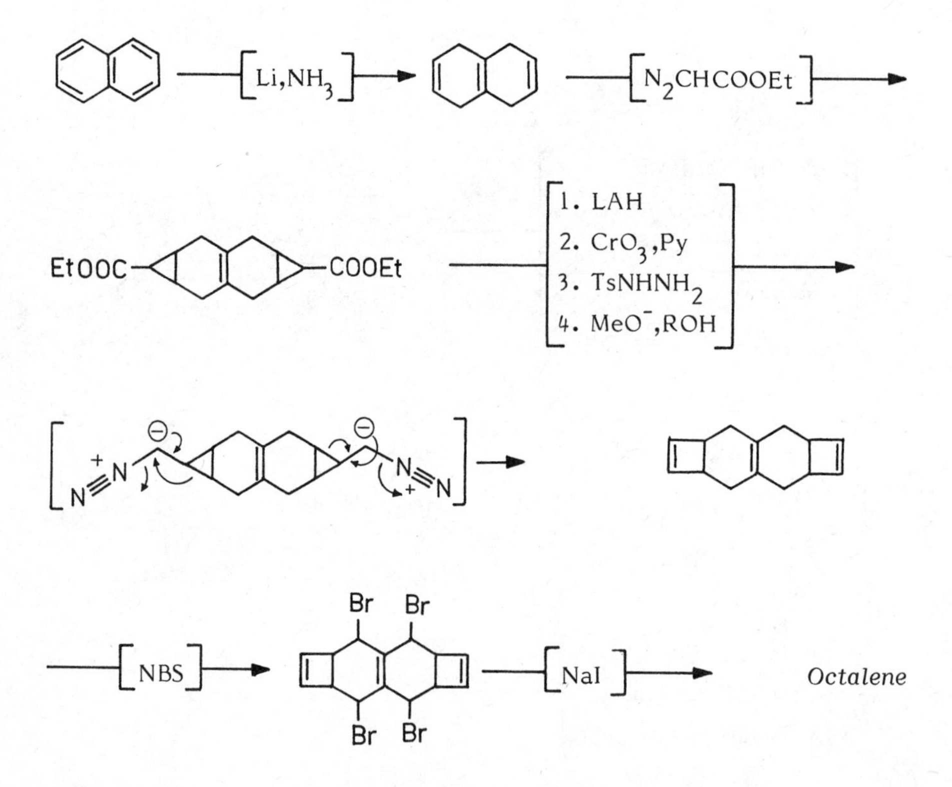

1. Vogel,E.; Runzheimer,H.V.; Hogrefe,F.; Baasner,B.; Lex,J. Angew. Chem. Int. Ed., 1977, 16, 871.

OUT-OUT, IN-IN BICYCLIC SYSTEMS

In [j.k.l]bicyclic systems, where j,k,l are small, the bridgehead hydrogens have to, necessarily, assume the out-out configuration. However, with increasing chain lengths all the three permissible configurations of the bridgehead hydrogens, namely, out-out, out-in and in-in become possible. The out-in and in-in isomers of the [8.8.8] bicyclic system have been prepared (1). The structures are established on the basis of ^{13}C-nmr and thermodynamic considerations. Interestingly, the in-in isomer, where the t-protons are not available for ready abstraction, is unreactive to bromine in contrast to the other isomers.

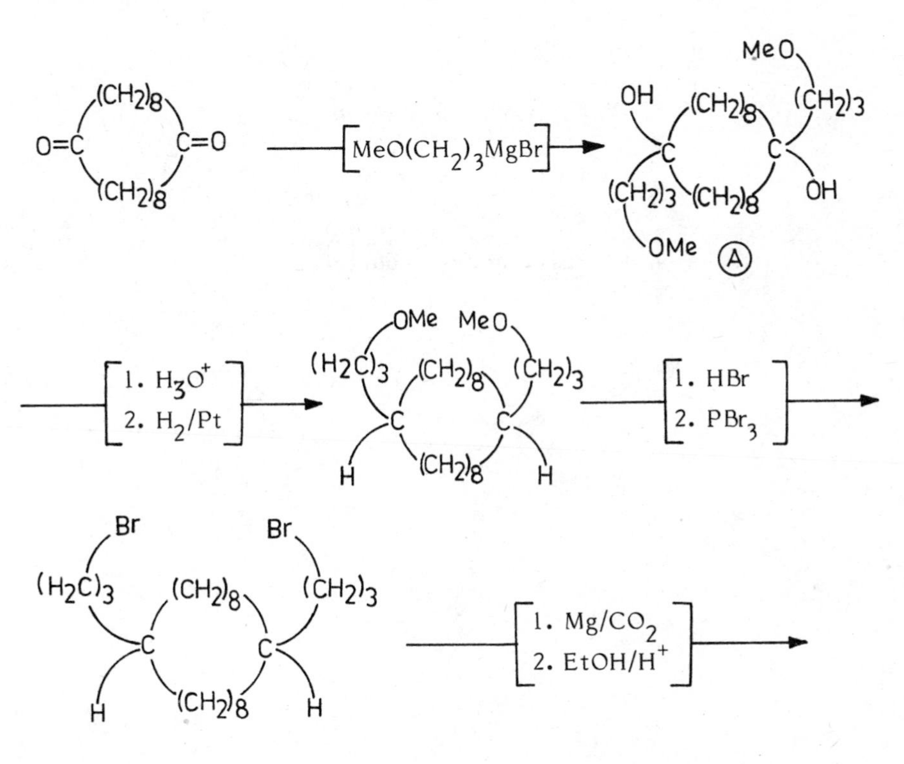

1. Park, C.H., Simmons, H.E. J. Am. Chem. Soc., 1972, 94, 7184.

2. The intermediate hydroxy-methyl ether (A) was obtained as a mixture of cis and trans isomers, but the isomers were separated at the stage of dibromides by fractional crystallisation, which were individually converted into the cis and trans di-esters.

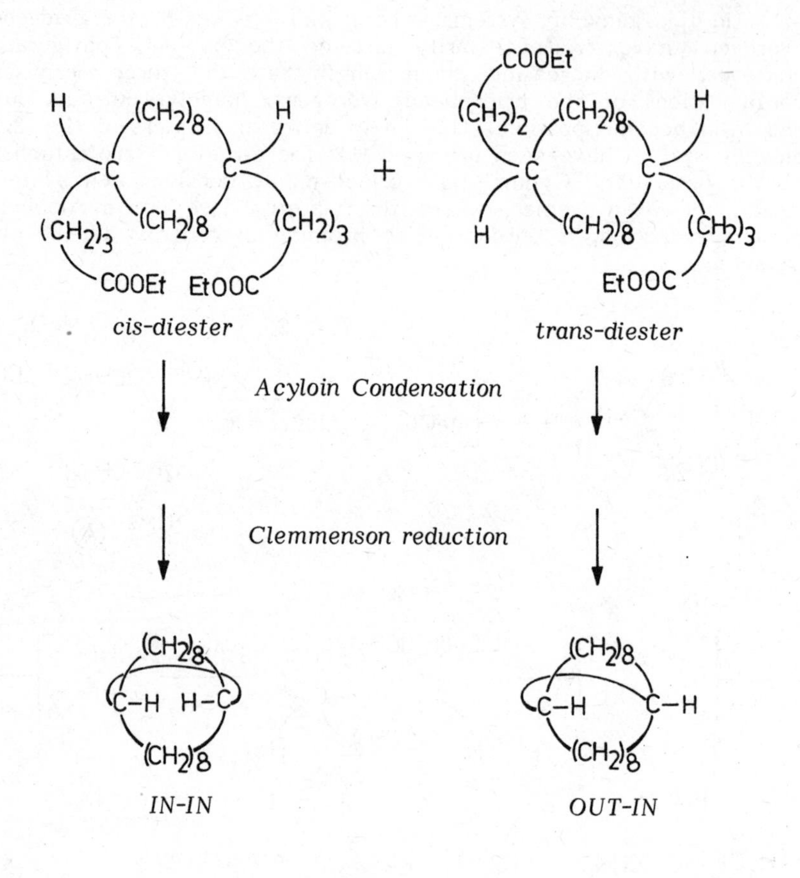
H
$(CH_2)_8$
H
C
C
+
$(CH_2)_3$
$(CH_2)_8$
$(CH_2)_3$
COOEt EtOOC
cis-diester
COOEt
$(CH_2)_2$
$(CH_2)_8$
H
C
C
H
$(CH_2)_8$
$(CH_2)_3$
EtOOC
trans-diester
Acyloin Condensation
Clemmenson reduction
$(CH_2)_8$
C–H H–C
$(CH_2)_8$
IN-IN
$(CH_2)_8$
C–H C–H
$(CH_2)_8$
OUT-IN

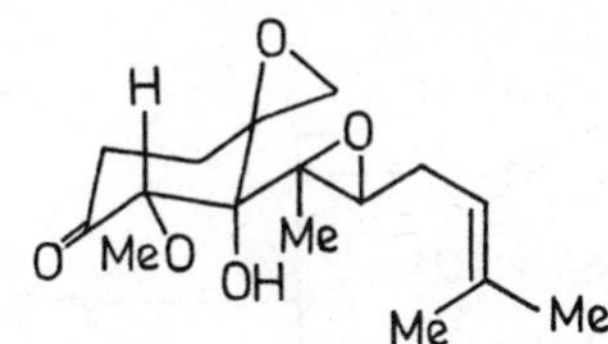

OVALICIN

Ovalicin, a sesquiterpene diepoxide antibiotic isolated from cultures of *Pseudorotum ovalis* Stolk(1),exhibits antimicrobial, anticancer and immunosupressive activities(2) and has been of much chemotherapeutic interest(3). Structurally ovalicin is related to fumagillin an antiprotozoal antibiotic which has been synthesised previously by the same authors(4). The stereocontrolled synthesis reported herein(5) follows a totally different strategy and synthetic methodology. One of the crucial steps was the selective reduction of 4,5 -double bond, which was achieved by using diimide(6). The method used for the synthesis of the side chain is a new approach to a stereospecific synthesis of E-trisubstituted 1,4-dienes.

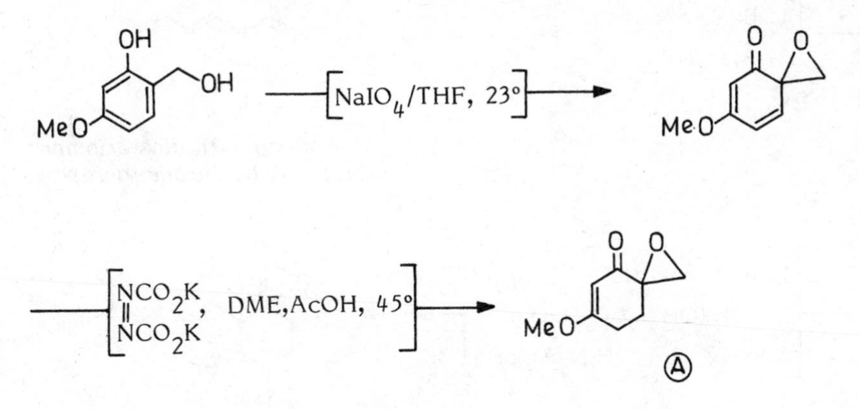

1. Sigg, H.P., Weber, H.P. Helv. Chim. Acta, 1968, 51, 1395.

2. Zimmerman, W.A., Hartman, G.R. Eur. J. Biochem., 1981, 118, 143.

3. Borel, J.F., Lazary, S., Staehelin, H. Agents & Actions, 1974, 4, 357.

4. Corey, E.J., Snider, B.B. J. Am. Chem. Soc., 1972, 94, 2549.

5. Corey, E.J., Dittami, J.P. J. Am. Chem. Soc.,1985, 107, 256-257.

6. Adam, W., Eggelte, H.J. J. Org. Chem., 1977, 42, 3987.

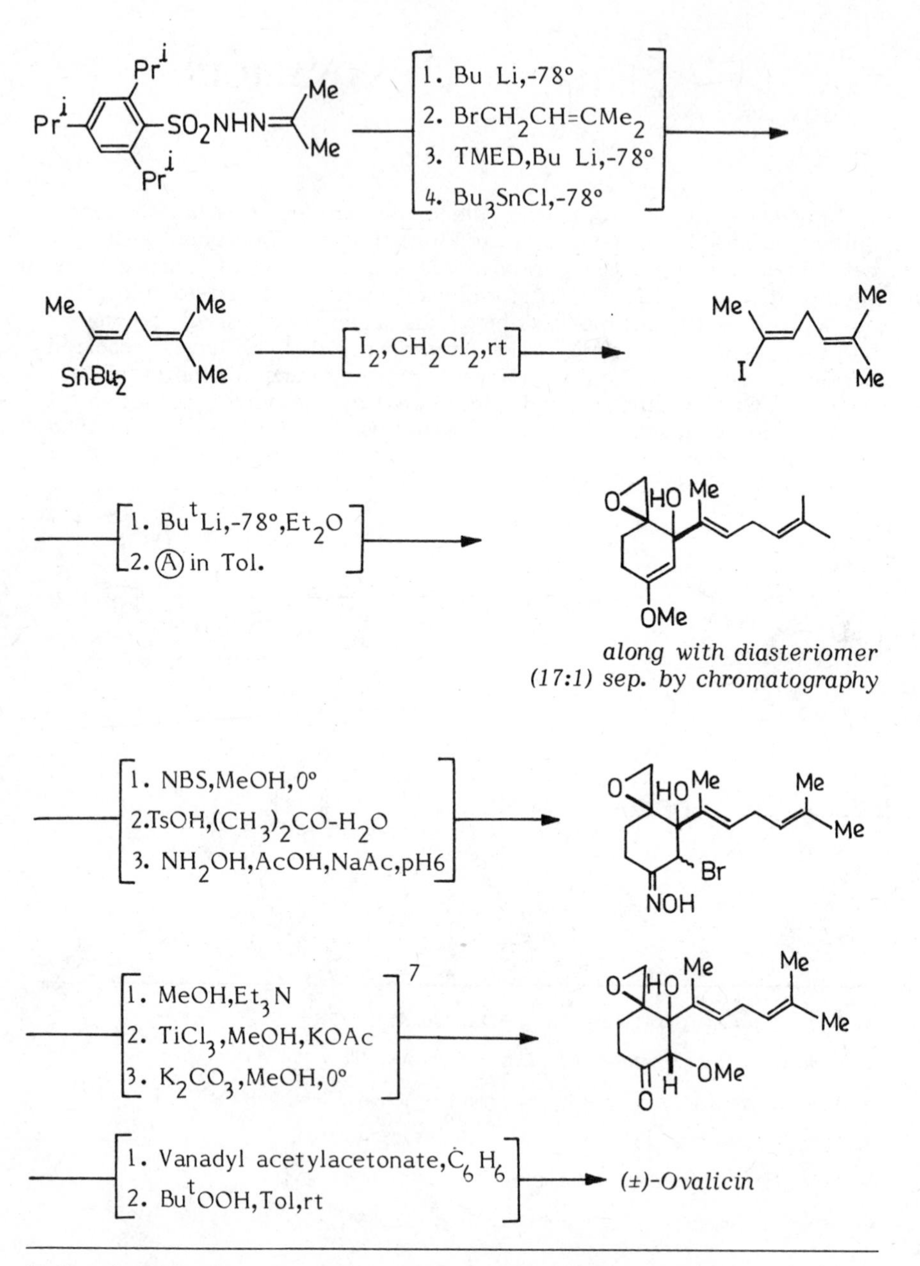

7. The above reactions gave a 1:1 mixture of diastereoisomers, which on this treatment underwent complete isomerisation to the right isomer.

PAGODANE

Eight five-membered rings radiating from a cyclobutane base generate the aesthetically pleasing pagodane constellation. Besides, the projection of the cyclobutane base outwards to pick up 4 hydrogens and the pair-wise union of the 4 peri methylenes with the loss of 4 hydrogens, would transform pagodane to dodecahedrane. Pagodane has been synthesised and its transformation to dodecahedrane demonstrated (1).

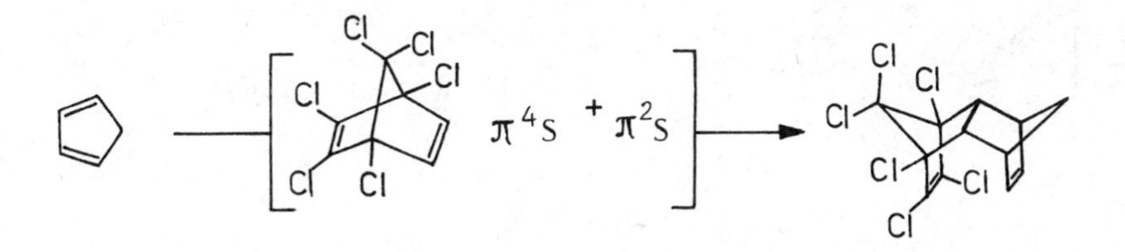

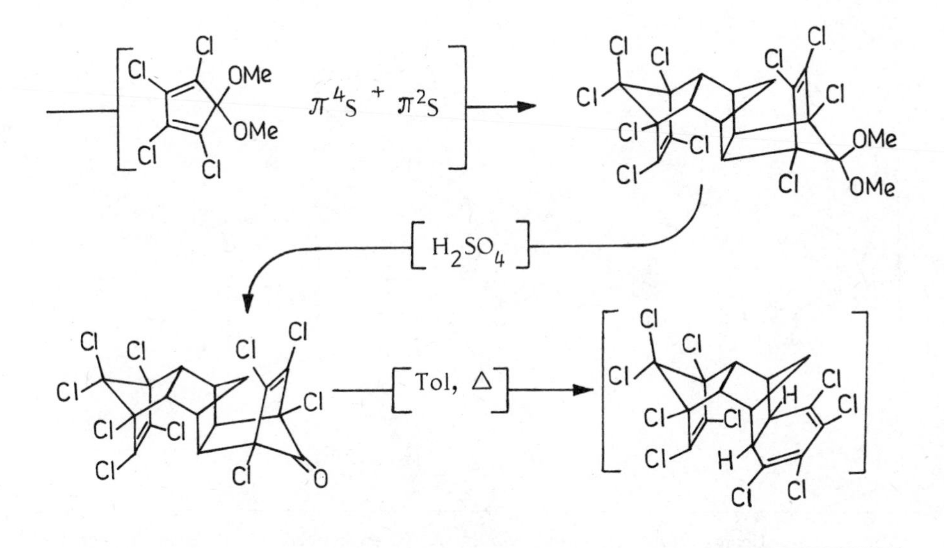

---

1. Fessner,W.D.; Prinzbach,H.; Rihs,G. Tetrahedron Lett., 1983, 24, 5857.

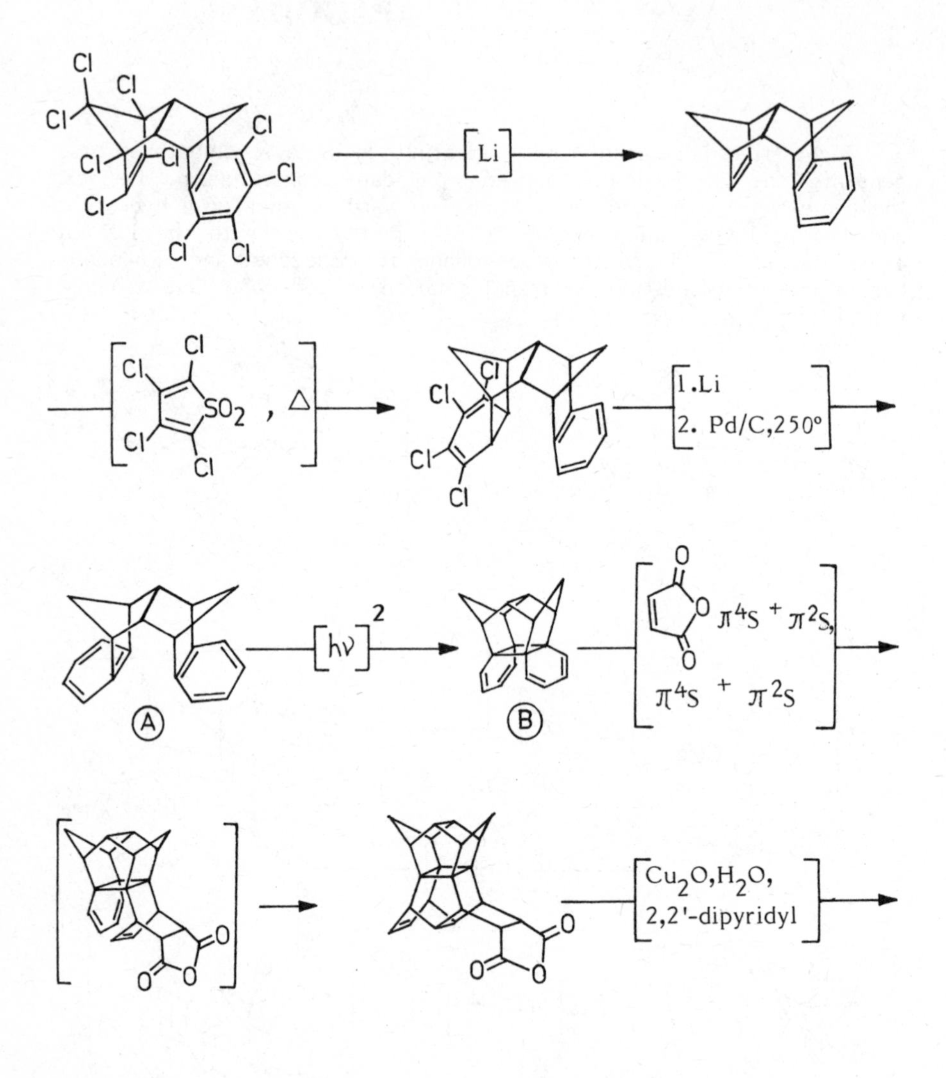

---

2. A photo-stationary state representing A:B::7:3 is established at 254 nm. B could be easily separated from the mixture. Thermolytic reversal is possible at 200°C in decalin.

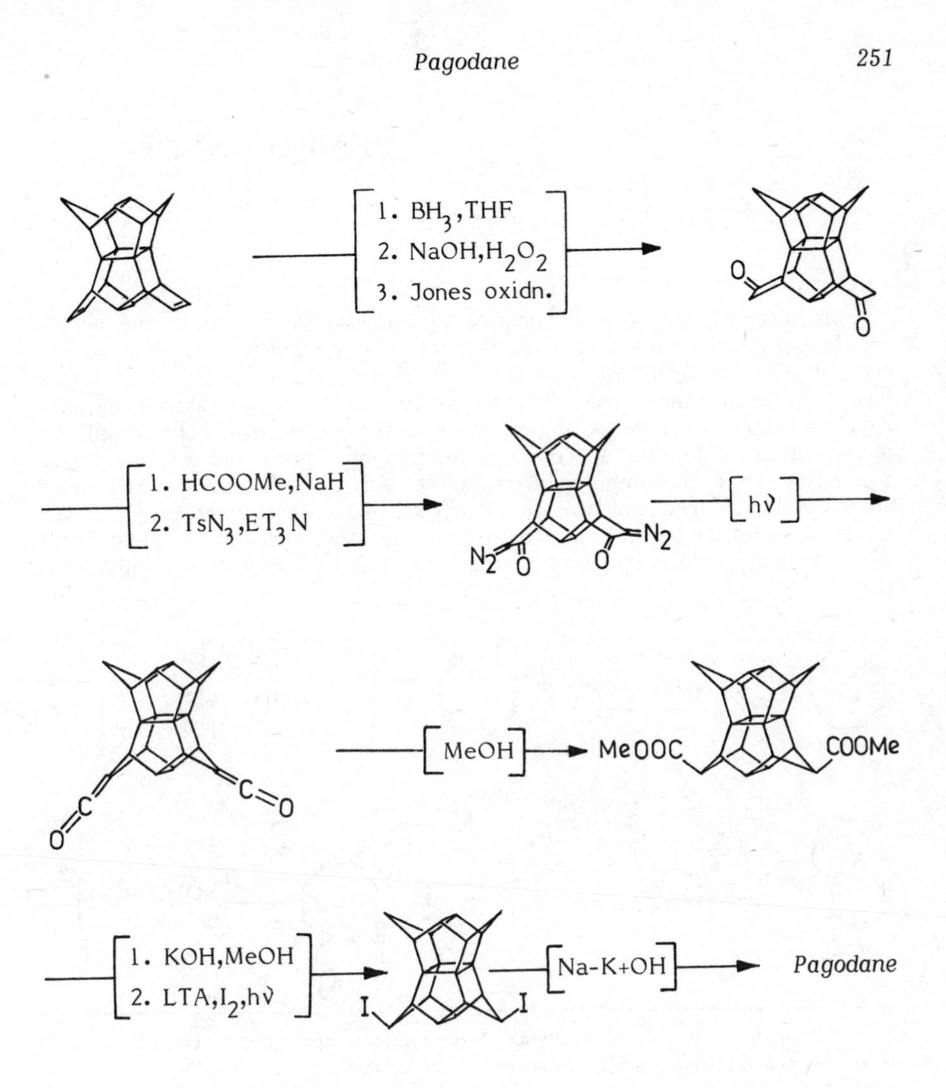
1. BH3,THF
2. NaOH,H2O2
3. Jones oxidn.
O
O
1. HCOOMe,NaH
2. TsN3,ET3N
N2
O
O
N2
hν
O=C
C=O
MeOH
MeOOC
COOMe
1. KOH,MeOH
2. LTA,I2,hν
I
I
Na-K+OH
Pagodane

PENTALENENE

Because of the wide occurence of tricyclo [6.3.0.0]undecane skeleton (angular triquinones in a number of sesterpenes and sesquiterpenoids the synthesis of natural products based on this carbon framework has been a popular target for the development of new synthetic approaches (1). The synthesis of pentalene described here in (2) exemplifies a general and flexible approach developed by Mehta and his associates for entry into the angular triquinones via a carbonium-ion-mediated transannulation reaction in the bicyclo [6.3.0.0]undecane system (3); which is based on taking cognisance of the marked propensity of the cycloacyl derivatives towards transannular cationic cyclisations (4).

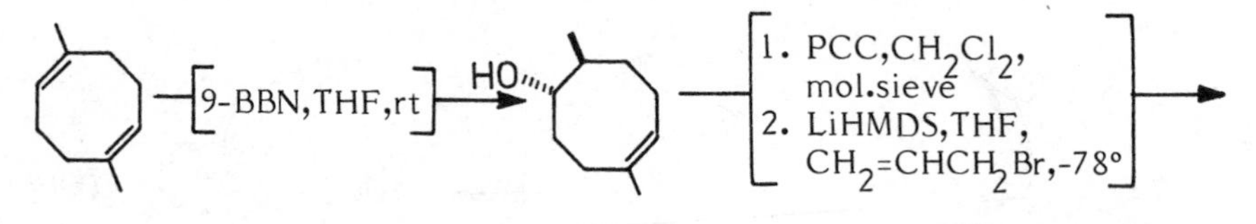

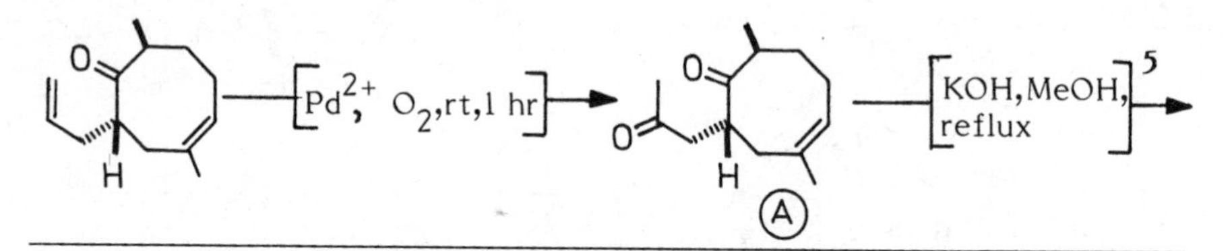

---

1. See references 11 & 12 of reference 2 for synthetic approches to tricyclic [6.3.0.0] undecanes and total synthesis of triquinane natural products.

2. Mehta,G.; Rao,K.S. J. Am. Chem. Soc., 1986, 108, 8015; J. Chem. Soc.; Chem. Commun., 1985, 1464.

3. For earlier synthesis of pentalenene see: Ohfune,Y.; Shirahama,H.; Matsumoto,T. Tetrahedron Lett., 1976, 2869; Misumi,S.; Ohtsuka,T.; Ohfune,Y.; Sugita,K.; Shirahama,H.; Matsumoto,T. Tetrahedron Lett., 1979, 31; Annis,G.D.; Paquette,L.A. J. Am. Chem. Soc., 1982, 104, 4504; Piers,E.; Karunaratne,V. J. Chem. Soc.; Chem. Commun., 1984, 959; Pattenden,G. Teague,S.J. Tetrahedron Lett., 1984, 3021; Mehta,G.; Rao,K.S. J. Chem. Soc., Chem. Commun., 1985, 1464; Crimmins,M.T.; DeLoach,J.A. J. Am. Chem. Soc., 1986, 108, 800; Crimmins,M.T.; Mascarella,S.W. J. Am. Chem. Soc., 1986, 108, 3435.

4. Cope,A.C.; Martin,M.M.; Mckerrey,M.A.Q. Rev. Chem. Soc., 1986, 20, 119.

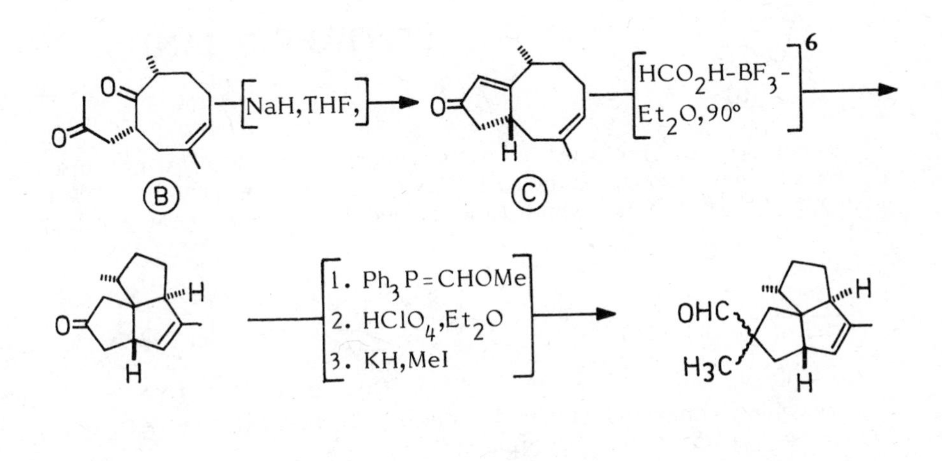

—[Wolff-Kis hner reduction]→ *(±)-Pentalenene*

From the trans-diketone Ⓐ by the same set of reactions (±) epi-pentalenene was obtained.

(±)-epi-Pentalenene

---

5. A 1:4 mixture of Ⓐ:Ⓑ was obtained in this equilibration; cf. Clark Still,W.; Galynker,I. Tetrahedron, 1981, 37, 3981.

6. Ⓒ was obtained along with C-2-epimeric enone in 4:1 ratio.

(III)

PENTAPRISMANE

Annulation of cyclobutane units generates an interesting series of hydrocarbons shown below with identical methine protons. Pentaprismane (III) over a period of time earned the reputation of a rather

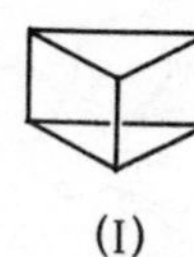

(I)

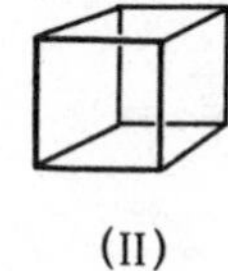

(II)

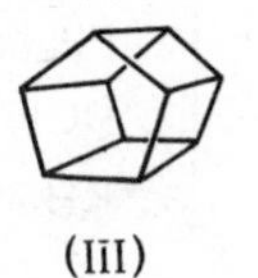

(III)

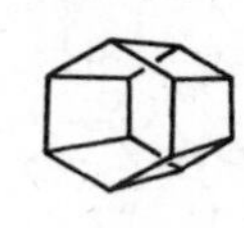

(IV)

"inaccessible" compound (1). Its synthesis (2) was achieved eventually via the key compound homopentaprismanone (A).

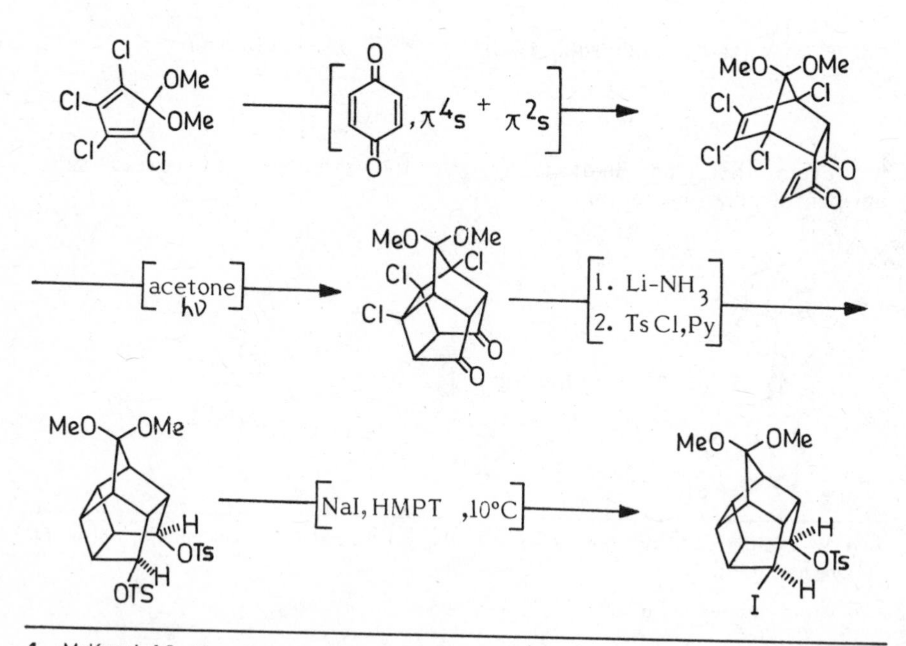

1. McKennis,J.S.; Brener,L.; Ward, J.S., Petit, R. J. Amer. Chem. Soc., 1971, 93, 4957; Paquette, L.A., Davis, R.F., James, D.R. Tetrahedron Lett., 1974, 1615.

2. Eaton, P.E., Or, Y.S., Branca, J.S. J. Amer. Chem. Soc., 1981, 103, 2134.

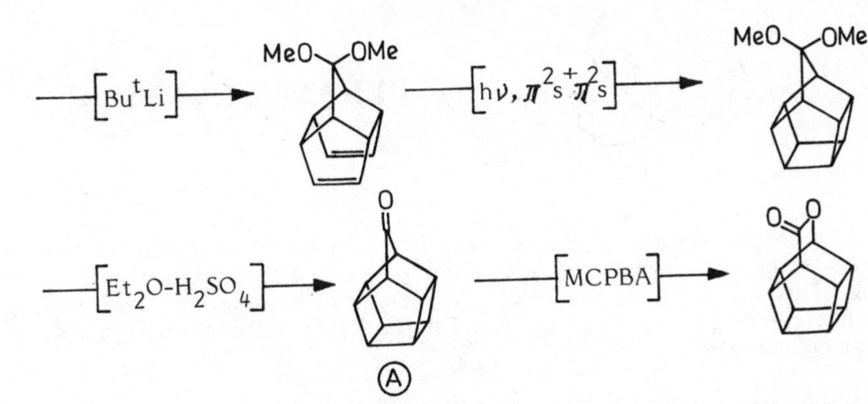

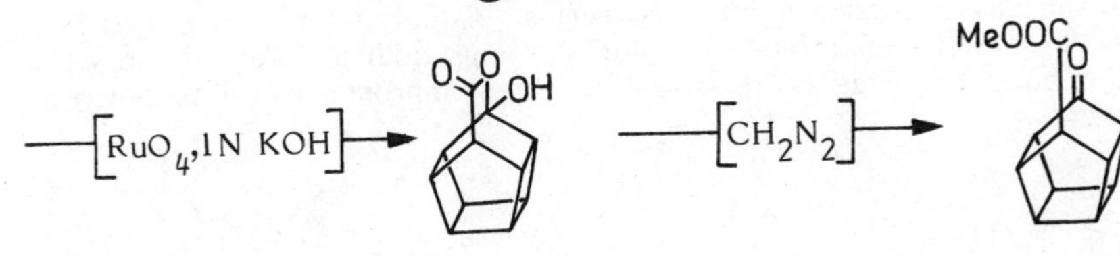

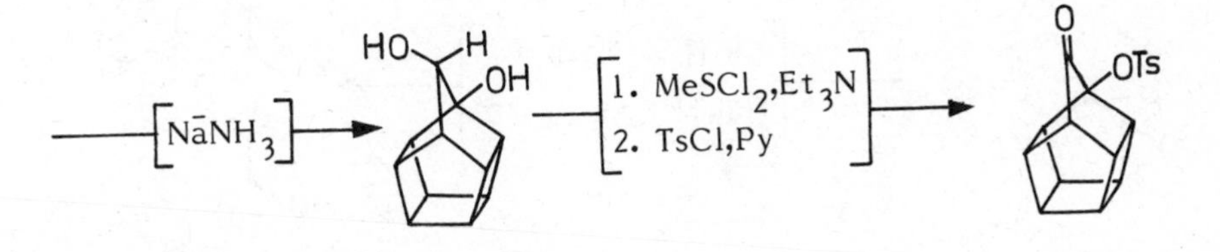

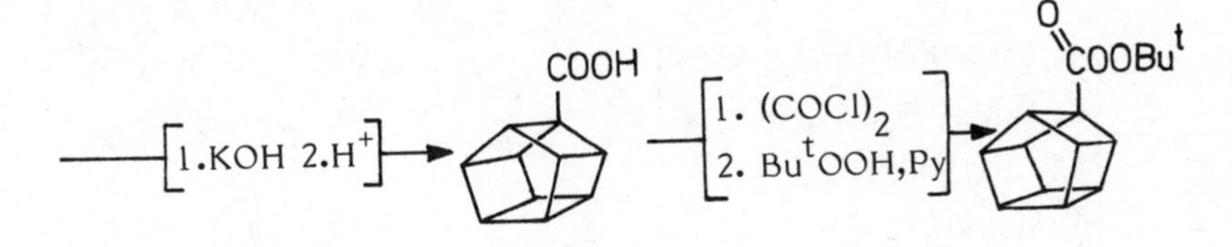

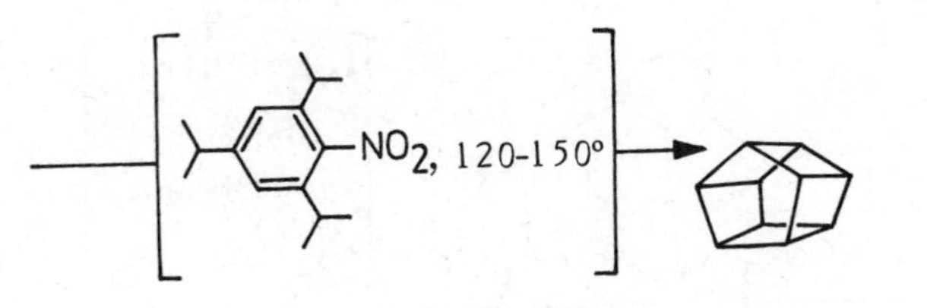

Pentaprismane

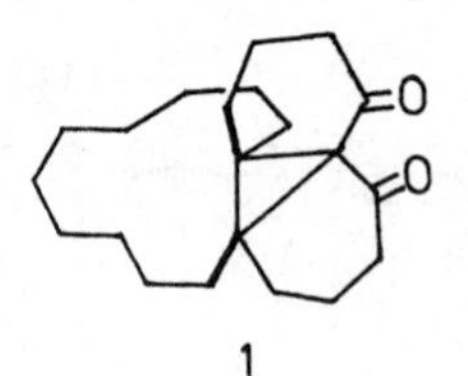

1

PERANNULANES

Perannulanes are perceived as fully annelated cycloalkanes in which rings of varying size are fused to each other to form a core ring. Their pleasing appearance belie complexities relating to stereochemistry and chirality, which would naturally increase with increasing perannulane size.

The base member of this series, namely, tri-annulane (I) has been prepared via intra-molecular carbene addition to the between-anene bond (1). This route also illustrates improved route to between-anenes.

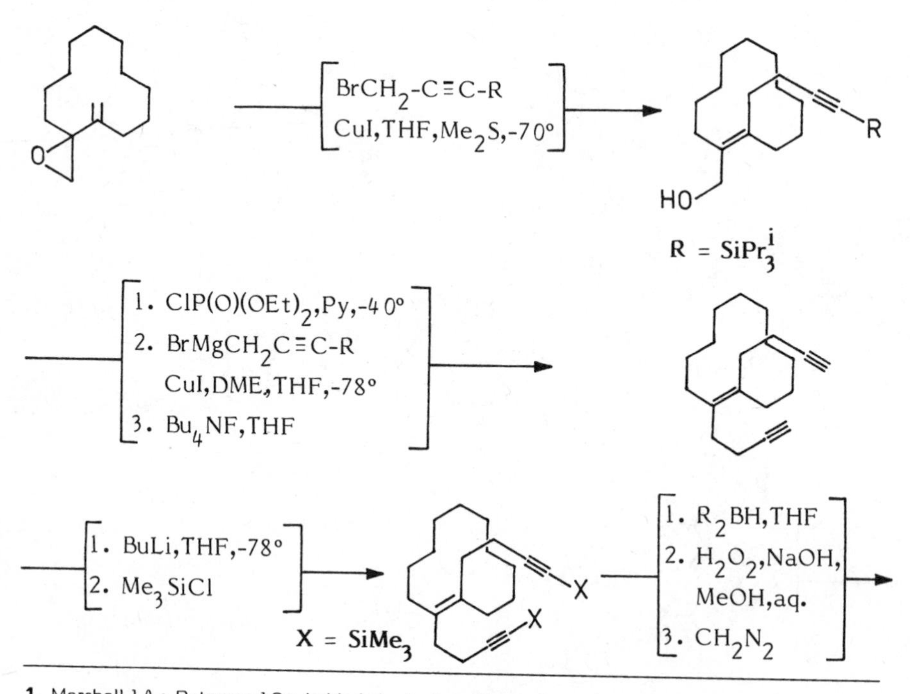

1. Marshall,J.A.; Peterson,J.C.; Lebioda,L. J. Am. Chem. Soc., 1983, 105, 6515.

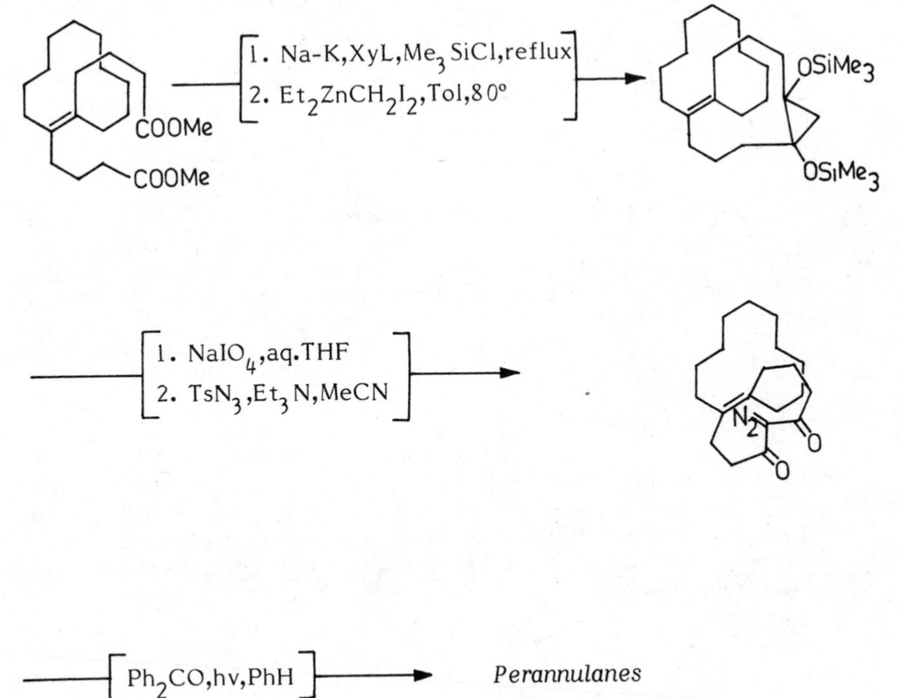

Perannulanes

trans, cis, cis-[10.4.4]
tri-annulene-16,18-dione

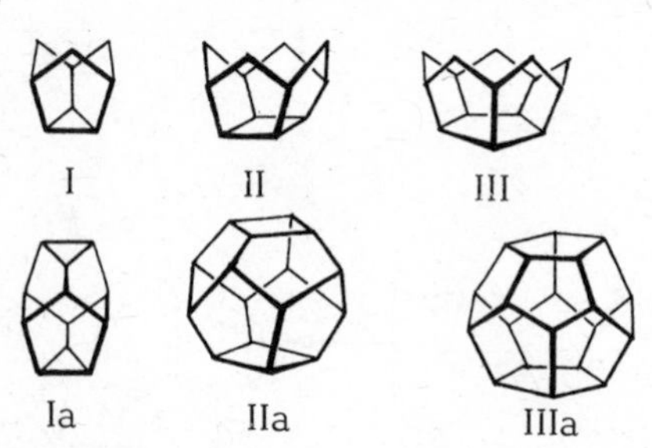

PERISTYLANES

The circumscription of five membered rings around a regular polygon generates a hemi-spherical cup (peristylane) topology such that the rim to base ratio is always 2:1. Their structures are aesthetically appealing, but even more pleasing are the spherical designs hypothetically arising from their dimerization. Thus [3]-, [4]- and [5]peristylanes can be represented as I, II, III and their dimers as Ia,IIa,IIIa. The synthesis of I to III has been achieved.

[3]Peristylane [Triaxane]

The 3 axial ligands of the chair cyclohexane describe a tripod and the union of these would result in [3]peristylane. When two of such ligands are united by a π bond a bicyclo[2.3.1]octane system is produced. The remaining axial ligands could be affixed to the π bond via cyclopropanation (1).

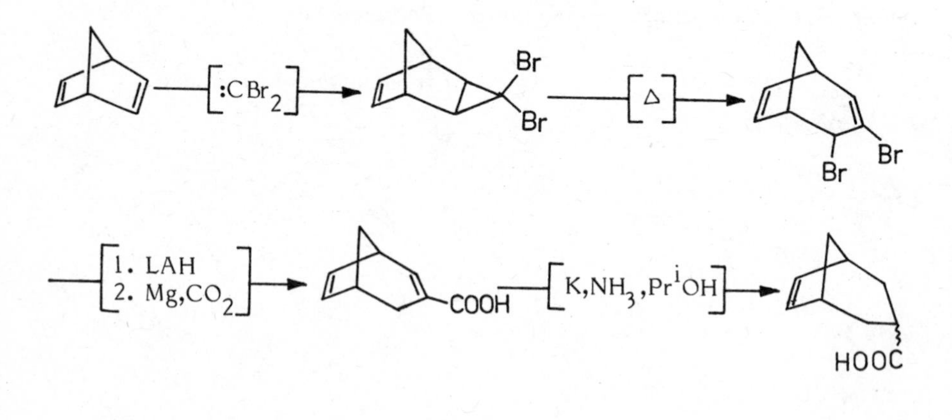

1. Garratt,P.J.; White,J.F. J. Org. Chem., 1977, 42, 1733.

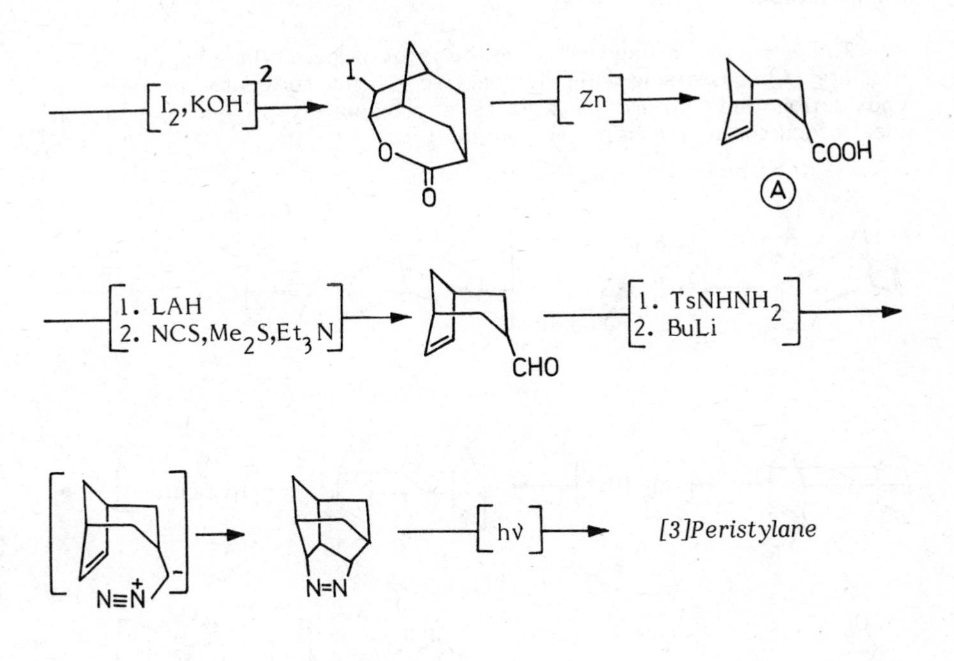

The intermediate A was also transformed to I, by an alternate pathway involving trans annular π participation.

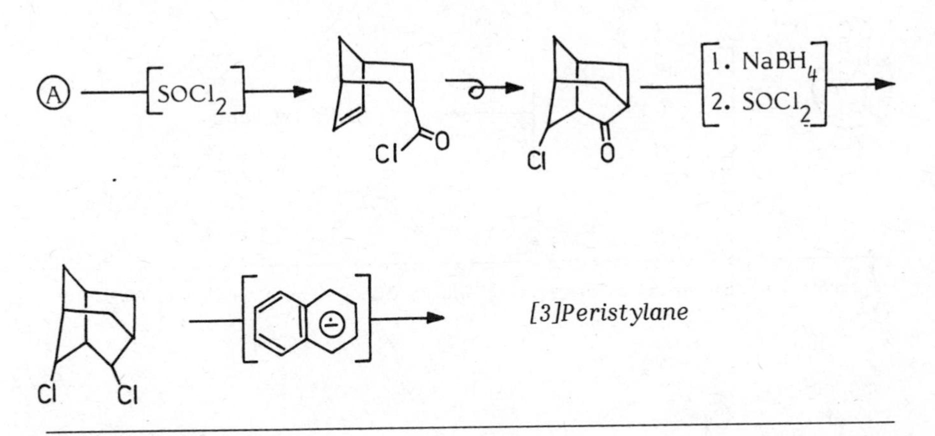

2. Iodolactonization-Zn cleavage sequence was used to separate the required endo acid.

[4]Peristylane

The core of an ingeneous synthesis of [4]peristylane is the intermediate (A), possessing a highly reactive π bond, suited for preferential epoxidation and rupture as well as a proximately aligned π pair, an ideal precursor to the base cyclobutane (3).

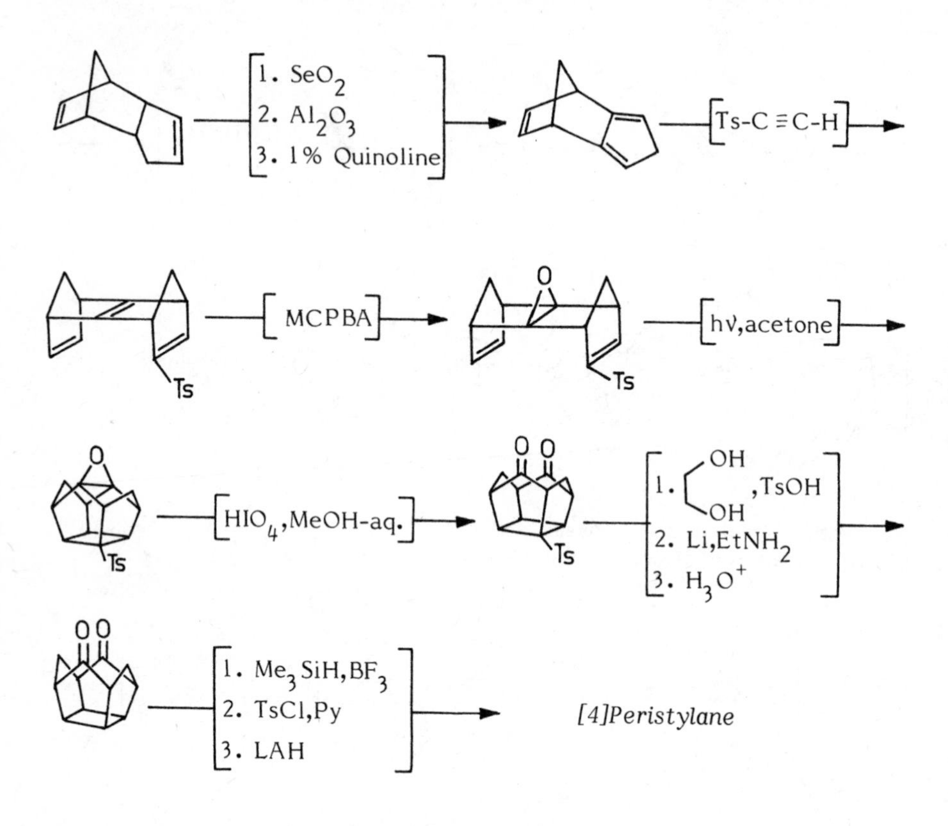

3. Paquette,L.A.; Browne,A.R.; Doecke,C.W.; Williams,R.V. J. Am. Chem. Soc., 1983, 105, 4113.

[5]Peristylane

The [5]peristylane synthesis (4) preceeded the intense endeavours to annelated cyclopentanes (5) and provided valuable methodologies in this direction. Indeed, the latest route to [5]peristylane involves the systematic cyclopentane annulation, starting from cyclopentenone.

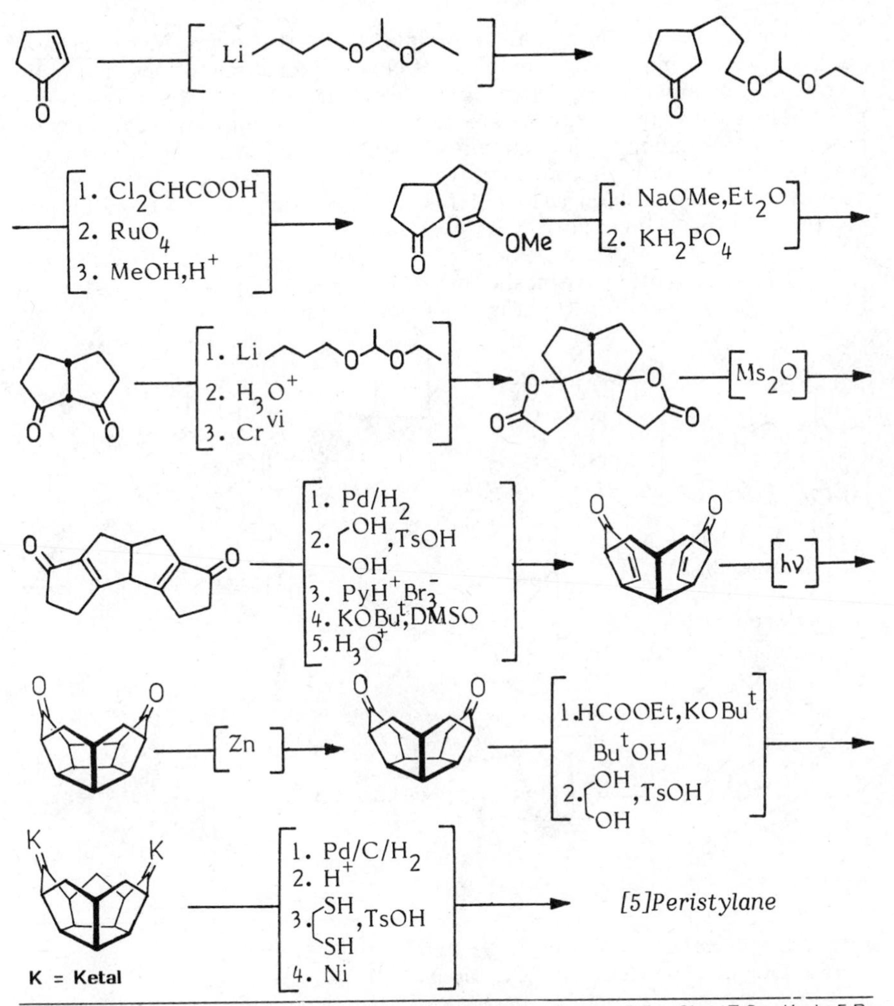

4. Eaton,P.E.; Mueller,R.H.; Carlson,G.R.; Cullison,D.A.; Copper,G.F.; Chou,T.C.; Krebs,E.P. J. Am. Chem. Soc., 1977, 99, 2751.
5. Paquette,L.A. "Recent Synthetic Developments in Polyquinane Chemistry", Springer-Verlag, 1984.

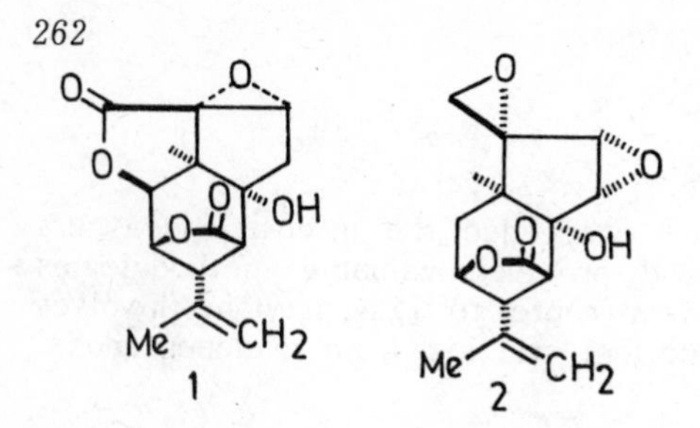

PICROTOXININ
CORIAMYRTIN

Picrotoxin, the poisonous principle from the plant Menispermum cocculus, is a molecular complex of toxic picrotoxinin and nontoxic picrotin. Coriamyrtin C, the toxin isolated from Coriaria species, also belongs to the picrotoxane group. Picrotoxinin and coriamyrtin posed a big challenge for chemical synthesis on account of delicate functionalities and high concentration of chiral centres in the molecule. Two synthesis have been reported for both picrotoxinin (1,3) and coriarmyrtin (2,3), each ingenious in its own way.

The first total synthesis for (-) picrotoxinin was reported in 1979 by Corey & Pearce (1) starting from (-) carvone.

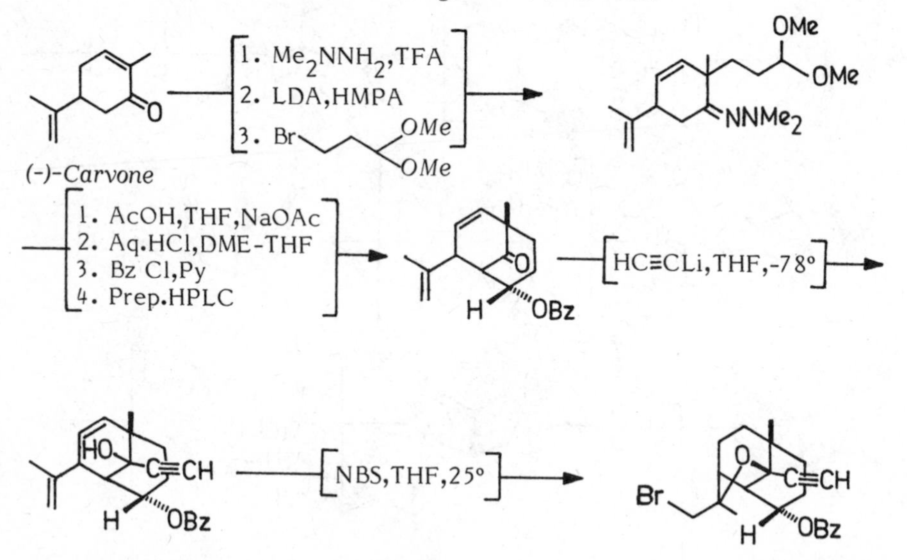

1. Corey,E.J.; Pearce,H.L. J. Am. Chem. Soc., 1979, 101, 5841.
2. Tanaka,K.; Uchiyama,F.; Sakamoto,K.; Inubishi,Y. J. Am. Chem. Soc., 1982, 104, 4965.
3. Niwa,H.; Wakamatsu,K.; Hida,T.; Niiyama,K.; Kigoshi,H.; Yamada,M.; Nagase,H.; Suzuki,M. Yamada,K. J. Am. Chem. Soc., 1984, 106, 4547.

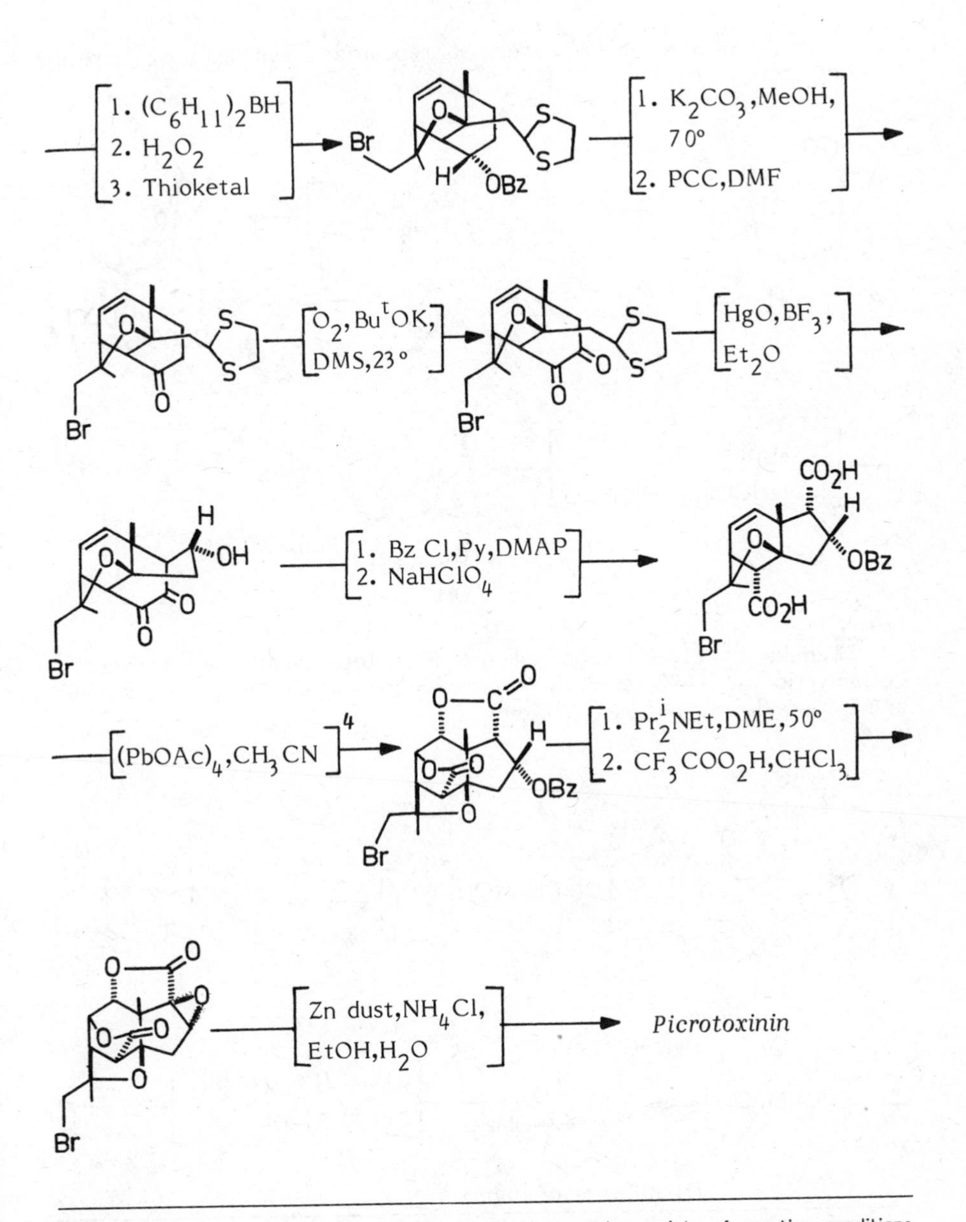

---

4. Halolactonisation processes to prepare the lactones under variety of reaction conditions failed, which may be due to the pronounced steric shielding of the double bond.

Corey and Pearce (5) have also reported a synthesis of picrotin starting from picrotoxinin.

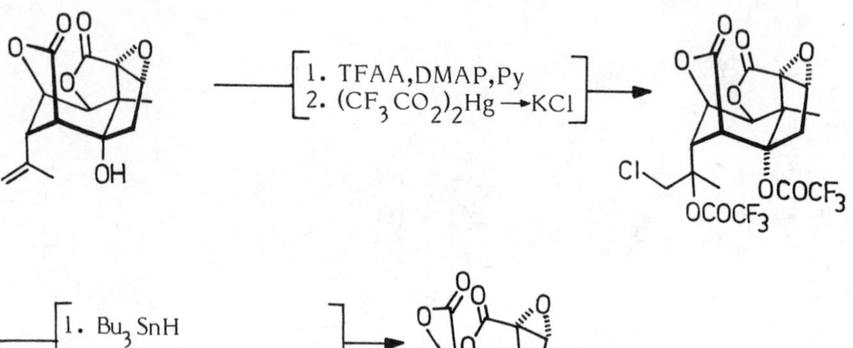

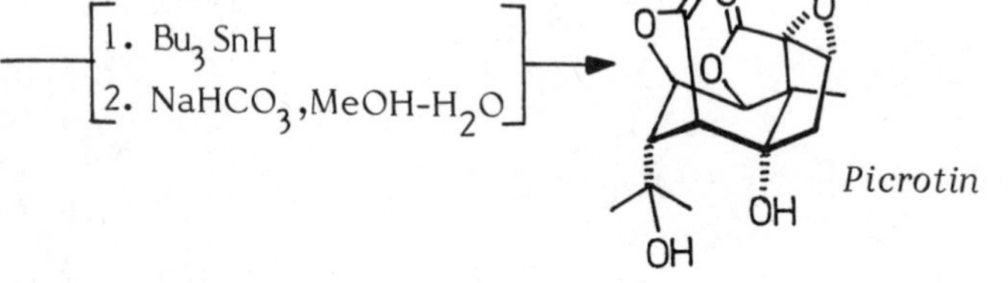

Picrotin

Tanaka et al. (2) reported the first total synthesis of racemic coriamyrtin in 1982 starting from readily available protoanemonin and 2-methylcyclopentane-1,3-dione.

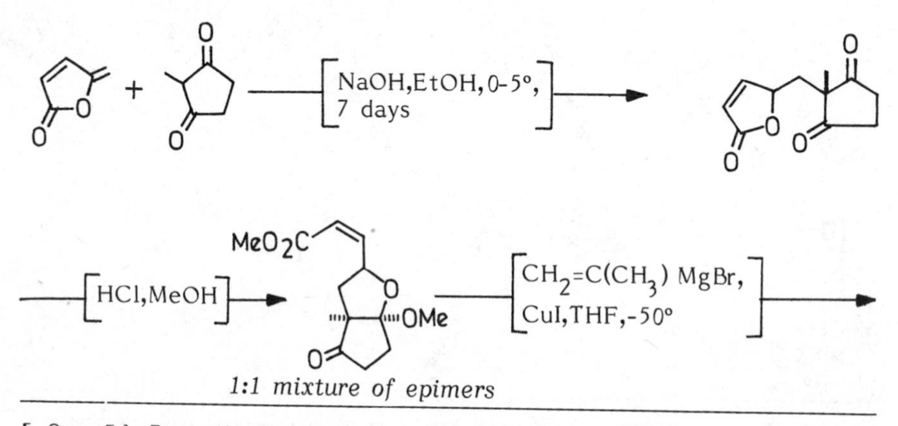

1:1 mixture of epimers

5. Corey,E.J.; Pearce,H.L. Tetrahedron Lett., 1980, 21, 1823.

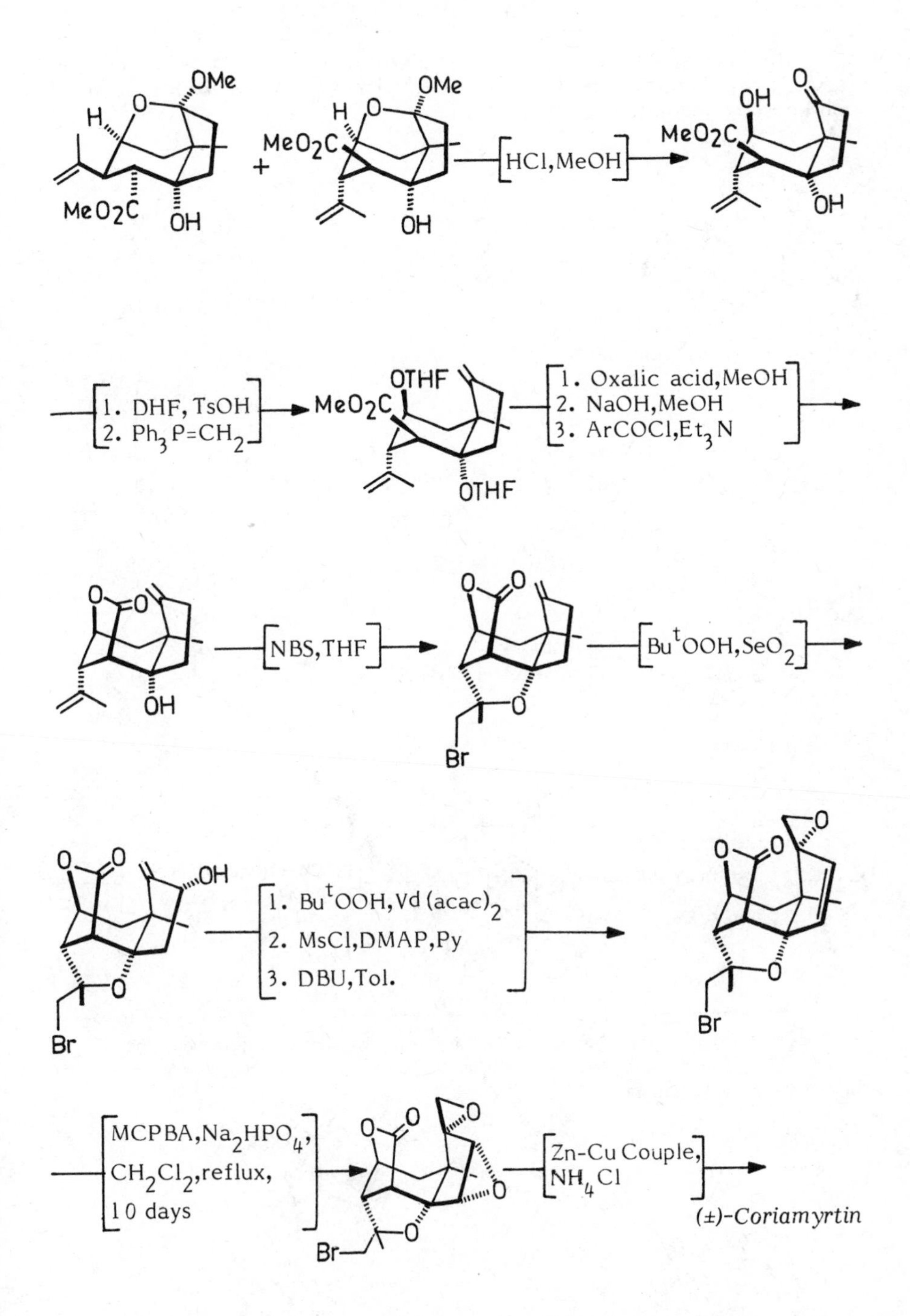

(±)-Coriamyrtin

The second synthesis of (-)-picrotoxinin and (+)-Coriamyrtin reported by Niwa et al. (3) utilises isotwistane compounds as common and key intermediates for their synthesis.

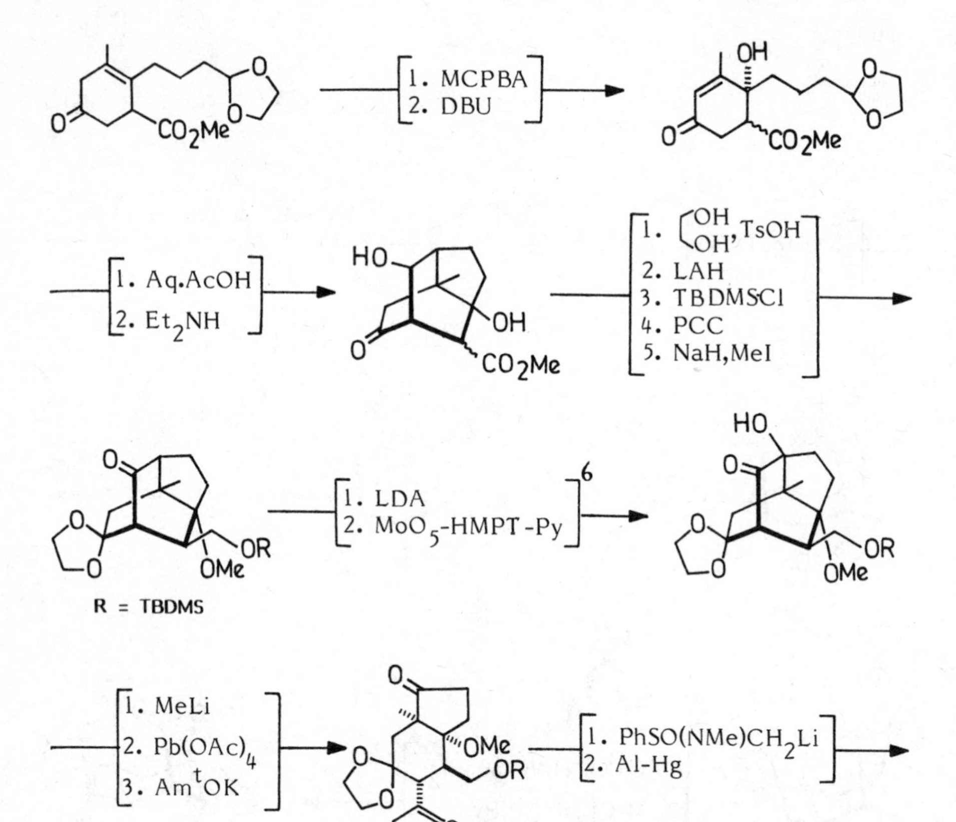

6. The success of bridgehead hydroxylation is due to the cyclohexanone ring in the bicyclo [3.2.1]octan-2-one being locked in the boat form making generation of the bridgehead enolate favourable.

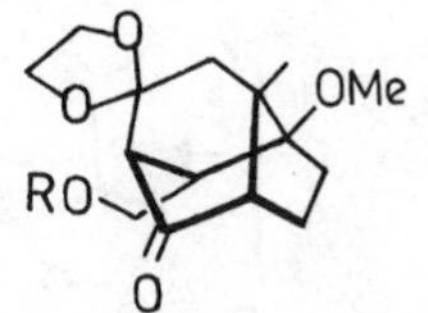

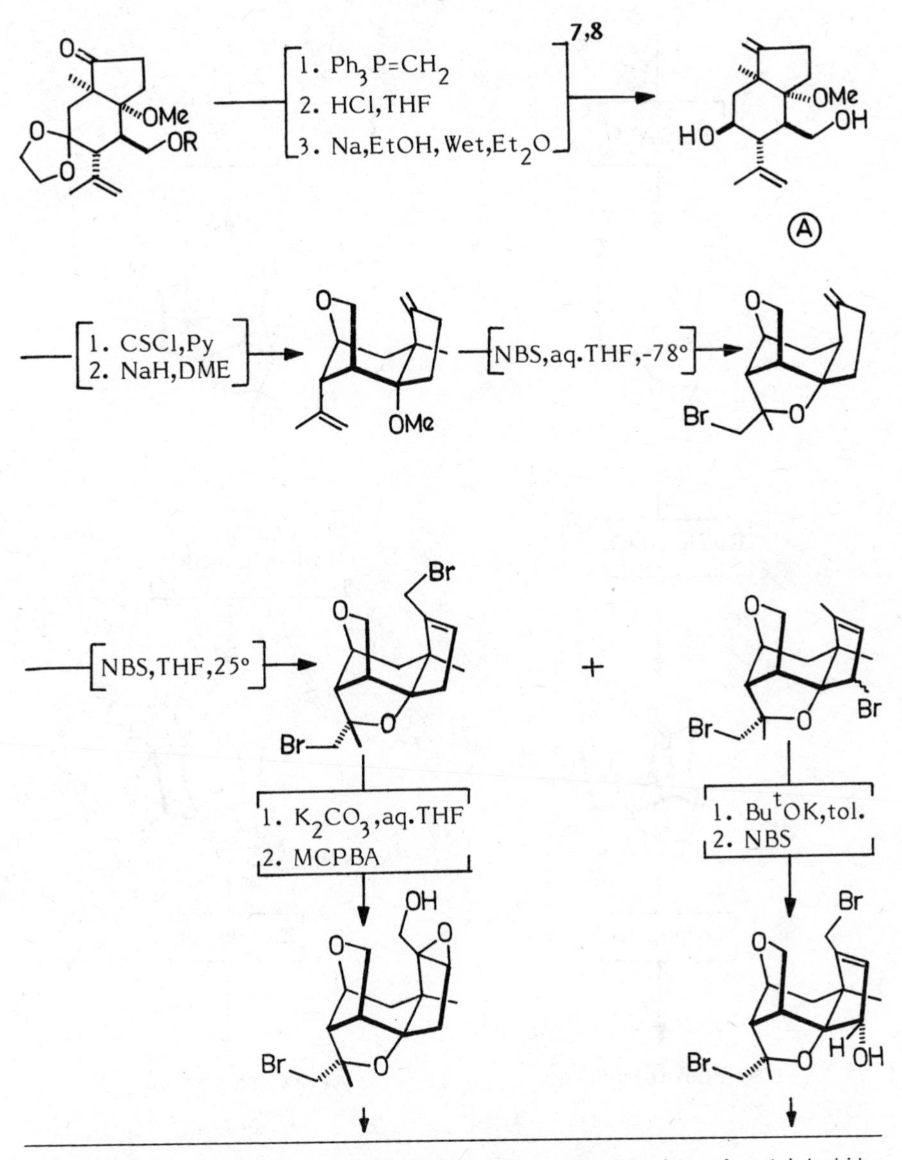

7. Reduction of the keto group after deacetylation with a variety of metal hydrides exclusively the epimer of (A) regarding the secondary hydroxyl group.
8. (A) was resolved through α-methoxy-α-trifluormethyl phenylacetyl ester formation followed by chromatographic separation and LAH reduction to give (+)-(A).

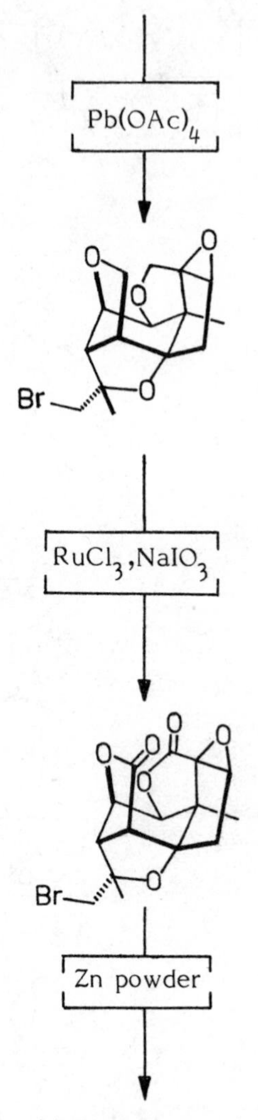

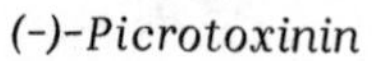

(-)-Picrotoxinin

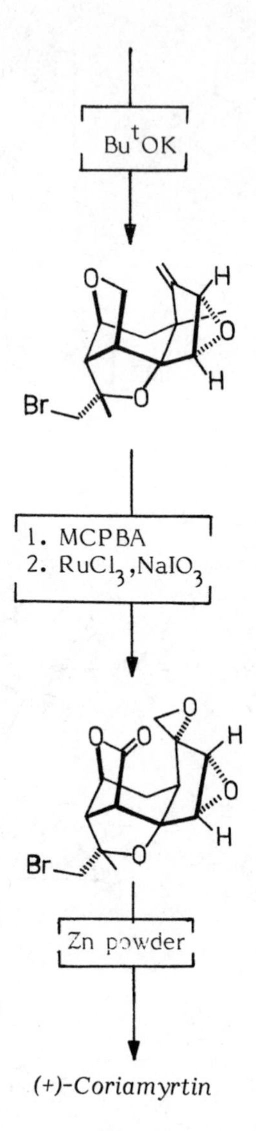

(+)-Coriamyrtin

PRISMANE

Inspite of the fact that prismane packs an additional 130 k cal mol^{-1} over benzene, this change, being forbidden by the Woodward-Hoffman rules, is sluggish with a half-life of 11 hours at 90°C. The synthesis of prismane (1) is remarkable and proceeds through yet another isomer of benzene, namely, benzvalene (2).

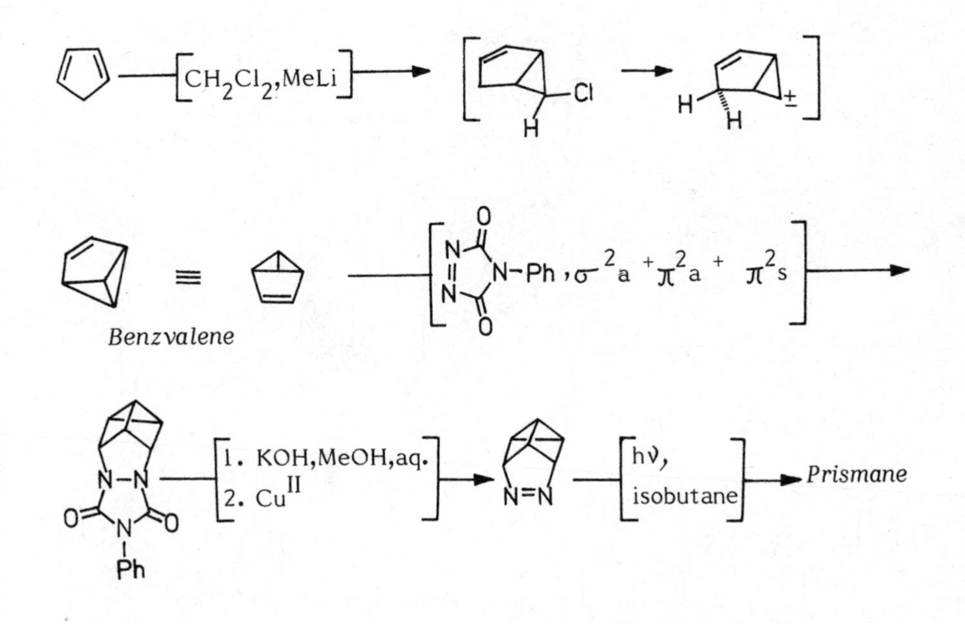

1. Katz,T.J.; Acton,N. J. Am. Chem. Soc., 1973, 95, 2738.
2. Katz,T.J.; Wang,E.J.; Acton,N. J. Am. Chem. Soc., 1971, 93, 3782.

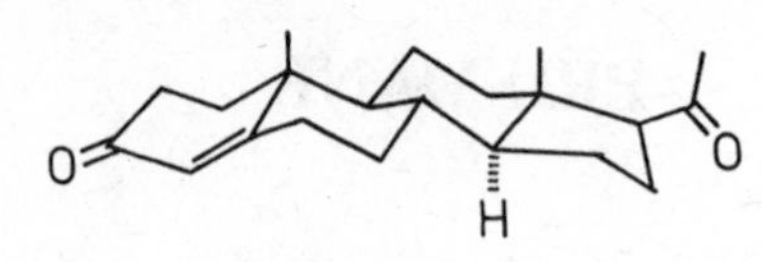

PROGESTERONE

The annelation reaction is an indispensable tool for the construction of polycyclic systems (1). In the synthesis of progesterone outlined below, Stork and McMurry (2) have illustrated the use of a new annelation reaction in which an isoxazole serves as a masked ketoalkyl function (4).

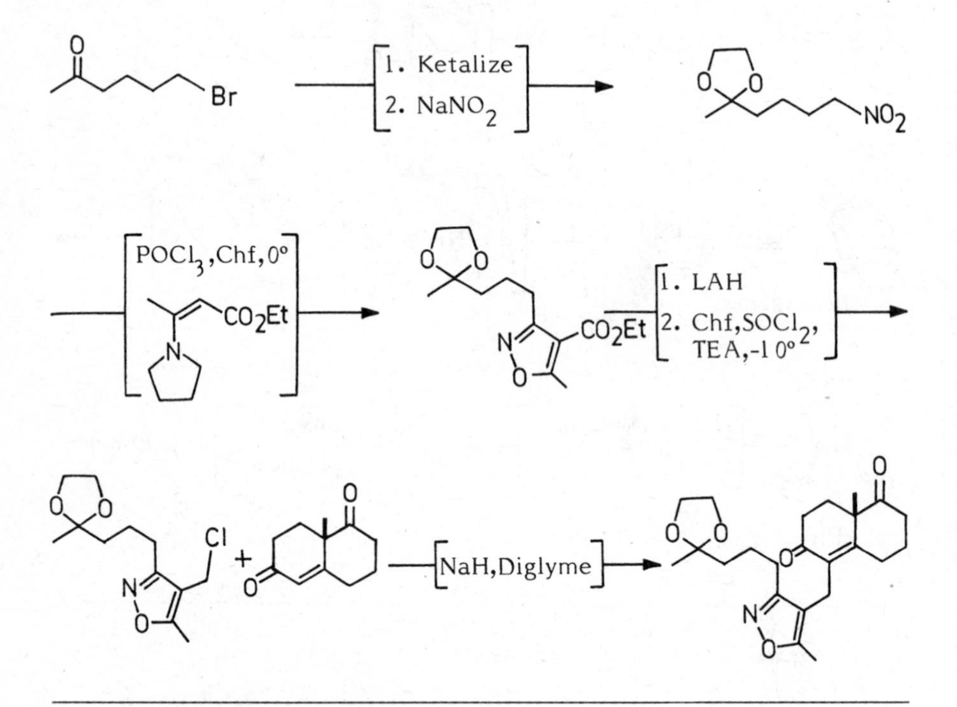

1. Stork,G. Pure Appl. Chem., 1964, 9, 931.

2. Stork,G.; McMurry,J.E. J. Am. Chem. Soc., 1967, 89, 5464.

3. This is a general synthesis of 4-isoxazolecarboxylic acids involving the addition of nitrile oxides (i) to derivatives of β-aminoacrylic esters (ii).

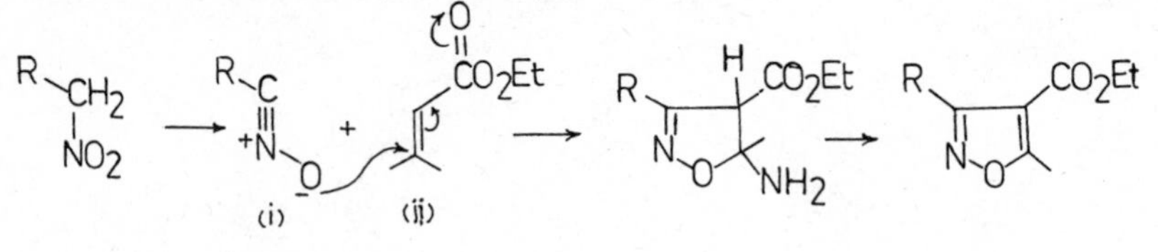

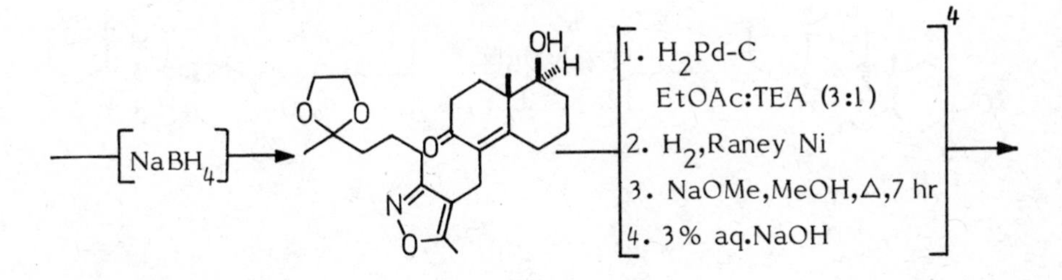

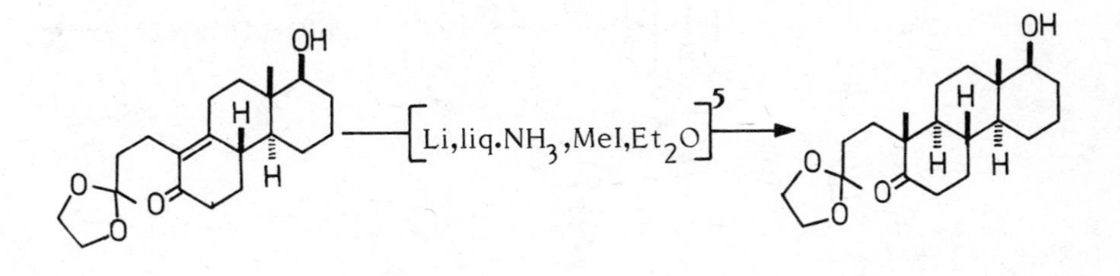

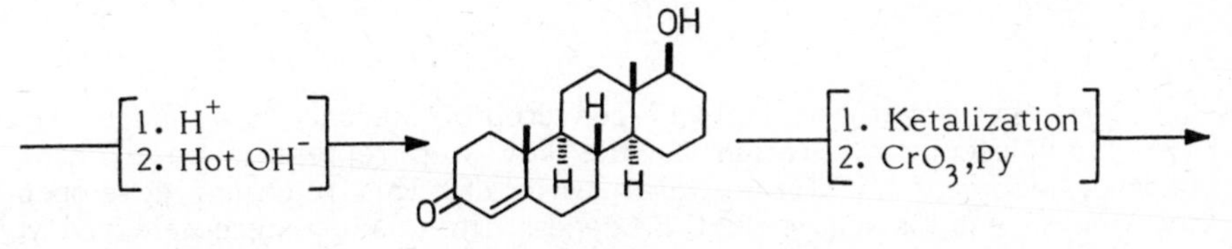

dl-D-Homotestosterone

---

4. This reaction is based on the lability of the N-O linkage of isoxazoles under special conditions, which makes possible their use as a potential ketoalkyl group. Hydrogenolysis with Raney nickel gives rise to a cyclic carbinolamine (i), which affords the annelated ketone on refluxing with aqueous alkali, Stork,G.; Danishefsky,S.; Ohashi,M. J. Am. Chem. Soc., 1967, 89, 5459.

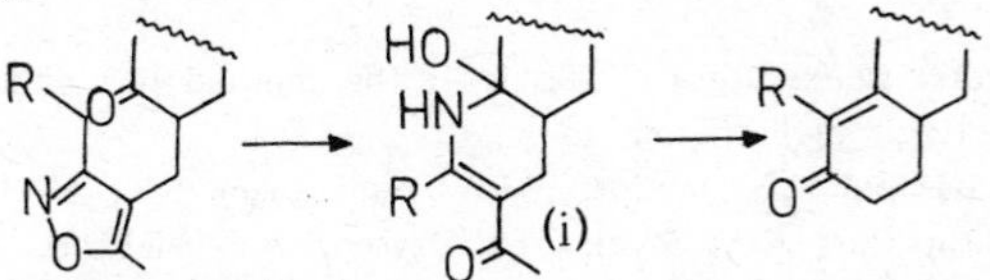

5. The alkylation-trapping method has been used for stereoselective introduction of the C-10 methyl group from the β-face, Stork,G.; Rosen,P.; Goldman,N.; Coombs,R.V.; Tsuji,J. J. Am. Chem. Soc., 1965, 87, 275.

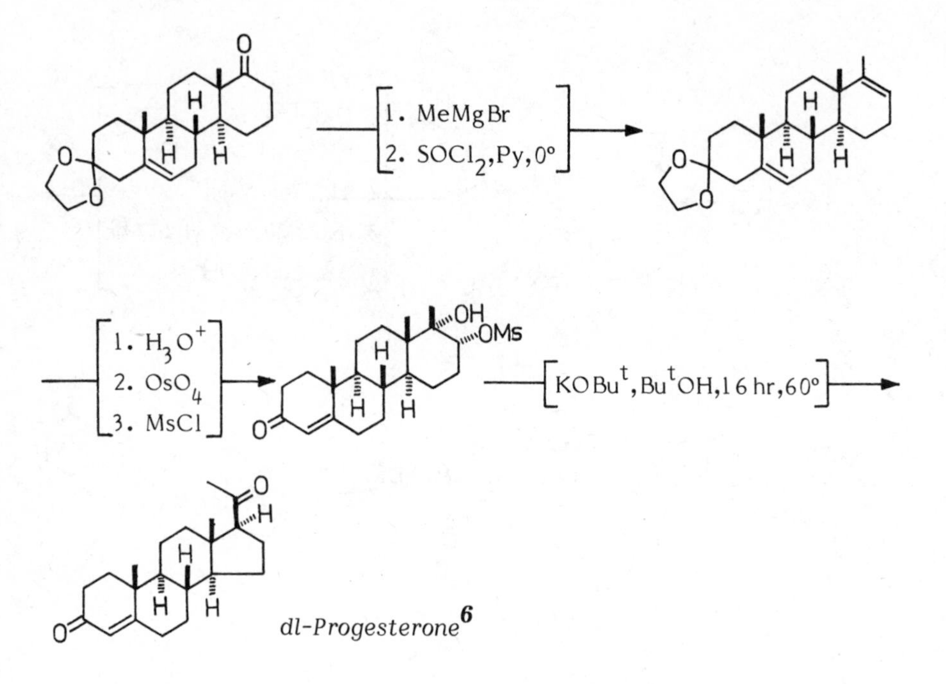

dl-Progesterone[6]

The recent synthesis of 16,17-dehydroprogesterone (7) by biogenetic type polyolefenic cyclisation as the key step represents an exciting new development in steroid total synthesis. This reaction, developed by Johnson and his school (8,9), proceeds with amazing stereoselectivity, all five steric centres of the required configuration being created in one step.

---

6. For a synthesis of progesterone using the "hydrochrysene approach", see: Johnson,W.S.; Marshall,J.A.; Keana,J.F.W.; Franck,R.W.; Martin,D.G.; Bauer,V.J. Tetrahedron, Supplement 8, 1966, Pt.II, 541.

7. Johnson,W.S.; Semmelhack,M.F.; Sultanbawa,M.U.S.; Dolak,L.A. J. Am. Chem. Soc., 1968, 90, 2994.

8. Johnson,W.S. Accts. Chem. Res., 1968, 1, 1; the method has already been illustrated in detail by the synthesis of Farnesol.

9. Johnson,W.S.; Gravestock,M.B.; McCarry,B.E. J. Am. Chem. Soc., 1971, 93, 4332; Johnson, W.S.; Gravestock,M.B.; Parry,R.J.; Myers,R.F.; Bryson,T.A.; Miles,D.H. J. Am. Chem. Soc., 1971, 93, 4330; Johnson,W.S.; Chen,Y.Q.; Kelogg,M.S. Biol. Act. Princ. Nat. Prod., 1984, 55; Johnson,W.S.; Chen.Y.Q.; Kellogg,M.S. Biol. Act. Princ. Nat. Prod., 1984, 55.

Preparation of Acetylenic Aldehyde (A)

Preparation of Progesterone

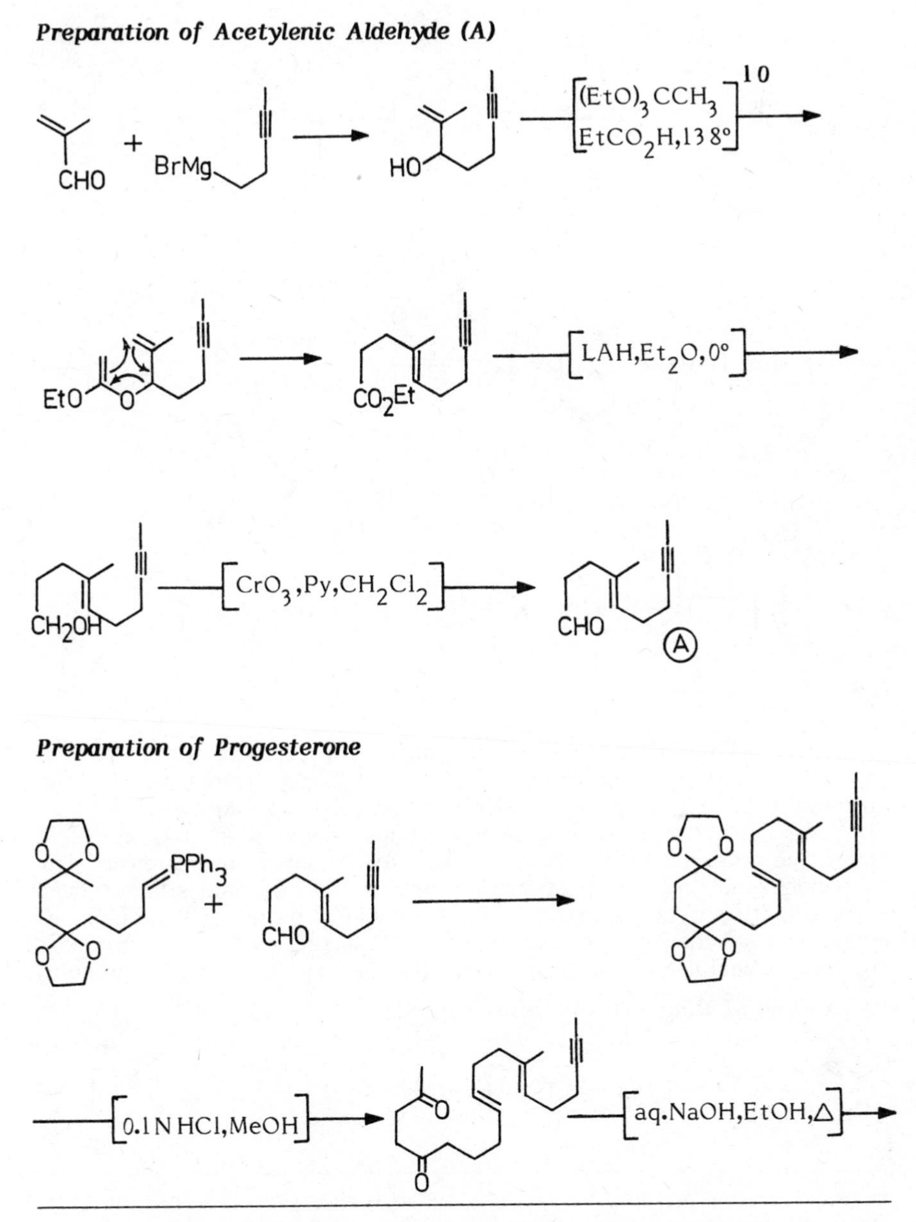

10. Ortho acetate Claisen reaction: Johnson,W.S.; Werthemann,L.; Bartlett,W.R.; Brocksom, T.J.; Li,T.; Faulkner,D.J.; Petersen,M.R. J. Am. Chem. Soc., 1970, 92, 741.

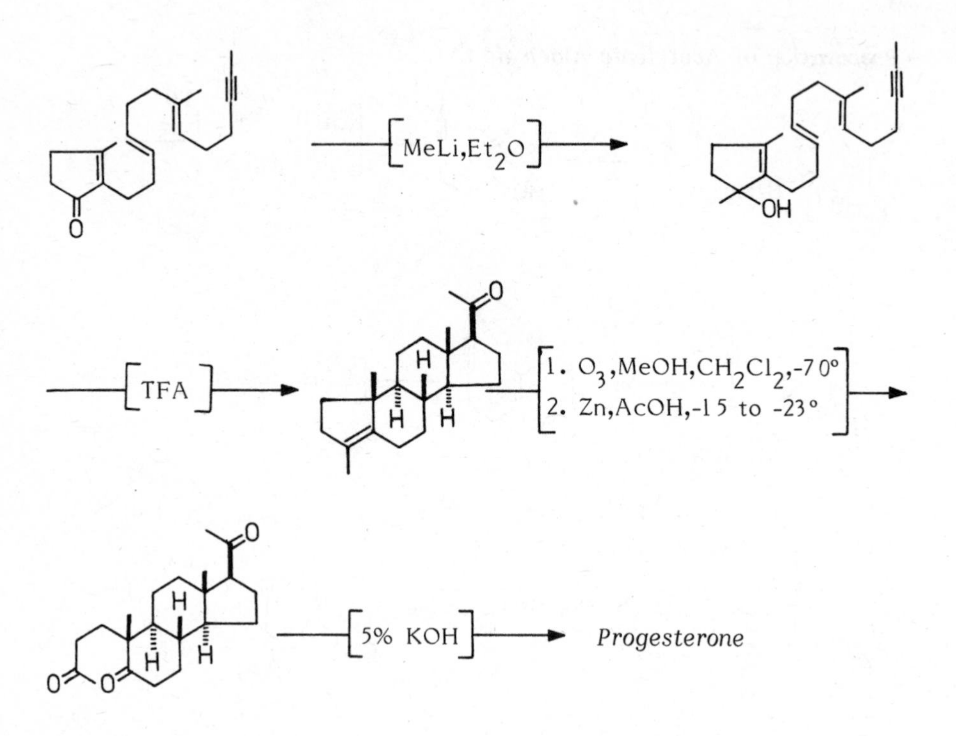

In a further refinement of this approach, Johnson and coworkers reported the direct construction of a complete steroid nucleus, in optically active form, by nonenzymic biogenetic-like cyclization (11,12). This was accomplished by incorporating, along with the terminal acetylenic residue into the key cyclization substrate, trienyol Ⓔ, a cyclohexenol moiety previously shown to be capable of initiating stereoselective cyclization to A/B cis decalin structures (13). Commencing with a substrate bearing a chiral C-5 the cyclization proceeds stereospecifically to yield an optically active tetracyclic product.

Preparation of Ring A/B Component

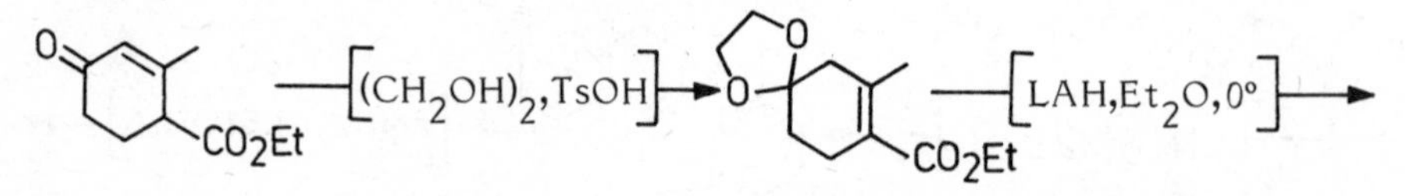

11. Markezich,R.L.; Willy,W.E.; McCarry,B.E.; Johnson,W.S. J. Am. Chem. Soc., 1973, 95, 4414.

12. McCarry,B.E.; Markezich,R.L.; Johnson,W.S. J. Am. Chem. Soc., 1973, 95, 4416.

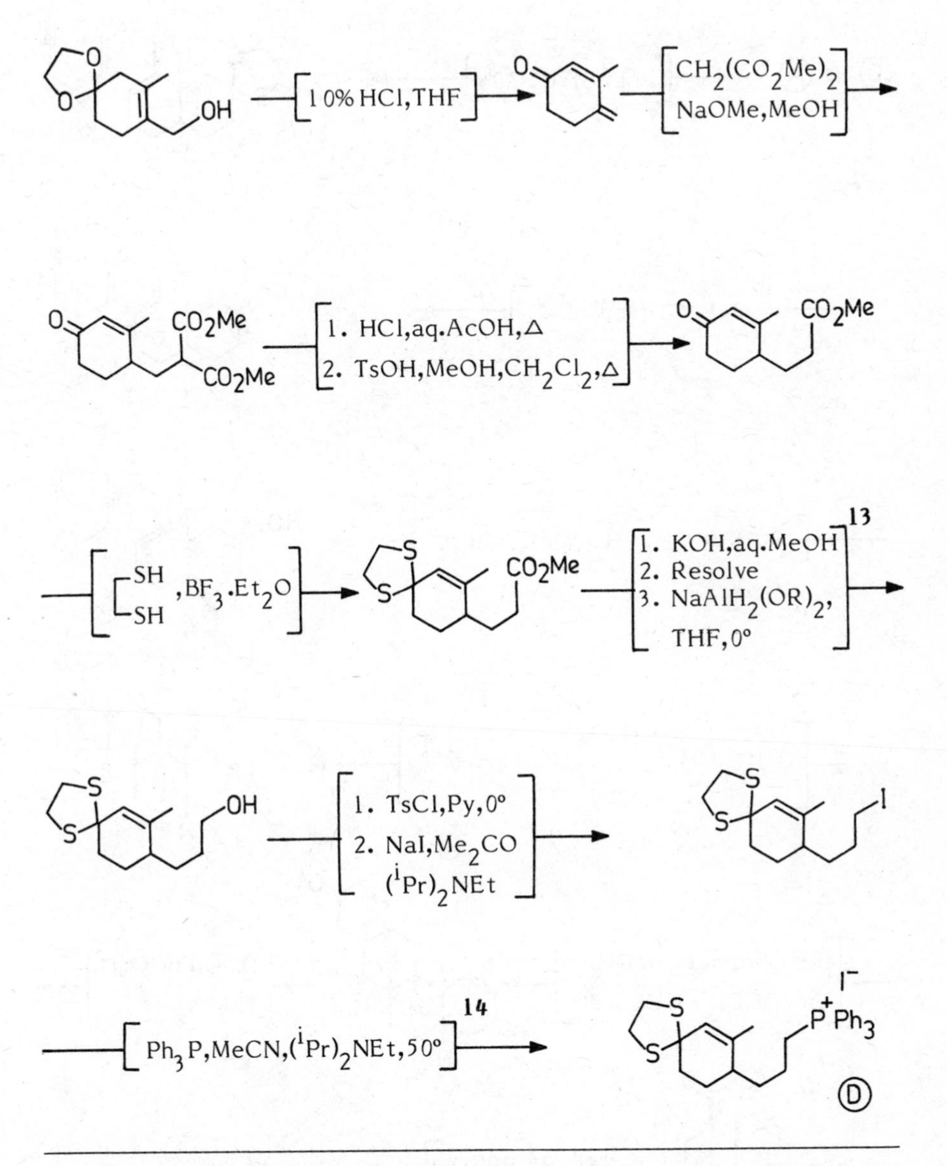

13. Johnson,W.S.; Neustaedter,P.J.; Schmiegel,K.K. J. Am. Chem. Soc., 1965, 87, 5148.
14. Diisopropylethylamine prevented migration of double bond to β,γ-position.

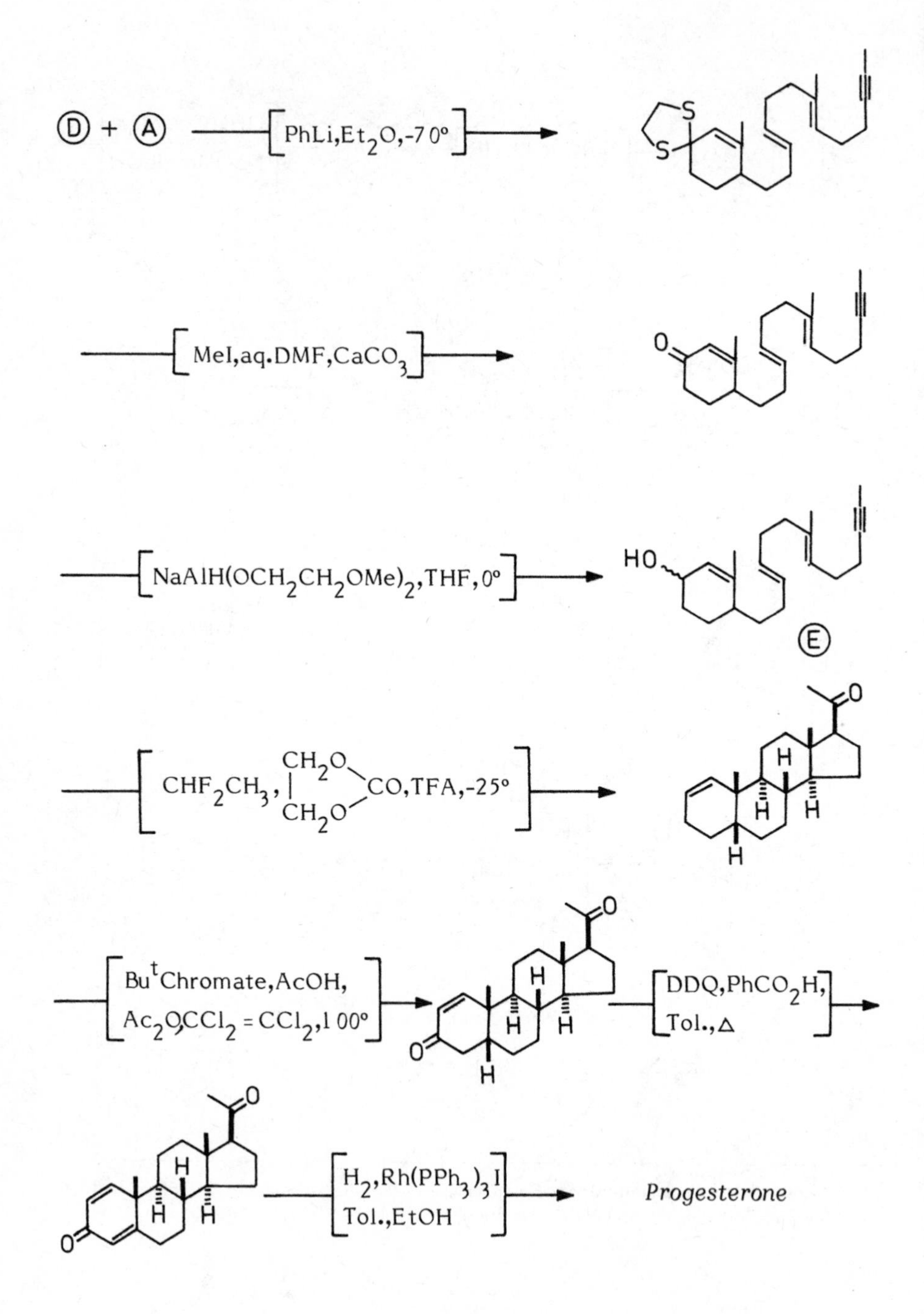
D + A
PhLi,Et2O,-70°
S
S
MeI,aq.DMF,CaCO3
O
NaAlH(OCH2CH2OMe)2,THF,0°
HO
E
CHF2CH3, CH2O CH2O CO,TFA,-25°
O
H
H
H
H
ButChromate,AcOH,
Ac2O,CCl2 = CCl2,100°
DDQ,PhCO2H,
Tol.,Δ
H2,Rh(PPh3)3I
Tol.,EtOH
Progesterone

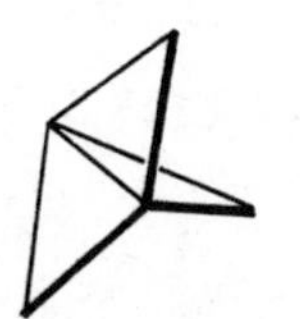

[1.1.1]PROPELLANE

The tetrahedral valency distribution of saturated carbon dictates that in any representation, only 3 of the ligands can be in the same hemisphere. Bending these ligands in the direction of the 4th by an "umbrella folding" operation, would steeply increase the strain. The fact that even the ultimate in such an operation, namely, [1.1.1]-propellane can be made (1), and is found stable, provides yet another example of the remarkable flexibility permitted in carbon constellations.

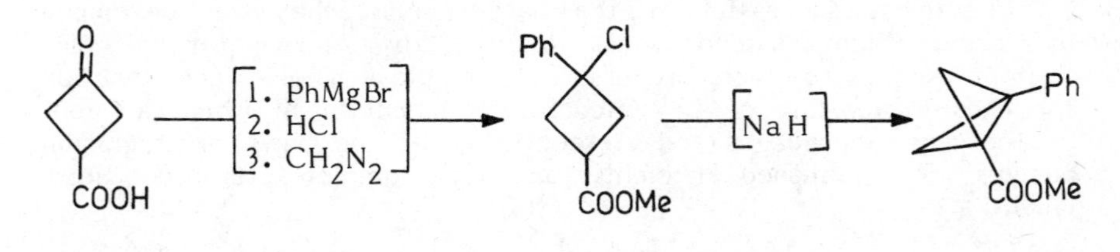

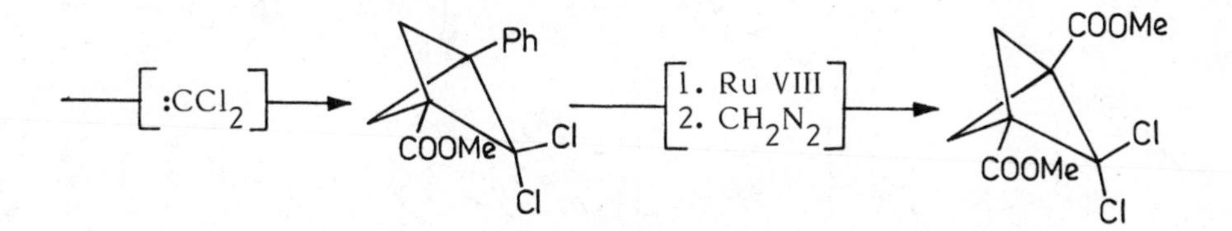

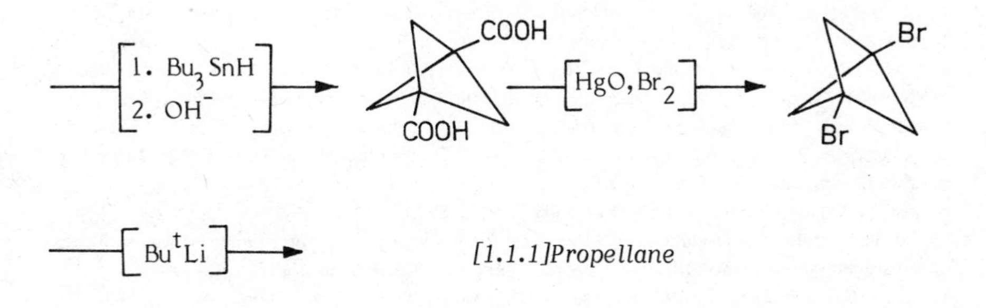

---

1. Wiberg,K.B. Acc. Chem. Res., 1984, 17, 379.

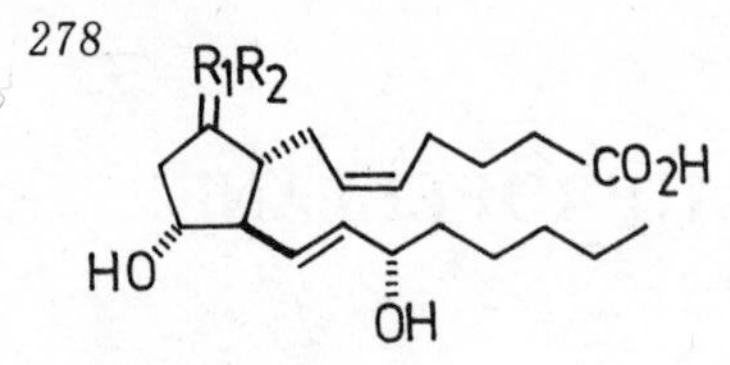

PROSTAGLANDINS

E_2, $R_1R_2 = O$

$F_2\alpha$, $R_1 = \alpha OH$, $R_2 = \beta H$

The number and variety of syntheses of these biologically important molecules which are now a legion (1,2) cover a wide span from the first almost 20 step synthesis of Just and Simonovitch (3) to the latest one pot, two step, greatly flexible synthesis of Suzuki et al. from an optically active 4-hydroxy-cyclopentenone (13).

The most successful of the earlier approaches was developed by Corey, which provided access to all of the primary prostaglandins from a single resolved precursor. It employs Diels-Alder reaction for construction of the key bicyclic intermediate in which all four of the ring appendages and stereocenters of the basic prostaglandin nucleus are established efficiently and with complete stereospecificity (4-6).

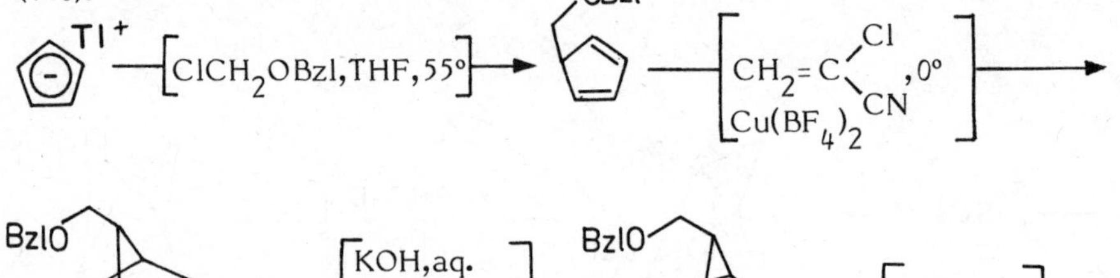

1. For a discussion of earlier synthetic strategies and classification of prostaglandins see: Bindra,J.S.; Bindra,R. Prostaglandin Synthesis, Academic Press, New York, 1977.
2. Reviews: Roberts,S.M.; Scheinmann,F. eds. "New Synthetic Routes to Prostaglandins and Thromboxanes", Academic Press, New York, 1982; Bindra,J.S. in "The Total Synthesis of Natural Products", Vol.4, ed., ApSimon,J. Wiley-Interscience, New York, 1981, p.353; Crabbe,P. ed. "Prostaglandin Research", Academic Press, New York, 1977; Caton,M.P.L.; Crowshaw,K. Progr. Med. Chem., 1978, 15, 357; Nicolaou,K.C.; Gasic,G.P.; Barnett,W. Angew. Chem. Int. Ed., 1978, 17, 293.
3. Just,G.; Simonovitch,C. Tetrahedron Lett., 1967, 2093.
4. For an excellent introduction to the strategy underlying the program of total synthesis of prostaglandins at Harvard, see: Corey,E.J. Ann. N.Y. Acad. Sci., 1971, 180, 24.
5. Corey,E.J.; Weinshenker,N.M.; Schaaf,T.K.; Huber,W. J. Am. Chem. Soc., 1969, 91, 5675.
6. Corey,E.J.; Schaaf,T.K.; Huber,W.; Koelliker,V.; Weinshenker,N.M. J. Am. Chem. Soc., 1970, 92, 397.

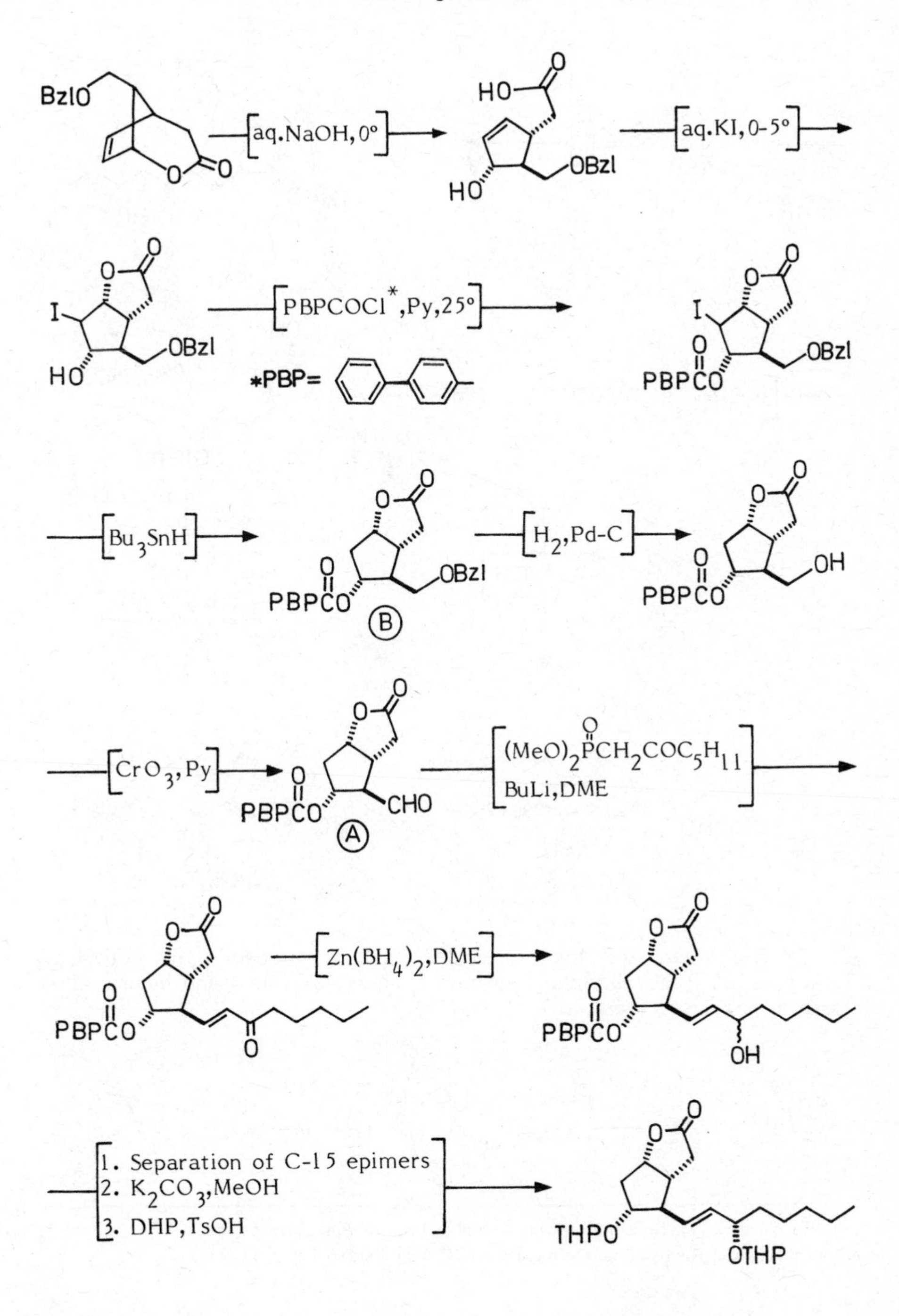
BzlO
aq.NaOH, 0°
HO
OBzl
aq.KI, 0-5°
I
PBPCOCl*, Py, 25°
*PBP=
PBPCO
Bu3SnH
B
H2, Pd-C
OH
CrO3, Py
CHO
A
(MeO)2PCH2COC5H11
BuLi, DME
Zn(BH4)2, DME
1. Separation of C-15 epimers
2. K2CO3, MeOH
3. DHP, TsOH
THPO
OTHP

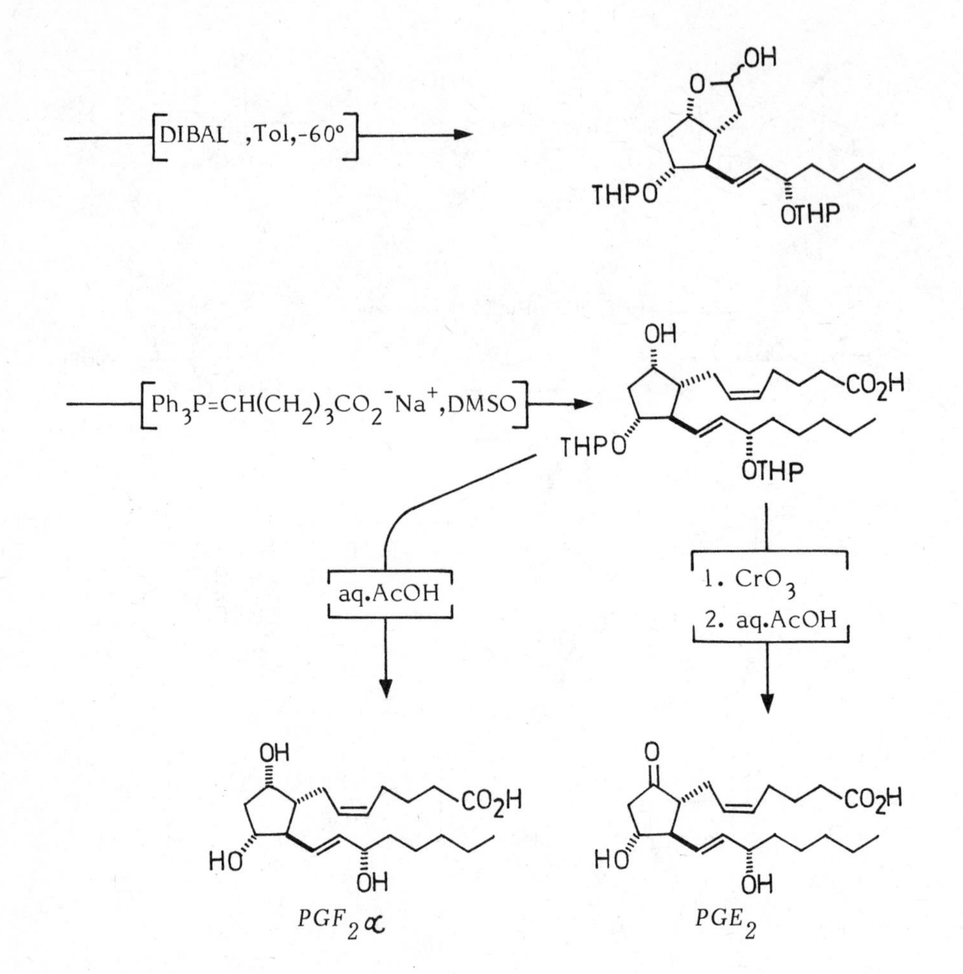

Ranganathan and his colleagues have developed a nitroethylene route to the Corey lactone B, which is short, practical and incorporates several novel features (7).

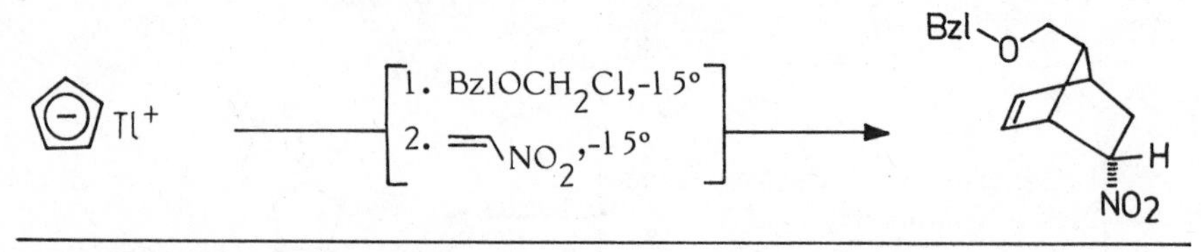

7. Ranganathan,S.; Ranganathan,D.; Mehrotra,A.K. J. Am. Chem. Soc., 1974, 96, 5261; Ranganathan,S.; Ranganathan,D.; Mehrotra,A.K. Tetrahedron Lett., 1975, 1215.

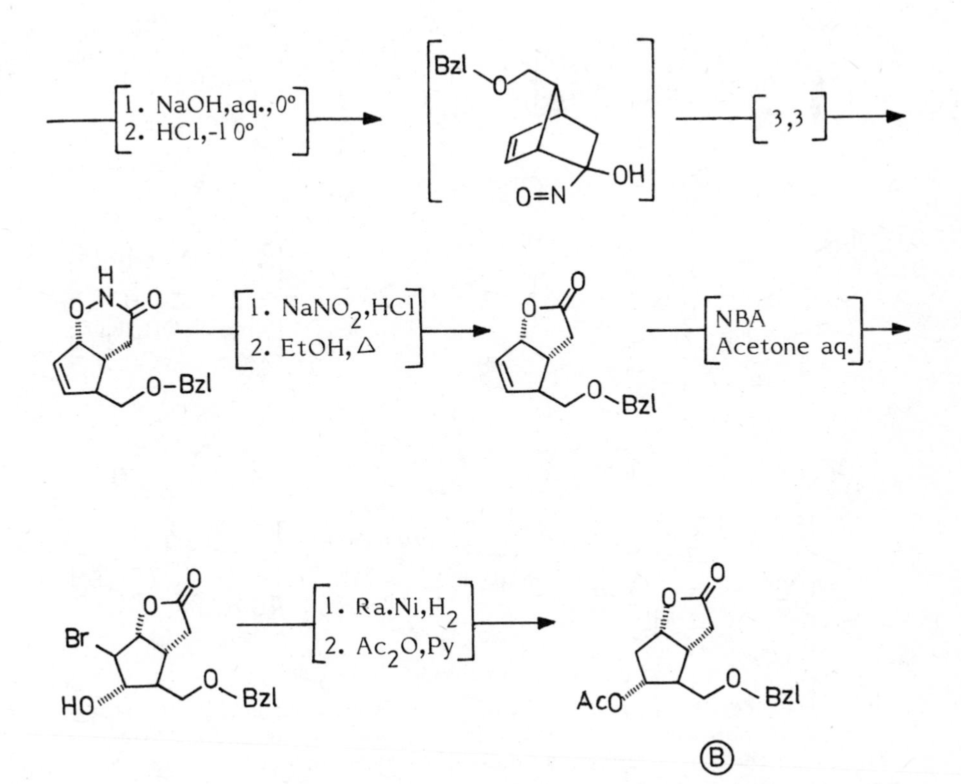

An alternative route to the Corey aldehyde (A), which was developed at Pfizer (8), utilizes an unusual Prins reaction on norbornadiene.

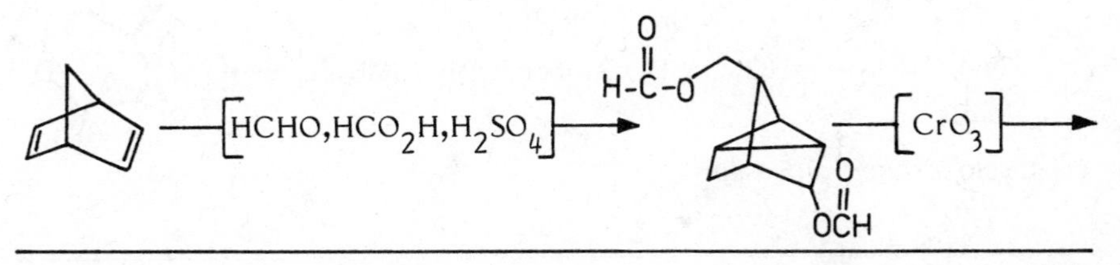

8. Bindra,J.S.; Grodski,A.; Schaaf,T.K.; Corey,E.J. J. Am. Chem. Soc., 1973, 95, 7522.

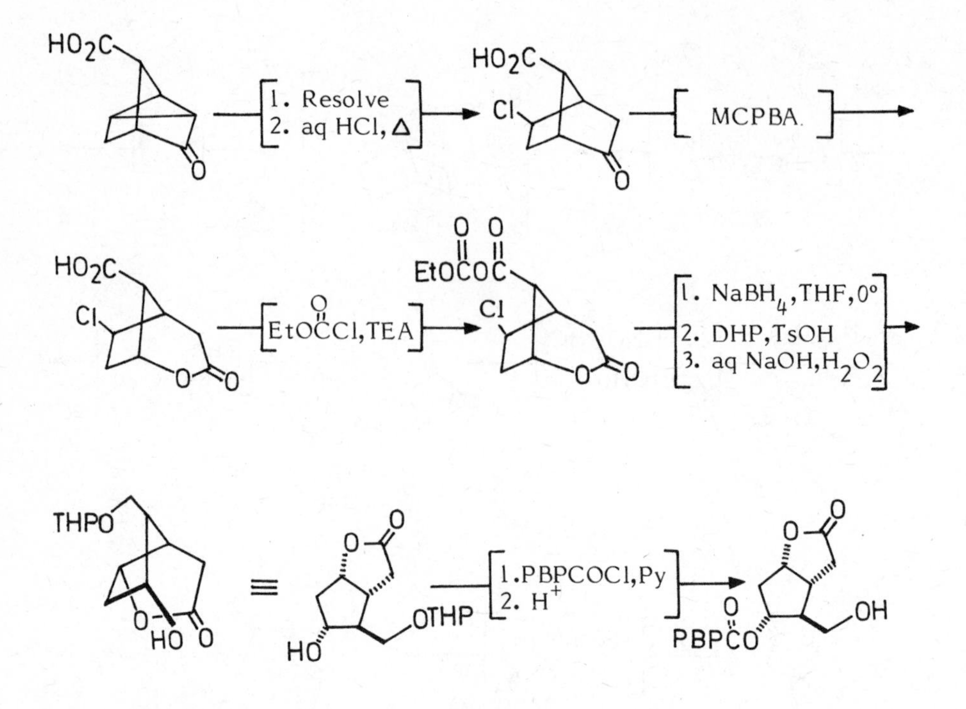

The stereospecific synthesis of $PGF_2\alpha$ by Woodward and his colleagues at Basel is noteworthy for its ingenious use of functional groups as internal protecting agents and their application to exert stereochemical control during synthetic operations. The synthesis commences with cyclohexane-1,3,5-triol, leading to a cyclohexyl amino diol Ⓐ which undergoes ring contraction to establish the prostanoid cyclopentane nucleus (9).

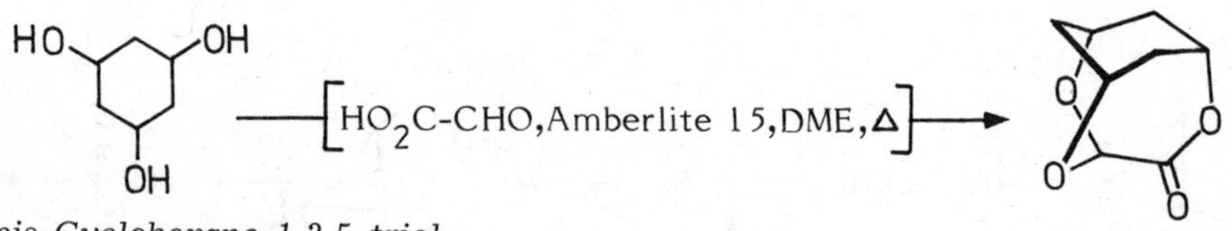

cis-Cyclohexane-1,3,5-triol

---

9. Woodward, R.B.; Gosteli, J.; Ernest, I.; Friary, R.J.; Nestler, G.; Raman, H.; Sitrin, R.; Suter, Ch.; Whitesell, J.K. J. Am. Chem. Soc., 1973, 95, 6853.

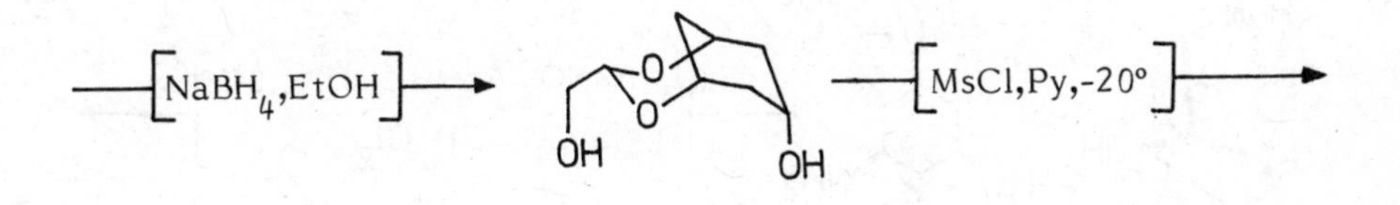
NaBH4,EtOH
O
O
OH
OH
MsCl,Py,-20°

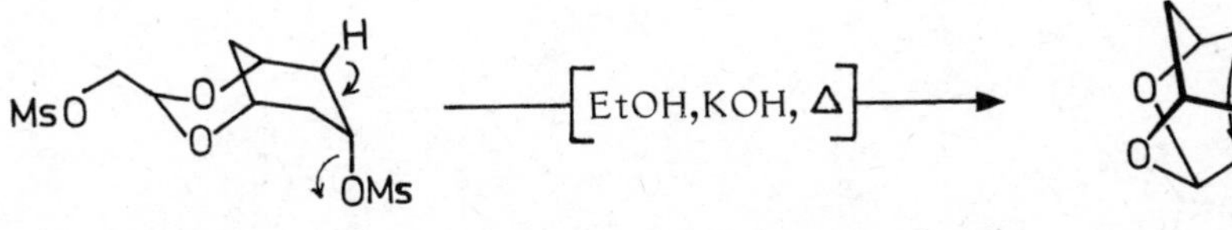
MsO
O
O
H
OMs
EtOH,KOH, Δ

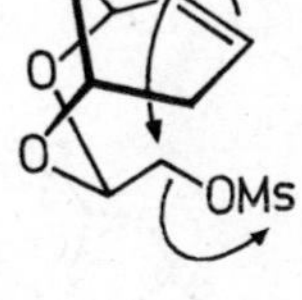
O
O
OMs

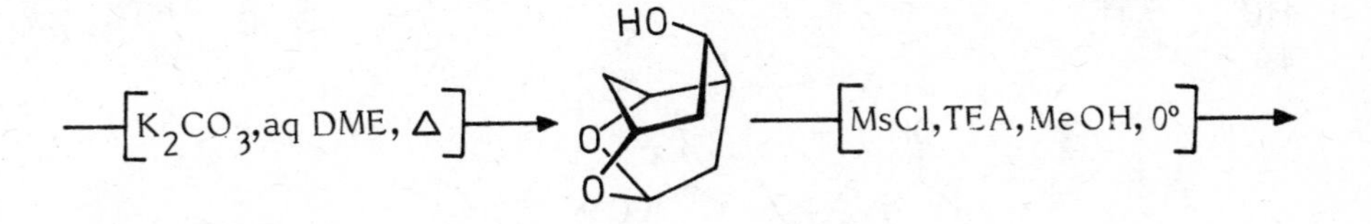
K2CO3,aq DME, Δ
HO
O
O
MsCl,TEA,MeOH, 0°

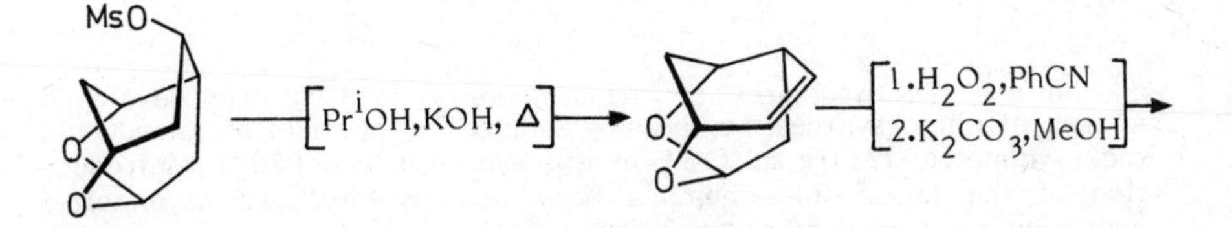
MsO
O
O
PriOH,KOH, Δ
O
O
1.H2O2,PhCN
2.K2CO3,MeOH

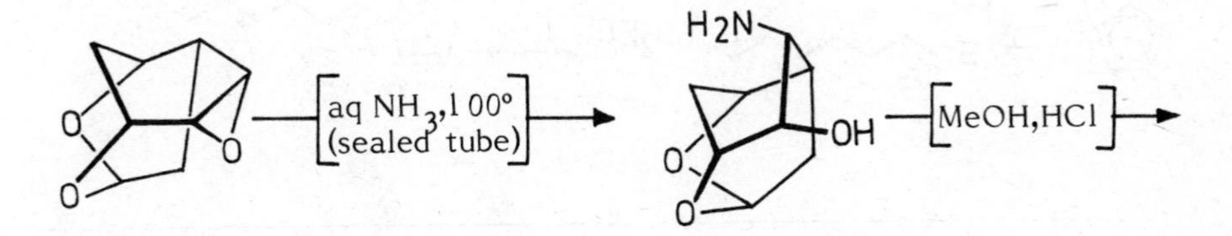
O
O
O
aq NH3,100°
(sealed tube)
H2N
OH
O
O
MeOH,HCl

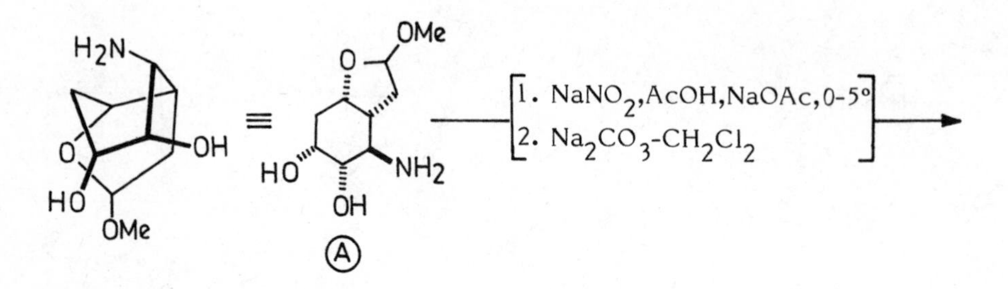

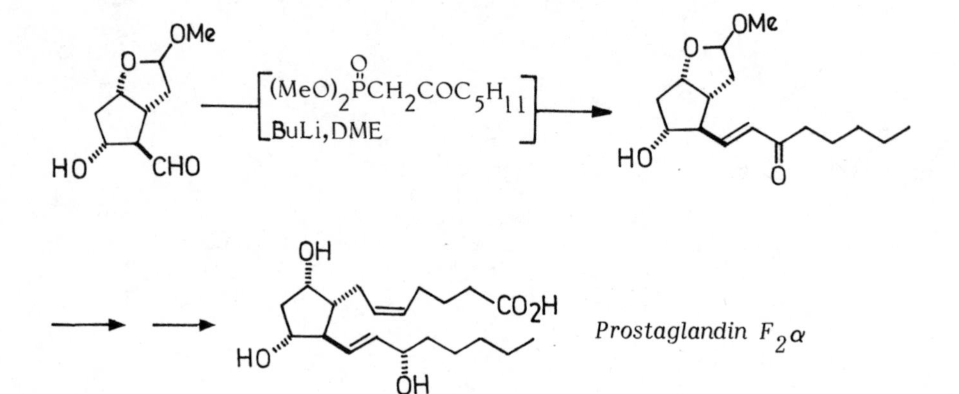

Sih and his associates at the University of Wisconsin developed a short efficient synthesis of prostaglandins. The problem of introducing the asymmetric centre at C-15 in this synthesis is solved by introduction of the lower side chain, already bearing the C-15 asymmetric centre by a 1,4-addition reaction (10).

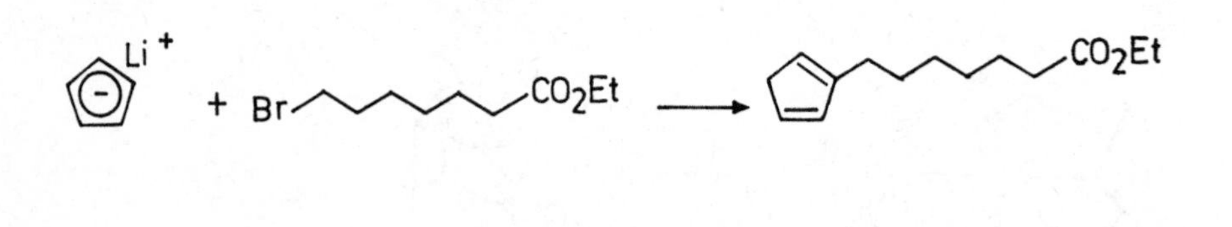

10.a) Sih,C.J.; Salomon,R.G.; Price,P.; Peruzzotti,G.; Sood,R. Chem. Commun., 1972, 240.
b) Sih,C.J.; Price,P.; Sood,R.; Salomon,R.G.; Peruzzotti,G.; Casey,M. J. Am. Chem. Soc., 1972, 94, 3643.

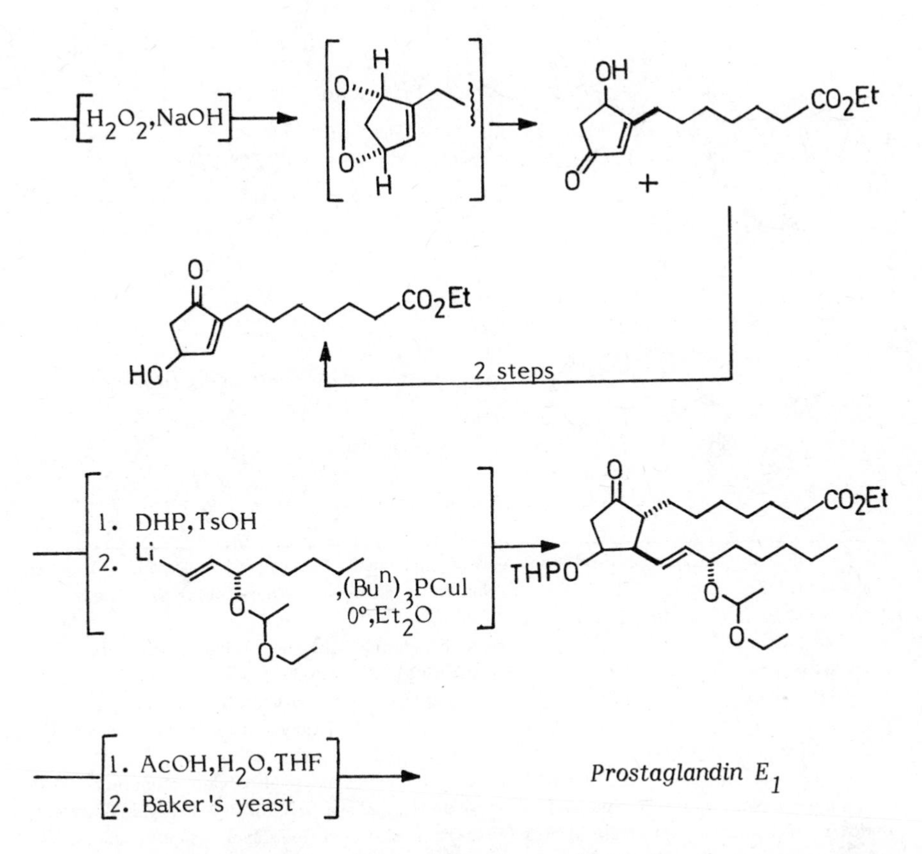

An 'ideal' approach to PG synthesis would be the three-component coupling process (11), conceived many years ago by the Syntex group (12) but realised only recently, which offers "an extremely short way to PG's" (13).

---

11. This was conceived as building the whole framework by coupling three units by the following strategy:

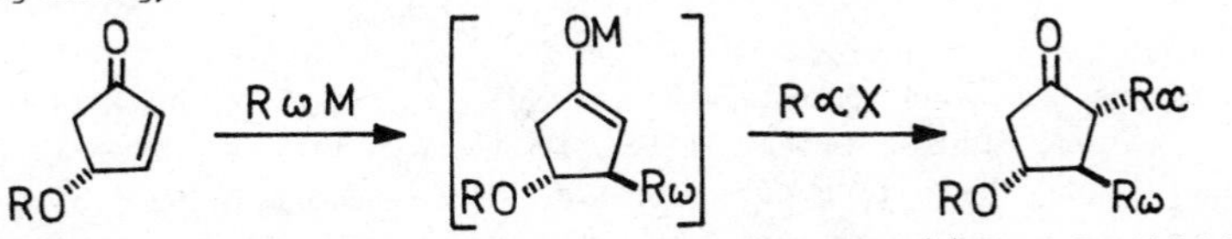

12. (a) Patterson,J.W.Jr.; Fried,J.H. J. Org. Chem., 1974, 39, 2506. (b) Davis,R.; Unich,K.G. ibid, 1979, 44, 3755.

13. Suzuki,M.; Yanagisawa,A.; Noyori,R. J. Am. Chem. Soc., 1985, 107, 3348.

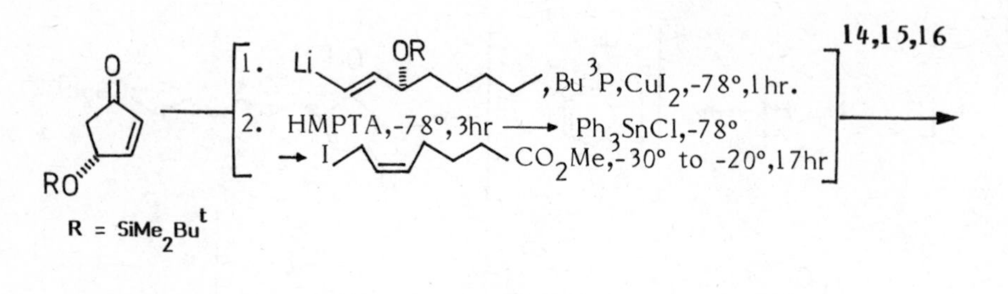

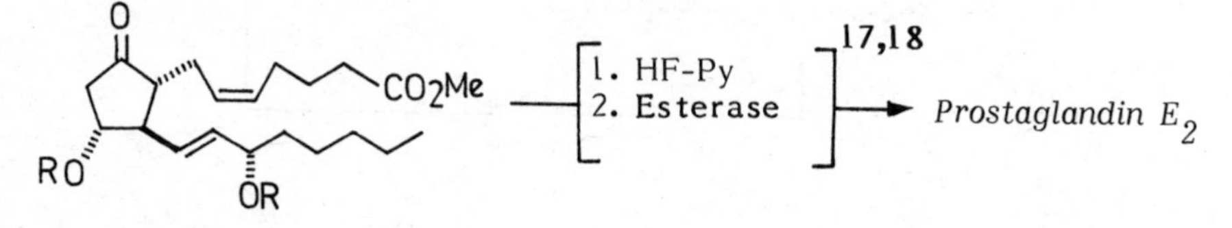

---

14. For optically active cyclopentenone and ω-side chain see: Noyori,R.; Suzuki,M. Angew. Chem. Intl. Ed. Engl., 1984, 23, 847; Noyori,R.; Tomino,I.; Yamada,M.; Nisizawa,M. J. Am. Chem. Soc., 1984, 106, 6717.

15. The conjugate addition proceeds in a completely stereoselective manner to give after quenching only the C-11/C-12 (PG numbering) trans product.

16. The success was due to the lithium (or copper) to tin transmetallation in the enolate stage: c.f. Tardella,P.A. Tetraahedron Lett., 1969, 117; Nishiyama,H.; Sakuta,K.; Itoh,K. ibid, 1984, 25, 223, 2487.

17. In a similar manner PGE, and PG's of the D&I series have also been prepared. Use of propargylic iodide as the α-side chain unit allowed the synthesis of the acetylenic compound given below in one step in 82% yield as a single stereoisomer which could serve as a common precursor of PG family.

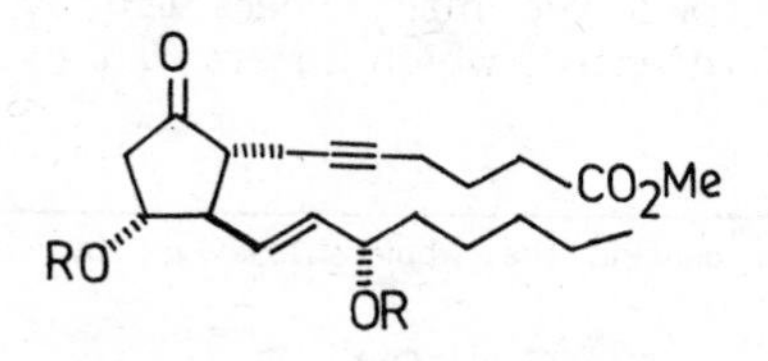

18. For other successful applications of the conjugate addition approach to the synthesis of PGF_2, see: Stork,G.; Isobe,M. J. Am. Chem. Soc., 1975, 97, 4745, 6260; Stork,G.; Kraus,G. J. Am. Chem. Soc., 1976, 98, 6747. The Columbia synthesis of $PGF_{2\alpha}$ from D-glucose is also particularly noteworthy, see: Stork,G.; Takahashi,T.; Kawamoto,I.; Suzuki, T. J. Am. Chem. Soc., 1978, 100, 8272.

287

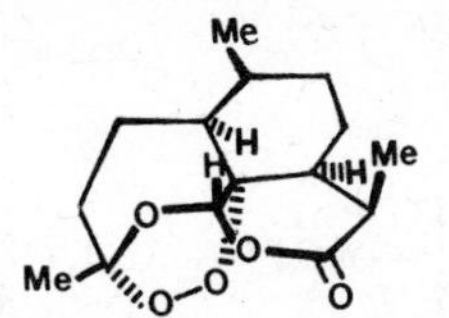

QINGHAOSU

Qinghaosu (also named artemesinin and arteannuin) isolated from Artemesia annua possessing an entirely novel structure for antimalarial activity, is perhaps the most important drug isolated in recent years from plants used in traditional systems of medicine. Its synthesis by Roche scientists (2) starts from (-)-isopulegone and the other chiral centres were generated by optical steric induction. The crucial acetal-lactone endoperoxide bridge was formed through an internal hydroperoxide cyclisation.

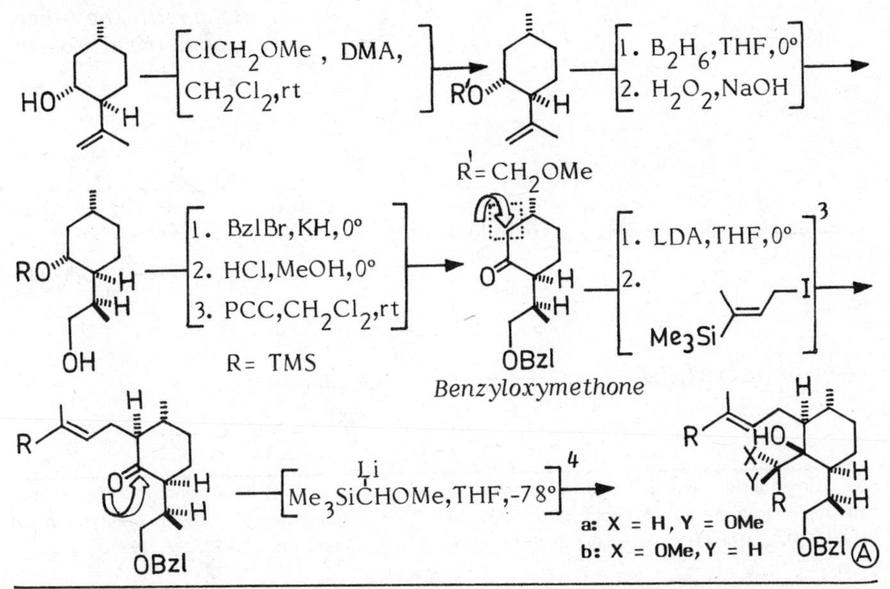

1. Liw,Jing-Ming; Ni,Mu-Yun; Yu-Fen; Tu,You-You; Wu,Zhao-Hua; Yu-Lin-Wu; Chou,Wei-Shan. Acta Chim. Sinica, 1979, 37, 129.

2. Schmid,G.; Hofheinz,W. J. Am. Chem. Soc., 1983, 105, 624.

3. This reaction provided a 6:1 excess of the required epimer.

4. The stereochemistry of (A) was assigned on the considerations that large nucleophiles attack cyclohexanones preferentially from the equatorial side and therefore, both epimers a and b must have the hydroxyl group in the axial position. With equimolar quantity of the reagent a 1:1 ratio of the epimers was obtained, but with a 10-fold excess of the reagent the ratio of a to b was shifted to 8:1, and the required epimer isolated in 89% yield. This stereoselectivity was considered to be the result of kinetic resolution of the organolithium reagent by the chiral ketone. The unambiguous assignment of configuration to the epimer followed from the result of the subsequent transformations.

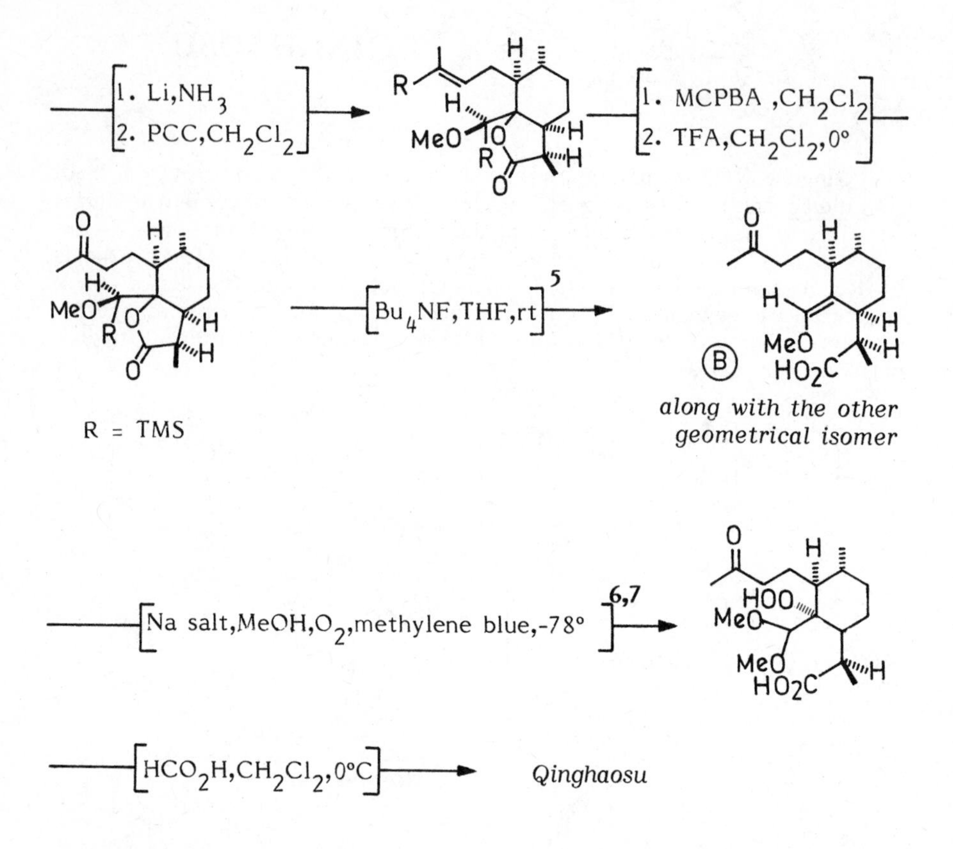

5. The fluoride ion induced β-elimination is stereospecific; it appears to be similar to acid catalysed synchronous anti-periplanar E_2 β-elimination of β-hydroxyalkyl silanes; Hudrlik,P.F.; Rona,R.J.; Misra,R.N.; Withers,G.P. J. Am. Chem. Soc., 1977, 99, 1993.

6. Asveld,E.W.S.; Kellogg,R.M. J. Am. Chem. Soc., 1980, 102, 3644.

7. A complex mixture of products was formed in this reaction; although none of the products could be identified, but from the course of the reaction of the next step which yielded qinghaosu in 30% yield the structure shown must be the main product. When this reaction was carried out at 0° in presence of methylene blue, Ⓒ shown below was formed.

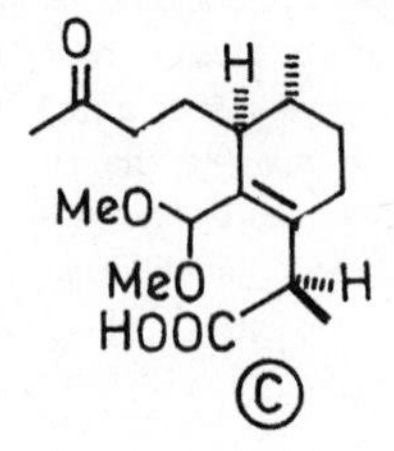

Chinese synthesis

In a somewhat different approach the Chinese scientists (8) have carried out the synthesis of qinghaosu starting from R(+)-citronellal (9); the intermediate Ⓐ, corresponding to the acid Ⓑ of Schmid & Hofheinz (2) was obtained in 19 steps, which on photo-oxidation to create the peroxide bridge followed by acid treatment gave qinghaosu. Hydroxylation of Ⓐ with OsO_4 yielded the corresponding ether, deoxy-arteannuin.

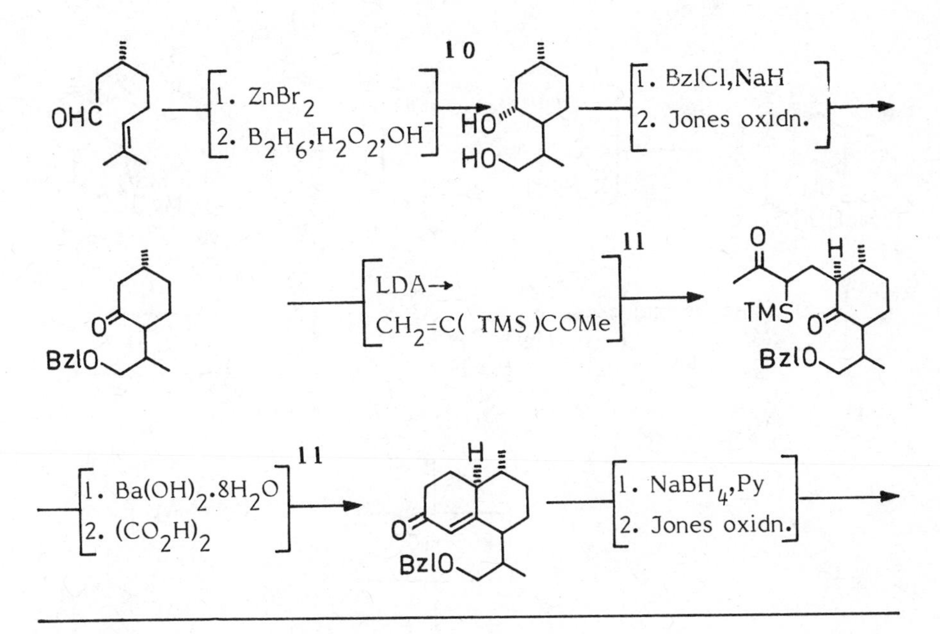

---

8. Xing-Xiang,Xu; Jie,Zhu; Da-Zhong,Huang; Wei-Shan,Zhou, Tetrahedron, 1986, 42, 819.
9. A review of the total synthesis of qinghaosu and related sesqinterpenes isolated from A. annua has been published; Zhou Wei-Shan, Pure & Appl. Chem., 1986, 58, 817.
10. Nakatani,Y., Kawashima,K.. Synthesis, 1978, 147; Schulte-Elte,K.H., Ohloff,G., Helv. Chim. Acta, 1967, 50, 153.
11. Use of NaOH in this cyclisation resulted in the inversion of configuration at C_7.

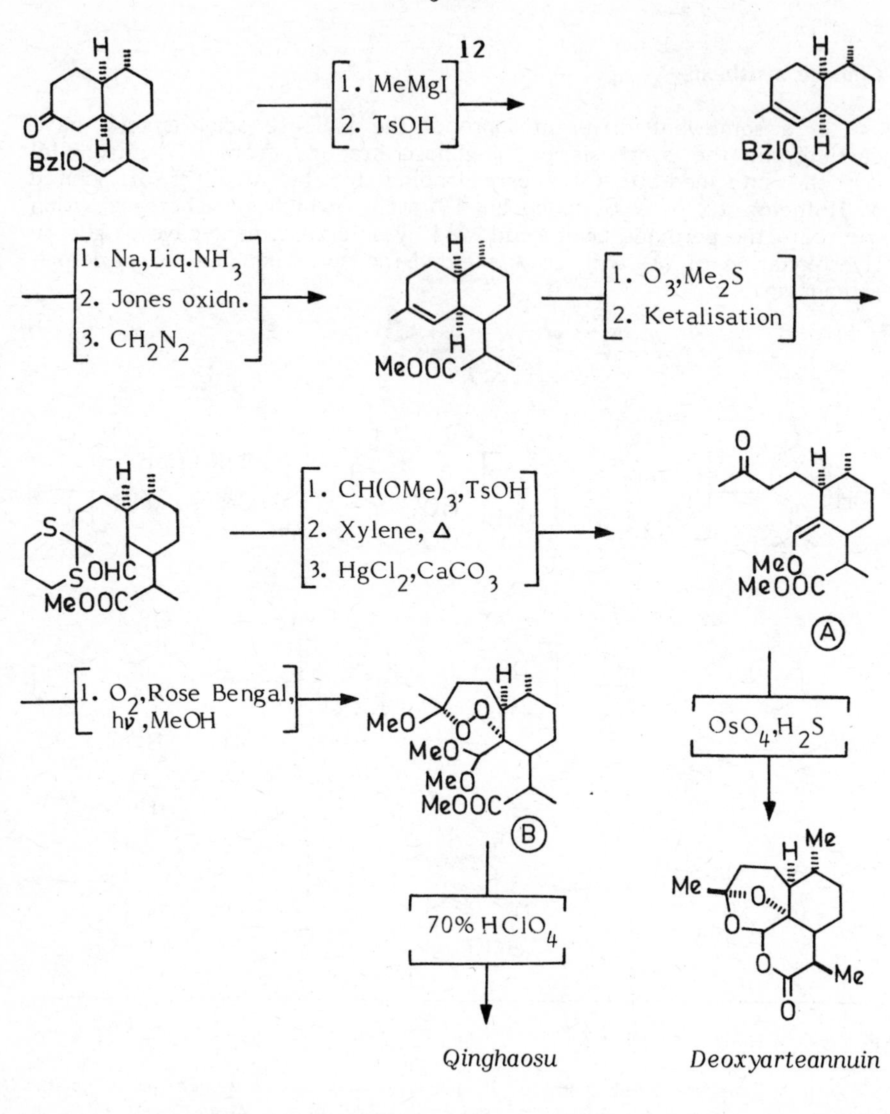

12. (A) 1:1 mixture of this and the Δ^3-isomer was obtained, which were separated by flash chromatography.

QUASSIN

Quassinoids, the bitter principles of the Simarubaceae family have attracted much attention recently because of the wide spectrum of biological activities shown by them, which include anticancer and antiparasitic activities (1). The total synthesis described herein of dl-quassin, the parent compound of this group, assembles a key intermediate Ⓐ by a Diels-Alder reaction which six of the seven chiral centres found in quassin (2).

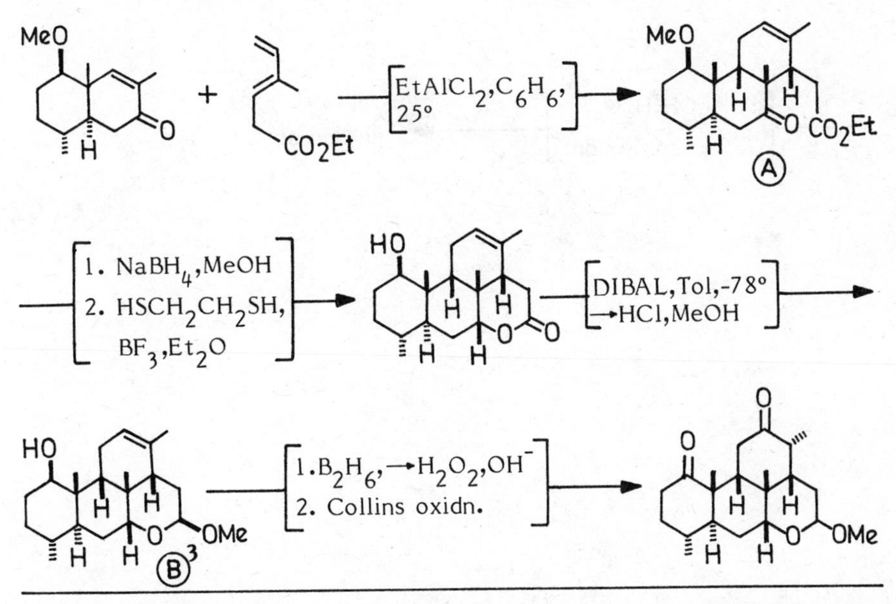

1. For comprehensive review of quassinoids see: Polonsky,J. in "Progress in the Chemistry of Natural Products", 1973, 30, 101; 1985, 47, 221.

2. Grieco,P.A.; Ferrino,S.; Vidari,G. J. Am. Chem. Soc., 1980, 102, 7586.

3. From the tetracyclic alcohol Ⓑ Grieco *et al.* have by similar synthetic reactions obtained dl-castelanolide: Grieco,P.A.; Lis,R.; Ferrino,S.; Jaw,J.Y. J. Org. Chem., 1982, 47, 601.

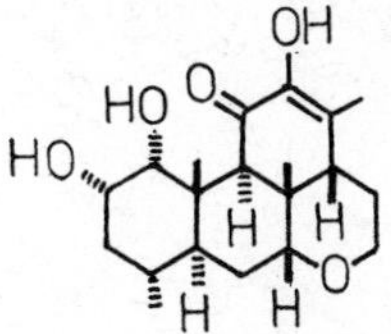

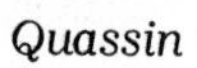

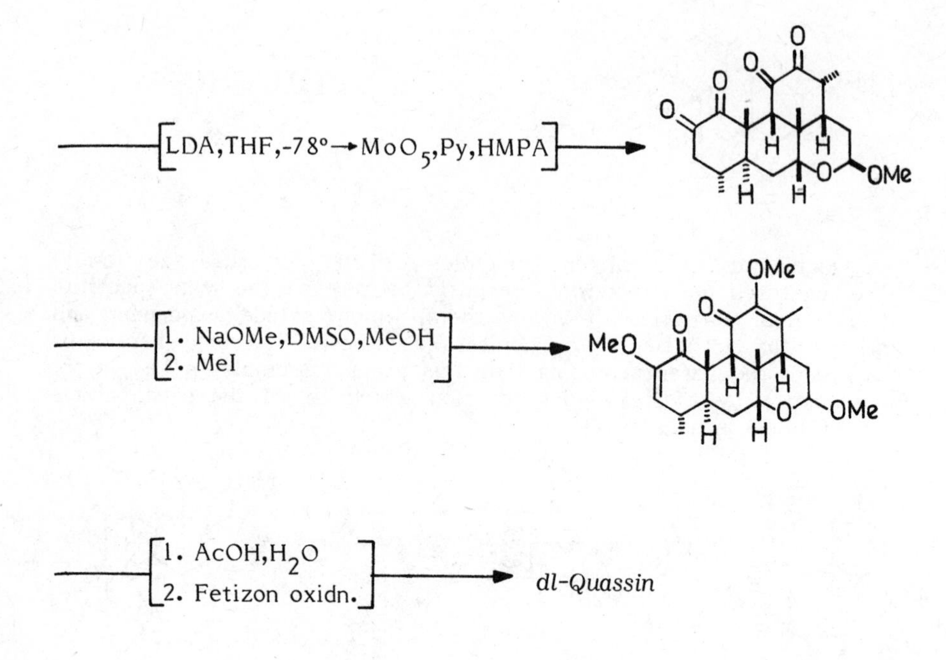
LDA,THF,-78°→MoO5,Py,HMPA
O
O
O
O
H
H
H
H
OMe
1. NaOMe,DMSO,MeOH
2. MeI
OMe
O
O
MeO
H
H
H
H
O
OMe
1. AcOH,H2O
2. Fetizon oxidn.
dl-Quassin

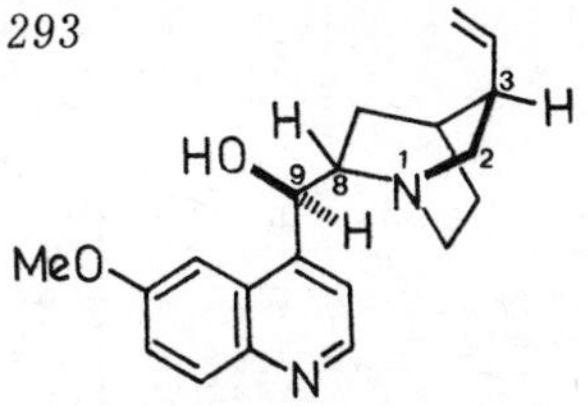

QUININE

The total synthesis of quinine by Woodward & Doering in 1944 (1) unshered in the modern era of the appreciation of the art in organic synthesis, when organic synthesis started being pursued not only to prove the structure of a natural product but for the creative elements in it and for its possible practical utility (2,3).

Rabe's synthesis of dihydroquinine and partial synthesis of quinine from quinotoxine (4,5) starting in 1910s brought into proper focus the problems involved and the direction for the synthesis of quinine. The intense effort expended over some forty years along various lines of investigations (3) culminated in the ingenious synthesis of quinotoxine by Woodward and Doering (1). The noteworthy feature of this synthesis is obtention of the cis-3,4-disubstituted piperidine Ⓐ from 7-hydroxyisoquinoline, with provision for generating the vinyl group by a Hofmann elimination.

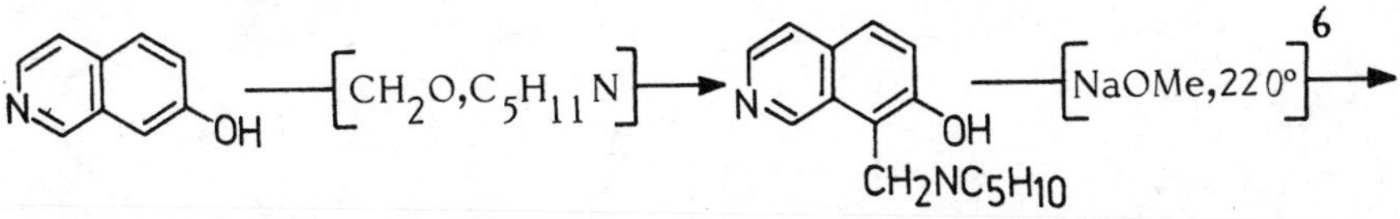

1. Woodward,R.B.; Doering,W.E. J. Am. Chem. Soc., 1944, 66, 849; 1945, 67, 860.
2. "Art in Organic Synthesis", Anand,N.; Bindra,J.S.; Ranganathan,S. Holden-Day Inc., San Francisco, 1970; "Creativity in Organic Synthesis", Bindra,J.S.; Bindra,R. Academic Press, 1975; Unique role played by physical chemistry in Chemical Sciences, Keniti,H.; Senryo to Yahukin, 1984, 29, 173.
3. For a review of the classic work on synthesis of Cinchona alkaloids, see: Turner,R.B.; Woodward,R.B. The Alkaloids, 1953, 3, 1; some of the later developments in this field have been reviewed by Uskokovic,M.R.; Grethe,G. The Alkaloids, 1973, 14, 181.
4. Rabe,P.; Kindler,K. Ber., 1918, 51, 466.
5. Rabe,P.; Huntenberg,W.; Schultze,A.; Volger,G. Ber., 1931, 64, 2487.
6. Hydride transfer from methoxide, cf. Cornforth,J.W.; Cornforth,R.; Robinson,R. J. Chem. Soc., 1942, 682.

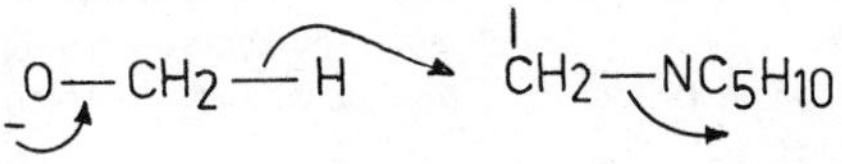

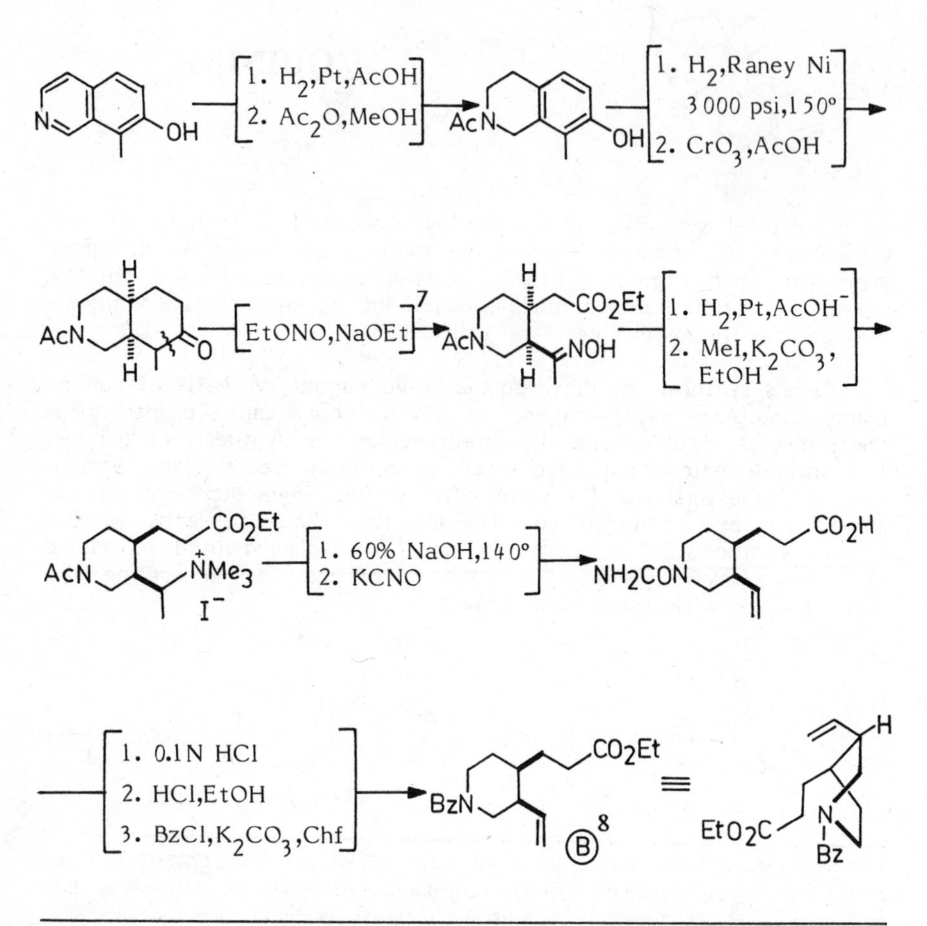

7. This reaction involves the heterolytic C-C fission in an intermediate tertiary nitroso compound (i).

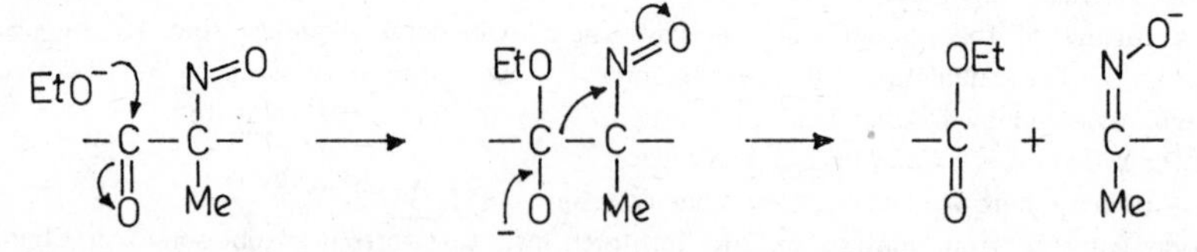

8. Uskokovic and his associates in the course of their extensive studies on the synthesis of cinchona alkaloids developed an alternative efficient synthesis for N-benzoylhomomeroquinene ethyl ester (12C) by employing Hofmann-Loffler-Freytag chlorination of side chain and formed the vinyl group by dehydrochlorination, which considerably improved the preparation of quinotoxine.

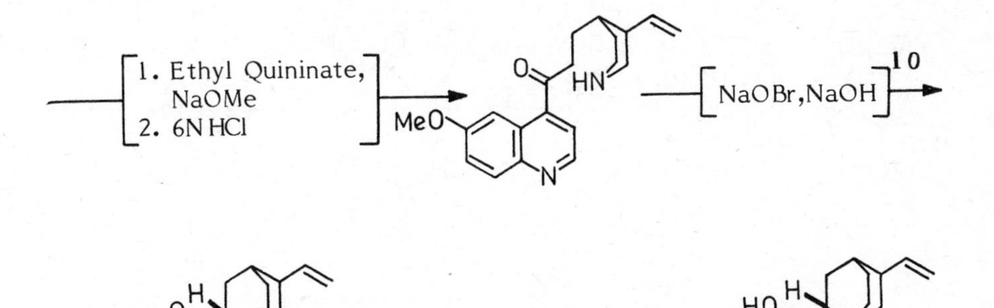

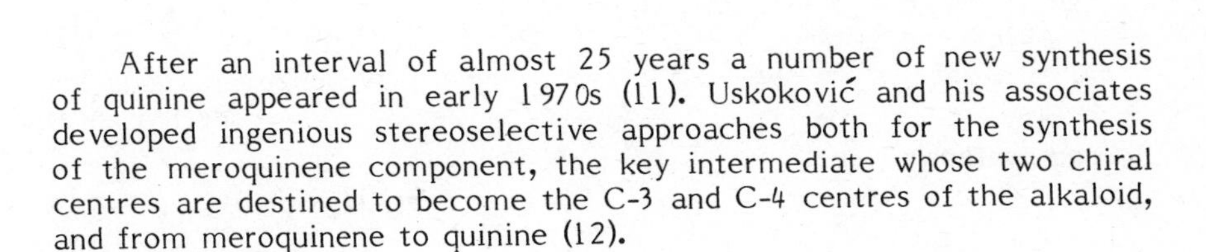

l-Quinine

After an interval of almost 25 years a number of new synthesis of quinine appeared in early 1970s (11). Uskoković and his associates developed ingenious stereoselective approaches both for the synthesis of the meroquinene component, the key intermediate whose two chiral centres are destined to become the C-3 and C-4 centres of the alkaloid, and from meroquinene to quinine (12).

N-Benzoylmeroquinene
First synthesis

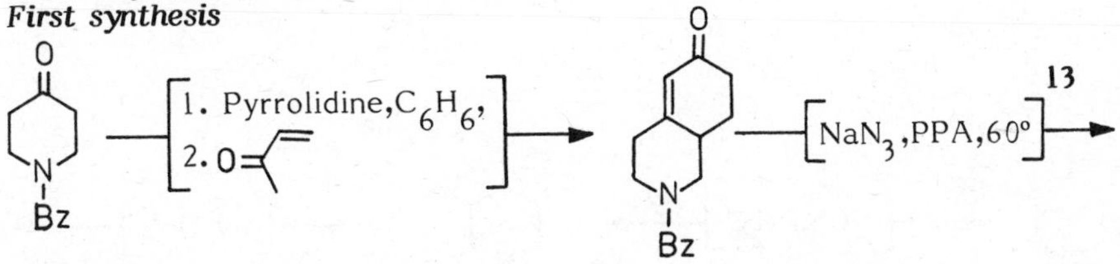

9. Resolved via dibenzoyl-d-tartrate.
10. First effected by Rabe and Kindler (5).
11. A common feature of all these syntheses is that they all proceed through meroquinene derived intermediate bearing a properly positioned functional group which facilitates the quinuclidine ring forming reaction (i→ii).

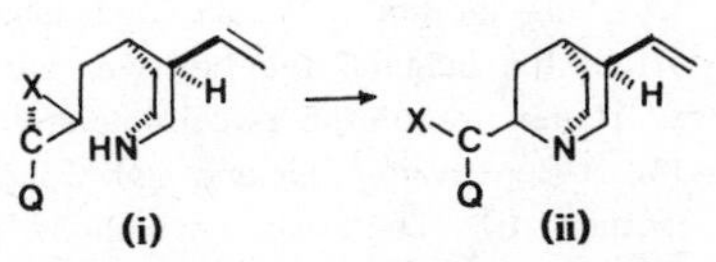

12. (a) Uskoković,M.R.; Gutzwiller,J.; Henderson,T. J. Am. Chem. Soc., 1970, 92, 203; (b) Uskoković,M.R.; Reese,C.; Lee,H.L.; Grethe,G.; Gutzwiller,J. ibid, 1971, 93, 5902; (c) Grethe,G.; Lee,H.L.; Mitt,T.; Uskoković,M.R. Helv. Chim. Acta, 1973, 56, 1485; (d) Uskoković, M.R.; Henderson,T.; Reese,C.; Lee,H.L.; Grethe,G.; Gutzwiller,J. J. Am. Chem. Soc., 1978, 100, 571; (e) Gutzwiller,J.; Uskoković,M.R. ibid, 1978, 100, 576.

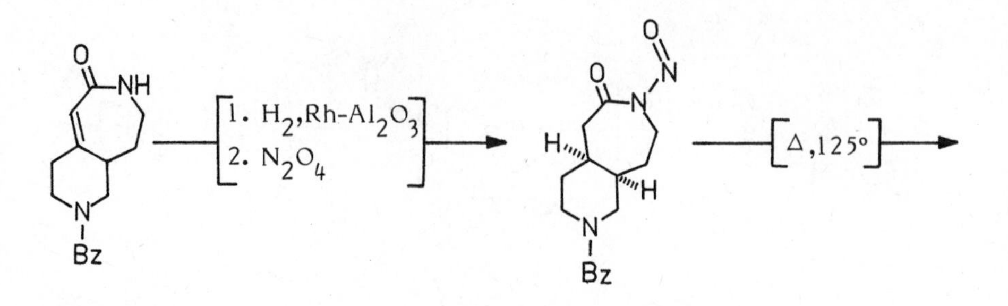

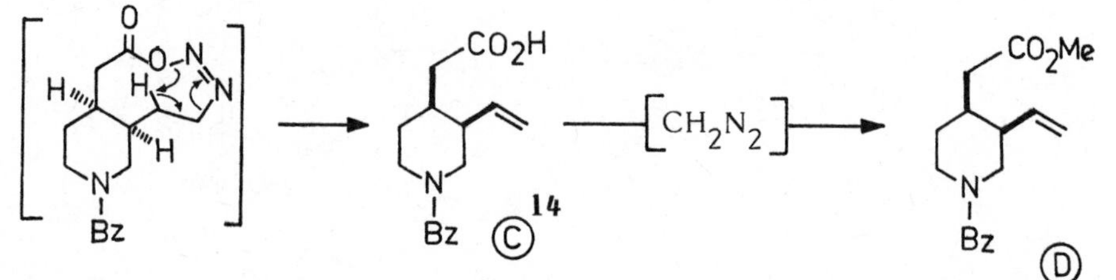

N-Benzoylmeroquinene Methyl Ester

Second synthesis

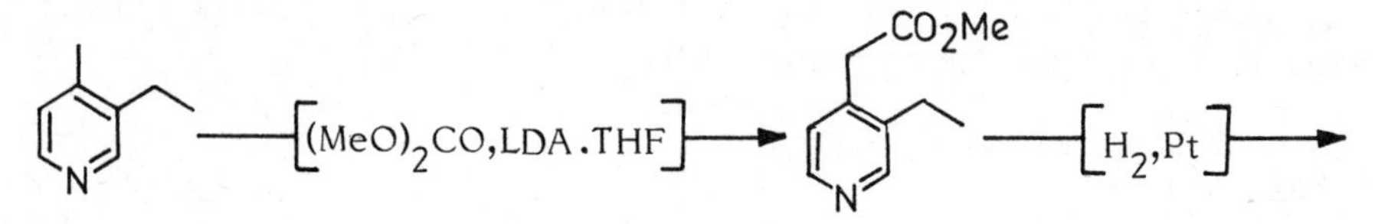

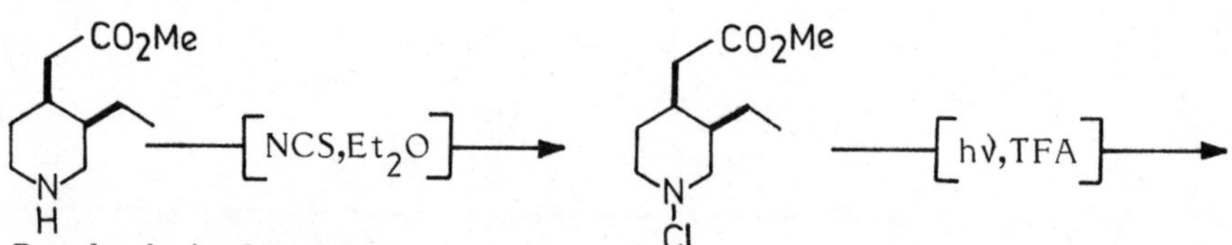

Resolved via d-tartrate

---

13. This lactam was obtained along with the isomeric enamine lactam in a ratio of 5:1, if the Schmidt reaction was carried out on the decahydroisoquinolone a 1:1 mixture of the two isomers.

14. N-Benzoylmeroquinene Ⓒ was accompanied by the seven-membered lactone (i). The two compounds were formed in 50 and 30% yield respectively. The latter is apparently formed by nucleophilic attack of one of the ester oxygens on the C-10 carbon. Lactone (i) could be converted to Ⓓ by two different routes.

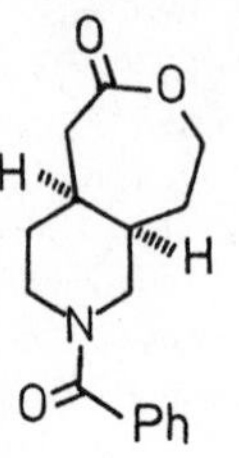

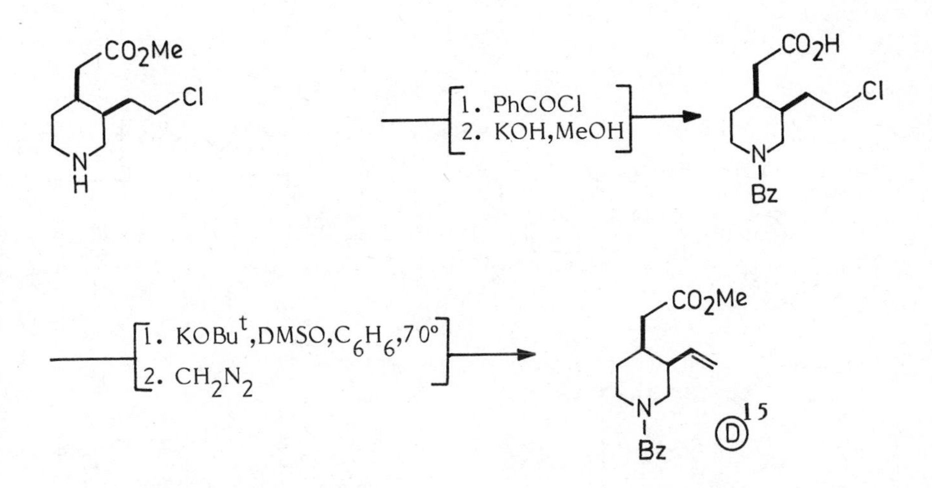

Preparation of Quinine

Uskokovic et al. explored various routes for the condensation of meroquinene component with the 6-methoxyquinoline unit (12e) and the two routes described below, though not completely stereospecific, proved relatively more efficient for stereoselectivity and yields.

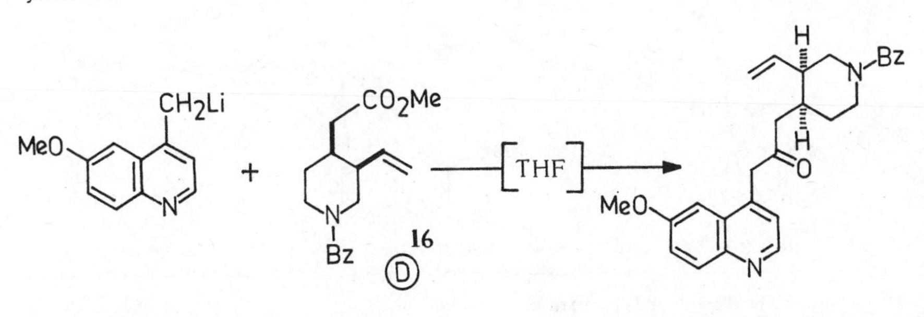

---

15. Imanishi et al. have described another alternative synthesis of meroquinine by a somewhat different approach starting from 3-hydroxy-Δ^4-tetrahydropyridine; Imanishi,T.; Inone,M.; Wada,Y.; Hanoaka,M. Chem. Pharm. Bull., 1982, 30, 1925; 1983, 31, 1551.

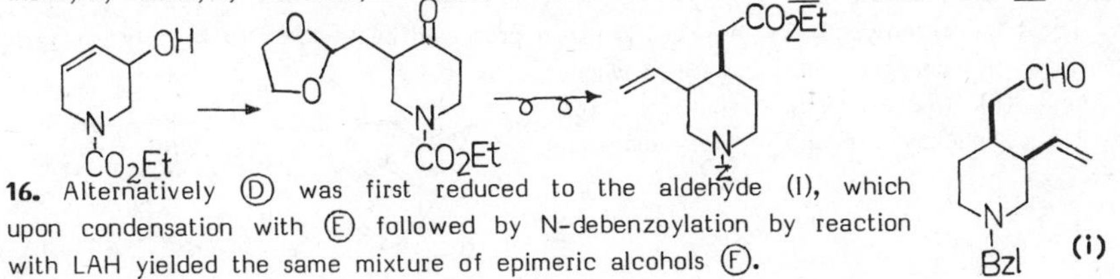

16. Alternatively (D) was first reduced to the aldehyde (i), which upon condensation with (E) followed by N-debenzoylation by reaction with LAH yielded the same mixture of epimeric alcohols (F).

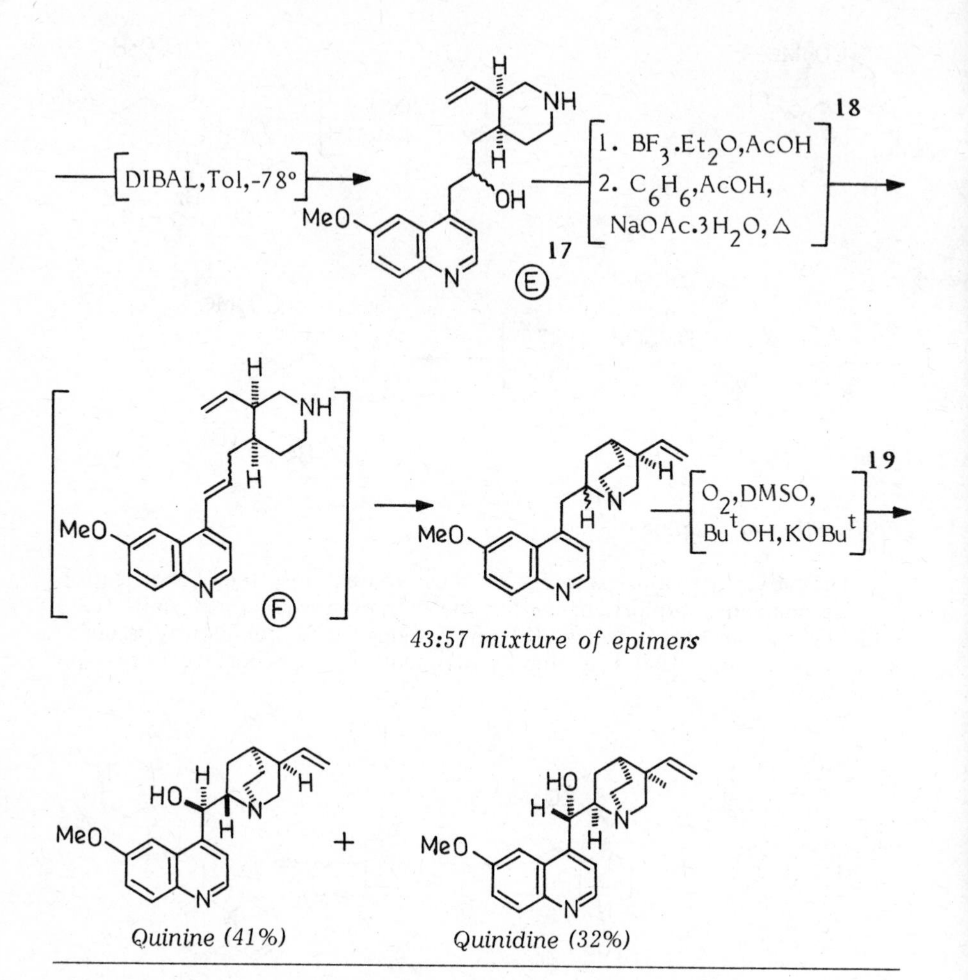

Quinine (41%)

Quinidine (32%)

---

17. Resolved via dibenzoyl (+)-tartrate.

18. Pure deoxyquinine and deoxyquinidine obtained from quinine and quinidine when treated under same cyclisation conditions resulted in each case in a similar mixture of epimers, thus indicating that vinyl quinoline to → deoxyquinine + deoxyquinidine is a reversible process.

19. The high stereospecificity may be due to a preferred back side attack of the oxygen radical anion on the intermediate radicals (i) and (ii) to avoid the repulsive force of the quinuclidine nitrogen free electron pair.

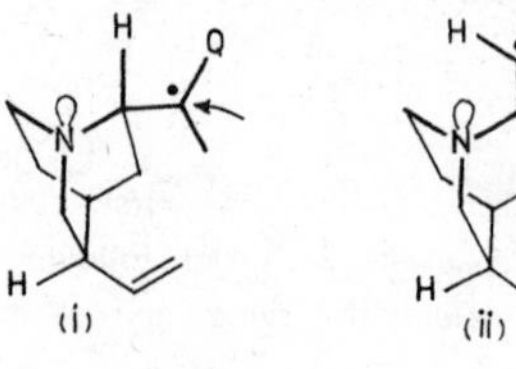

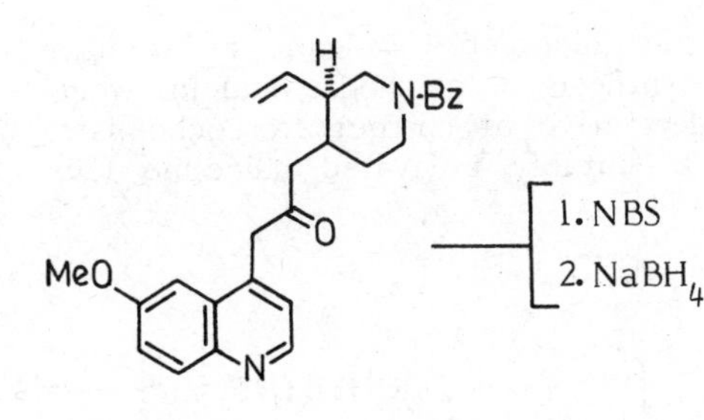

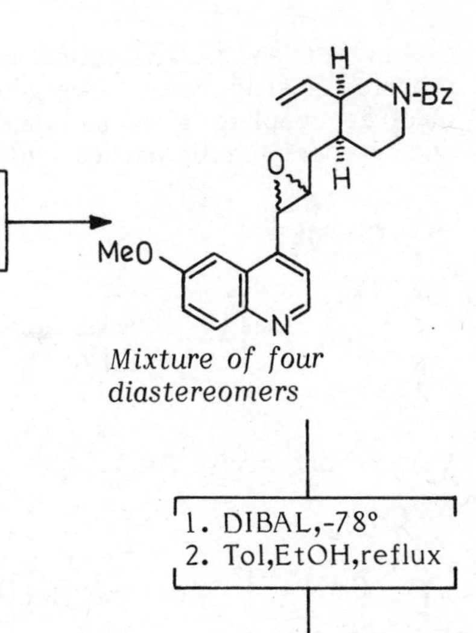

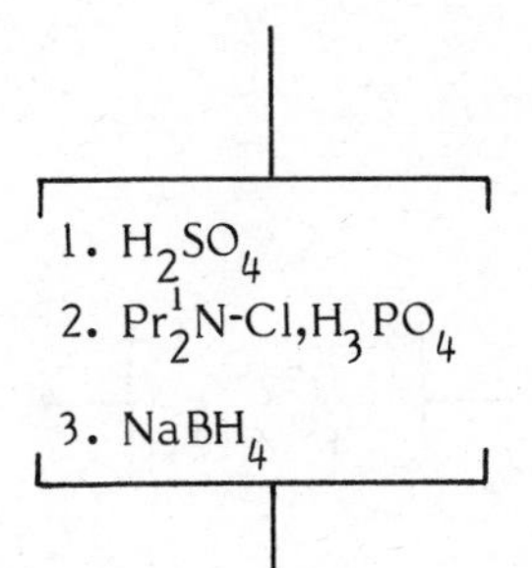

13% quinine, 24% quinidine
18% epiquinine, 18% epiquinidine

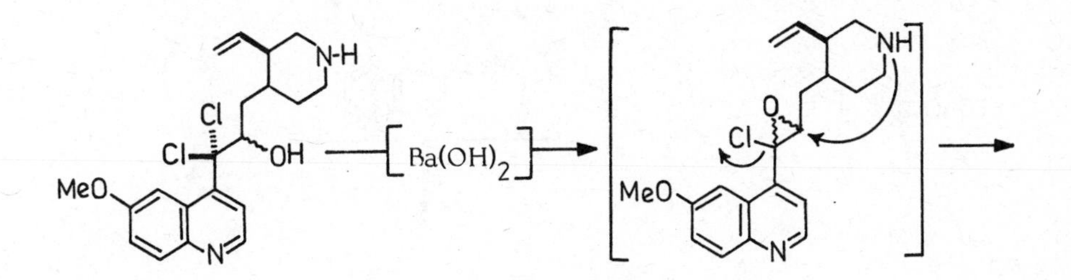

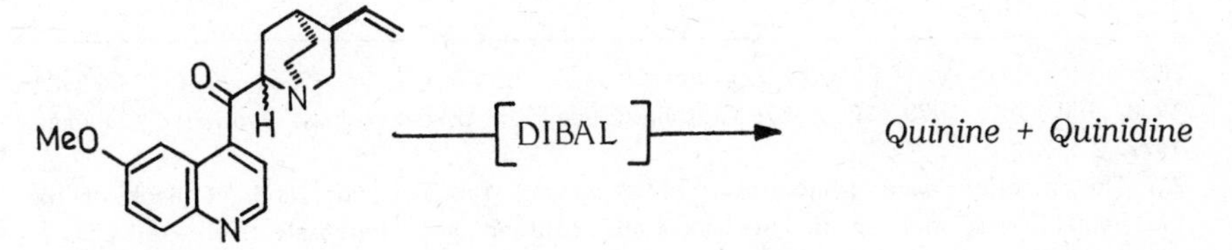

Quininone +
Quinidinone

In yet another variation of this approach Uskoković et al. have reported a still more convergent synthesis of cinchona alkaloids which involves coupling a quinuclidine derivative of correct stereochemistry and proper functionalities with a suitably activated quinoline (20).

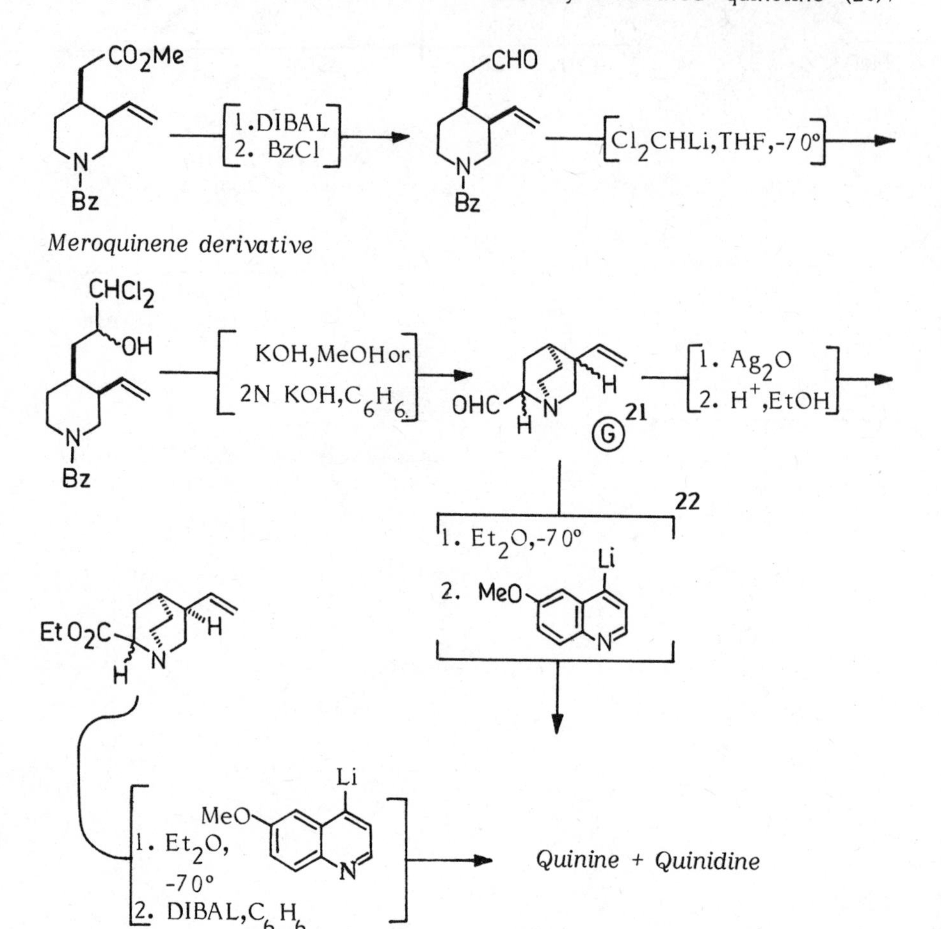

20. Grethe,G.; Lee,H.L.; Mitt,T.; Uskoković,M.R. J. Am. Chem. Soc., 1971, 93, 5904; 1978, 100, 581; 1978, 100, 589; Grethe,G.; Lee,H.L.; Uskoković,M.R. Synthetic Commun., 1972, 2, 55.

21. This reaction gave a mixture of 13% quinine, 15% quinidine, and 5% each of the two epi-isomers; the epi-isomers could be oxidised and stereoselectively reduced to the desired erythro compounds.

22. This aldehyde has also been used by Uskokovic et al. for the synthesis of dihydrocinchonamine starting from N-Li o-toluidine and quinuclidine ester.

In an alternative procedure developed by Gates et al. the vinylpiperidine to quinoline union, which precedes the quinuclidine ring formation, is brought about by a Wittig reaction; the key quinuclidine cyclisation step is non-stereospecific (23).

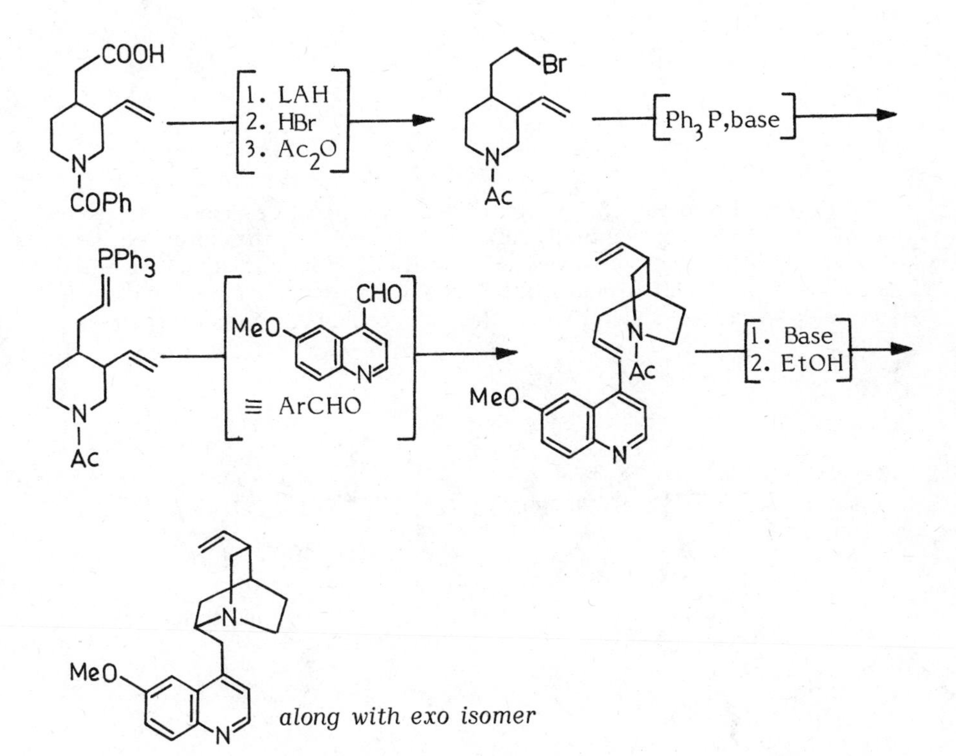

In another variation of this approach Taylor and Martin synthesised the piperidino-methylvinylquinoline Ⓗ from meroquine aldehyde and quinoline Wittig reagent (24).

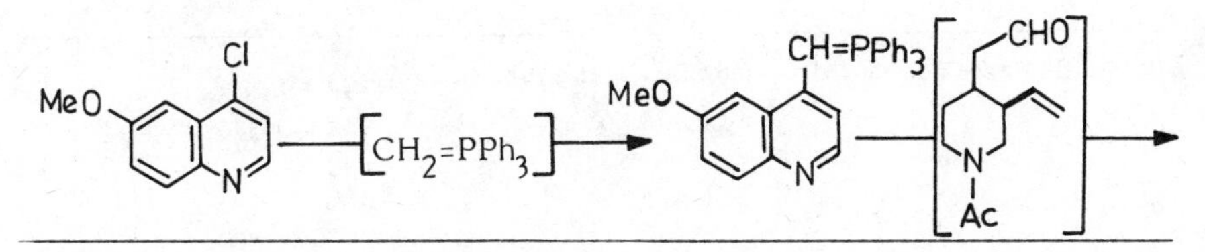

23. Gates,M.; Sugavanam,B.; Schreiber,W.L. J. Am. Chem. Soc., 1970, 92, 205.
24. Taylor,E.C.; Martin,S.F. J. Am. Chem. Soc., 1972, 94, 6218.

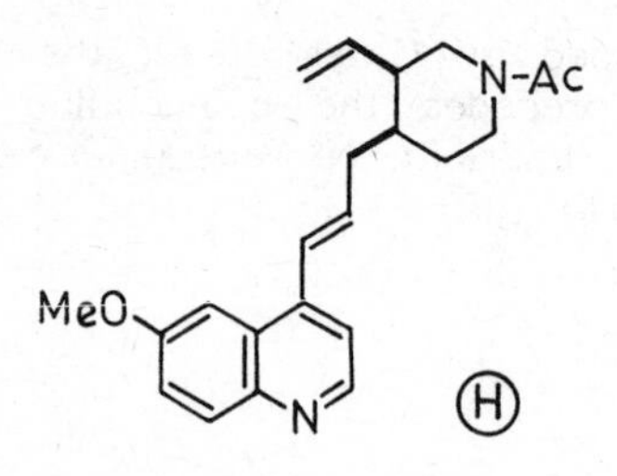

Stotter, Friedman & Minter have recently described, on some model compounds, conceptually a new approach to the total synthesis of quinine (25), which involves as a critical step a diastereoselective aldol condensation of a quinuclidone enolate with an aromatic aldehyde, to yield an amino-alkanol with the required erythro stereochemistry.

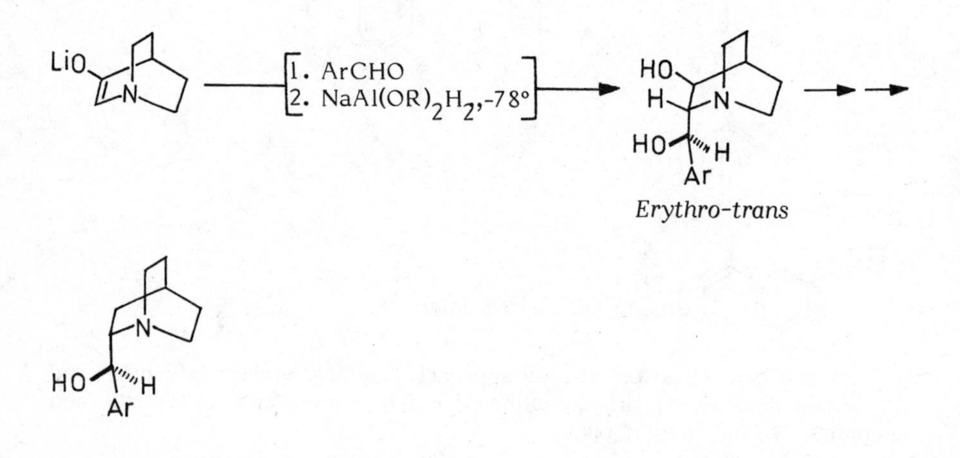

Erythro-trans

---

25. Stotter,P.L.; Friedman,M.D.; Minter,D.E. J. Org. Chem., 1985, 50, 29.

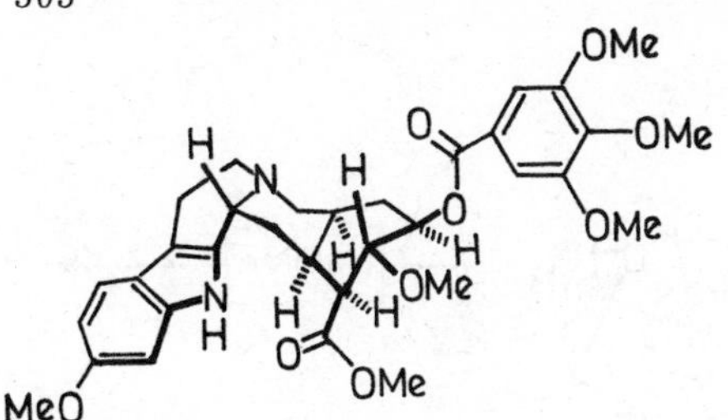

RESERPINE

The discovery of reserpine, pharmacologically the most important alkaloid of Rauwolfia serpentina, in 1952 was a big event in drug research as it sharply focussed attention on the possibility of finding major drugs from plants based on knowledge of the traditional systems of medicine(1)Its structure, with six chiral centres, five in one ring, and all stereochemically not in the most stable orientation offered a big challenge to synthetic ingenuity. Four synthesis have been reported all of which first constructed the essential features of D/E ring with all the five chiral centres correctly disposed but by very different synthetic strategies, each ingenious in its own way.

The synthesis by Woodward and his associates (2) reported in 1956 created all the five chiral centers of E ring in three steps in a highly stereoselective manner (3). The remaining task of completing the pentacyclic system was accomplished by an oxidative scission of the enone Ⓑ to permit insertion of N-4 and completion of the synthesis.

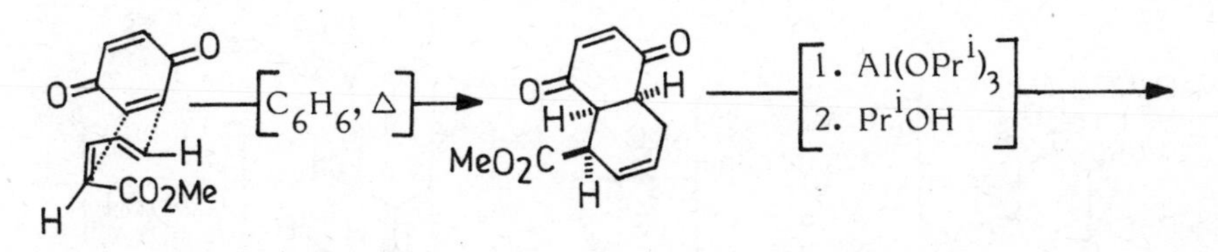

1. Woodson,R.E.; Younken,H.W.; Schlittler,E.; Schneider,J.A. "Rauwolfia: Botany, Pharmacognosy, Chemistry and Pharmacology". Little Brown & Co., Boston.

2. Woodward,R.B.; Bader,F.E.; Bickel,H.; Frey,A.J.; Kierstead,R.W. J. Am. Chem. Soc., 1956, 78, 2023, 2657; Tetrahedron, 1958, 2, 1.

3. Diels-Alder reaction gave the adduct with the cis-fused D/E ring junction and three chiral centres correctly set up as a result of the preference of endo-addition; thereafter the new centres of asymmetry were introduced stereoselectively under direction of existing centres of asymmetry. The high degree of stereochemical control evident in transformations of the cis-decalin is due to the folded structure of the molecule in which the side carrying the cis bridgehead hydrogens, the convex face of the boat-shaped molecule, is readily accessible to attacking reagents; see: Woodward,R.B. in "Pointers and Pathways in Research", Hofteizer,G. for Ciba of India Ltd., Bombay, 1963,23.

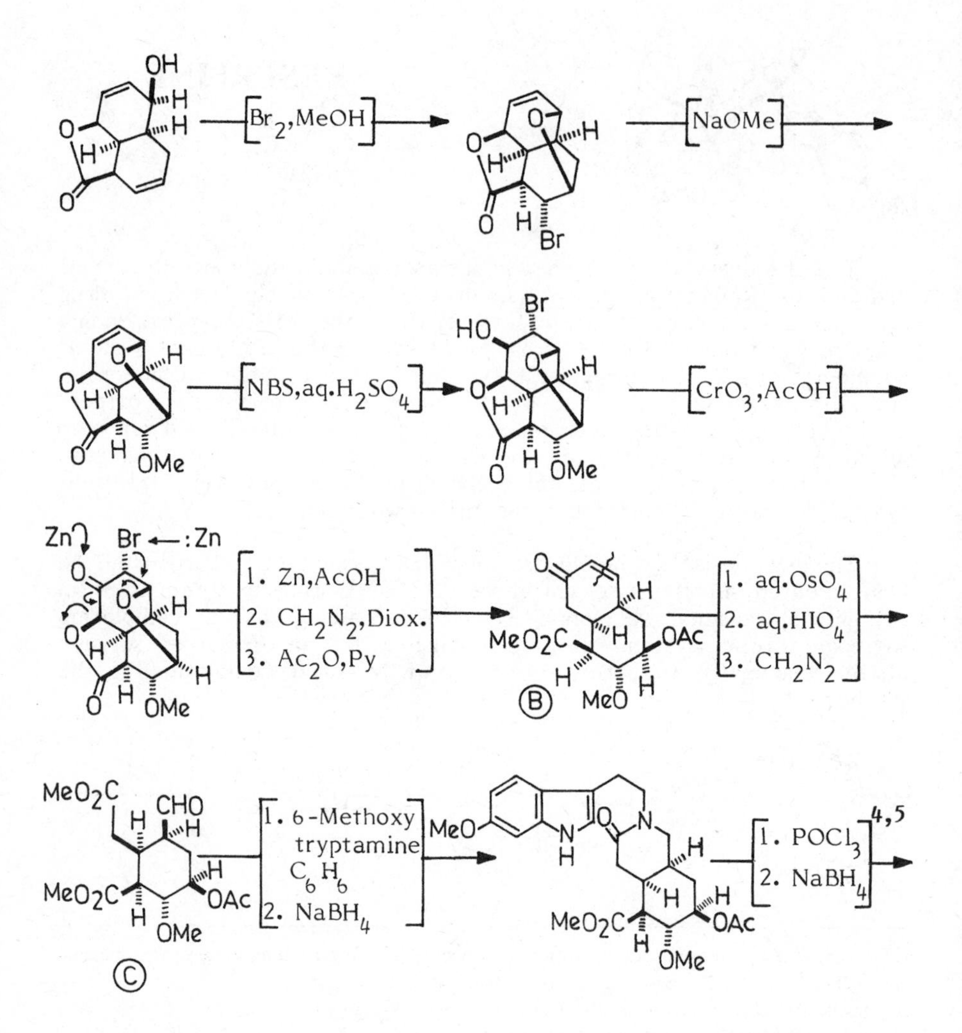

4. Reduction of the intermediate immonium salt directly to a reserpic acid derivative (C-3 hydrogen, cf.ref.5) may be accomplished with zinc and acetic acid, Weisenborn,F.L.; Diassi,P.A. J. Am. Chem. Soc., 1950, 78, 2022. Other modifications of the synthesis by later workers are all based on Woodward's intermediate carbocyclic aldehyde (C) and the synthesis is now a commercial feasibility. For a review, see: Schlittler,E. in "The Alkaloids", Vol.VIII, ed. R.H.F. Manske, Academic Press, New York, 1965, 287.

5. An isoreserpic acid derivative is obtained in which C-2,3 linkage and all large groups attached to rings D/E are equatorially disposed.

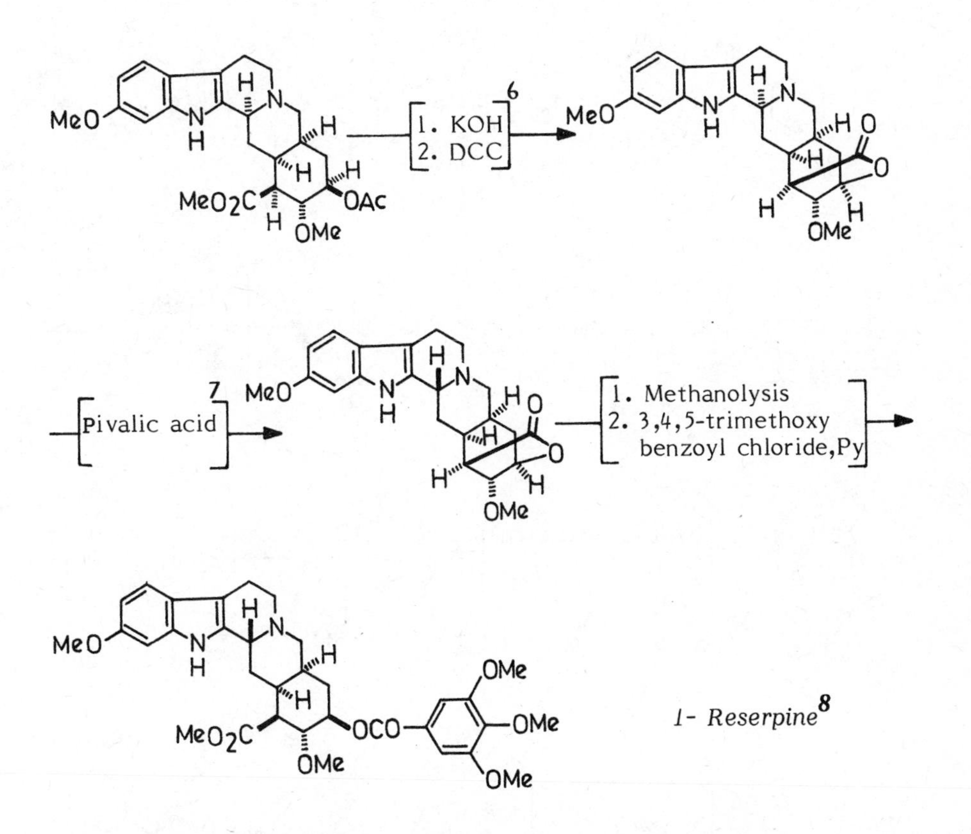

Following a gap of almost 20 years the second synthesis of reserpine was reported by Pearlman (9) which uses a novel deMayo reaction with formyl acetic ester to place a vicinal carboxaldehyde and acetic ester appendages onto double bond to get the key Woodward precursor Ⓐ (10).

---

6. Lactonization forces the carboxyl at C-16 to become axial, and in the process conformation of entire substituents on the alicyclic system of iso-reserpine is inverted.

7. C-2,3 linkage which is now axial, is smoothly equilibrated to the more stable equatorial conformation; pivalic acid was chosen as the equilibrating agent because it is too weak a nucleophile to effect side-reactions such as opening of the lactone ring.

8. Resolved via d-camphor-10-sulfonate.

9. Pearlman,B.A. J. Am. Chem. Soc., 1979, 101, 6404.

10. For a discussion of the philosophy of this approach and its broad implications in natural product synthesis: see Pearlman,B.A. J. Am. Chem. Soc., 1979, 101, 6398.

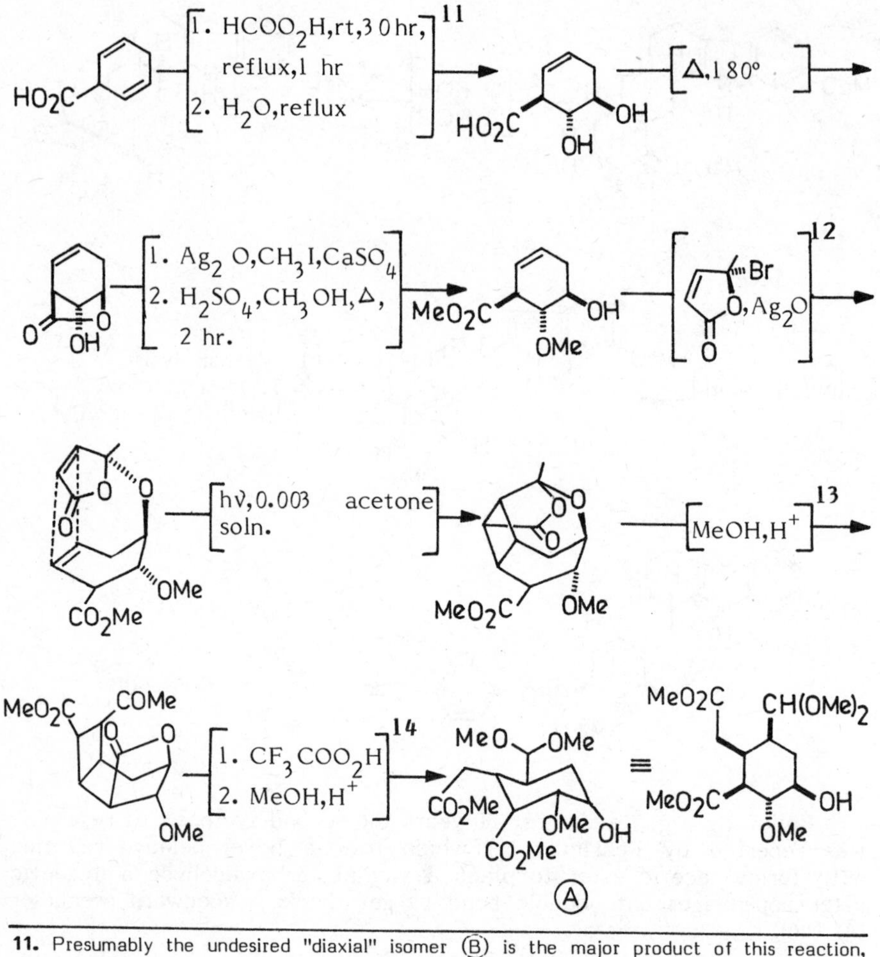

11. Presumably the undesired "diaxial" isomer Ⓑ is the major product of this reaction, but the required diequatoral diol is formed in sufficiently high yield to make the route attractive.

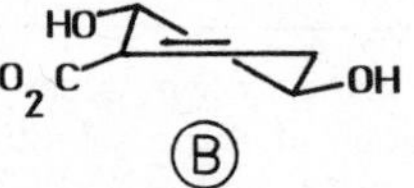

12. The product was a 1:1 mixture of Ⓒ and its diastereomer, which were separated by column chromatography; the undesired isomer was converted back to the parent alcohol by methanolysis.

13. The ketal bridge underwent methanolysis, and the liberated methyl ketone epimerised to the exo configuration, and the liberated hydroxyl group formed a lactone with the carbomethoxy group of the cyclohexane ring.

14. During methanolysis the cyclobutane ring opened quantitatively by retroaldolisation, and the lactone ring underwent methanolysis by about 56%; the crude dimethyl acetal wastrimethoxybenzoylated and characterised by conversion to 3-epireserpine by known method.

The third synthesis by Wender, Schaus and White (15) is based on a general method for the synthesis of cis-hydroisoquinolines.

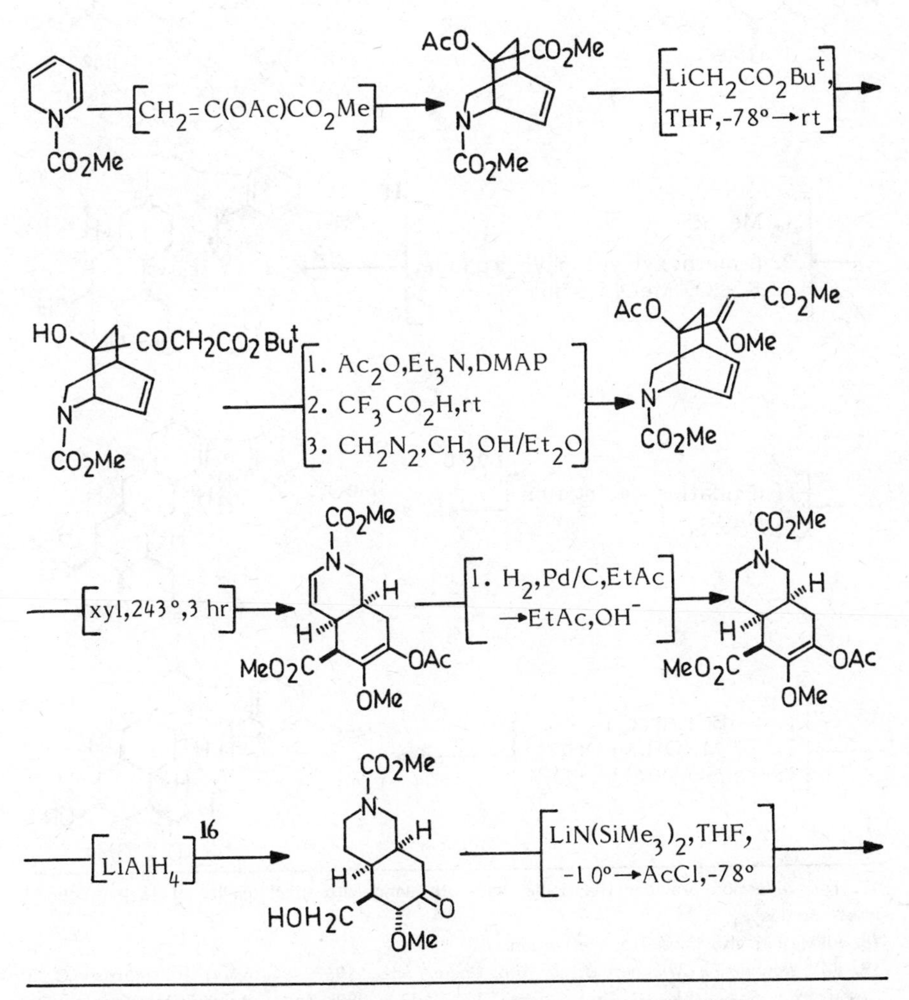

---

15. Wender,P.A.; Schaus,J.M.; White,A.W. J. Am. Chem. Soc., 1980, 102, 6159.

16. The ketone was obtained alongwith a small quantity of 18α & 18β-OH compounds. Since this epimerisation of (D) with $NaOCH_3$ produced a 6:1 mixture of the 17α:17β epimer, the formation of single ketone in the reaction appears to be a result of kinetic protonation of the enol aluminate intermediate.

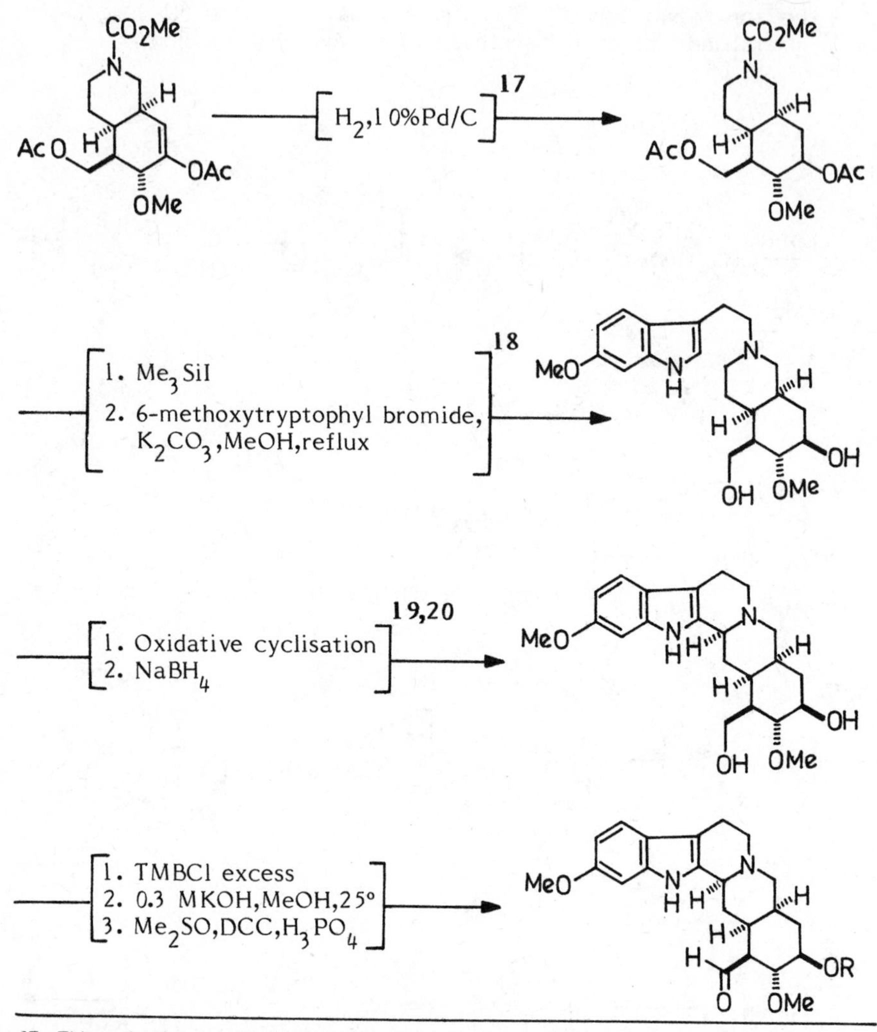

17. This reduction provided the single diacetate along with small quality of 18-hydrogenolysed product.

18. Alkylation also caused deacetylation.

19. (a) Wenkert,E.; Wickberg,B. J. Am. Chem. Soc., 1962, 84, 4914; (b) Morrison,G.C.; Cetenko,W.; Shavel,J.Jr. J. Org. Chem., 1967, 32, 4089; (c) Gutzwiller,J.; Pizzolato,G.; Uskokovic,M. J. Am. Chem. Soc., 1971, 93, 5908; (d) Stork,G.; Guthikonda,N. ibid, 1972, 94, 5110; (e) Aimi,N.; Yamanaka,E.; Endo,J.; Sakai,S.; Haginiwa,J. Tetrahedron Lett., 1972, 1081; (f) Tetrahedron, 1973, 29, 2015.

20. The isoreserpine diol was obtained alongwith a 4.5:3 ratio of inside reserpinediol.

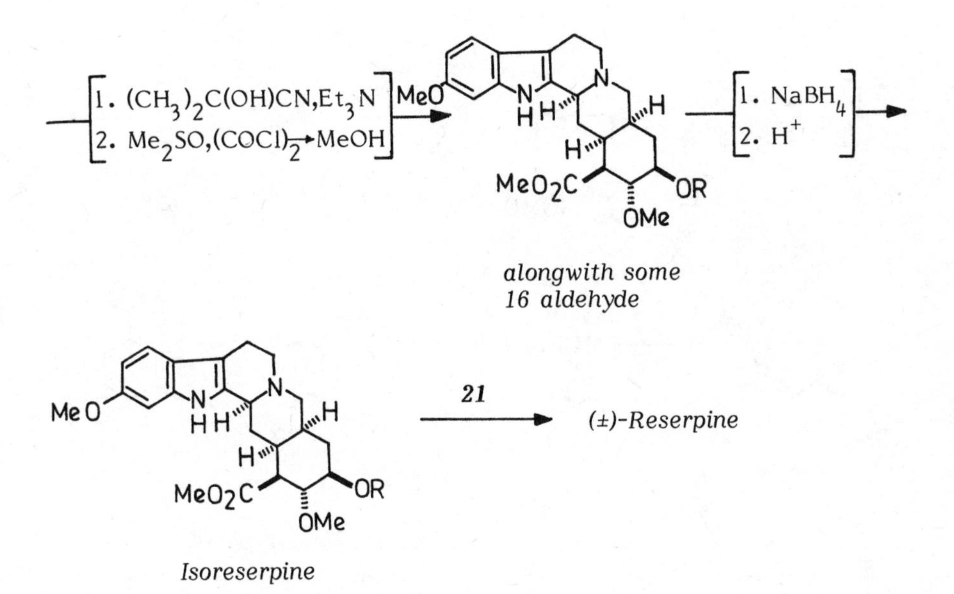

The fourth synthesis by Martin et al (22) is based on an intramolecular Diels-Alder cycloaddition as the key step for the facile construction of the highly functionalised hydroisoquinoline component followed by catalytic hydrogenation which gave the ring D/E with all the stereocenters correct.

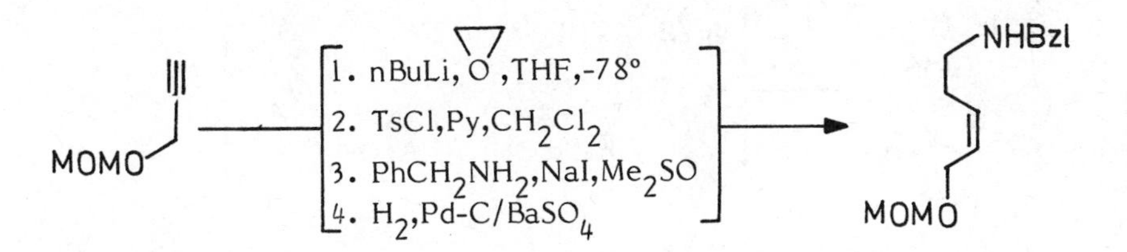

---

21. Huebner,C.F.; Kuehne,M.E.; Korzun,B.; Schlitter,E. Experientia, 1956, 12, 249; Weisenborn F.L.; Diassi,P.A. J. Am. Chem. Soc., 1956, 78, 2022.

22. Martin,S.F.; Grzejszczak,S.; Rueger,H.; Williamson,S.A. J. Am. Chem. Soc., 1985, 107, 4072.

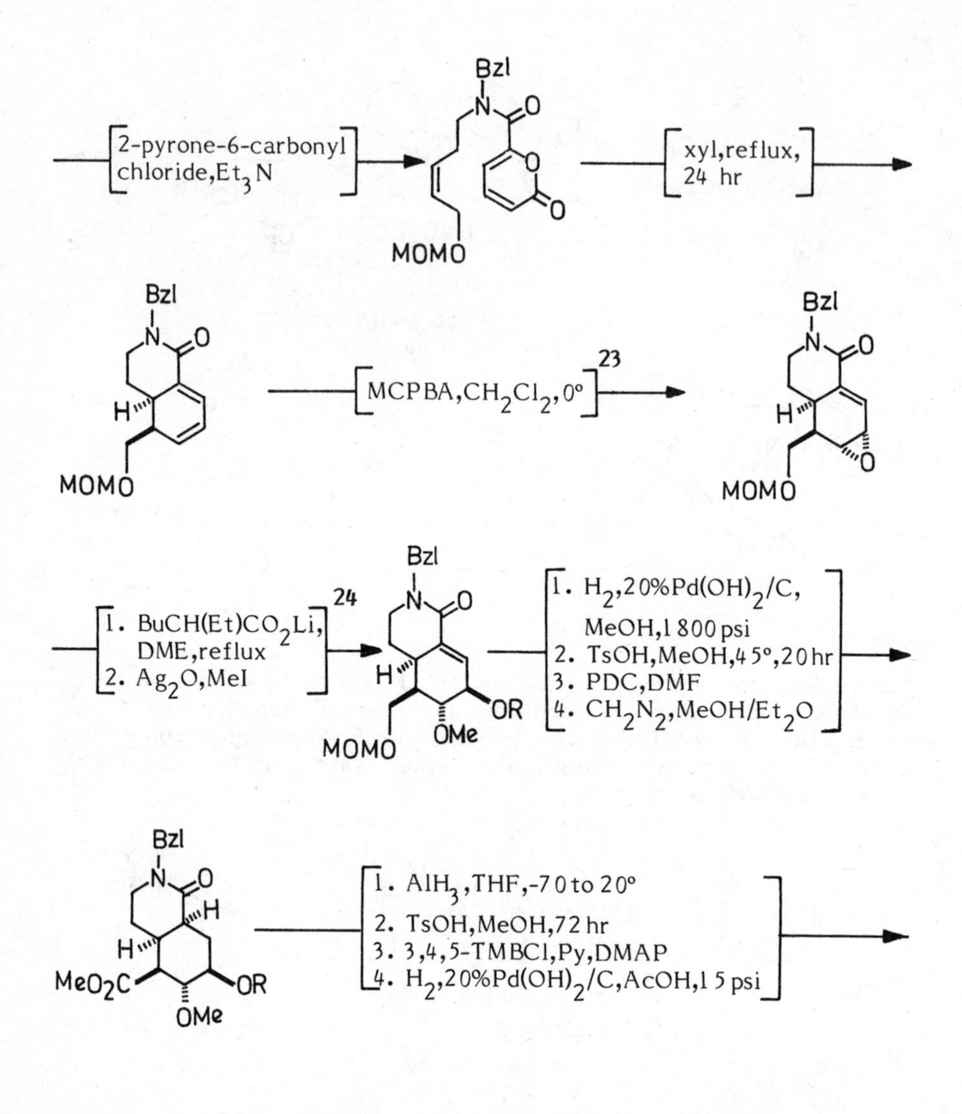

---

23. This epoxidation proceeded with a high degree of stereoselectivity from the less encumbered α-face.

24. The opening of the epoxide occurred exclusively at the allylic terminus.

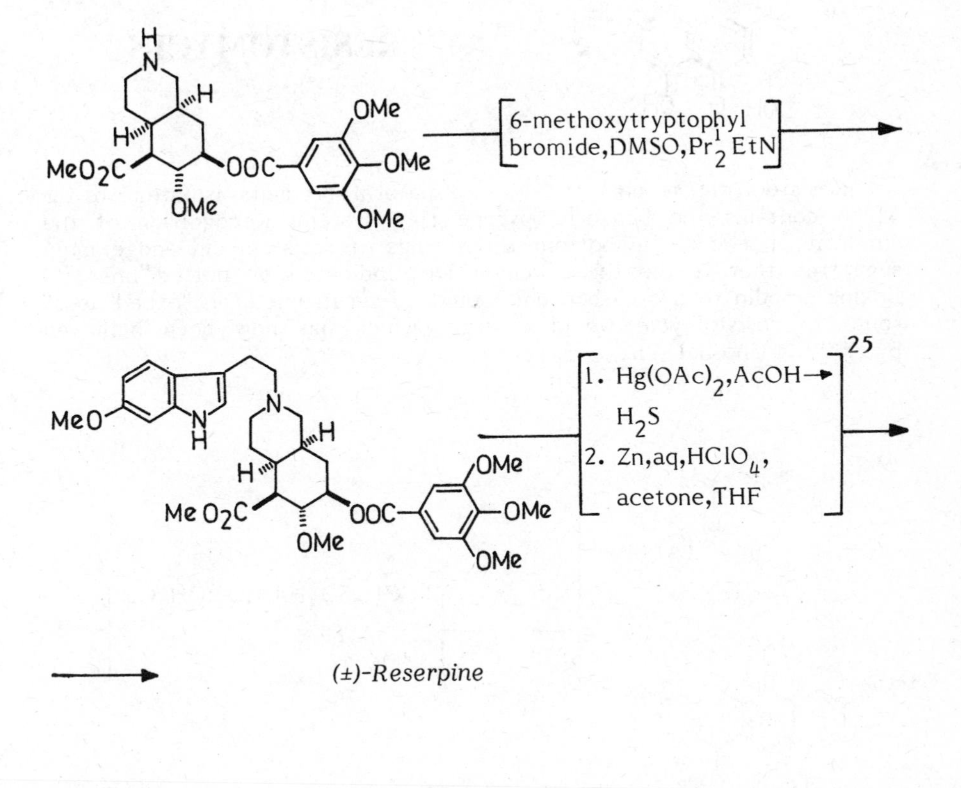

(±)-*Reserpine*

---

25. This conversion produced reserpine in 35% yield along with isoreserpine 8% and two corresponding inside derivatives in 18 and 4% yields.

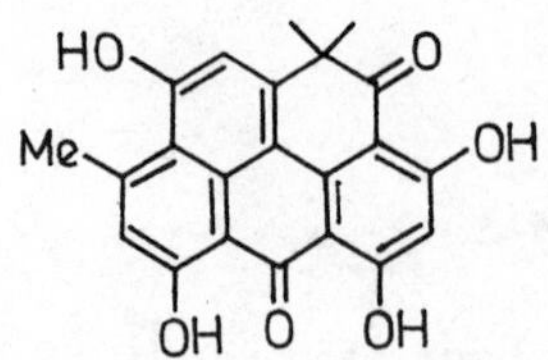

RESISTOMYCIN

Resistomycin is one of the two natural products isolated so far which contains the benzo [cd]pyrene ring system. Recognition of the similarity between the bottom three rings of resistomycin and emodin suggested that if the three connections indicated by dotted lines (1) linking emodin to a 5 carbon unit could be constructed, an "expeditious" route to resistomycin would emerge, which has now been achieved by Kelly & Ghoshal (2).

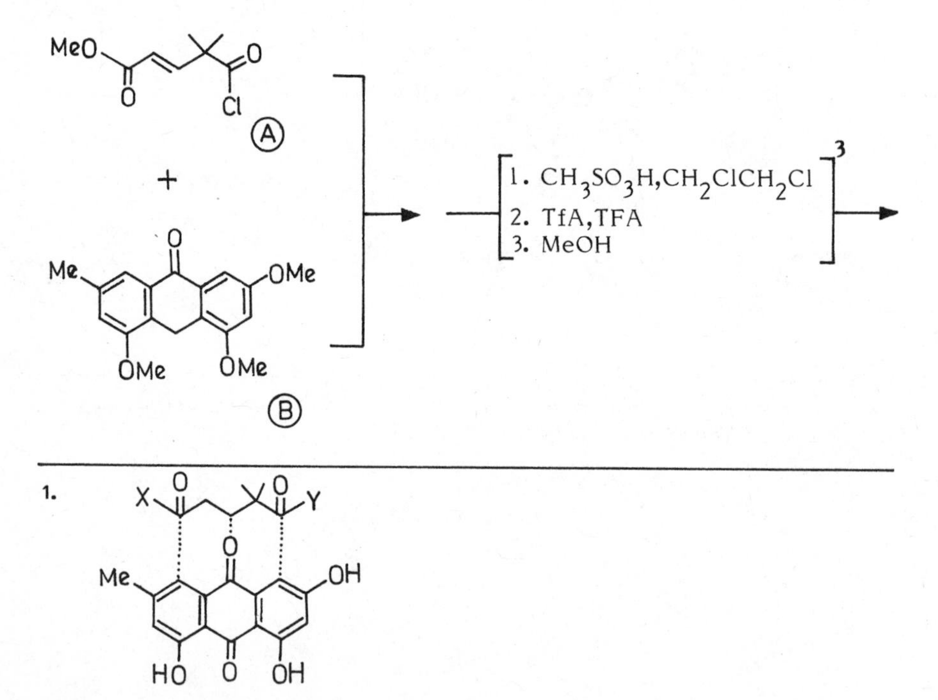

2. Kelly,T.R.; Ghoshal,M. J. Am. Chem. Soc., 1985, 107, 3879.

3. The sequence of events of the fusion of (A) & (B) has been examined by the authors in some detail and two- and three-bond formed intermediates isolated, which established that the reaction proceeded as indicated. Regio-selectivity seems to be determined by the acidity of the reaction conditions. Mechanistic details of the reaction are discussed in the paper.

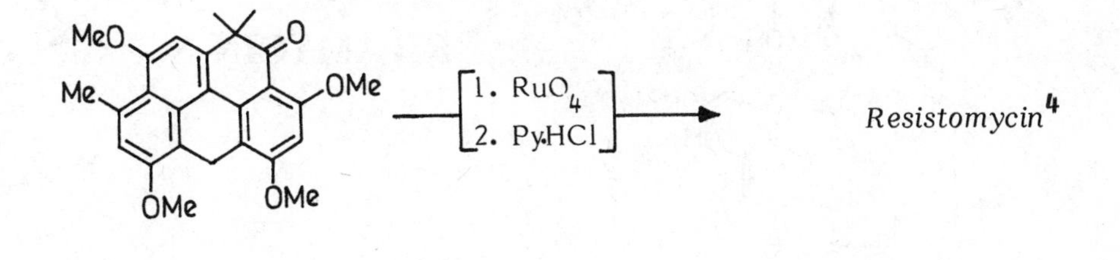

---

4. The only other synthesis of resistomycin was on classical lines and employed an intramolecular Diels-Alder reaction of an insitu generated isobenzofuran as the key constructive step to get (A);[Keay,B.A.; Rodrigo,R. J. Am. Chem. Soc., 1982, 104, 4725.];(A) on treatment

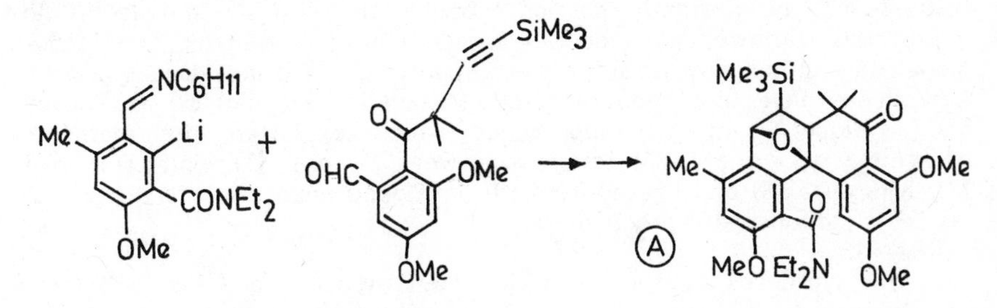

with pyridine hydrochloride underwent demethylation, desilylation, aromatisation and cyclisation to yield resistomycin.

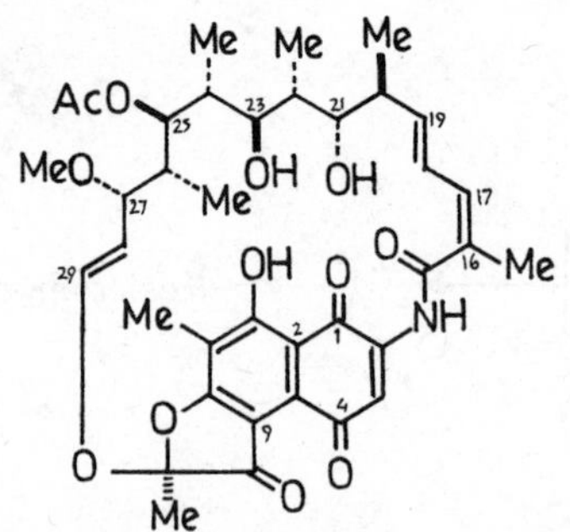

RIFAMYCIN S

Rifamycins, isolated from the fermentation broth of Norcardia mediterranei, were the first examples of ansamycins (1) characterised by an aliphatic bridge linking two non-adjacent positions of an aromatic nucleus. The first and the only total synthesis of rifamycin S by Kishi and his associates (2) builds separately the aliphatic chain (the ansa bridge) and the aromatic portion, couples the two to first form the acetal link followed by macro-lactamisation and adjustment of functionalities. Other synthesis for the aliphatic chain using various acyclic approaches have also been reported; Masamune (8), Still (9) and Corey (5) developed synthetic approaches which exploited the elements of symmetry in the aliphatic chain between C_{20}-C_{26} (3), while Hanesian (7), Kinoshita (8) and Fraser-Reid (9) developed enantioselective synthesis from carbohydrate precursors.

The syntheses of the aliphatic chain by Kishi and his associates are based on the appreciation of the elements of symmetry around C_{23}, and that fragment C_{20}-C_{26} consists of four repeating structural units which led to a strategy amenable to repetitive condensation based on same synthetic operations (4). The first synthesis reported was of the racemic compound (2a); in the next modification (2d) optically pure β-benzyloxy-isobutyraldehyde was used as the starting aldehyde and the enantiomerically correct aliphatic segment prepared

---

1. For reviews on rifamycins see: Prelog,V. Pure & Appl. Chem., 1963, 7, 551; Reinhart,K.L. Jr. Acc. Chem. Res., 1972, 5, 57; Sensi,P. Pure & Appl. Chem., 1975, 41, 15; Reinhart,K.L. Jr.; Shidd,L.S. Prog. Chem. Org. Nat. Prod., 1976, 33, 231; Patterson,I.; Mansuri,M.M. Tetrahedron, 1985, 41, 3569.

2. (a) Nagaoka,H.; Rutsch,W.; Schmid,G.; Lio,H.; Johnson,M.R.; Kishi,Y. J. Am. Chem.Soc., 1980, 102, 7962; (b) Lio,H.; Hagaoka,H.; Kishi,Y. J. Am. Chem. Soc., 1980, 102, 7965; (c) Nagaoka,H.; Schmid,G.; Lio,H.; Kishi,Y. Tetrahedron Lett., 1981, 22, 89; (d) Nagaoka,H.; Kishi,Y. Tetrahedron, 1981, 37, 3873; (e) Johnson,M.R.; Kishi,Y. Tetrahedron Lett., 1979, 4347.

3.

26 23 20

OH OH OH

using the previously established route (5). Further improvement in overall stereoselectivity was achieved by use of a sequence of three Sharpless asymmetric epoxidation reactions for controlling the six asymmetric centres at C_{20}-C_{23}, C_{25} and C_{26} and the stereo-selectivity at C_{27} also substantially improved by using an allyltin (II) reagent for coupling instead of the allylzinc reagent (2d). In the final and most efficient route described below the construction of the C_{19}-C_{27} segment was further simplified by conducting two of the two-carbon chain extensions with a crotyl-chromium reagent, which served to set up correctly two new asymmetric centres at the same time (2d).

Aliphatic Segment C_{15}-C_{29}

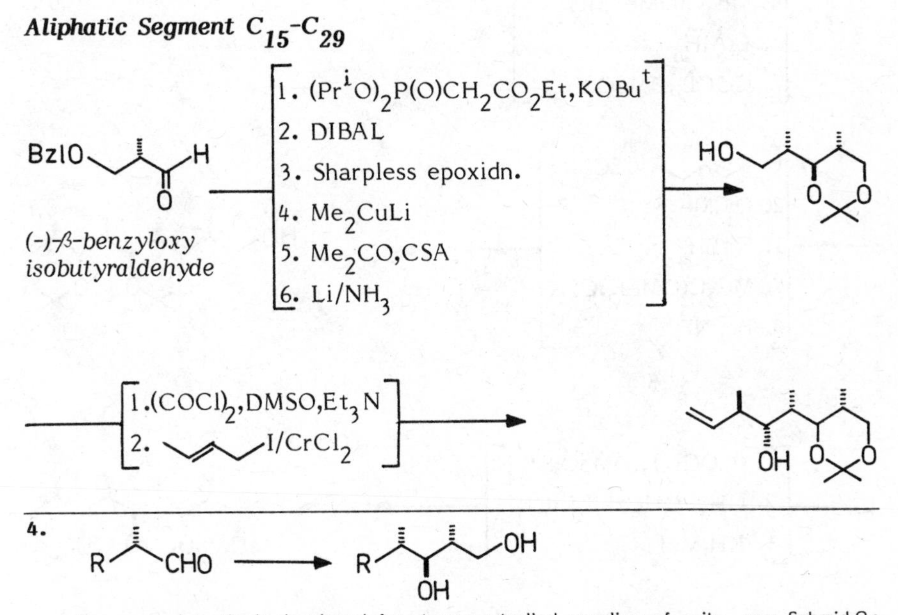

For general methods developed for stereocontrolled coupling of units see: Schmid,G.; Fukuyama,T.; Akasaka,K.; Kishi,Y.; J. Am. Chem. Soc., 1979, 101, 259; Johnson,M.R.; Nakata,J.; Kishi,Y. Tetrahedron Lett., 1979, 4343; Hasan,I.; Kishi,Y. ibid, 1980, 4229.

5. In view of the presence of the aldehyde group adjacent to the asymmetric centre the effect of different reagents and reactions on optical purity of the starting aldehyde and products has been studied in great detail and such reaction conditions selected which had minimal effect on optical purity; Nagaoka & Kishi (2d); changes of reagent were found to markedly affect the optical purity e.g. in oxidation of primary alcohol while Swern's oxidation (Mancuso,A.J.; Jung,S.L.; Swern,D. J. Org. Chem., 1978, 43, 2480) maintained optical purity in the resulting aldehyde; with pyridinium chlorochromate there was considerable loss of optical purity.

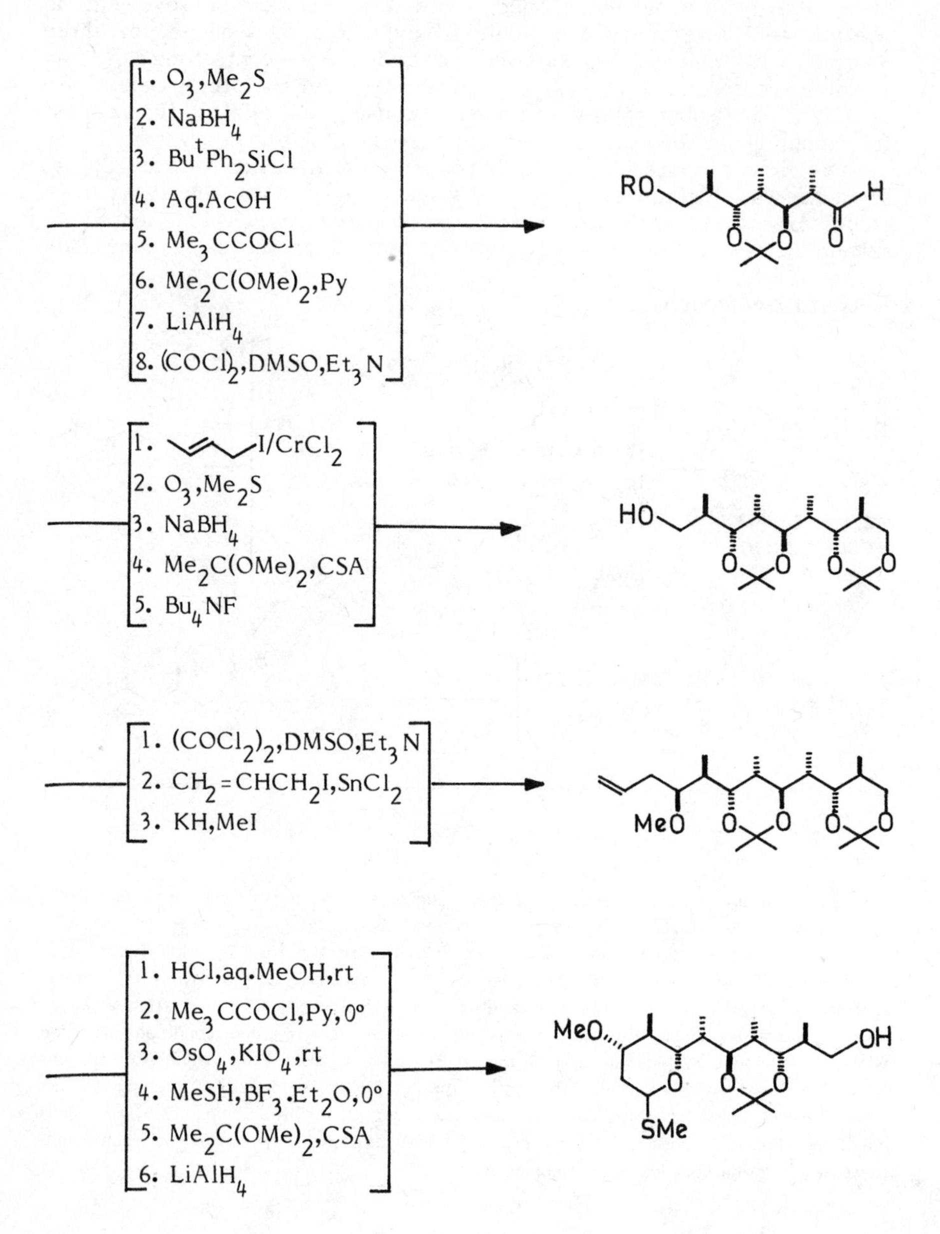
1. O3,Me2S
2. NaBH4
3. ButPh2SiCl
4. Aq.AcOH
5. Me3CCOCl
6. Me2C(OMe)2,Py
7. LiAlH4
8. (COCl)2,DMSO,Et3N
RO
H
O
O
O
1. I/CrCl2
2. O3,Me2S
3. NaBH4
4. Me2C(OMe)2,CSA
5. Bu4NF
HO
O
O
O
O
1. (COCl2)2,DMSO,Et3N
2. CH2=CHCH2I,SnCl2
3. KH,MeI
MeO
O
O
O
O
1. HCl,aq.MeOH,rt
2. Me3CCOCl,Py,0°
3. OsO4,KIO4,rt
4. MeSH,BF3.Et2O,0°
5. Me2C(OMe)2,CSA
6. LiAlH4
MeO
OH
O
O
O
SMe

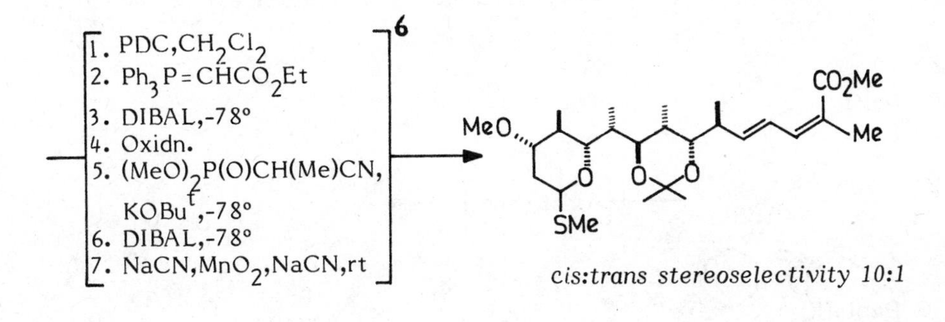

cis:trans stereoselectivity 10:1

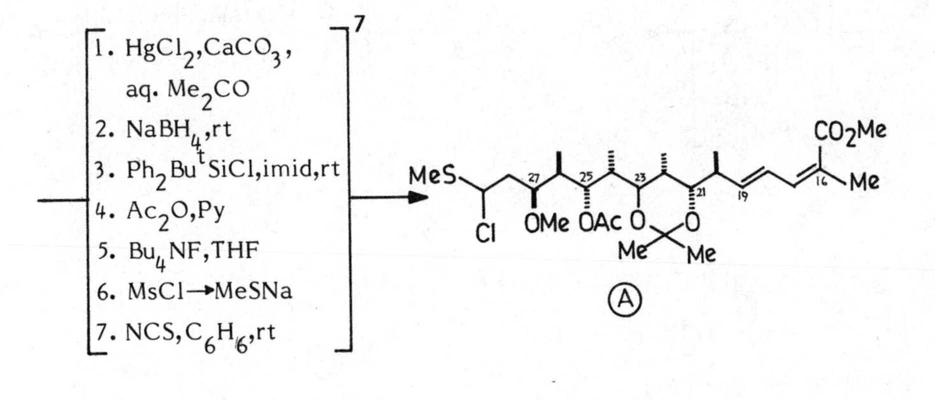

---

6. Use of methyl ester in the second Wittig reaction gave instead almost exclusively the trans olefine, and the best cis: trans ratio (10:1) was obtained with the cyano Wittig reagent.

7. The N-chlorosuccimide (NCS) chlorination gave a diastereomeric mixture of α-chlorosulfides.

Masamune's (8) highly convergent asymmetric synthesis of the aliphatic fragment described herein assembles seven of the eight asymmetric centres by four directed aldol condensation reactions; the eighth centre, C_{23}, was created by a stereoselective reduction.

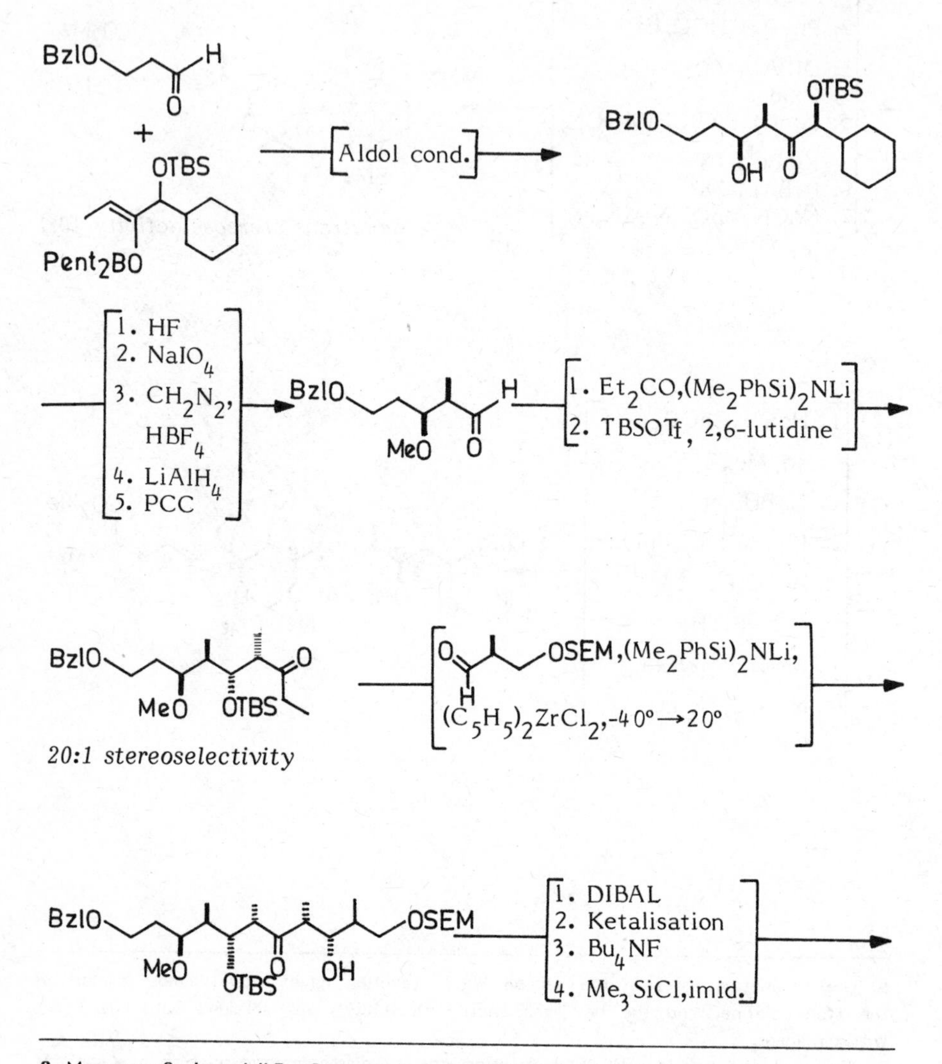

8. Masamune,S.; Imperiali,B.; Garvey,D.S. J. Am. Chem. Soc., 1982, 104, 5528.

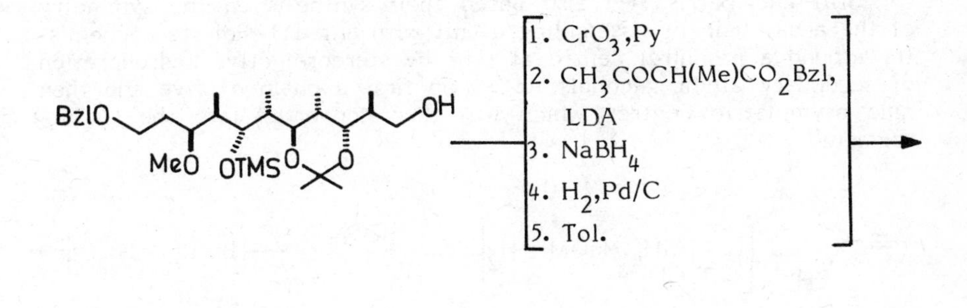
BzlO
MeO
OTMS
O
O
OH
1. CrO3,Py
2. CH3COCH(Me)CO2Bzl,
LDA
3. NaBH4
4. H2,Pd/C
5. Tol.

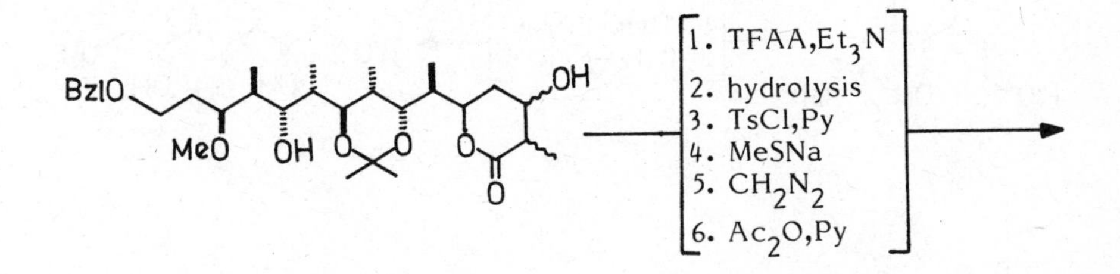
BzlO
MeO
OH
O
O
O
O
OH
1. TFAA,Et3N
2. hydrolysis
3. TsCl,Py
4. MeSNa
5. CH2N2
6. Ac2O,Py

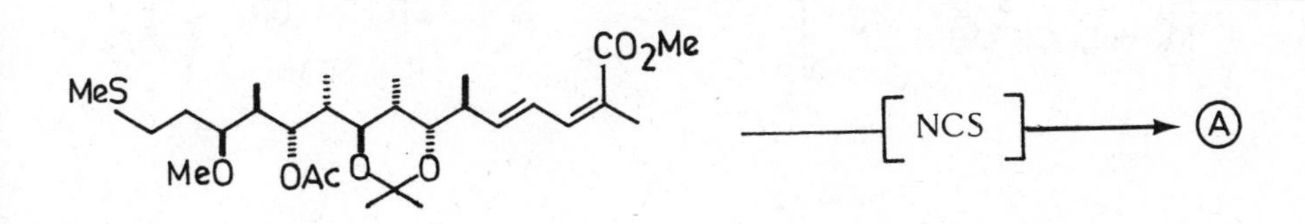
MeS
MeO
OAc
O
O
CO2Me
NCS
A

Still and Barrish (9) also based their synthesis on the symmetry of the ansa chain by assembling a unit with anti-1,3-diol stereochemistry around a prochiral centre at C_{23} by stereoselective hydroboration of secondary allylic alcohols to set up first a chain of five and then nine asymmetric centres which was then converted into the C_{17}-C_{28} segment.

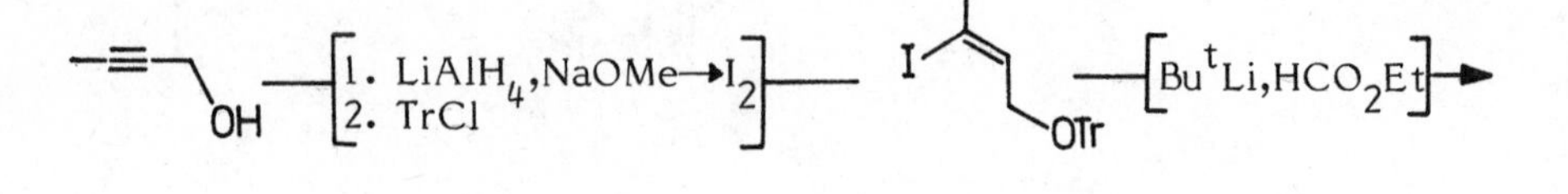

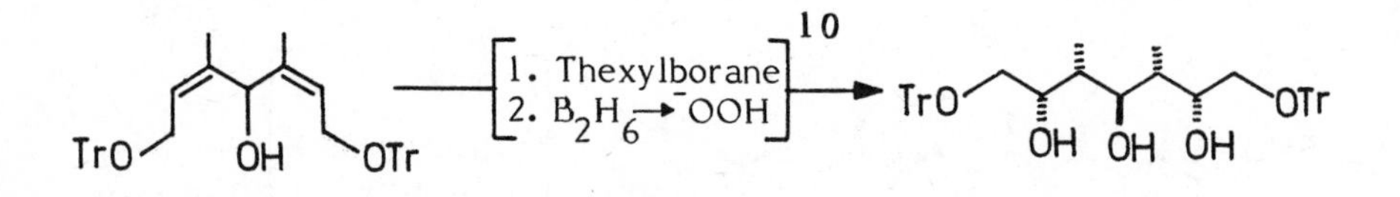

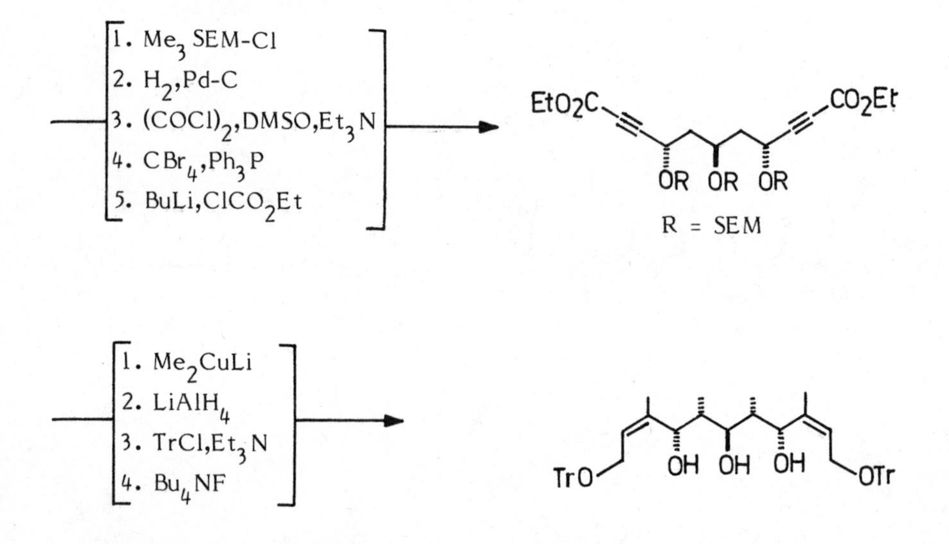

---

9. Still,W.C.; Barrish,J.C. J. Am. Chem. Soc., 1983, 105, 2487.
10. This reaction gave a 5:1 ratio of meso- and dl-triols.

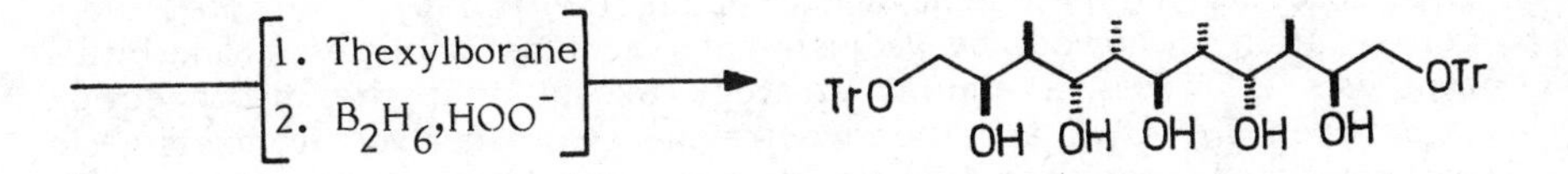

Corey and Hase (11) noting the antipodal relationship of 5 carbon atom fragments around C_{23} developed a highly stereoselective synthesis of fragments (C) & (D) from enantiomeric but otherwise identical 6-carbon precursors (B) based on application of the halolactonisation reaction to acyclic systems. Condensation of B&C with deprotonated acetonitrile as the C_{23} nucleophilic carbonyl equivalent was used to complete the C_{19}-C_{27} fragment. Corey and Clark (11) have also developed a method for the formation of the macrolactam ring based on a relay substrate.

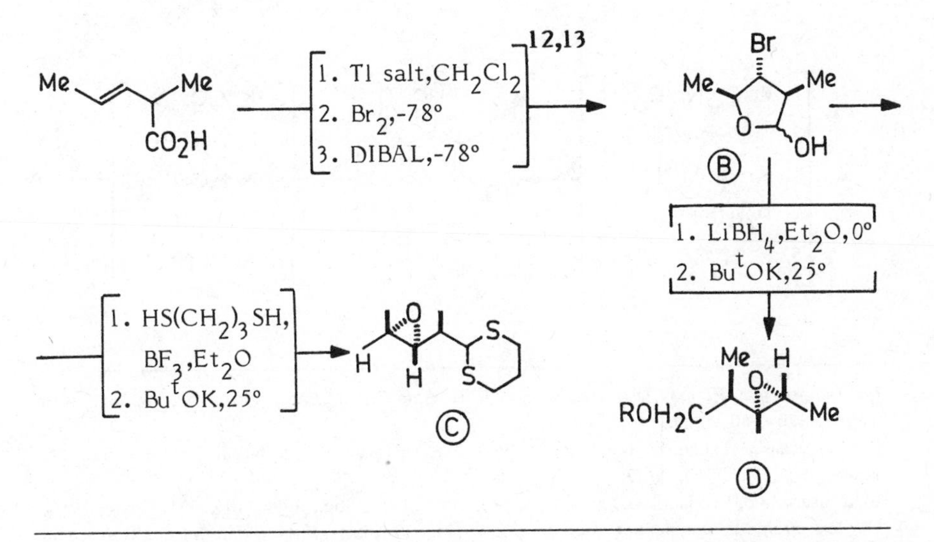

---

11. Corey,E.J.; Hase,T. Tetrahedron Lett., 1979, 335; Corey,E.J.; Clark,D.A. Tetrahedron Lett., 1980, 2045.

12. A trans relationship in halolactonisation is indicated by the known mode of these reactions and was supported by PMR studies; this is a result of kinetically controlled process.

13. The corresponding iodo lactal was much less stable and therefore bromo lactal was used.

Hanesian's carbohydrate approach (14) is based on constructing enantiomerically pure C_{19}-C_{24} and C_{25}-C_{29} fragments from a common D-glucose derived precursor and coupling the two by a directed aldol condensation followed by adjustment of functionalities. Kinoshita's synthesis (15) uses a similar strategy except that the disconnection is between C_{23}-C_{24}, and the two fragments are derived from epimeric pyranosides. Fraser-Reid (16) developed a novel ring cleavage approach from levoglucosan, which avoided the coupling of two sugar derived fragments, and generated the C_{19}-C_{28} chain.

Aromatic fragment

Kishi Synthesis[17]

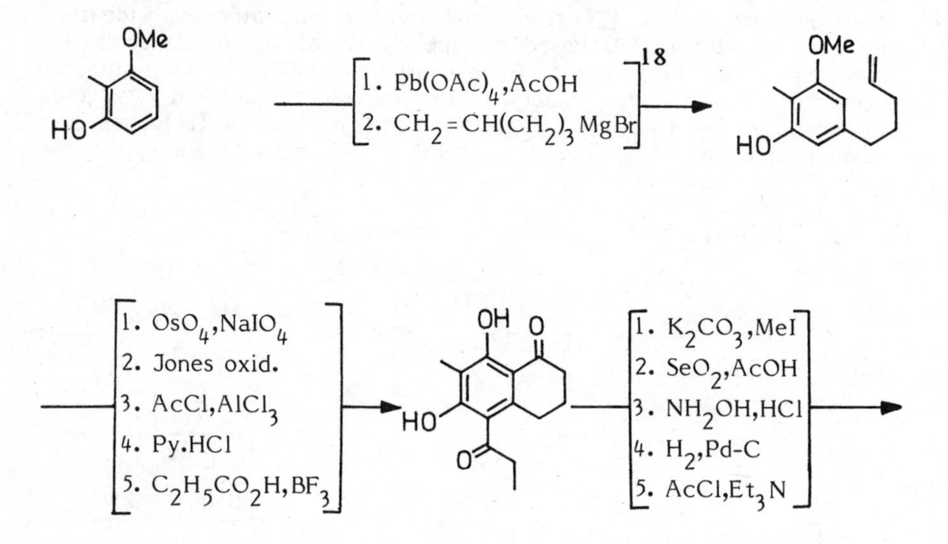

---

14. Hanesian,S.; Pougny,J.R.; Boessenkool,I. J. Am. Chem. Soc., 1982, 104, 6164; Tetrahedron, 1984, 40, 1289.

15. Nakata,M.; Takao,H.; Ikeyama,Y.; Sakai,T.; Tatsuta,K.; Kinoshita,M. Bull. Chem. Soc. Japan, 1981, 54, 1743, 1749.

16. Fraser-Reid,B.; Magdinski,L.; Molino,B. J. Am. Chem. Soc., 1984, 106, 731.

17. The aromatic component used in the original synthesis was prepared from the aromatic fragment obtained by hydrolysis of rifamycin S (2b), and the synthesis described herein was reported subsequently; Nagaoka,H.; Schmid,G.; Lio,H.; Kishi,Y. Tetrahedron Lett., 1981, 22, 899.

18. $Pb(OAc)_4$ reaction helped to create a nucleophilic site through the intermediate.

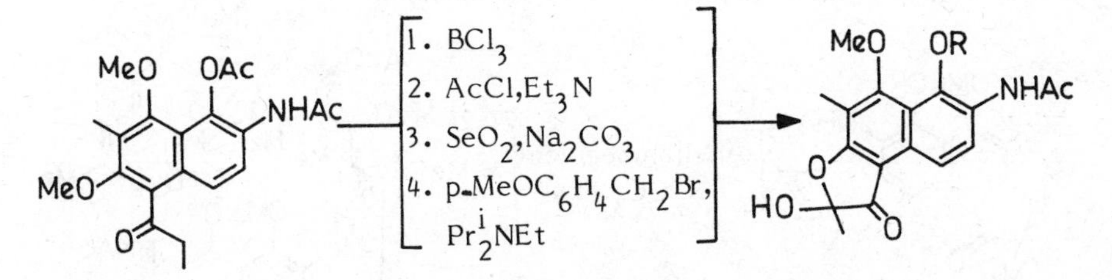

Kelly et al. (19) prepared the aromatic component with all the substituents assembled by a regiospecific Diels-Alder reaction between a suitable diene and a benzoquinone dienophile.

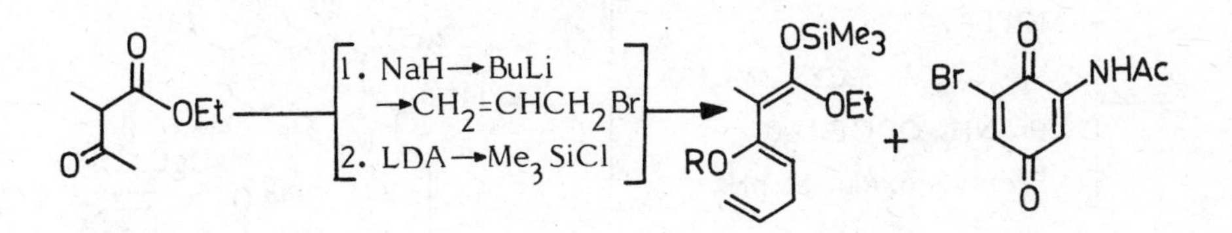

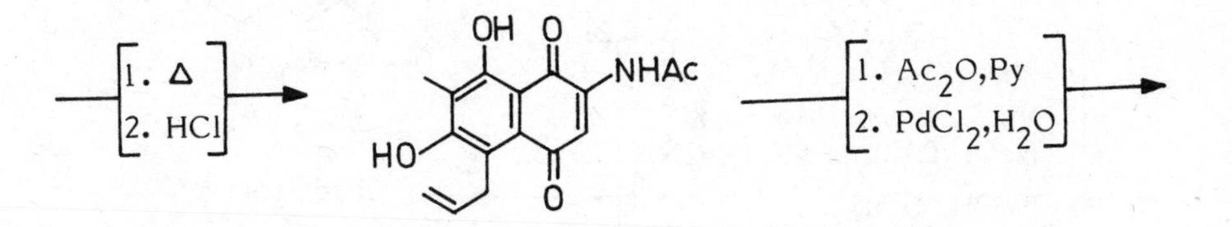

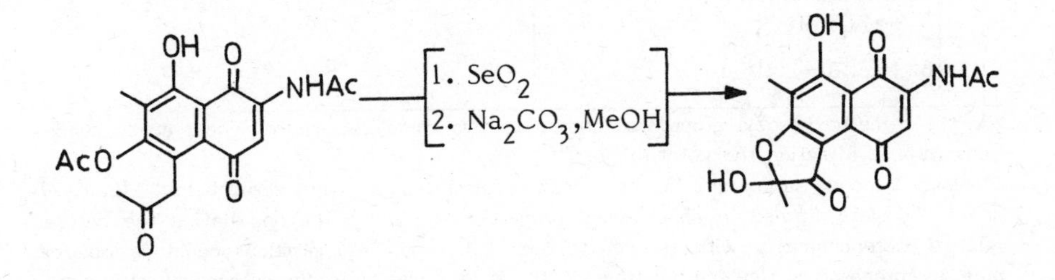

19. Kelly,T.R.; Behforouz,M.; Echavarren,A.; Vaga,J. Tetrahedron Lett., 1983, 23, 2331.

Condensation of aliphatic & aromatic Components

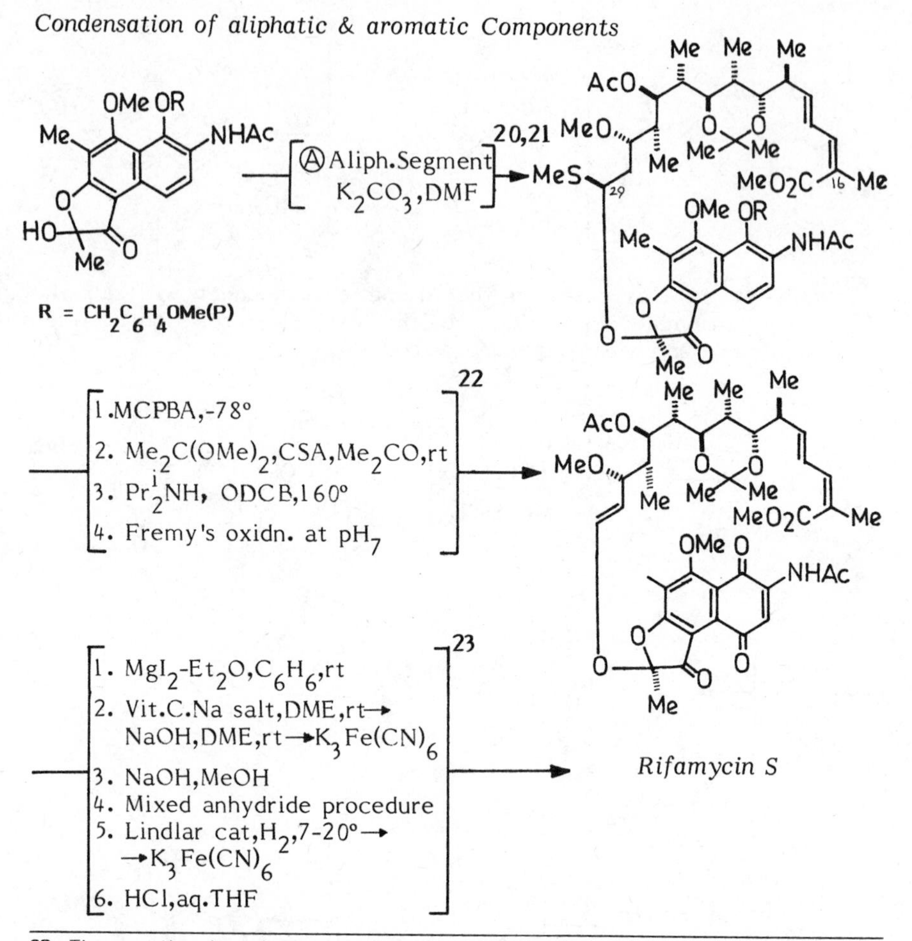

20. The p-methoxybenzyl group was chosen as this could be removed under acidic conditions without affecting the ketal group.

21. In this condensation all the four possible diastereomers with respect to C-12, C-27 and C-29 were formed in about equal proportion, of which the one having the natural relative stereochemistry with respect to the C-12 and C-27 positions could be isolated pure by preparative tlc; the mixture of these two was used for subsequent operations.

22. Olefin formed by sulfoxide elimination was a 1:1 mixture of the trans and cis olefins, which were separated at the next step after oxidation by preparative tlc, and the required trans isomer used for further conversions.

23. Lindlar catalyst at low temperature was used to reduce the naphthoquinone moiety (and in consequence to increase the nucleophilicity of the C-2 amino group in lactam formation) without affecting the olefinic bonds.

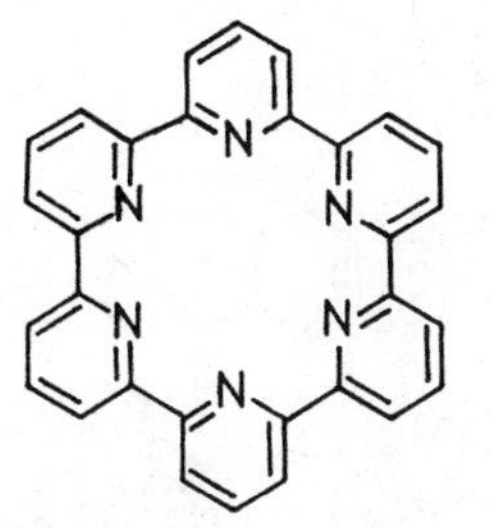

SEXIPYRIDINE

Sexipyridine offers an irresistible electron rich cavity and its synthesis has been vainly pursued since 1932. A synthetic strategy incorporating initial macrocyclization to give a flexible poly-functional intermediate which can be irreversibly rigidized to build the remaining pyridine rings, has resulted in a brilliant synthesis of sexipyridine.

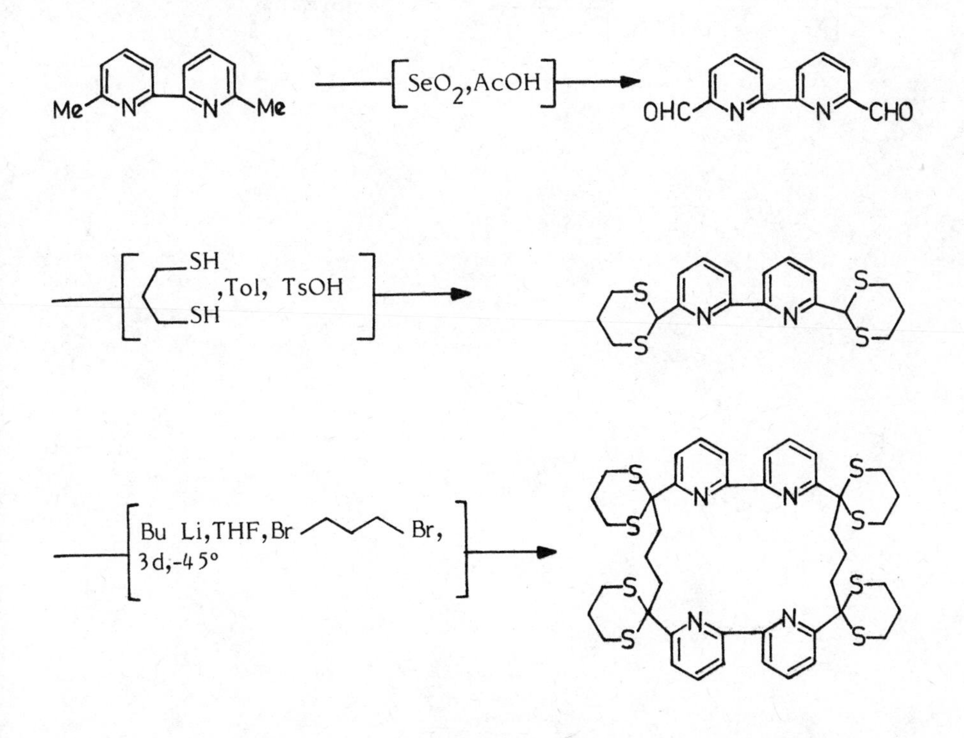

1. Newkome,G.R.; Lee,H.W. J. Am. Chem. Soc., 1983, 105, 5956.

[NBS,THF,MeOH] ⟶

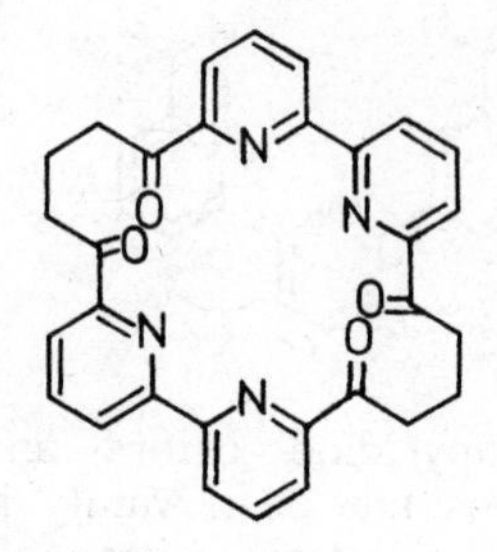

[NH_2OH,HCl
AcOH,reflux,12 h.] ⟶ *Sexipyridine*

SIGMA DIRECTED pi-SYSTEMS

I

The reversal of the usual π-σ-orientation of aromatic systems would lead to extra-ordinary structures where the π component would be directed towards the core of the molecule as illustrated with II-IV. In appropriate cases the electron rich cavity thus generated can harbor metal ions. The computer generated x-ray crystal structure of the IV-Ag^+ complex is represented by I. The strain arising from the holding back of the σ plane is, to a measure, compensated by the increase of the s character of the π bond. This is pronounced in the case of II.

Compound II arises by the dimerization of the extra-ordinarily strained compound, bicyclo [2.2.0]hex-1-ene Ⓐ (1).

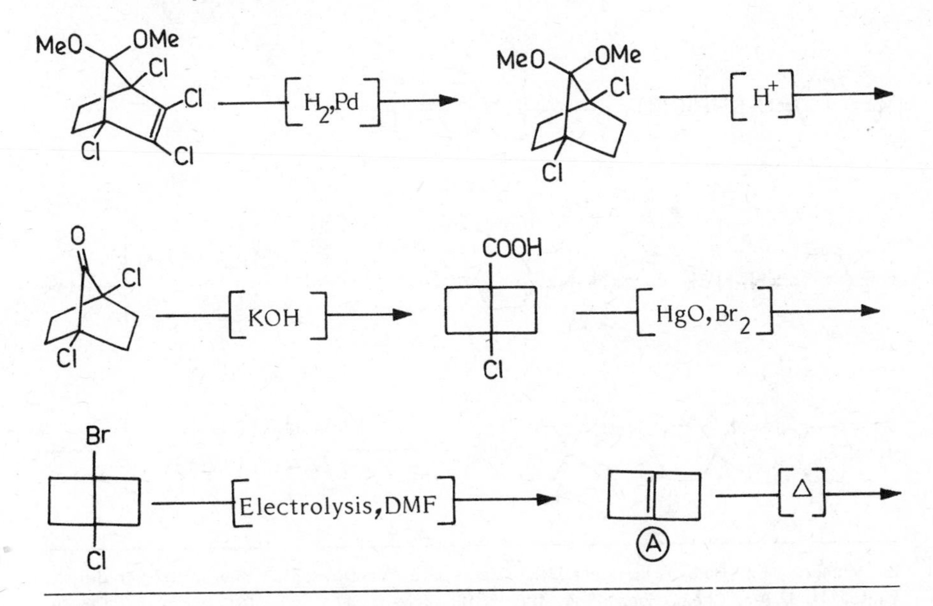

1. Wiberg,K.B.; Matturro,M.G.; Okarma,P.J.; Jason,M.E.; Dailey,W.P.; Burgmaier,G.J.; Bailey, W.F.; Warner,P. Tetrahedron, 1986, 42, 1895.

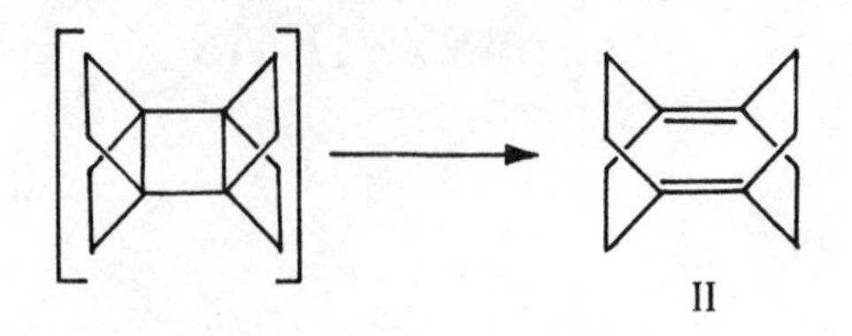

Synthesis of III & IV

Compounds III and IV were prepared by a generalized procedure for cyclohexane-1,4-dione unit elongation incorporating a nitrogen and sulfur extrusion sequence. The linear units are cyclized with Ti° (2).

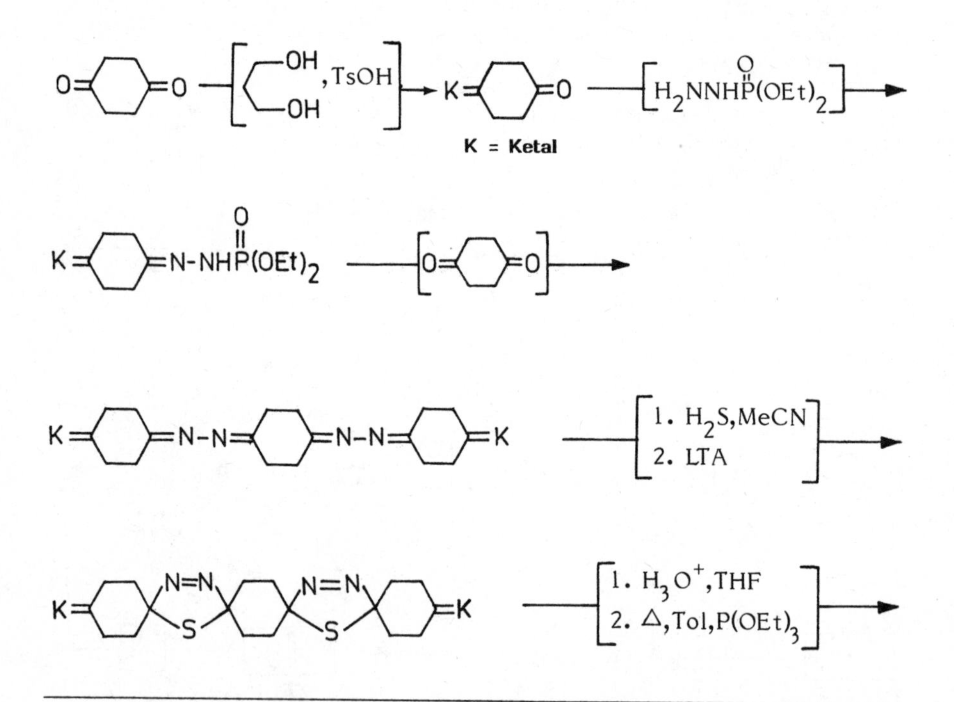

2. McMurry,J.E.; Haley,G.J.; Matz,J.R.; Clardy,J.C.; VanDuyne,G.; Gleiter,R.; Schafer,W.; White,D.H. J. Am. Chem. Soc., 1984, 106, 5018; McMurry,J.E.; Haley,G.J.; Matz,J.R.; Clardy, J.C.; Mitchell, J. J. Am. Chem. Soc., 1986, 108, 515.

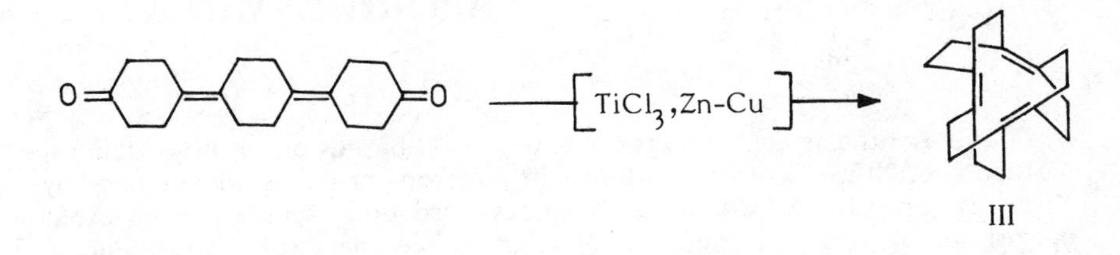

Synthesis of IV

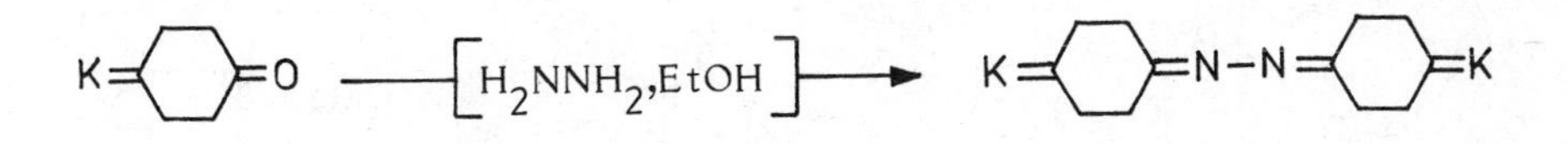

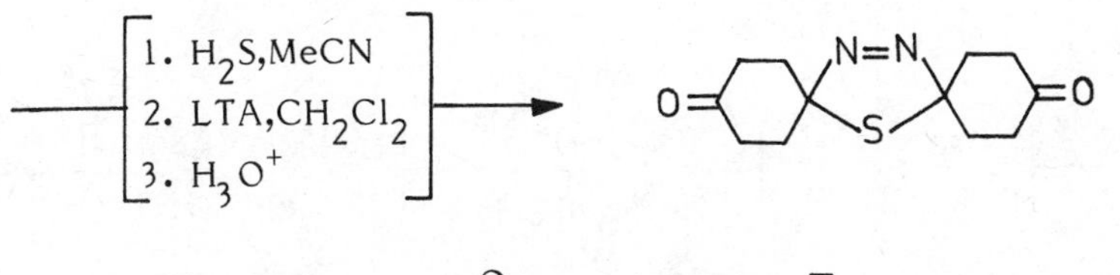

[K=⟨cyclohexylidene⟩=N-NH-P(=O)-$(OEt)_2$,NaH,THF] →

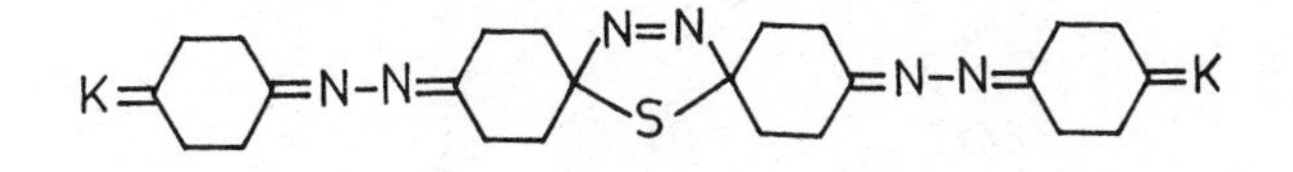

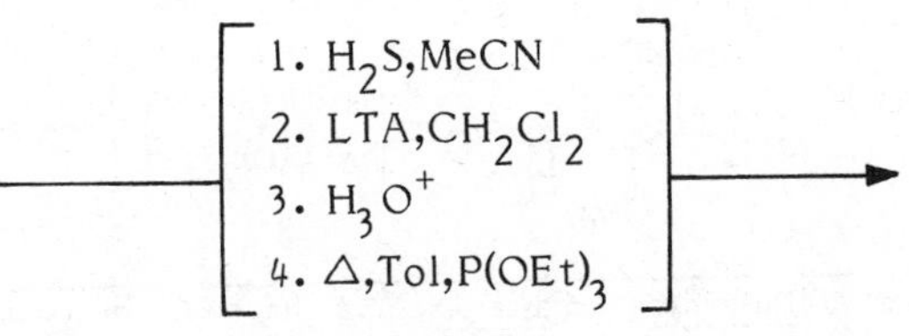

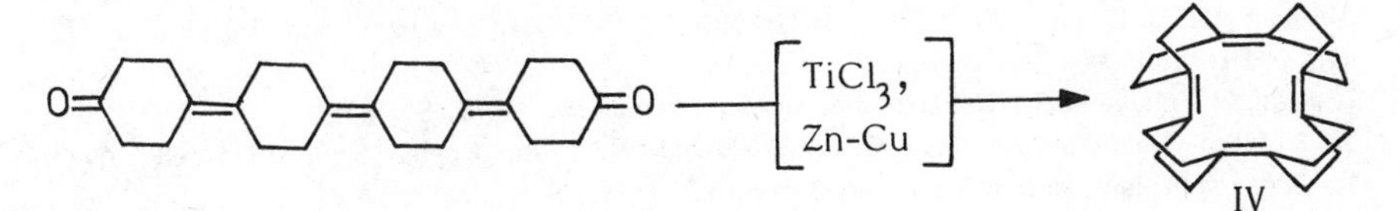

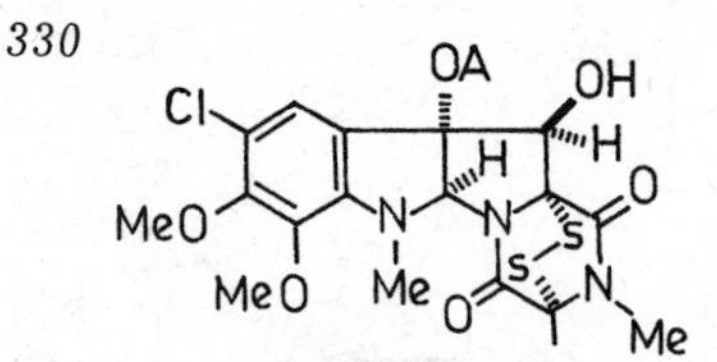

SPORIDESMIN-A

The synthesis of sporidesmin-A (1) is based on a new general method for the synthesis of epidithioketopiperazines developed by Kishi et al. (2), in which a thioacetal grouping serves as a latent episulfide bridge. Introduction of the indole part of the molecule has been carried out by acylation of the bridgehead monocarbanion derived from Ⓐ followed by oxidative ring closure of ring C Ⓒ→Ⓓ (3).

Piperazinedione part:

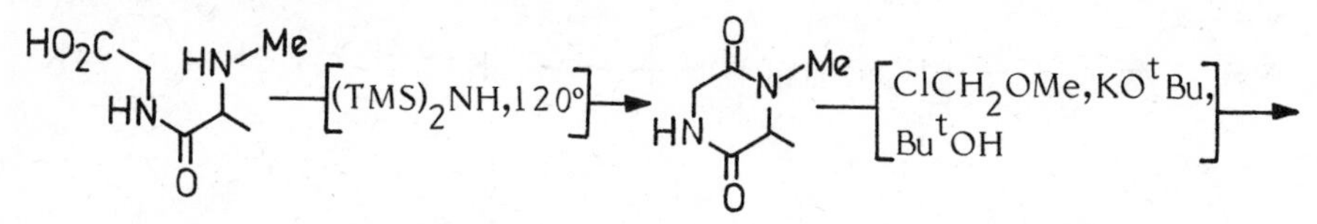

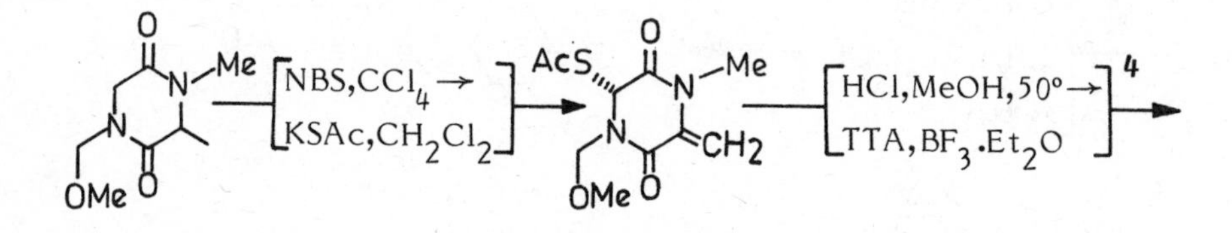

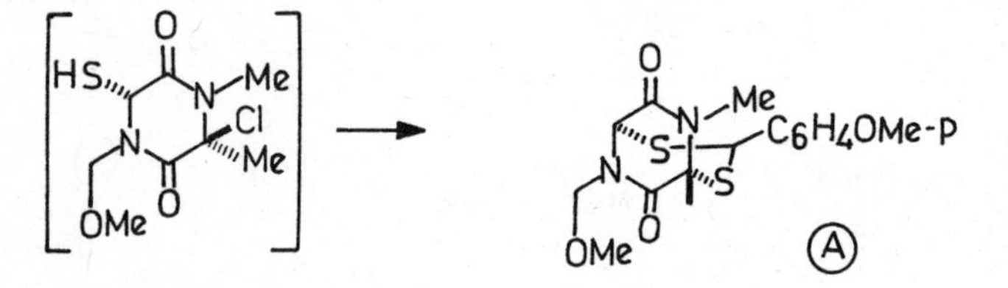

Indole part:

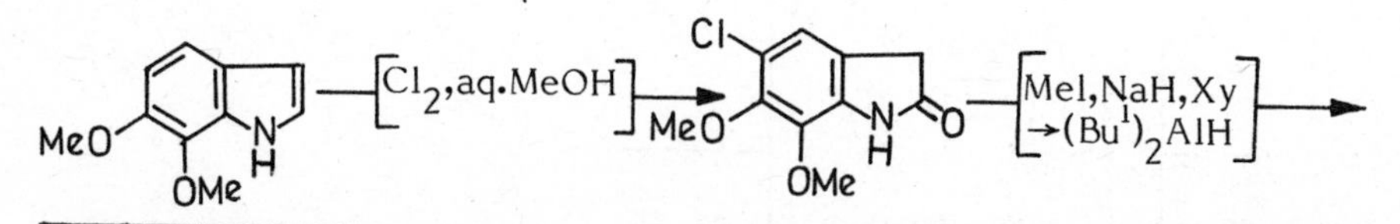

1. Sporidesmins are toxic metabolites of Pithomyces chartarum, and the cause of facial eczema and serious hepatoxic disease in sheep; Ronaldson,J.W.; Taylor,A.; White,E.P.; Abraham,R.J. J. Chem. Soc., 1963, 3172; Safe,S.; Taylor,A. J. Chem. Soc. Perkin I, 1972, 470 and references cited therein.

2. Kishi,Y.; Fukuyama,T.; Nakatsuka,S. J. Am. Chem. Soc., 1973, 95, 6491.

3. Kishi,Y.; Nakatsuka,S.; Fukuyama,T.; Havel,M. J. Am. Chem. Soc., 1973, 95, 6493.

4. TTA = Trithiane derivative of anisaldehyde.

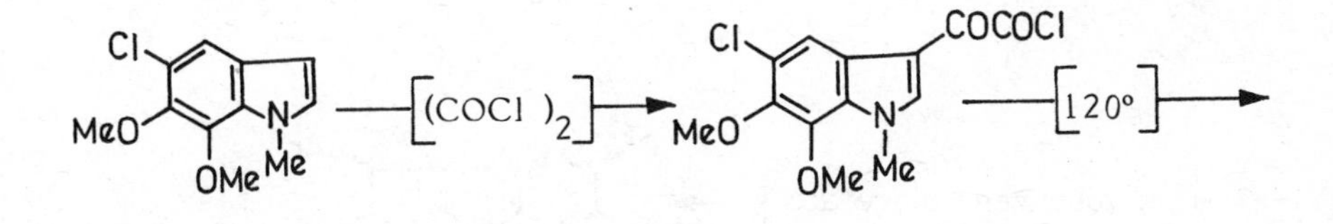

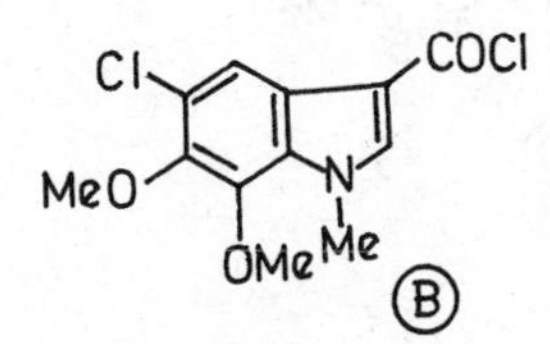

Sporidesmine-A:

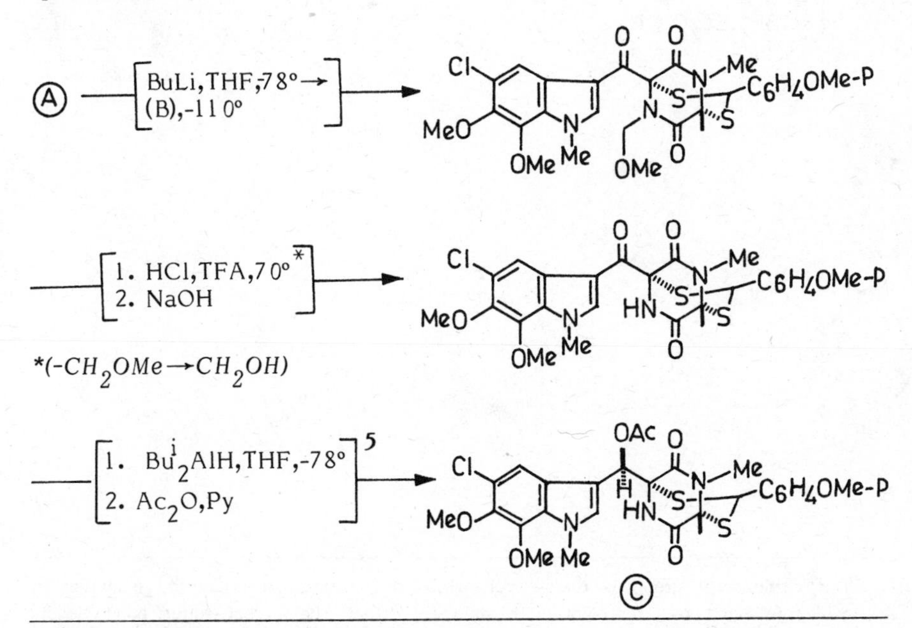

5. This reduction is rationalized as involving complex formation between the amide N-H group and the reducing agent. The crucial intramolecular hydride transfer then occurs from the α-side of the diketopiperazine which stays as far away as possible from the bulky indole unit (i).

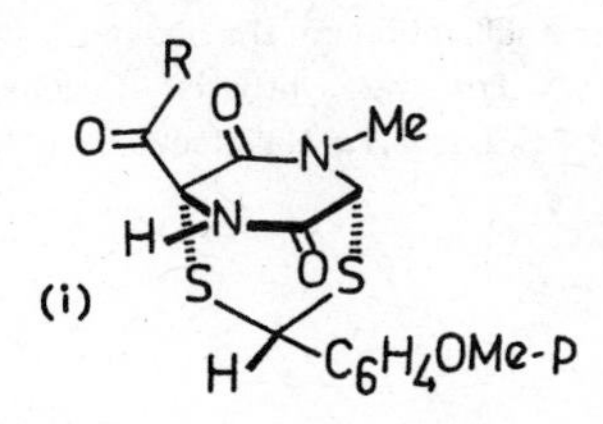

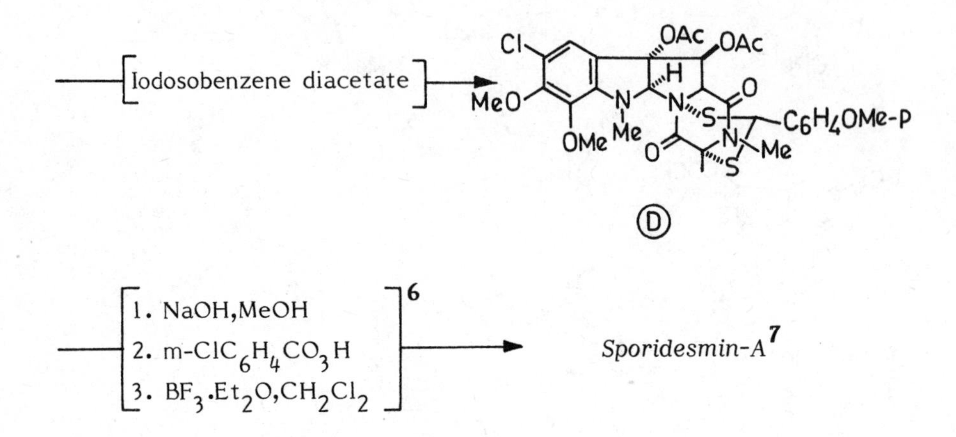

6. The dithioacetal group is readily converted to the disulfide bridge by oxidation to sulfoxide, followed by treatment with acid to effect the carbon-sulfur bond fission. The p-methoxybenzene group is essential for this cleavage and facilitates resonance stabilization of the incipient carbonium ion.

7. For the synthesis of dehydrogliotoxin, a related epidithioketopiperazine, see: Kishi,Y.; Fukuyama,T.; Nakatsuka,S. J. Am. Chem. Soc., 1973, 95, 6492.

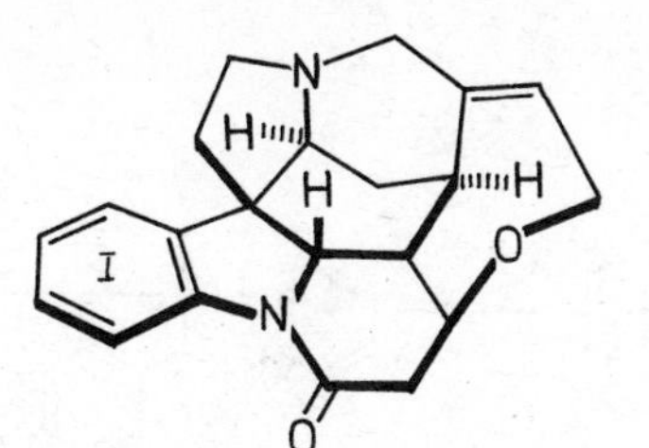

STRYCHNINE

The synthesis of strychnine (1) illustrates how biogenetic considerations may be employed advantageously in planning the synthesis of a complex natural product (2). In cognizance of the natural process Woodward and his colleagues constructed ring V in strychnine by the condensation of an aldehyde between the β-position of a 2-veratryltryptamine (4) and the side-chain nitrogen. Another crucial step in the synthesis was the oxidative cleavage of a carbocyclic ring to provide the right structural elements for cyclizing to a pyridone Ⓐ capable of transformation to the vital intermediate pentacyclic acid Ⓑ.

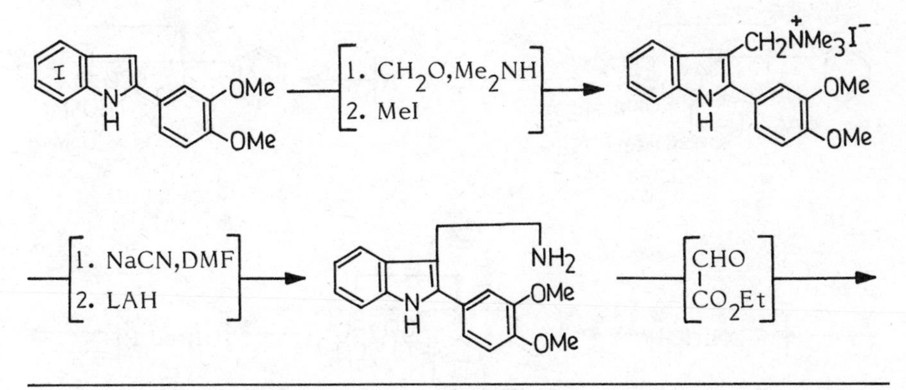

---

1. Woodward,R.B., Cava,M.P., Ollis,W.D., Hunger,A., Daeniker,H.U., Schenker,K. J.Am. Chem. Soc., 1954, 76, 4749; Tetrahedron, 1963, 19, 247.

2. The history of strychnine is singularly deficient in attempts directed towards its total synthesis. Apart from the synthesis by Woodward one need only mention Robinson's efforts in this connection (ref.3), notably his attempt to obtain the Wieland-Gumlich aldehyde through an ingenious biogenetically patterned intramolecular cyclization, Robinson,R., Saxton,J.E. J. Chem. Soc., 1953, 2598. See also Van Tamelen, E.E., Dolby, L.J., Lawton,R.G., Tetrahedron Lett., 1960, 30 for a more recent and successful one-step reproduction of the essential features of strychnine molecule patterned along similar lines.

3. For two excellent reviews on the synthesis of strychnine, see Hendrickson,J.B. in "The Alkaloids", Vol.VI, ed. R.H.F. Manske, Academic Press, New York, (1960), p.211; and G.F. Smith in Vol.VIII, 1965, p.591.

4. The veratryl group serves an additional purpose in blocking the reactive α-position of the indole nucleus.

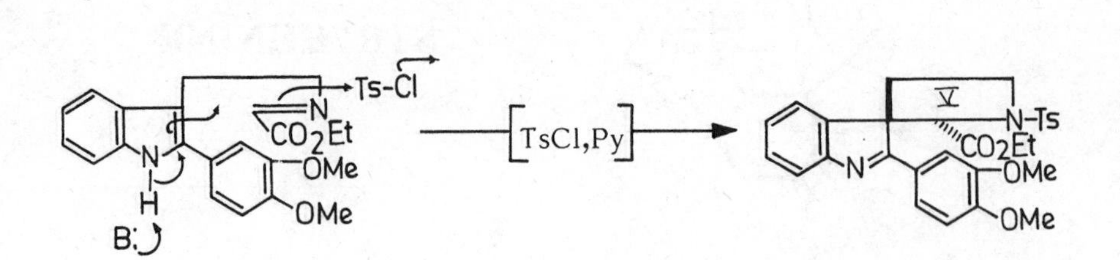
Ts-Cl
N
CO2Et
OMe
OMe
N
H
B:
TsCl,Py
V
N-Ts
CO2Et
N
OMe
OMe

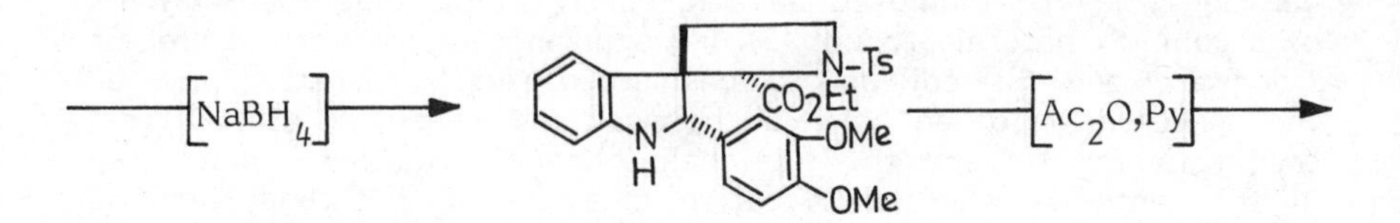
NaBH4
N-Ts
CO2Et
N
H
OMe
OMe
Ac2O,Py

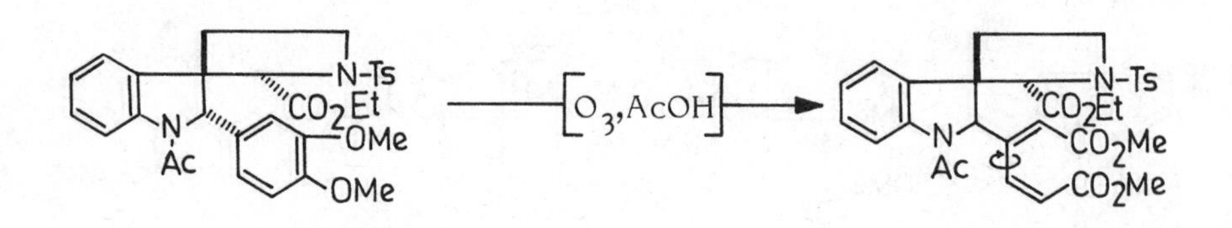
N-Ts
CO2Et
N
Ac
OMe
OMe
O3,AcOH
N-Ts
CO2Et
CO2Me
CO2Me
N
Ac

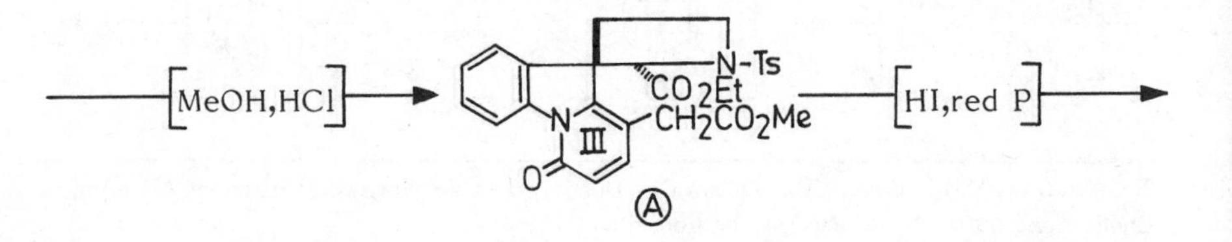
MeOH,HCl
N-Ts
CO2Et
CH2CO2Me
N
III
O
A
HI,red P

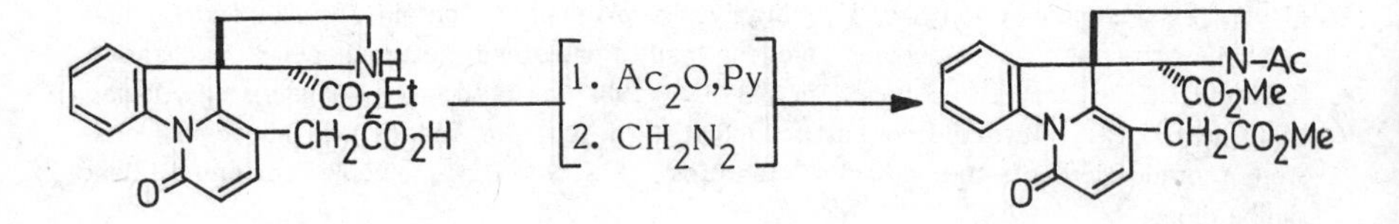
NH
CO2Et
CH2CO2H
N
O
1. Ac2O,Py
2. CH2N2
N-Ac
CO2Me
CH2CO2Me
N
O

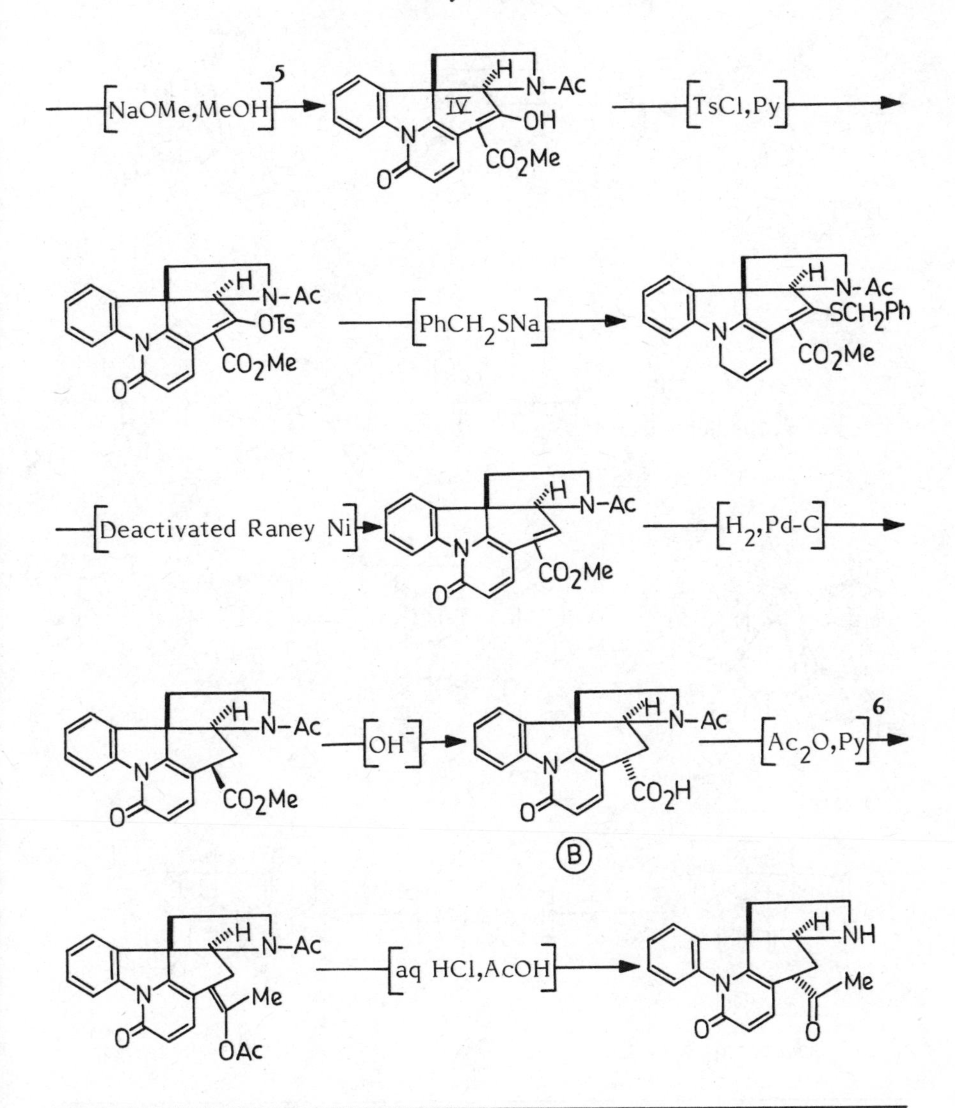

5. With inversion at C-16.

6. This unusual formation of a ketone is obviously due to II-14 being ionized, the steps leading to formation of ketone from the resulting ion are then easily explainable:

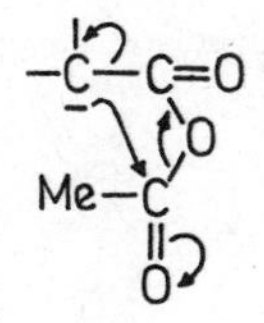

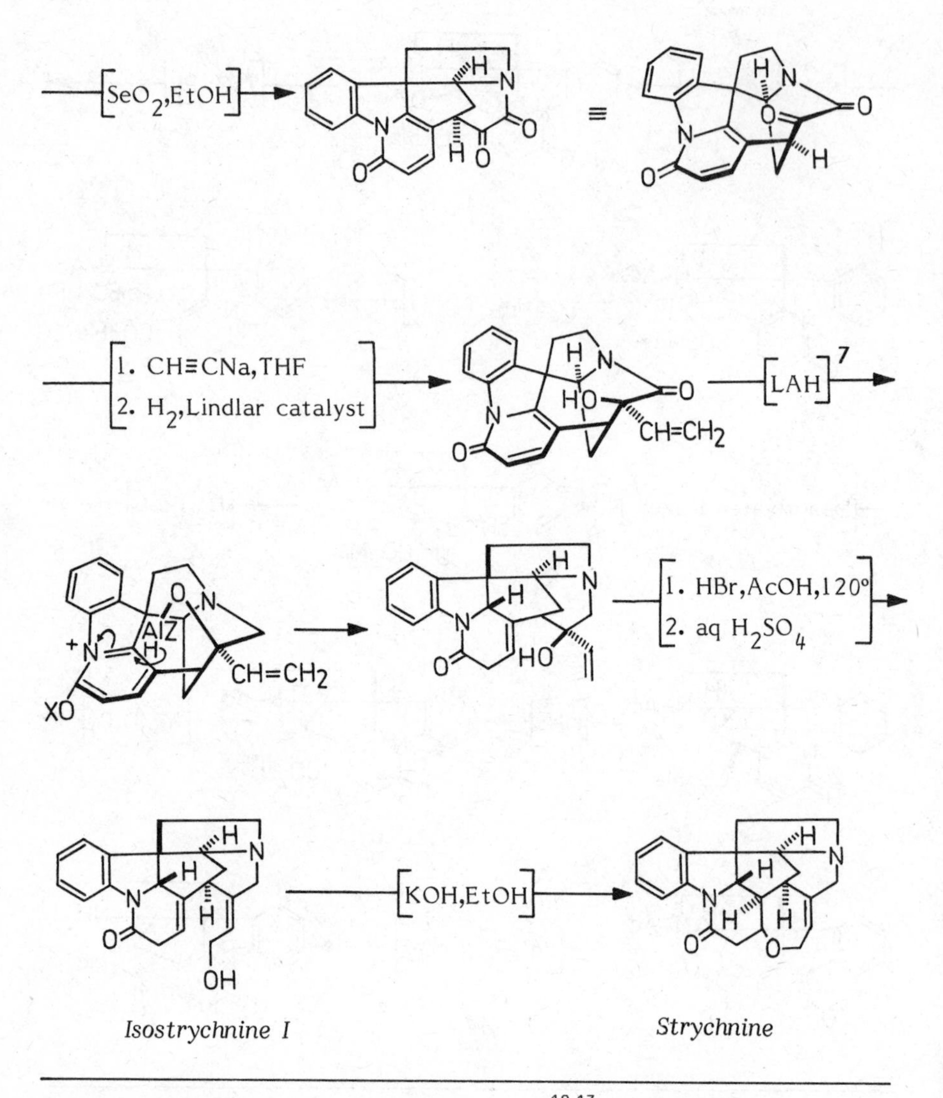

---

7. Concomitant reduction of the α-pyridone to $\Delta^{12,13}$-dihydro-α-pyridone oxidation level occurs from the more hindered concave side and is explicable only by assembling an intermediate capable of specific intramolecular delivery of the hydride ion to C-8.

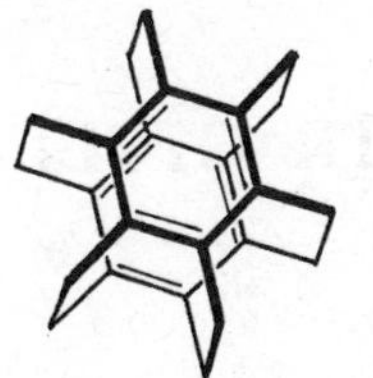

SUPERPHANE

Cyclophanes have been the subject of many studies as models to get answers to questions of bonding, strain-energy and π-π electron interaction. Superhane, which was the ultimate goal in this field, has been synthesised recently based on the ability of benzocyclobutenes to dimerise to form multibridged cyclophanes (1,2).

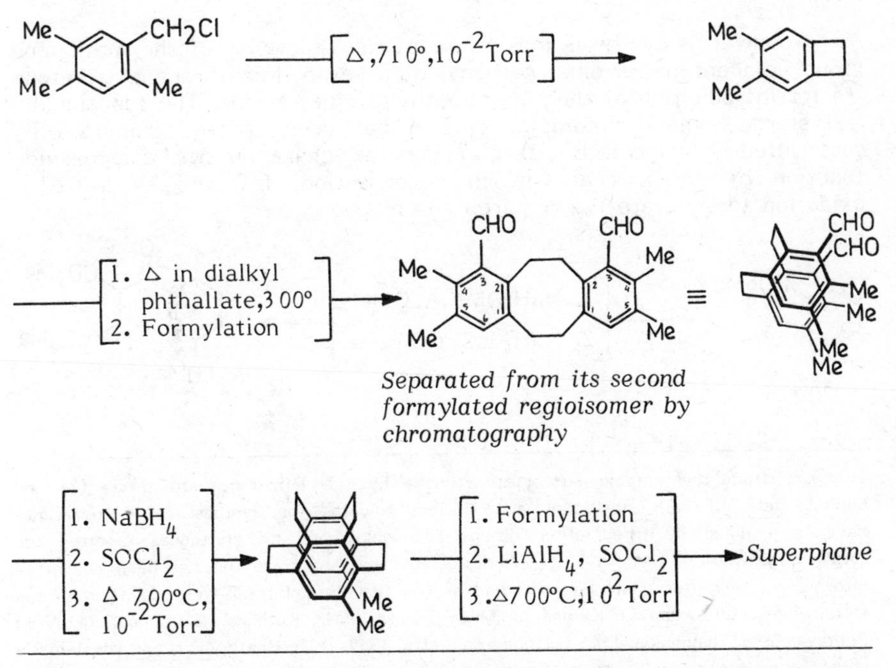

Separated from its second formylated regioisomer by chromatography

1. For review of earlier work see: Smith,B.H., "Bridged Aromatic Compounds", Academic Press, New York, 1964; Vogtle,F.; Neumann,P. Angew. Chem. Int. Ed. Engl., 1972, 11, 73; Cram,D.J.; Cram,J.M. Acc. Chem. Res., 1971, 4, 204; Misumi,S.; Otsubo,T. ibid, 1978, 11, 251; Lindner,H.J. Tetrahedron, 1976, 32, 753.

2. (a) Schirch,P.F.T.; Boekelheide,V. J. Am. Chem. Soc., 1979, 101, 3125; (b) Sekine,V.; Brown,M.; Boekelheide,V. ibid, 1979, 101, 3126.

3. Rieche,I.; Gross,H.; Hoft,E. Chem. Ber., 1960, 93, 88.

4. PMR spectrum of superhane exhibited a singlet at 2.98, while its ^{13}CNMR exhibited (proton decoupled) singlets at 144.2 and 32.3.

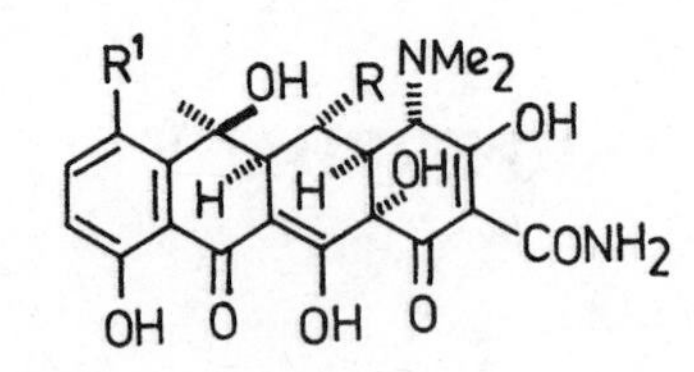

Tetracycline; R=R$_1$=H
Terramycin; R=OH, R$_1$=H
Aureomycin; R=H, R$_1$= Cl

TETRACYCLINES

Ever since elucidation of the structures of aureomycin and terramycin in 1952, the total synthesis of tetracyclines has engaged the attention of several groups of workers (1), culminating, almost a decade later, in the synthesis of 6-demethyl-6-deoxytetracycline by Woodward and his collaborators (2) and of terramycin by Muxfeldt et al. (5,6).

Woodward's synthesis makes use of the reactivity of the methylene group adjacent to terminal carbonyl of a key hydroanthracene triketone (A) for introduction of the N,N-dimethylglycine residue. The functionally and stereochemically complex ring A has been created from a fully substituted intermediate (B) carrying the characteristic carboxamide function by an internal Claisen condensation, followed by a Ce^{+++} oxidation to generate the required chromophore.

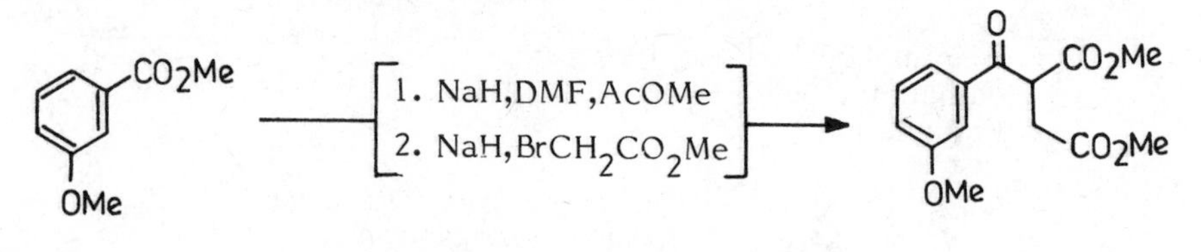

---

1. Apart from the notable successes wrought by the Pfizer-Harvard group (2) and Muxfeldt et al. (5,6), significant contributions towards the synthesis of tetracyclines have been made by three other groups: Shemyakin and his colleagues reported the total synthesis of 12α-deoxy-5 ,6-anhydrotetracycline, and since this has been converted into tetracycline, this could be construed as the first formal total synthesis of a natural tetracycline; Gurevich,A.I.; Karapetyan,M.G.; Kolosov,M.N.; Korobko,V.G.; Onoprienko,V.V.; Popravko,S.A.; Shemyakin,M.M. Tetrahedron Lett., 1967, 131; Booth et al. at the Lederle Laboratories carried out the synthesis of dediethylamino-12α-deoxy-6-demethylanhydrochlorotetracycline; Booth,A.H.; Kende,A.S.; Fields,T.L.; Wilkinson,R.G. J. Am. Chem. Soc., 1959, 81, 1006; Barton & his colleagues in a series of papers have described the synthesis of 6-methylpretetramid, a naphthacene precursor which can be microbiologically converted into tetracyclines; Barton,D.H.R.; Magnus,P.D.; Hase,T. J. Chem. Soc. (C), 1971, 2215, and accompanying series of papers for their general approach to the total synthesis of tetracyclines.

2. Conover,L.H.; Butler,K.; Johnston,J.D.; Korst,J.J.; Woodward,R.B. J. Am. Chem. Soc., 1962, 84, 3222; Woodward,R.B. Pure Appl. Chem., 1963, 6, 651; Korst,J.J.; Johnston,J.D.; Butler,K.; Bianco,E.J.; Conover,L.H.; Woodward,R.B. J. Am. Chem. Soc., 1968, 90, 439.

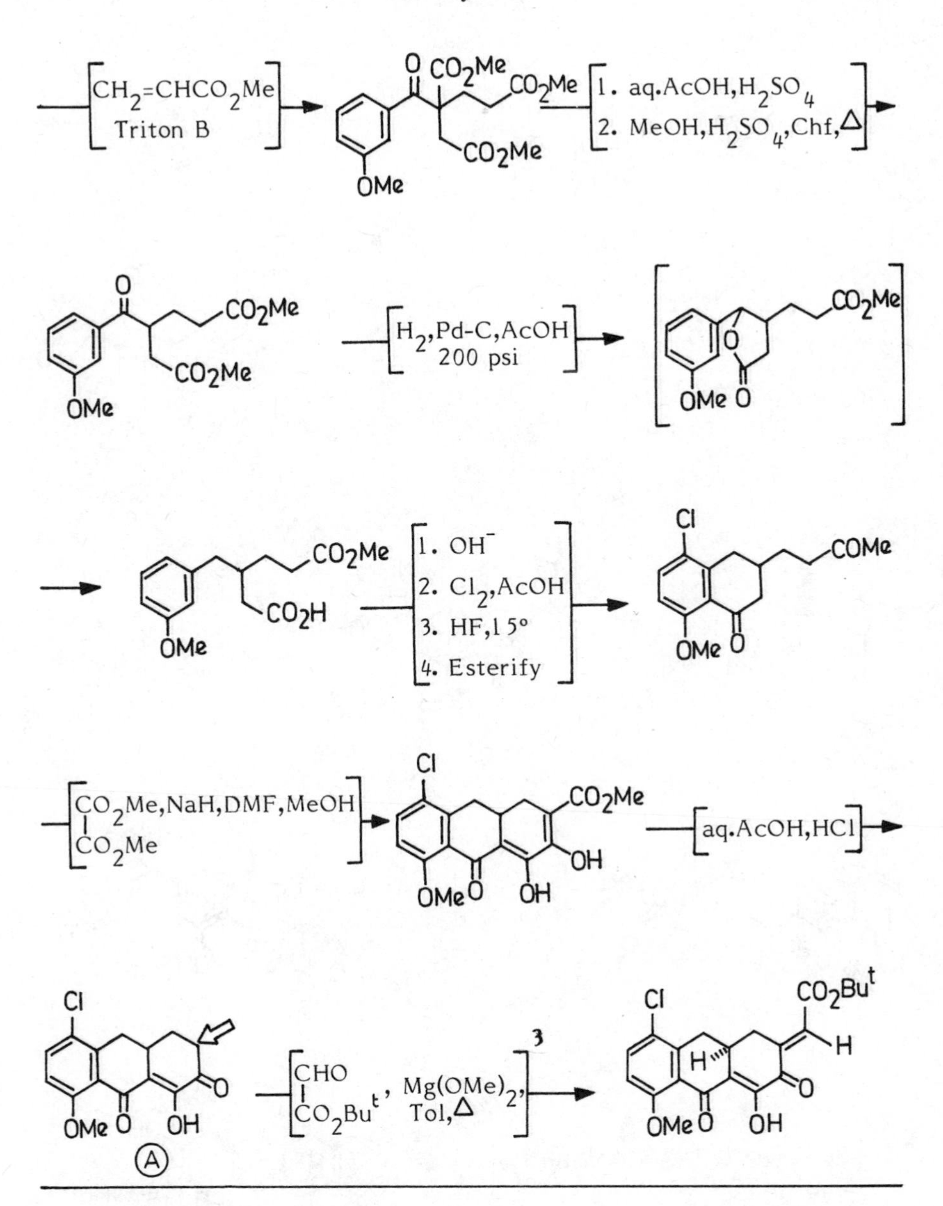

3. Choice of magnesium methoxide as the condensing agent was made with a view to protecting the β-dicarbonyl system against base-cleavage.

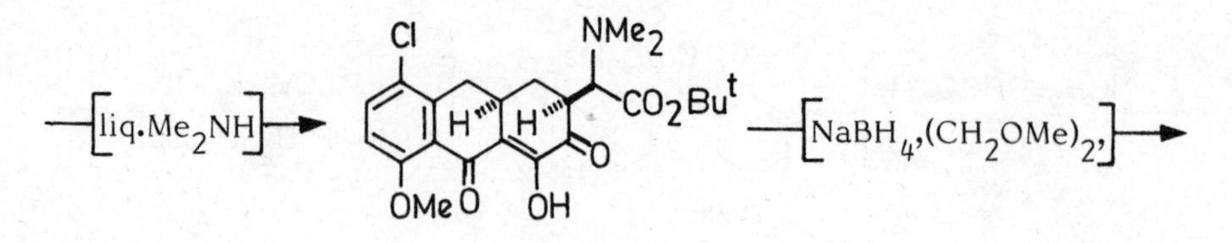
liq.Me2NH
Cl
NMe2
CO2But
H
H
O
OMe
O
OH
NaBH4,(CH2OMe)2,

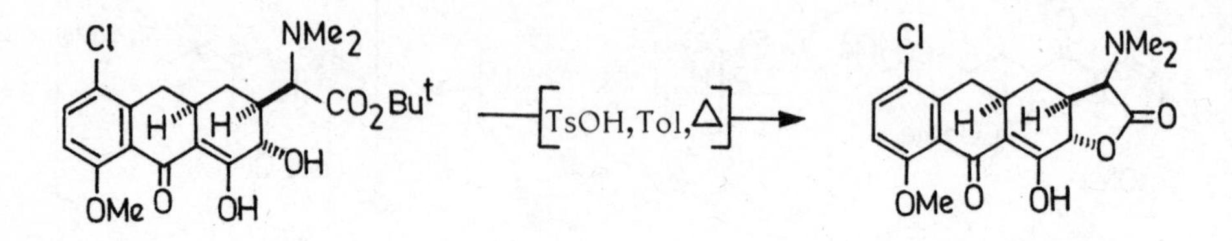
Cl
NMe2
CO2But
H
H
OH
OMe
O
OH
TsOH,Tol,Δ
Cl
NMe2
H
H
O
O
OMe
O
OH

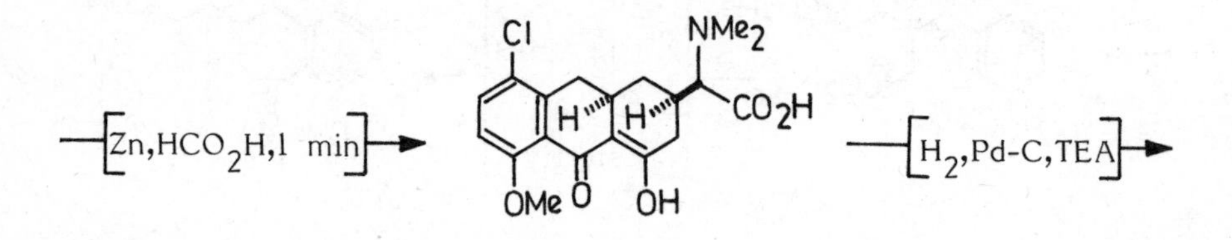
Zn,HCO2H,1 min
Cl
NMe2
H
H
CO2H
OMe
O
OH
H2,Pd-C,TEA

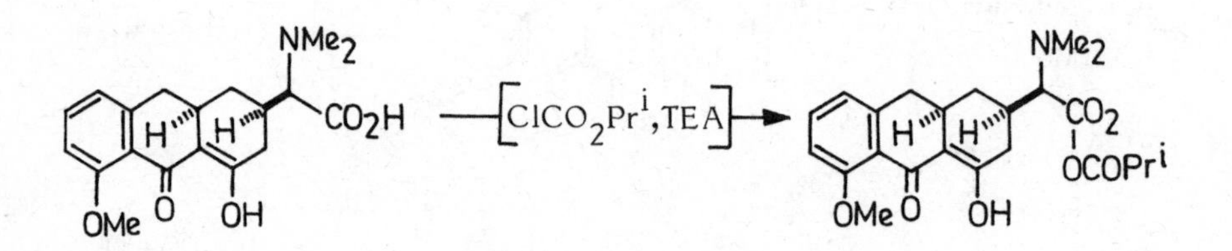
NMe2
H
H
CO2H
OMe
O
OH
ClCO2Pri,TEA
NMe2
H
H
CO2
OCOPri
OMe
O
OH

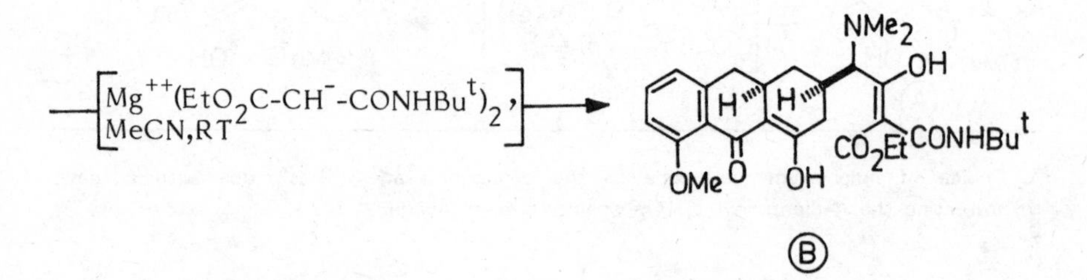
Mg++(EtO2C-CH−-CONHBut)2,
MeCN,RT
NMe2
OH
H
H
CO2Et
CONHBut
OMe
O
OH
B

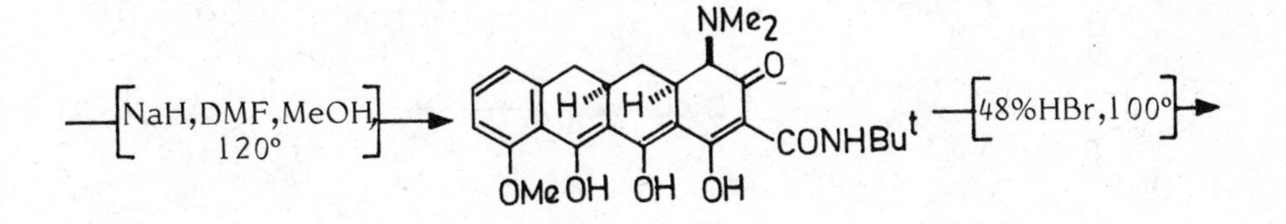

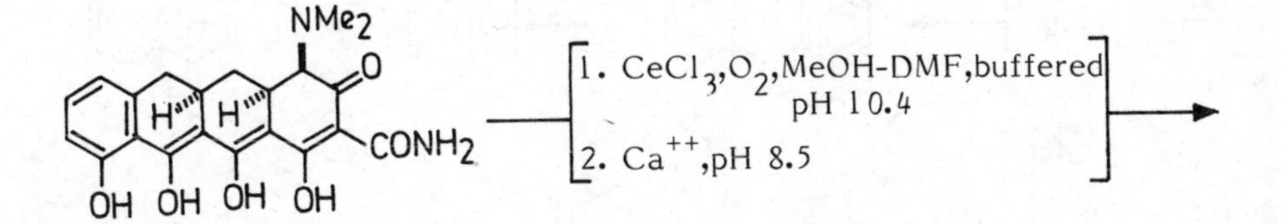

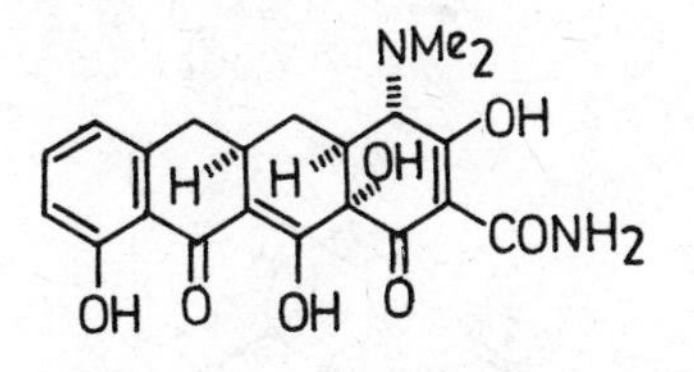

dl-6-Demethyl-6-deoxytetracycline[4]

Using a different approach fo the hydronaphthacene nucleus, Muxfeldt and his collaborators (5), have accomplished an elegant synthesis of terramycin (6), the first total synthesis of a natural tetracycline. Noteworthy feature of this synthesis are the stereospecific elaboration of rings C and D as the key bicyclic aldehyde Ⓐ. introduction of the 4-amino function as a thiazolidinyl group and completion of rings A and B using an oxoglutamate.

---

4. For another synthesis of this compound, see: Muxfeldt,H.; Rogalski,W. J. Am. Chem. Soc., 1965, 87, 933.

5. Muxfeldt,H. Angew. Chem., Internat. Ed., 1962, 1, 372; ibid, Angew. Chem., 1962, 74, 825; Muxfeldt,H.; Hardtmann,G. Ann., 1963, 669, 113.

6. Muxfeldt,H.; Hardtmann,G.; Kathawala,F.; Vedejs,E.; Mooberry,J.B. J. Am. Chem. Soc., 1968, 90, 6534. For an engrossing account of the total work on the chemical synthesis of terramycin by the Muxfeldt group see: Muxfeldt,H.; Haas,G.; Hardtmann,G.; Kathawala, F.; Mooberry,J.B.; Vedejs,E. J. Am. Chem. Soc., 1979, 101, 689.

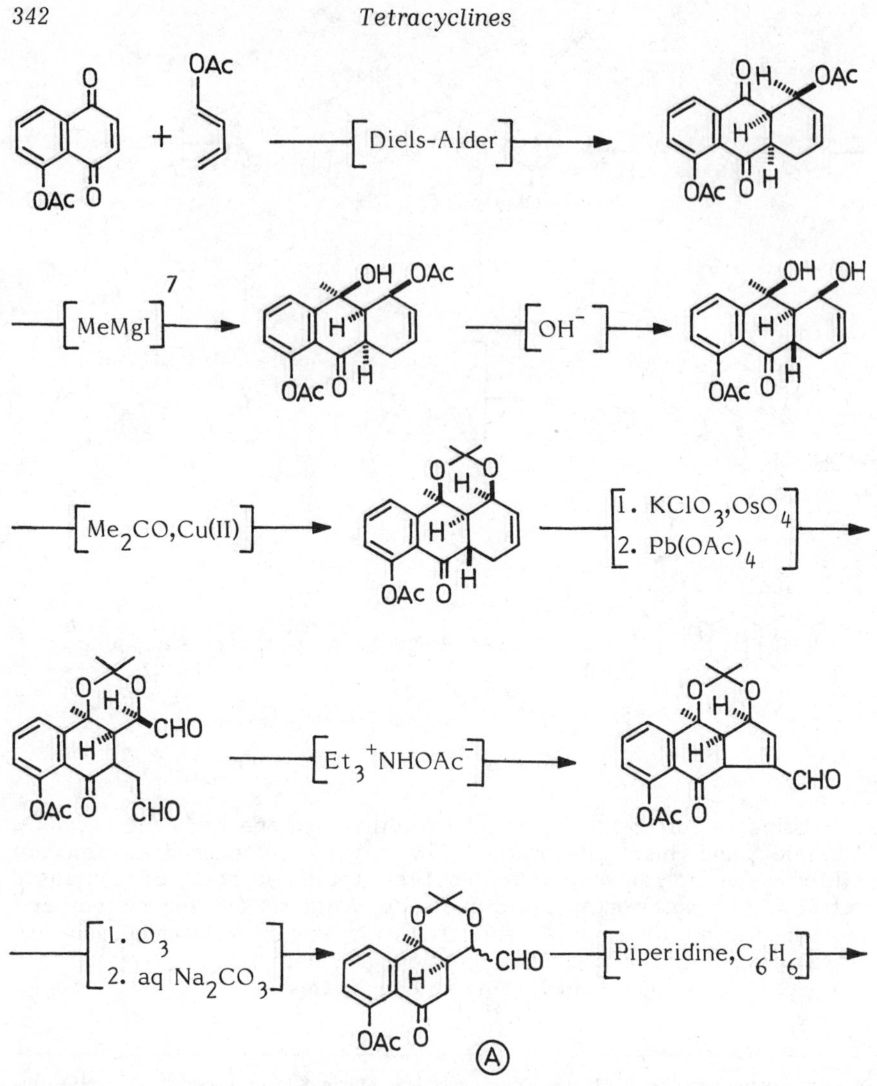

7. Selective attack of the grignard reagent on the carbonyl situated adjacent to the acetoxy group can be explained if it is assumed that the acetoxy group takes part in the reaction as depicted:

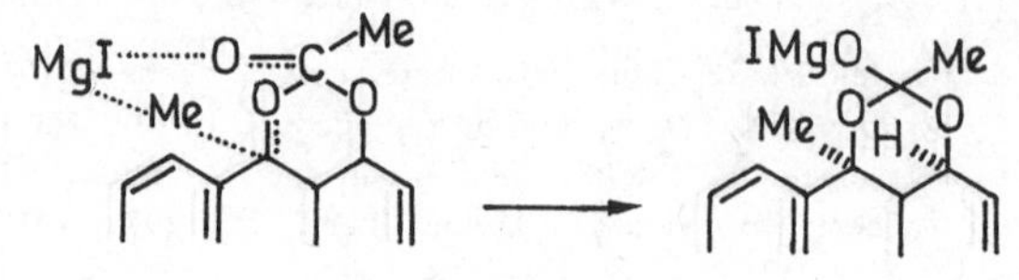

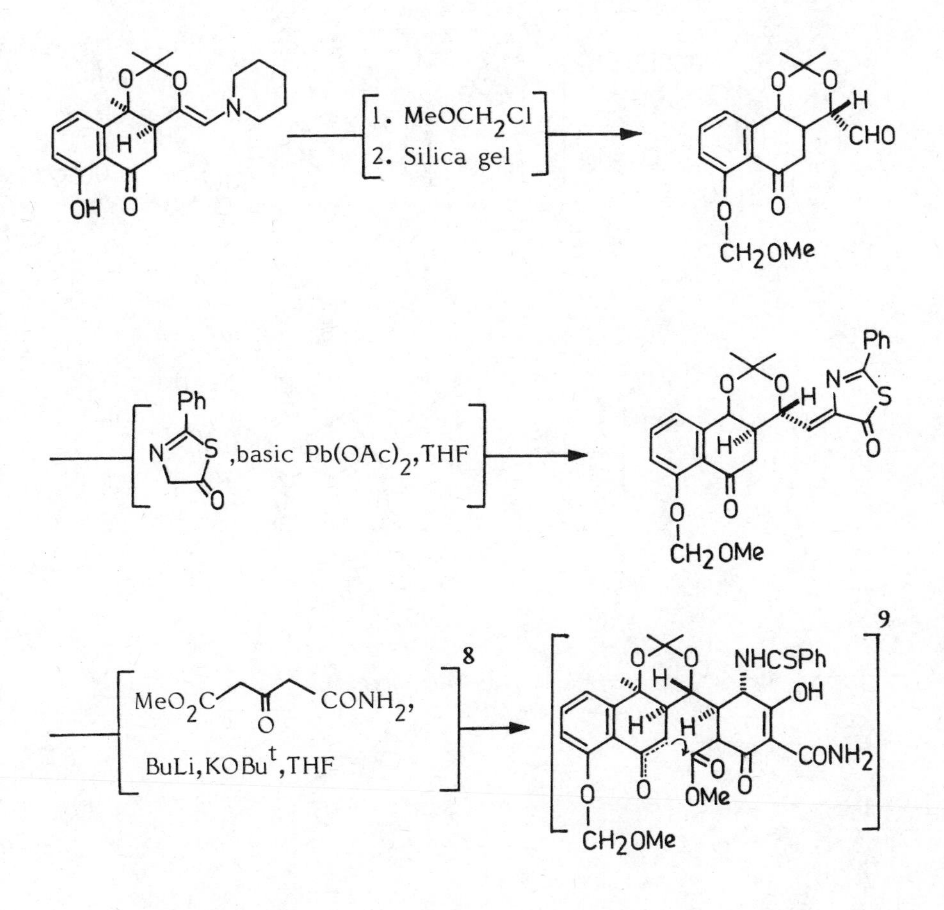

8. This is a new general reaction between azlactones and derivatives of acetonedicarboxylic esters leading to amino dihydroresorcinol derivatives.

9. There are two possible modes of cyclization of the tricyclic ACD-intermediate and are shown in (i) and (ii); the final configuration at 4a is determined by the fact that conformationally (i) is the most stable intermediate (half-chair) on the pathway leading to the tetracyclic system.

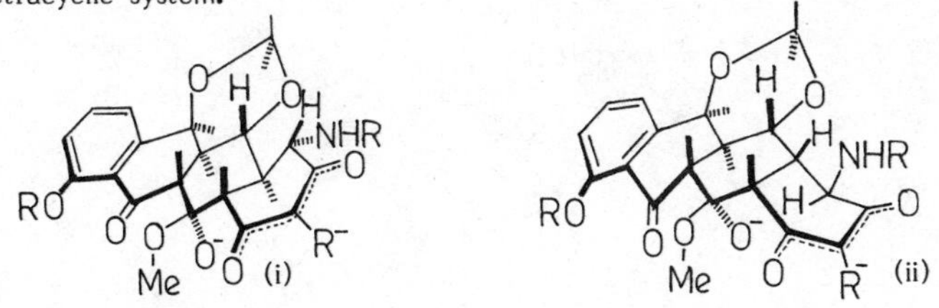

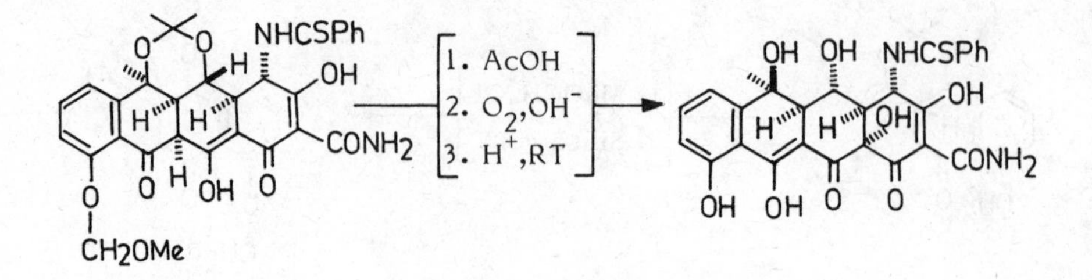

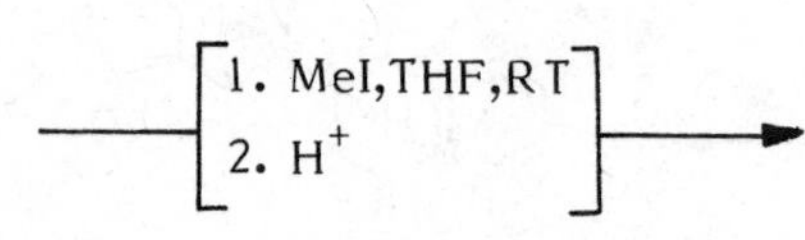

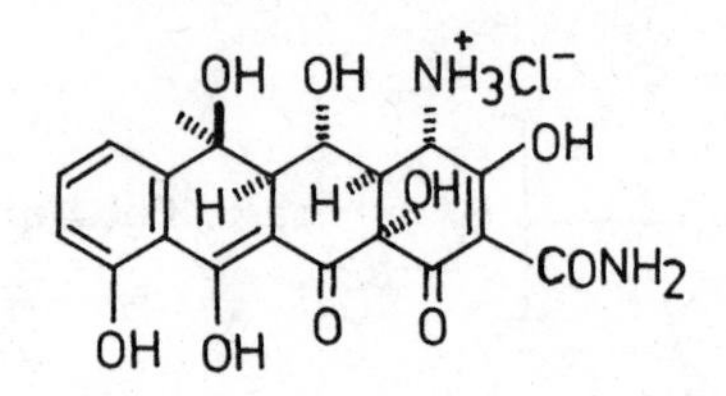

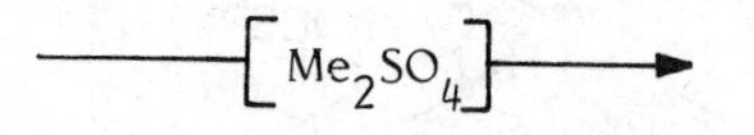

dl-Terramycin

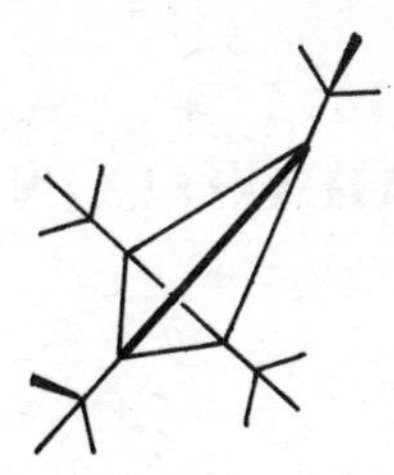

TETRAHEDRANE, TETRA-tert-BUTYL

Tetrahedrane represents the very ultimate in carbon angle distortion and this compound has been the synthetic objective of endeavours spanning well over five decades. The highly substituted tetrahedrane has now been made in a truly spectacular manner, inorporating, inter alia, the photochemical extrusion of carbon dioxide and carbon monoxide (1).

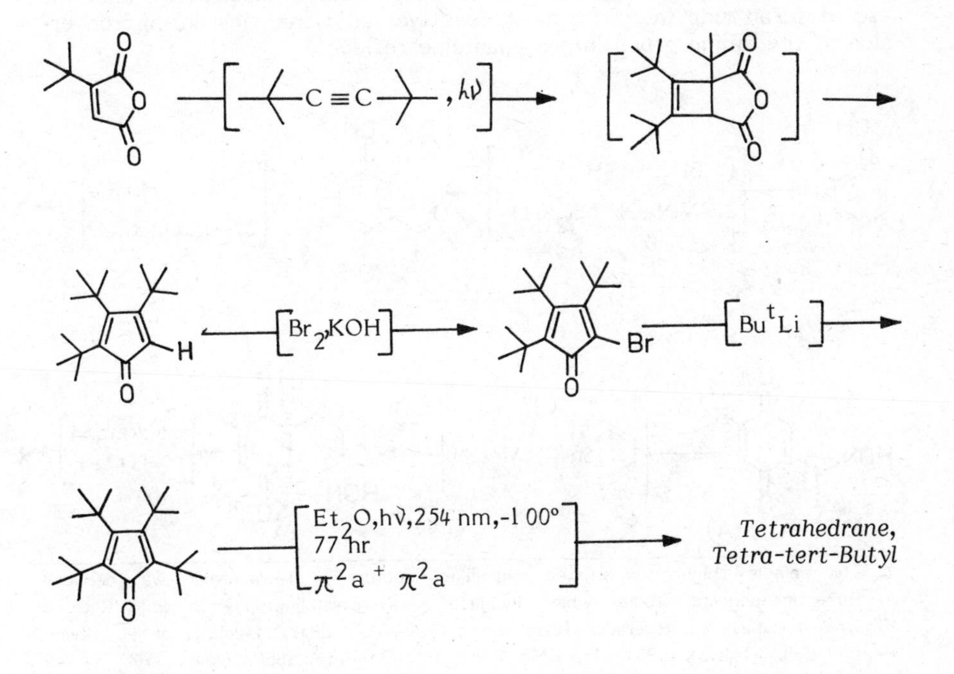

1. Maier,G.; Pfriem,S. Angew. Chem. Int. Ed., 1978, 17, 519; Maier,G.; Pfriem,S.; Shafer,U.; Matusch,R. ibid, 1978, 17, 520.

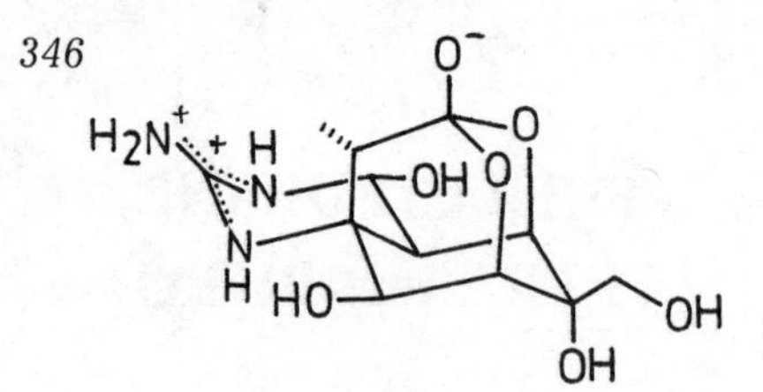

TETRODOTOXIN

Tetrodotoxin, isolated from the ovaries and liver of the puffer fish (Japanese: fugu) (1), is one of the most toxic low molecular weight poisons known. A noteworthy feature of the only synthesis of tetrodotoxin reported by Kishi's group (2-4) is the aggrandization of six chiral centers on the cyclohexane nucleus in the key tetrodamine intermediate Ⓑ by effective use of proximity effects. Other important features of the synthesis are the use of Lewis acid in Diels-Alder reaction of Ⓐ, which introduced the desired regioselectivity by enhancing the electron withdrawal of the α-oximinoalkyl substituent and the use of furan ring in Ⓓ to mask reactive functionalities during conversion of the amino group into a guanidine residue.

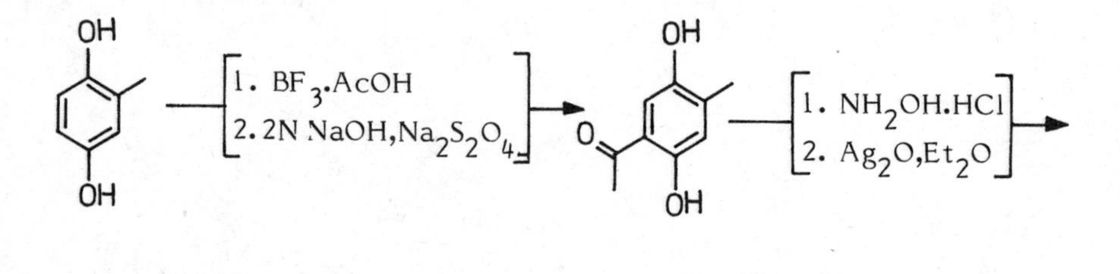

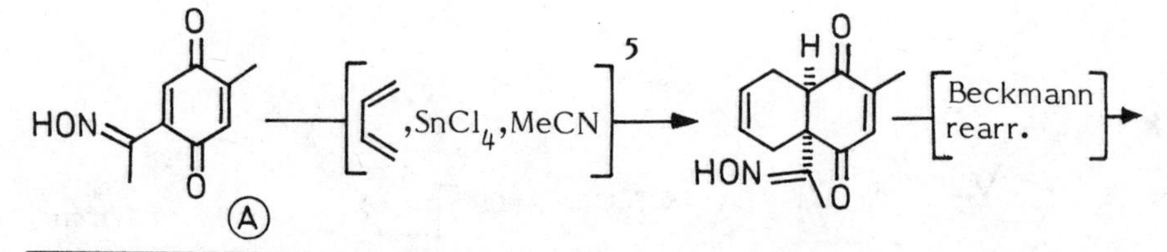

1. The complex highly oxygenated cage-like structure of tetrodotoxin was deduced in three independent studies which included X-ray crystallography as well; Goto,T.; Kishi,Y.; Takahashi,S.; Hirata,Y. Tetrahedron, 1965, 21, 2059; Tsuda,K. el al. Chem. Pharm. Bull. (Tokyo), 1964, 12, 634; Woodward,R.B. Pure Appl. Chem., 1964, 9, 49.

2. Kishi,Y.; Nakatsubo,F.; Aratani,M.; Goto,T.; Inoue,S.; Kakoi,H.; Sugiura,S. Tetrahedron Lett., 1970, 5127.

3. Kishi,Y.; Nakatsubo,F.; Aratani,M.; Goto,T.; Inoue,S.; Kakoi,H. Tetrahedron Lett., 1970, 5129.

4. Kishi,Y.; Aratani,M.; Fukuyama,T.; Nakatsubo,F.; Goto,T.; Inoue,S.; Tanino,H.; Sugiura,S.; Kakoi,H. J. Am. Chem. Soc., 1972, 94, 9217, 9219.

5. The required regioselectivity was the result of the enhanced electron withdrawal from α-oximinoalkyl substituent induced by the Lewis acid.

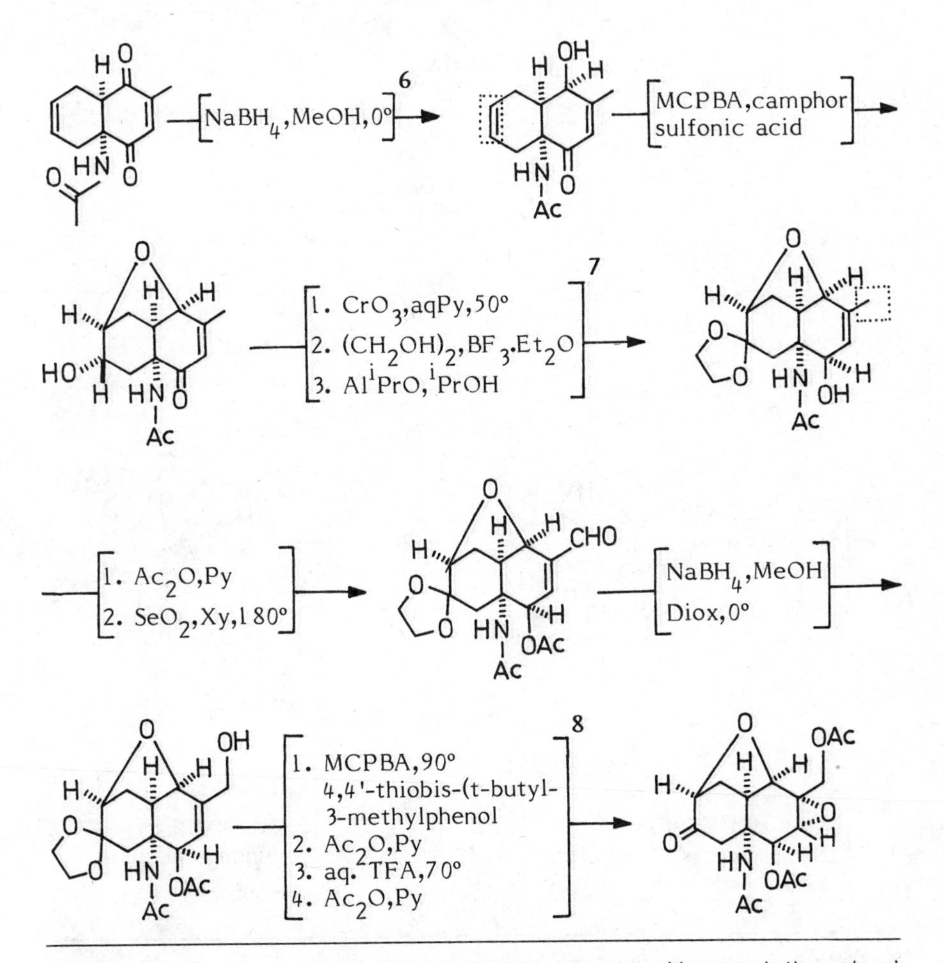

6. The reduction was highly stereoselective since the reagent could approach the carbonyl group only from the outer side of the cage-like molecule, and regiospecific since the carbonyl at C-8 was protected by presence of the vicinal axial acetamido group at C-8a.

7. The reducing agent approaches from outer side of the cage-like molecule.

8. Addition of a radical inhibitor prevented thermal decomposition of the peracid, thus allowing epoxidation of the otherwise poorly reactive olefin to take place at elevated temperatures, Kishi,Y.; Aratani,M.; Tanino,H.; Fukuyama,T.; Goto,T.; Inoue,S.; Sugiura,S.; Kakoi,H. Chem. Commun., 1972, 64.

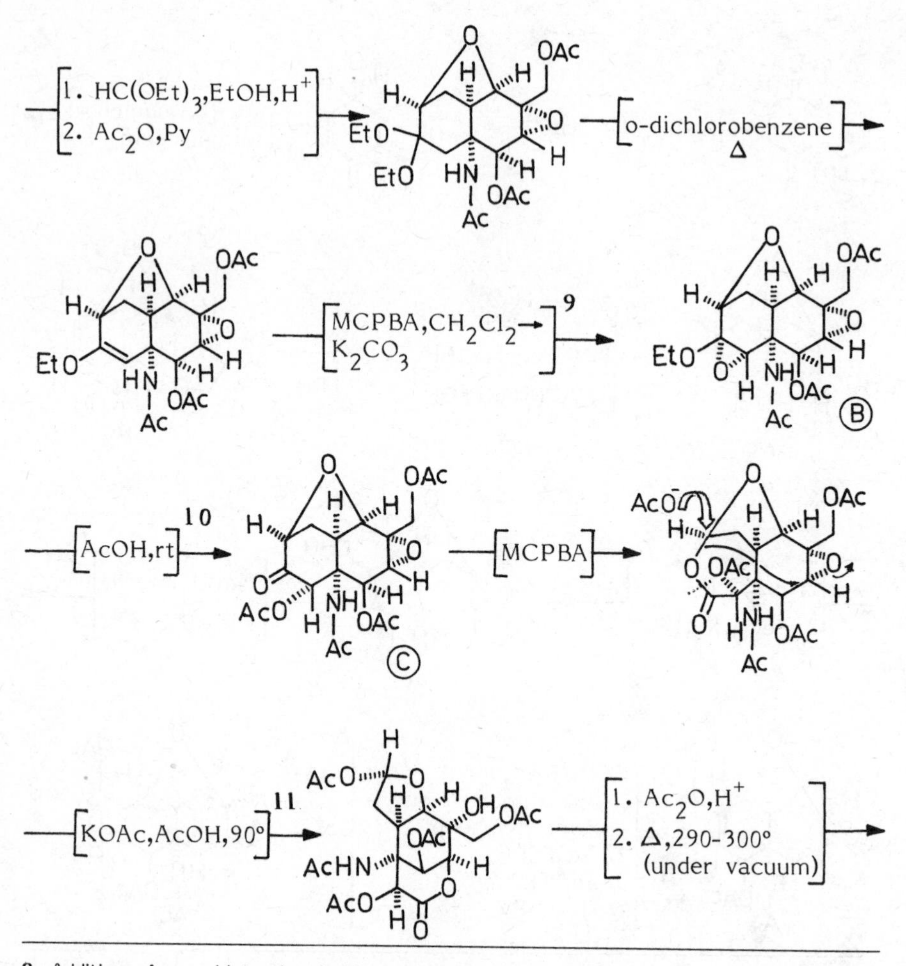

9. Addition of peracid to the double bond occurs from the α-side due to steric reasons.
10. Stereospecific opening of the epoxy ether by acetic acid may be rationalized as depicted below: (Cf. Stevens, C.L.; Dykstra, S.J. J. Am. Chem. Soc., 1953, 75, 5975)

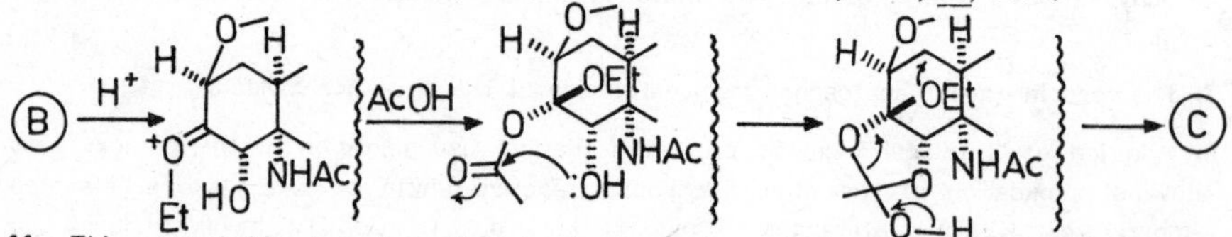

11. This remarkable transformation involves opening of the seven-membered lactone ring, to generate at one terminus a carboxylate group which attacks the epoxide ring by an intramolecular SN2 reaction, while at the other terminus the oxonium ion is attacked by acetate ion from the less hindered side.

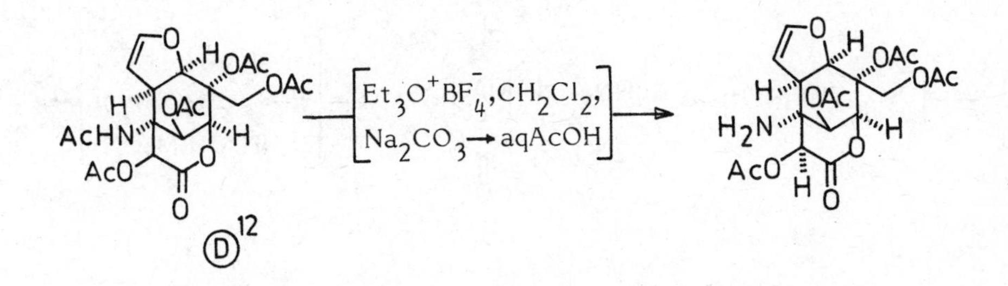

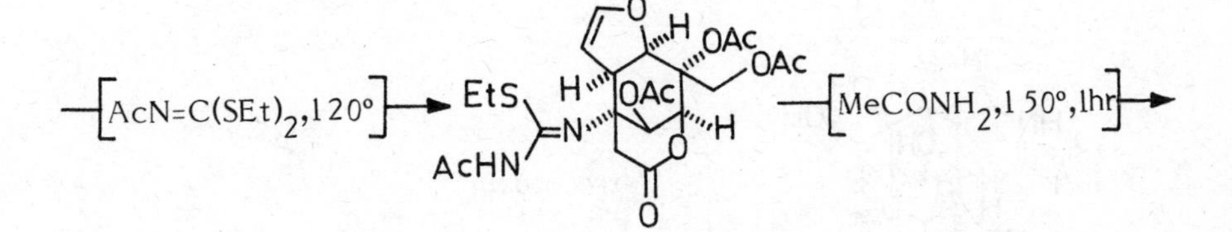

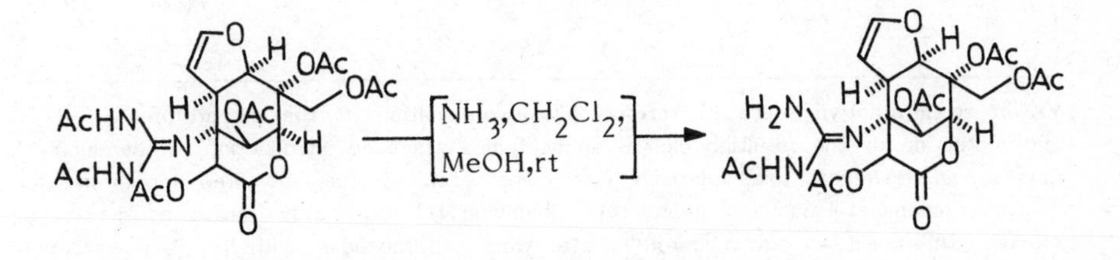

---

12. Because of the rather facile elimination (i→ii) generation of the aldehyde group at C-4a had to be deferred until after the guanidino group had been introduced at the C-8a position.

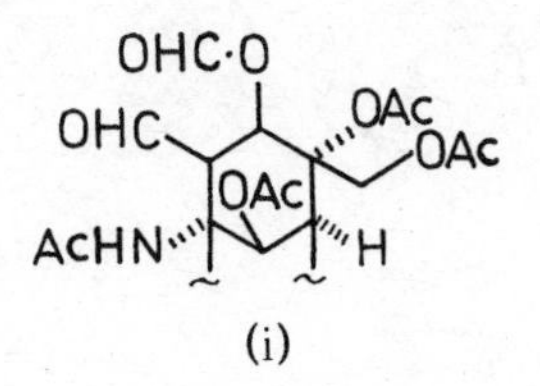

(i)

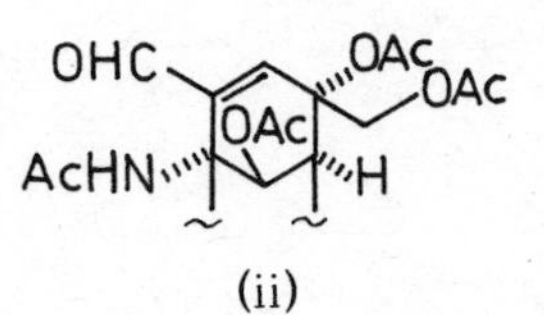

(ii)

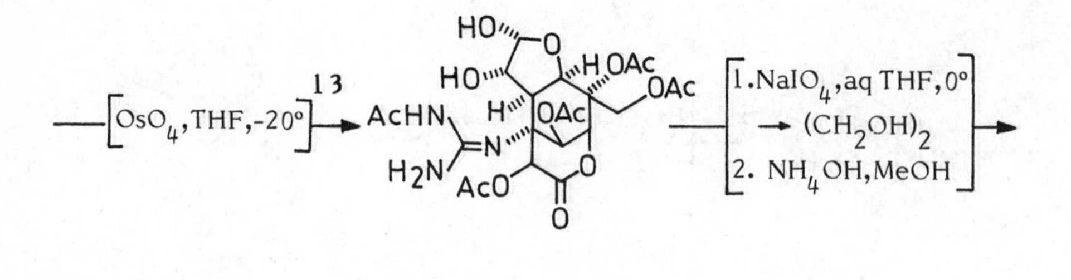

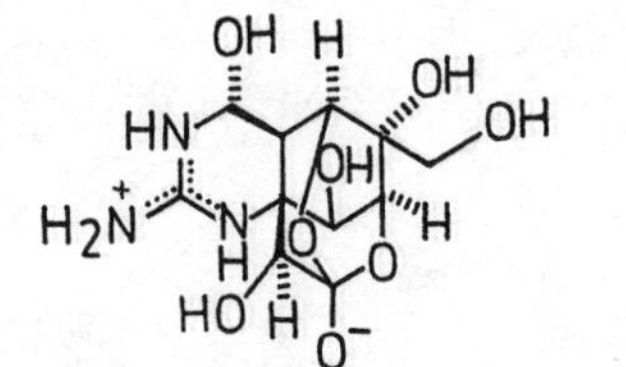

Tetrodotoxin[13]

---

13. In an alternative route to tetrodotoxin, hydroxylation of the dihydrofuran in (D) and protection of the resulting diol as an acetonide preceded introduction of the guanidino group. However, this approach was complicated by the unwanted formation of a cyclic monoacetylguanidino group (cf i) involving the C_9-acetoxy group as the only product. This has been circumvented in later work by proceeding with the C_9-p-anisoate derivative of (D); Tanino,H.; Inoue,S.; Aratani,M.; Kishi,Y. Tetrahedron Lett., 1974, 335.

(i)

351

THIENAMYCIN

The discovery of carbapenem of microbial origin has been a major development in β-lactam antibiotics, as these have a novel chemical structure, exceptional potency, unusually wide antibacterial spectrum and high stability to β-lactamases (1); thienamycin, isolated in 1976 is an important member of this class. Their fermentation yields are, however, low. Their synthesis has, therefore, attracted much interest, both for structure-activity relationship studies as also to provide practical methods of total synthesis. Although a number of synthesis of thienamycin have been reported (2), the one described below by the Merck group (3) was the first stereocontrolled synthesis of (+)-thienamycin, and employ a highly efficient carbene insertion reaction to produce the bicyclic nucleus by formation of the N-C_3 bond.

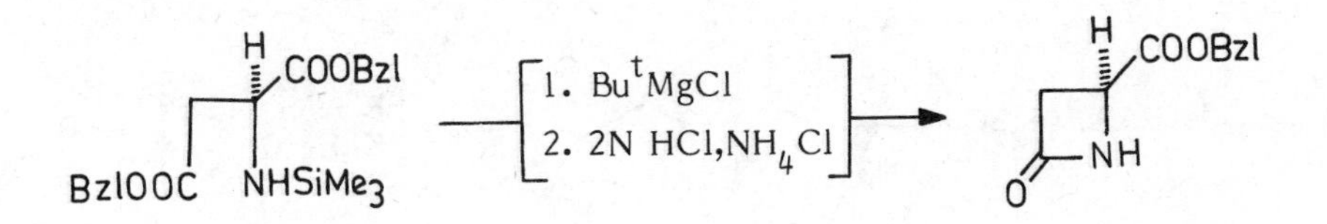

1. For recent reviews see: (a) Brown,A.G.; Roberts,S.M. Recent Advances in the Chemistry of β-lactam Antibiotics; The Royal Society of Chemistry, Burlington House, London, 1984; (b) Kametani,T. Heterocycles, 1982, 17, 463; (c) Labia,R.; Morin,C. J. Antibiot., 1984, 37, 1103.

2. For other synthetic approaches to thienamycin and related carbapenems see: (a) George,G.I.; Kant,J.; Gill,H.S. J. Am. Chem. Soc., 1987, 109, 1129; (b) From penicillin: Kerady,S.; Amats,J.S.; Reemer,R.A.; Weirstock,L.M. J. Am. Chem. Soc., 1981, 103, 6765; from 6-APA: Maruyama,H.; Hiraoka,T. J. Org. Chem., 1986, 51, 399; (c) from chiral sugar templates: Ikota,N.; Yoshino,O.; Koga,K. Chem. Pharm. Bull., 1982, 30, 1929; (d) Okano,K.; Izawa,T.; Ohno,M. Tetrahedron Lett., 1983, 24, 217.

3. Salzmann,T.N.; Ratcliffe,R.W.; Christensen,B.G.; Bouffard,F.A. J. Am. Chem. Soc., 1980, 102, 6161.

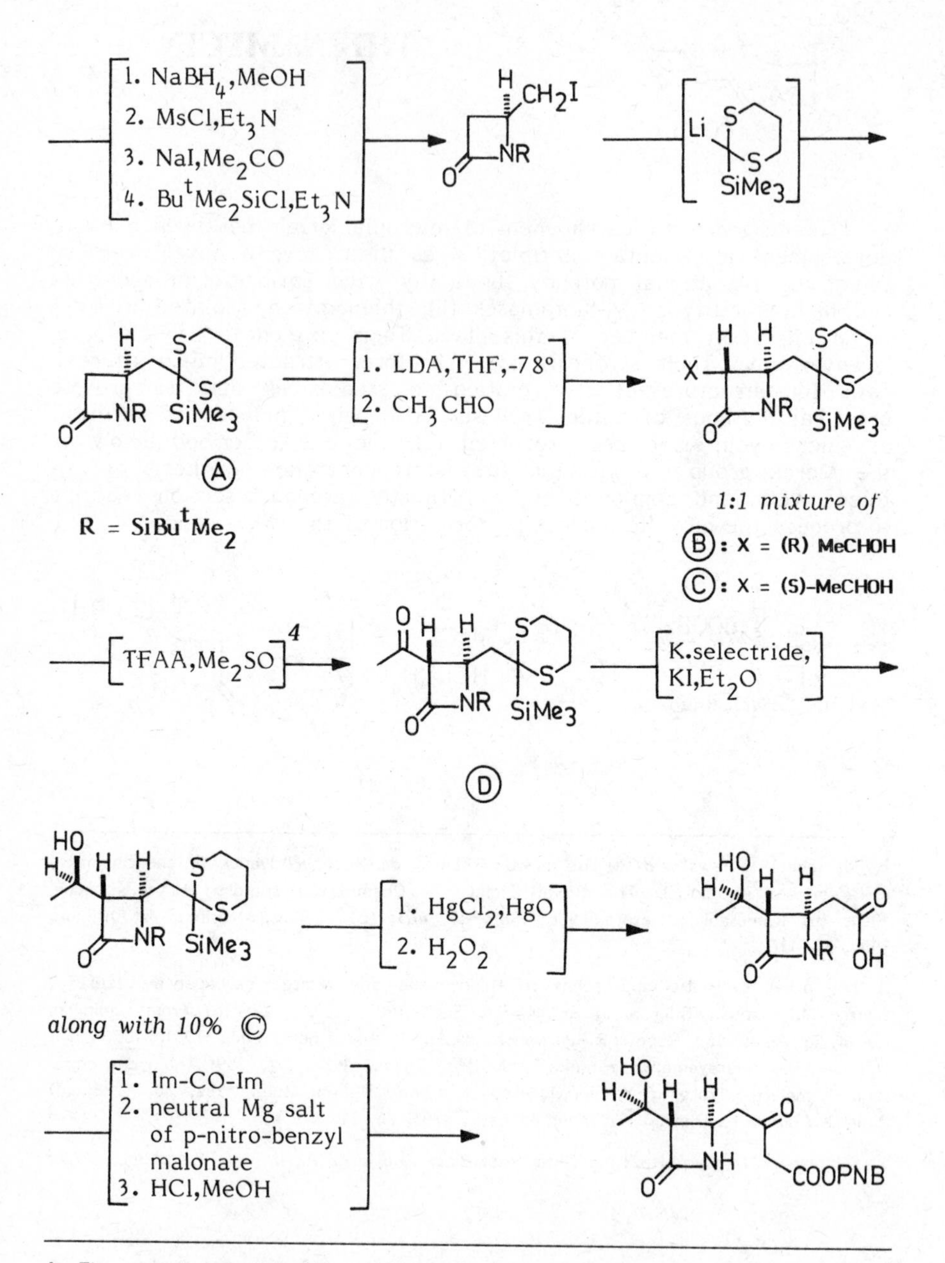

4. The acetyl derivative (D) could also be prepared directly from (A) treatment of its enolate with N-acetylimidazole.

5. Brooks,D.W.; Lu,D.L.L.; Masamune,S. Angew. Chem. Int. Edn., 1979, 18, 72.

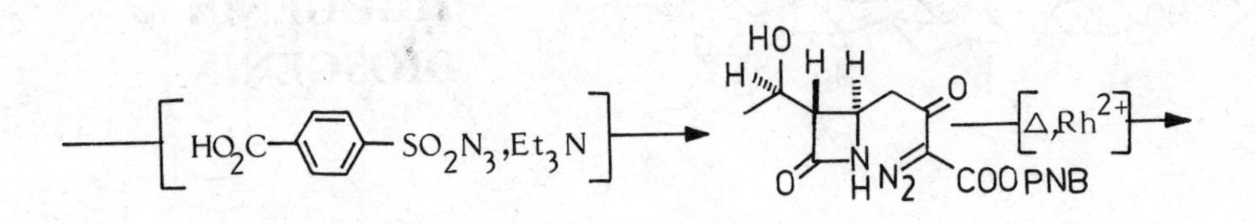
HO₂C–C₆H₄–SO₂N₃, Et₃N
HO
H
H H
O
O
N
H
N₂
COOPNB
Δ, Rh²⁺

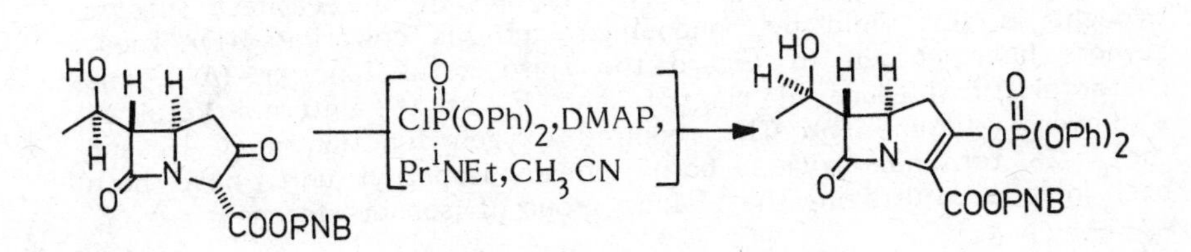
HO
H H
H
O
N
=O
COOPNB
ClP(O)(OPh)₂, DMAP,
PrⁱNEt, CH₃CN
OP(O)(OPh)₂
COOPNB

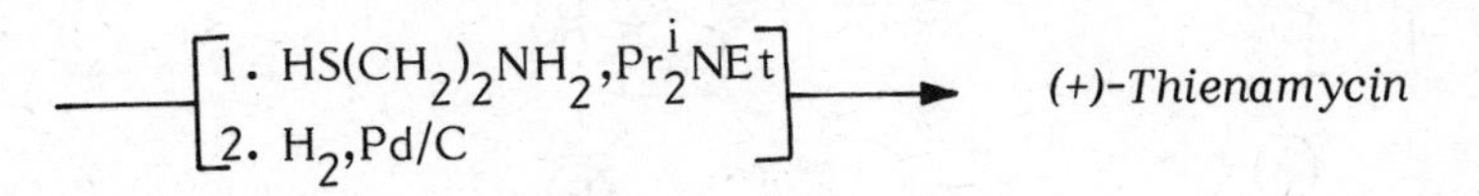
1. HS(CH₂)₂NH₂, Prⁱ₂NEt
2. H₂, Pd/C
(+)-Thienamycin

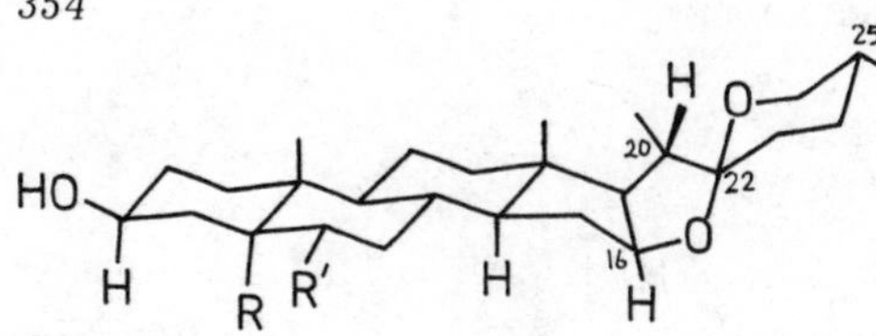

TIGOGENIN
DIOSGENIN

The total synthesis of steroids having been secured, steroidal sapogenins became an obvious target for synthetic attack. The route to tigogenin outlined below represents one of the earliest success wrought in this field by Sondheimer and his coworkers (1). These authors have chosen to regard the spiro ketal function Ⓐ , the characteristic feature of rings E and F, as the internal ketal of a dihydroxy-ketone Ⓑ. The required oxygen function at C-16 and the C-26 terminal oxygen, borne on an isopropyl unit, have both been introduced utilizing the 17-keto group of isoandrosterone.

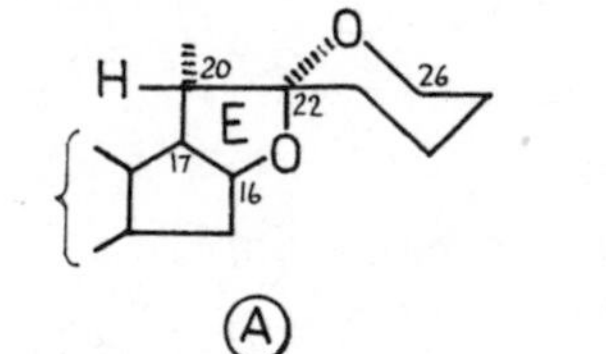

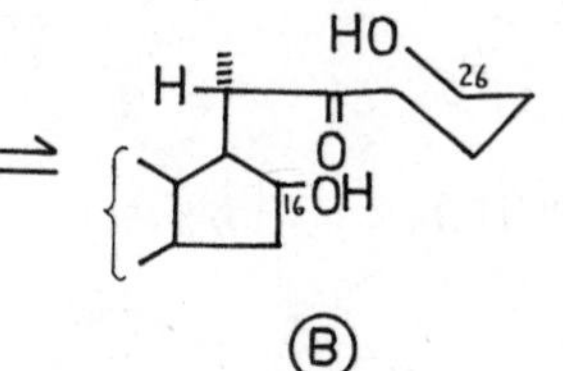

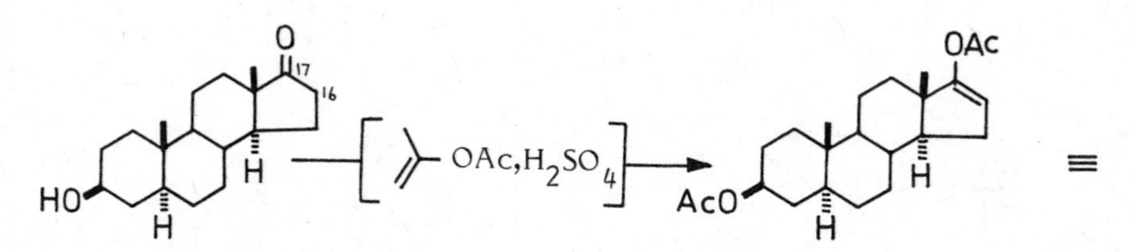

3β-Hydroxyandrostan-17-one

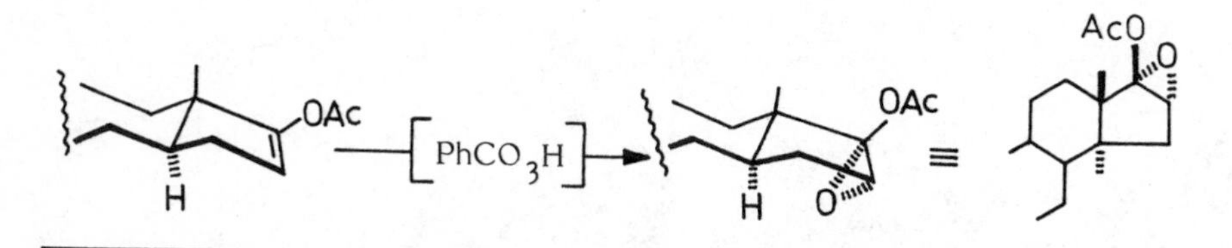

1. Mazur,Y., Danieli,N., Sondheimer,F., J. Am. Chem. Soc., 1960, 82, 5889.

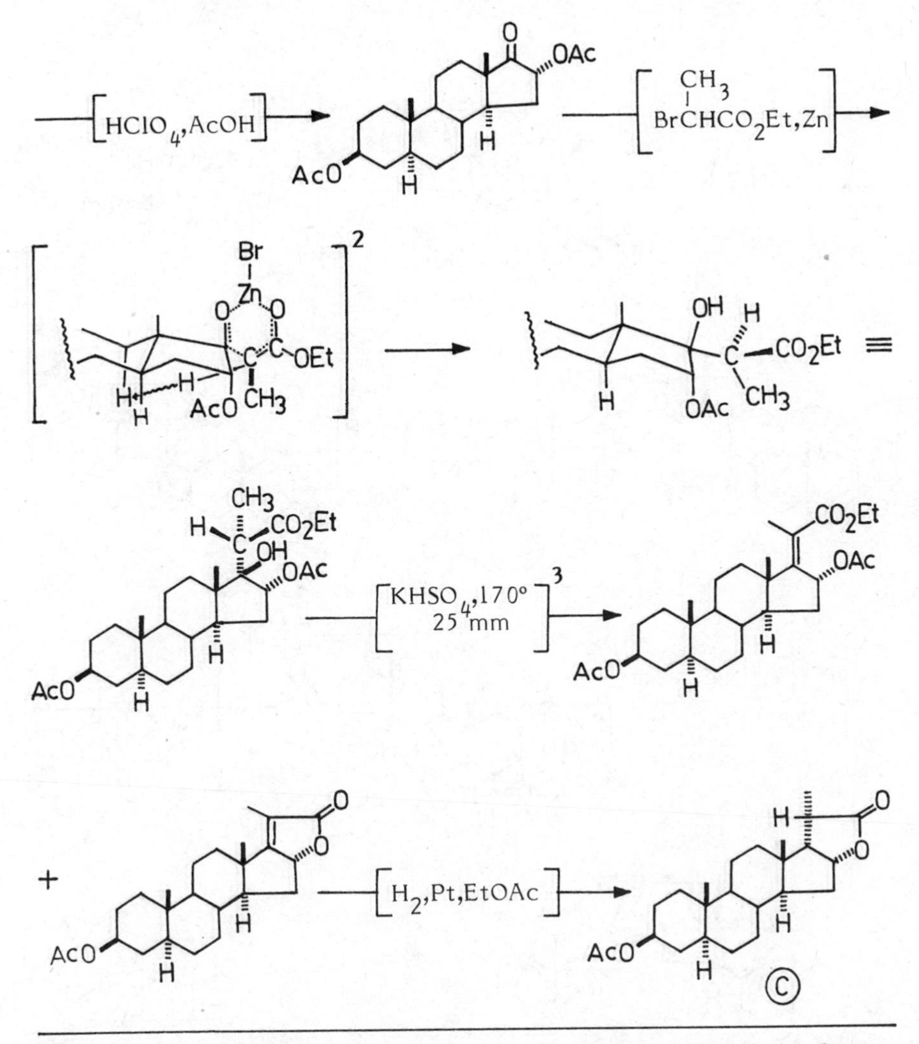

2. If the opposite configuration is formulated at C-20, there is considerable interference between the 12-α hydrogen and the 20-methyl group.

3. A mixture of two products is obtained. Each of these can be transformed independently into Ⓒ using slightly different routes. The route from the α,β-unsaturated γ-lactone is only shown.

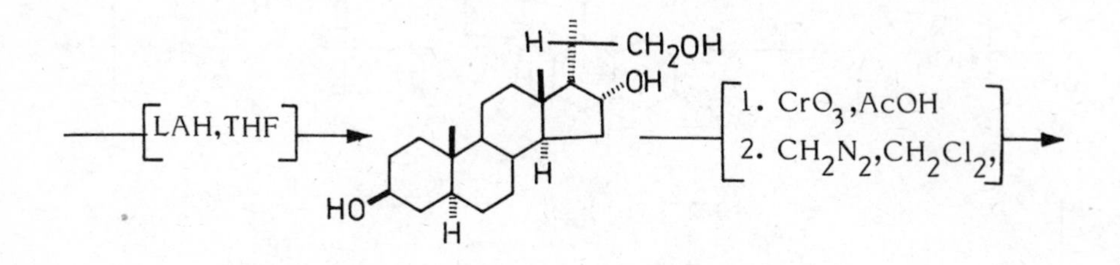
LAH,THF
H
CH2OH
OH
HO
H
1. CrO3,AcOH
2. CH2N2,CH2Cl2

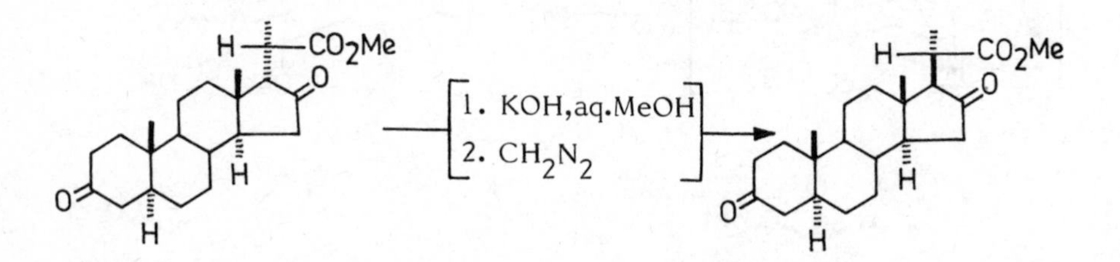
H
CO2Me
O
1. KOH,aq.MeOH
2. CH2N2

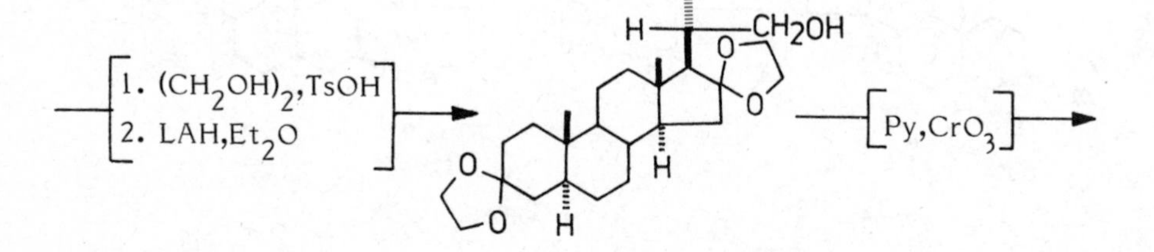
1. (CH2OH)2,TsOH
2. LAH,Et2O
CH2OH
Py,CrO3

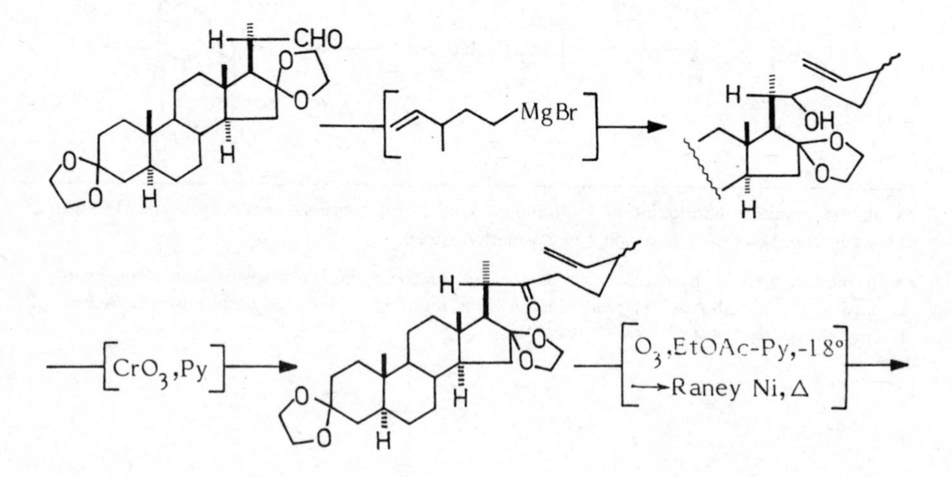
CHO
MgBr
OH
CrO3,Py
O3,EtOAc-Py,-18°
Raney Ni,Δ

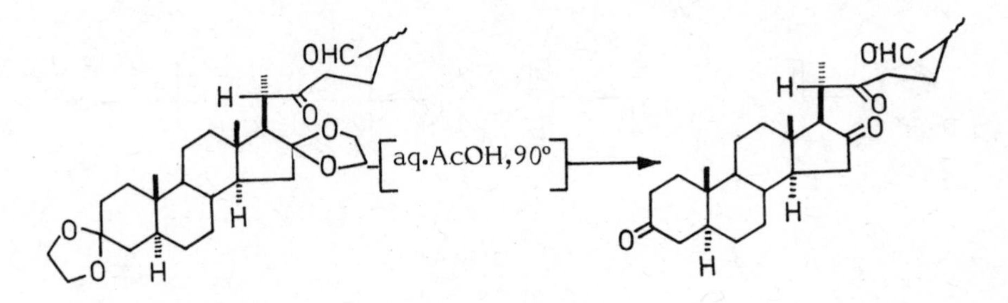

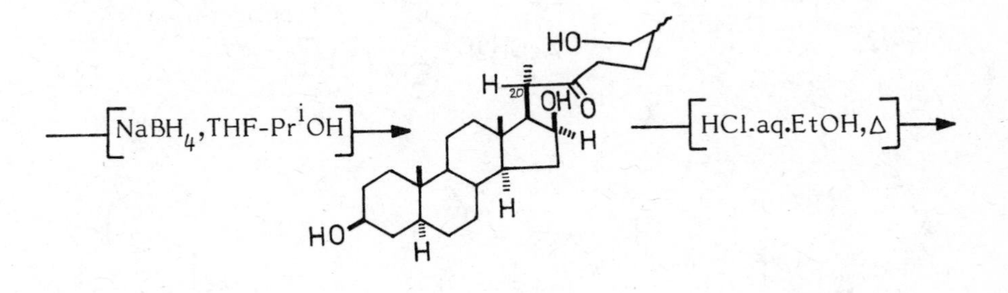

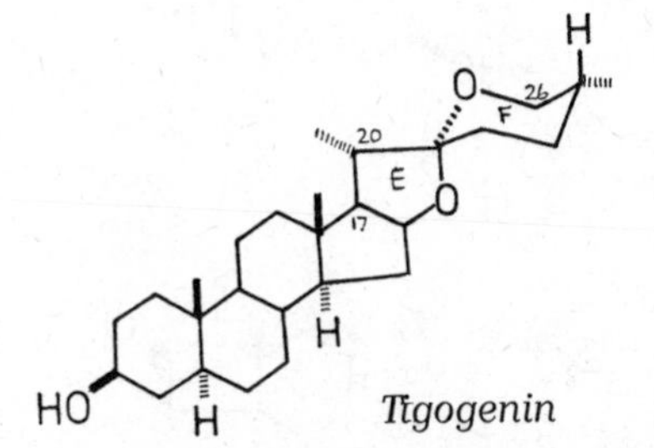

Tigogenin

A versatile new approach to steroidal sapogenins has been developed by Kessar et al. (4) and is exemplified in the synthesis of diosgenin outlined below. The key step in this synthesis is the introduction of elements of rings E and F by Michael addition of 1-acetoxy-5-nitro-2-methylpentane Ⓐ to the α,β-unsaturated ketone Ⓑ.

---

4. Kessar,S.V.; Gupta,Y.P.; Mahajan,R.K.; Joshi,G.S.; Rampal,A.L. Tetrahedron, 1968, 24, 899.

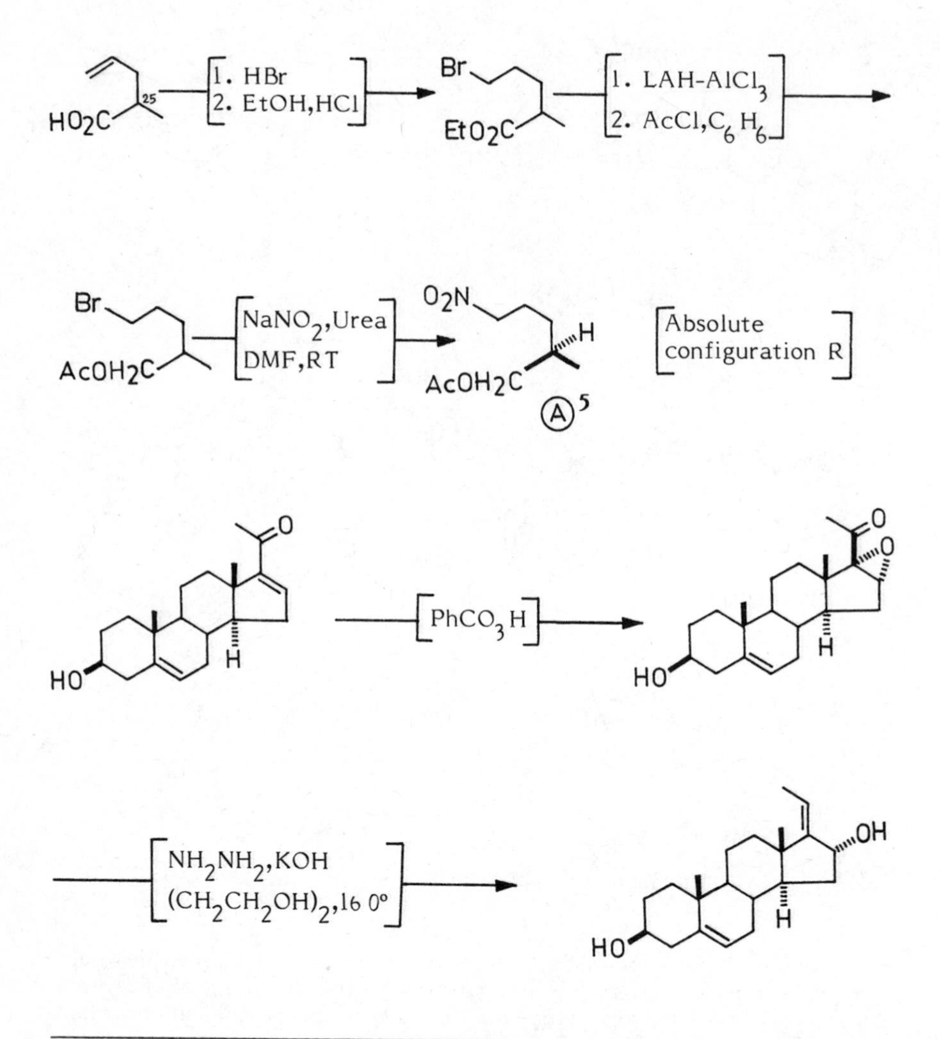

5. The stereochemistry at C-25 was controlled by selecting a nitroacetate possessing the correct configuration.

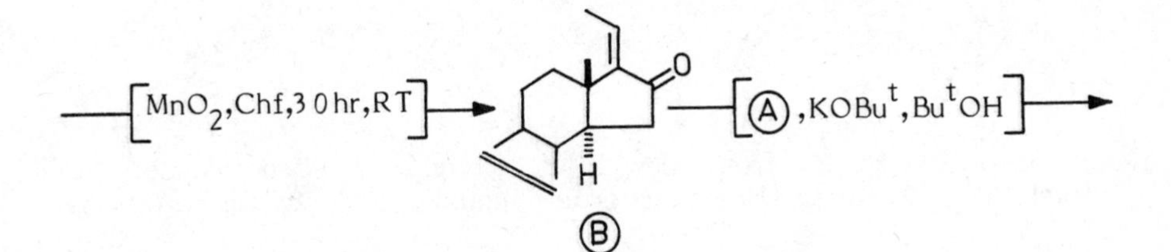
[MnO2,Chf,30 hr,RT]
O
H
B
[A,KOBut,ButOH]

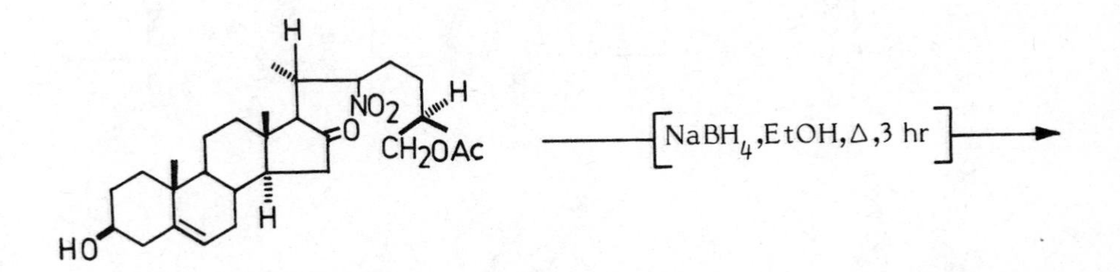
H
NO2
H
O
CH2OAc
H
HO
[NaBH4,EtOH,Δ,3 hr]

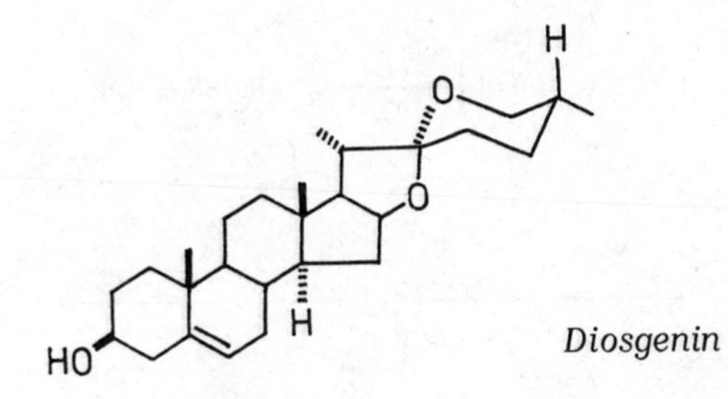
H
O
O
H
HO

Diosgenin

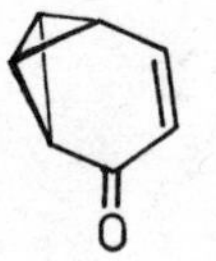

TROPAVALENE

The synthesis of tropavalene (1) highlights the confidence with which the Woodward-Hoffmann rules could be used for synthesis.

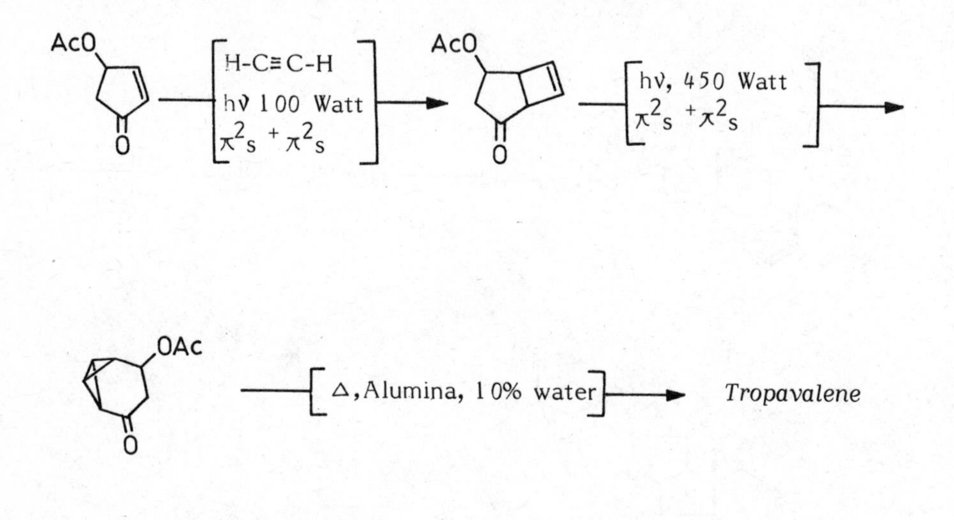

1. Sugihara,Y.; Morokoshi,N.; Murata,I. Tetrahedron Lett., 1977, 3887.

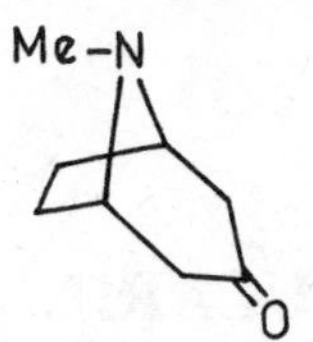

TROPINONE

Robinson's (1) synthesis of tropinone in 1917 was a landmark in the development of the art of organic synthesis. It demonstrated in a spectacular manner the power of the retro-synthetic analysis (2). The stunning simplicity and efficiency of the synthesis led Robinson to propose that plants use similar reactions for biosynthesis of this group of alkaloids, and to his general speculations about biogenesis of natural products (3). This synthesis was the fore-runner for biomimetic synthesis (although this term was not used by Robinson).

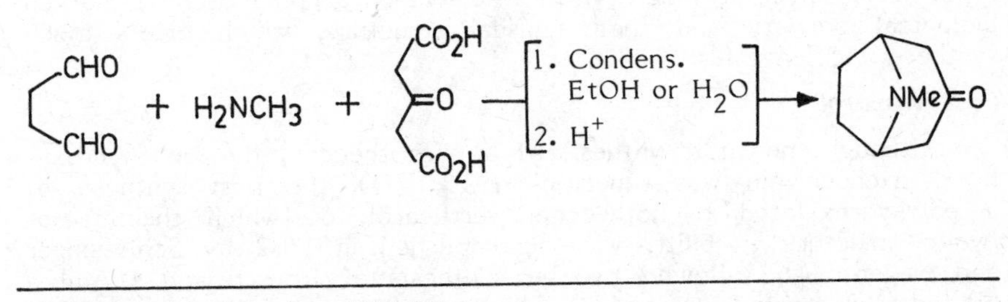

1. Robinson,R., J. Chem. Soc., 1917, 762.

2. Robinson described his thinking and his course of action thus: "By imaginary hydrolysis at the points indicated by the dotted line, the substance may be resolved into succindialdehyde, methylamine & acetone, and this observation suggested a line of attack of the problem which has resulted in a direct synthesis. It was found that tropinone was formed in small yield by condensation of succinaldehyde with acetone and methylamine in aq. solution. An improvement followed by the replacement of acetone by a salt of acetone dicarboxlic acid.

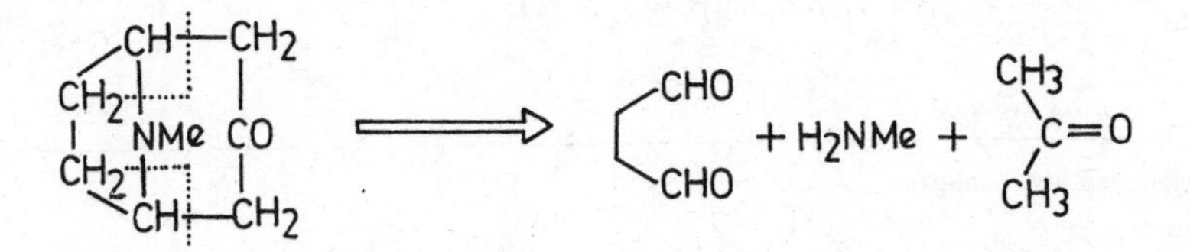

3. Robinson,R., J. Chem. Soc., 1917, 876.

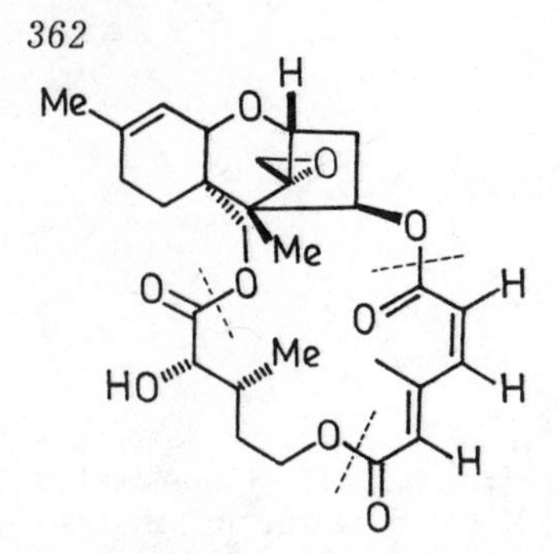

VERRUCARIN A

The trichothecenes are a group of closely related tetracyclic sesquiterpenoid epoxides produced by various species of imperfect fungi (1), and about 80 natural trichothecenes have been characterised todate. One important structural type of trichothecenes possess a macrocycle on the trichothecene skeleton, to which verrucarins belong. These have attracted much synthetic effort on account of their marked biological activity and their unusual structure, which offers much synthetic challenge (2).

(±)-Verrucarol

Although the first synthesis of a trichothecene, the monohydroxylated trichodermin was reported in 1971 (3), the first synthesis of a polyhydroxylated trichothecene, verrucarol, on which the macrocyclic structure is built, was accomplished in 1982 by Schlessinger and Nugent (4), followed by the synthesis of Rousch and D'Ambra (5) and by Trost & McDougal (7), which are described below.

Schlessinger and Nugent synthesis followed a biomimetic approach, and starts from a hydrindenone, carrying some of the centres in the configuration in which they appear in the final molecule.

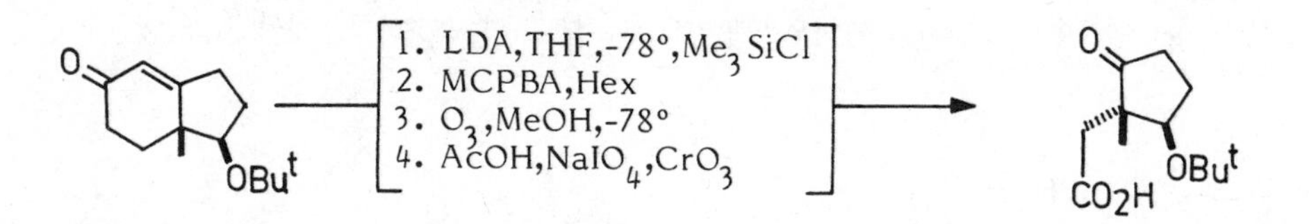

1. Trichothecane skeleton.

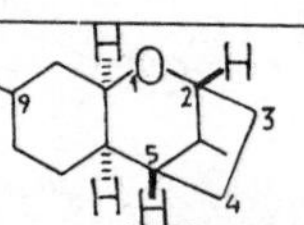

2. For a comprehensive review see: McDougal,P.G.; Schmuff,N.R. Prog. Chem. of Natural Products, 1985, 47, 153.

3. Colrin,E.W.; Raphael,R.A.; Roberts,J.S. Chem. Comm., 1971, 858; Colrin,E.W.; Malchenko,S.; Raphael,R.A.; Roberts,J.S. J. Chem. Soc. Perkins. Trans., 1973, 1989.

4. Schlessinger,R.H.; Nugent,R.A. J. Am. Chem. Soc., 1982, 104, 1116.

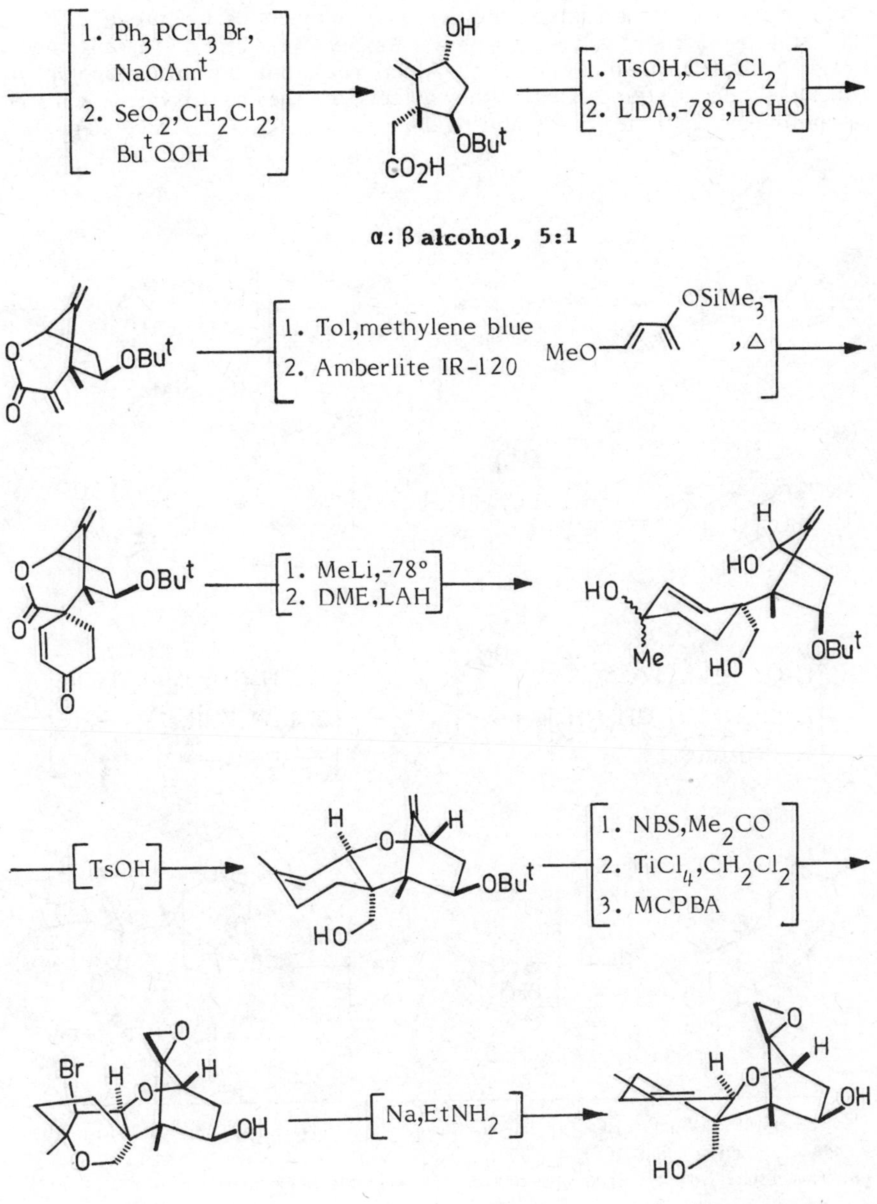
1. Ph3PCH3 Br, NaOAmt
2. SeO2,CH2Cl2, ButOOH
OH
OBut
CO2H
1. TsOH,CH2Cl2
2. LDA,-78°,HCHO
α: β alcohol, 5:1
O
O
OBut
1. Tol,methylene blue
2. Amberlite IR-120
MeO
OSiMe3
, Δ
O
O
OBut
O
1. MeLi,-78°
2. DME,LAH
H
HO
HO
Me
HO
OBut
TsOH
H
O
H
OBut
HO
1. NBS,Me2CO
2. TiCl4,CH2Cl2
3. MCPBA
Br
H
O
O
H
OH
O
Na,EtNH2
O
H
H
O
OH
HO
Verrucarol

Rousch and D'Ambra (5) in an alternative approach to synthesis of verrucarol constructed the critical bicyclo[2.2.1]heptene Ⓐ by a stereoselective silyl-controlled Wagner-Meerwein rearrangement; ring A was formed by a Diels-Alder reaction, and the topology of the bicyclic system directs the attack of the acetoxydiene on the α-methylene lactone to establish the required C_5,C_6 stereochemistry.

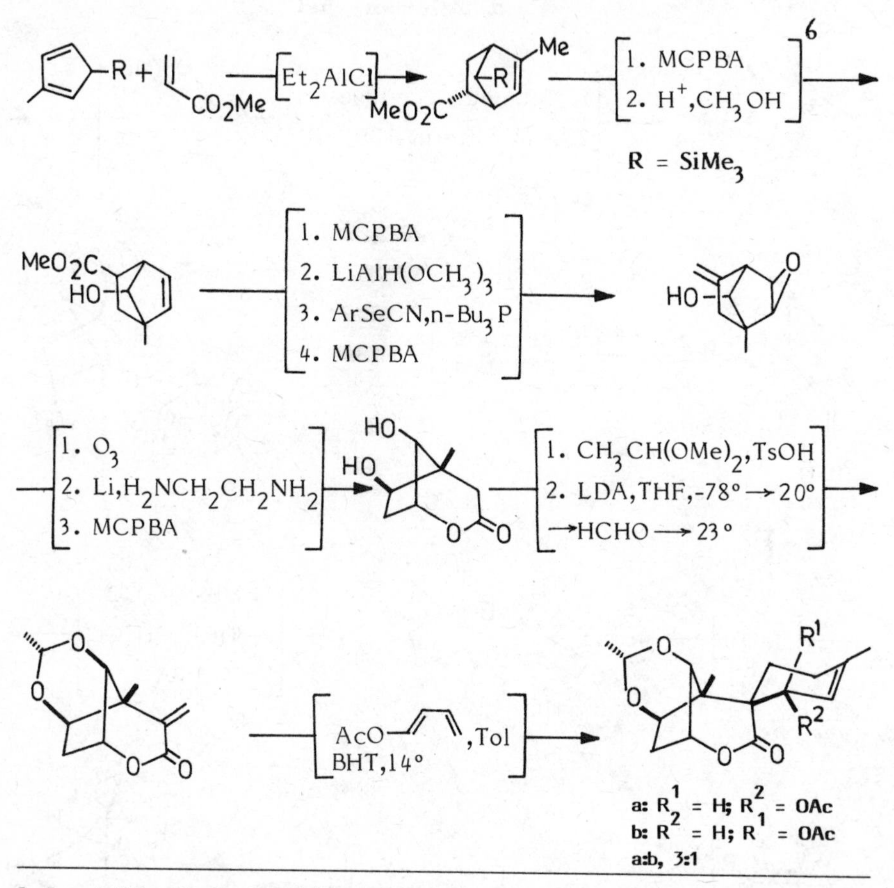

5. Rousch,W.R.; D'Ambra,T.E. (a) J. Am. Chem. Soc., 1983, 105, 1058; (b) J. Org. Chem., 1980, 45, 3927; (c) ibid, 1981, 46, 5045.

6. This results from a unique silyl-controlled Wagner-Meerwein arrangement.

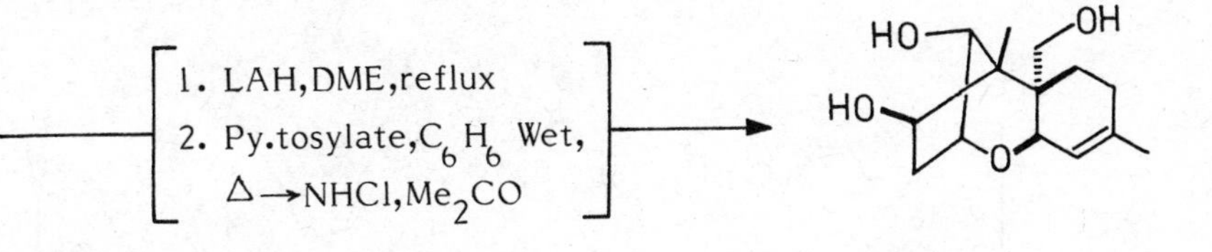
1. LAH,DME,reflux
2. Py.tosylate,C6H6 Wet,
Δ→NHCl,Me2CO
HO
OH
HO
O

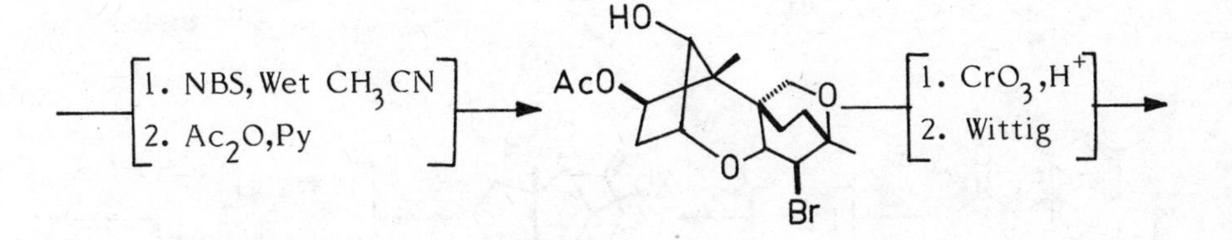
1. NBS,Wet CH3CN
2. Ac2O,Py
HO
AcO
O
O
Br
1. CrO3,H+
2. Wittig

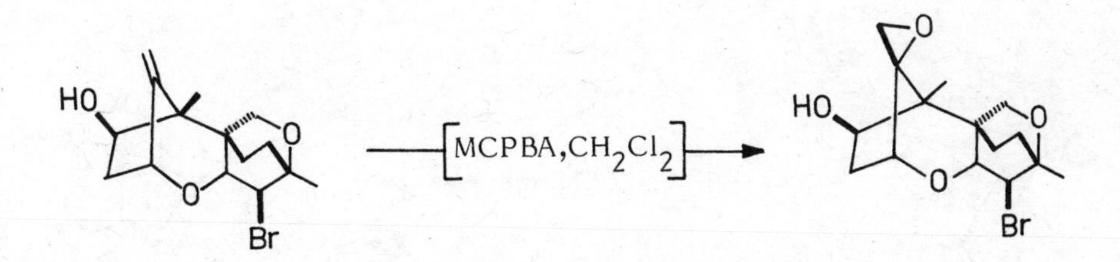
HO
O
O
Br
MCPBA,CH2Cl2
O
HO
O
O
Br

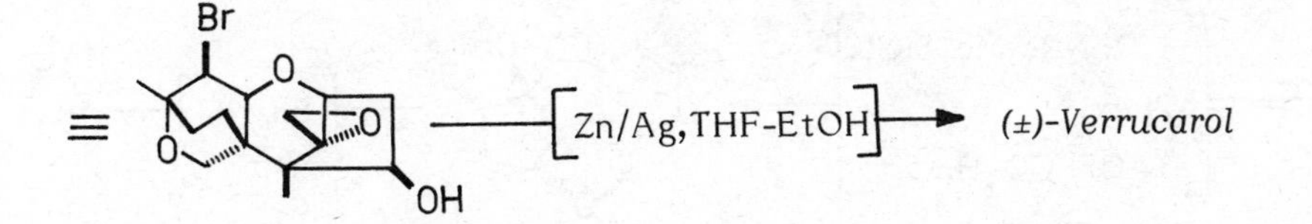
≡
Br
O
O
O
OH
Zn/Ag,THF-EtOH
(±)-Verrucarol

Trost and McDougal (7) developed yet another biomimetic route to verrucarol described below starting from 2-methyl-1,3-cyclopentadione.

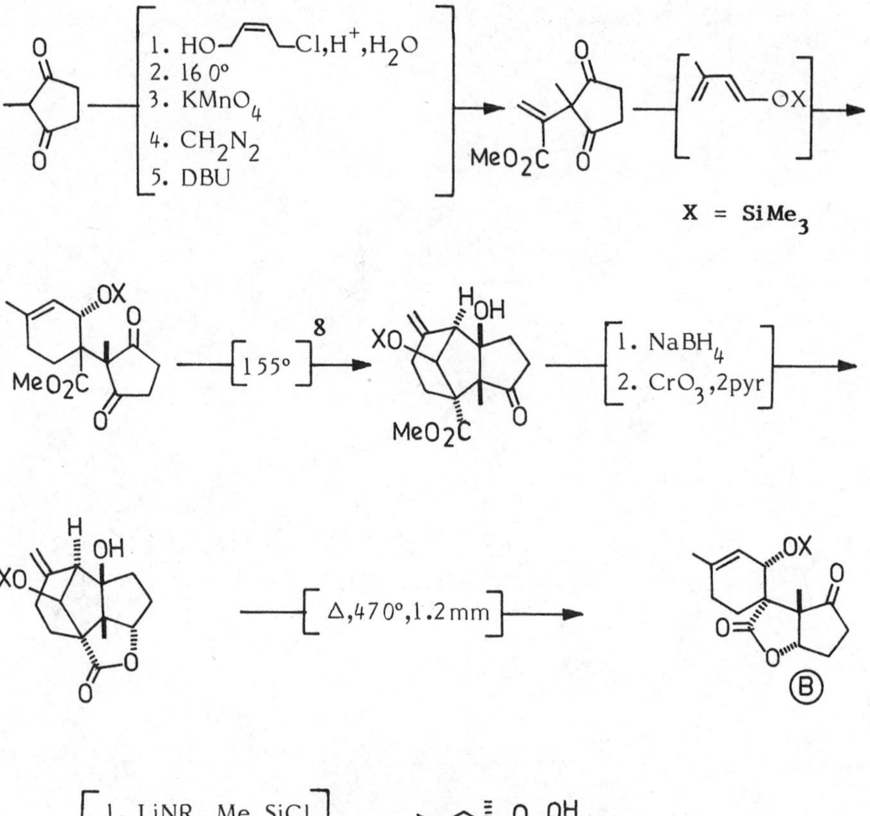

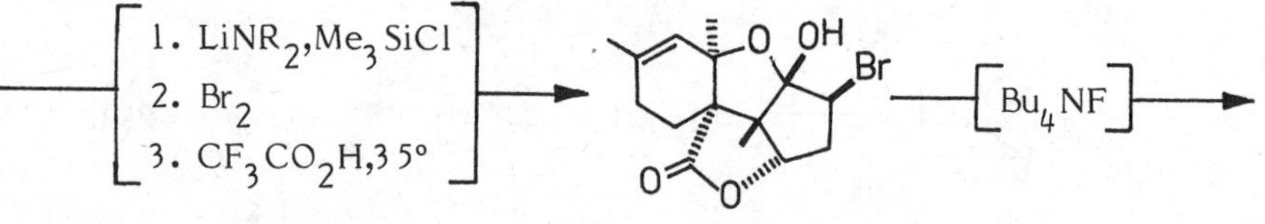

---

7. Trost,B.M.; McDougal,P.G. J. Am. Chem. Soc., 1982, 104, 6110.

8. Only one of the two keto groups in the dione could align itself in the right position to undergo the enone reaction, which thus serves to differentiate the two ketones; this differentiation is maintained in the subsequent step when through a retroene reaction a modified Diels-Alder adduct (B) is obtained.

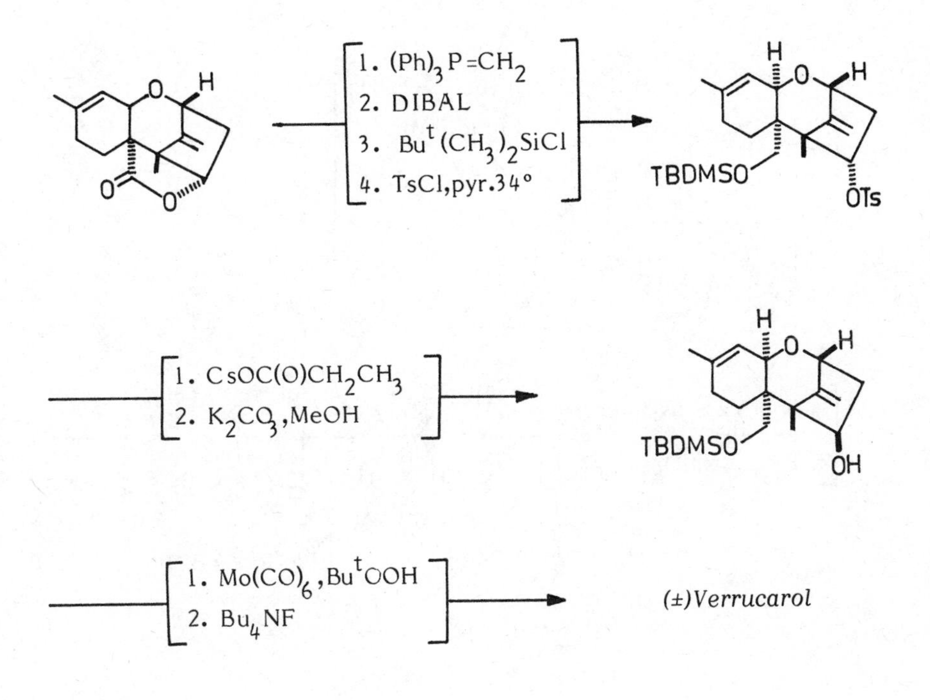

Verrucarin A

Still & Ohmizu were the first to synthesise a naturally occuring macrocyclic trichothecene, verrucarin A, in 1981 (9). An alternative synthesis of verrucarin A following a similar strategy was reported soon after this by Tamm and his associates (10); the synthesis of a number of related macrocycles has since then been reported (2,11). The synthesis by Still and Ohmizu (9) described below constructs the bridge as two units, one of them obtained in optically pure form through a chiral epoxidation of an olefin precursor.

---

9. Still,W.C.; Ohmizu,H. J. Org. Chem., 1981, 46, 5244.

10. Mohr,P.; Tori,M.; Grossen,P.; Harold,P.; Tamm,Ch. Helv. Chim. Acta, 1982, 65, 1412.

11. Most of the reported synthesis share two common strategies: (a) the syntheses are convergent, with acyclic units, mostly two, assembled prior to their attatchment to verrucarol; (b) a common retrosynthetic connection has been at the ester linkages both for the attatchment of the acyclic units and for the formation of the macrocycle, as shown by dotted lines in the formula of verrucarin A.

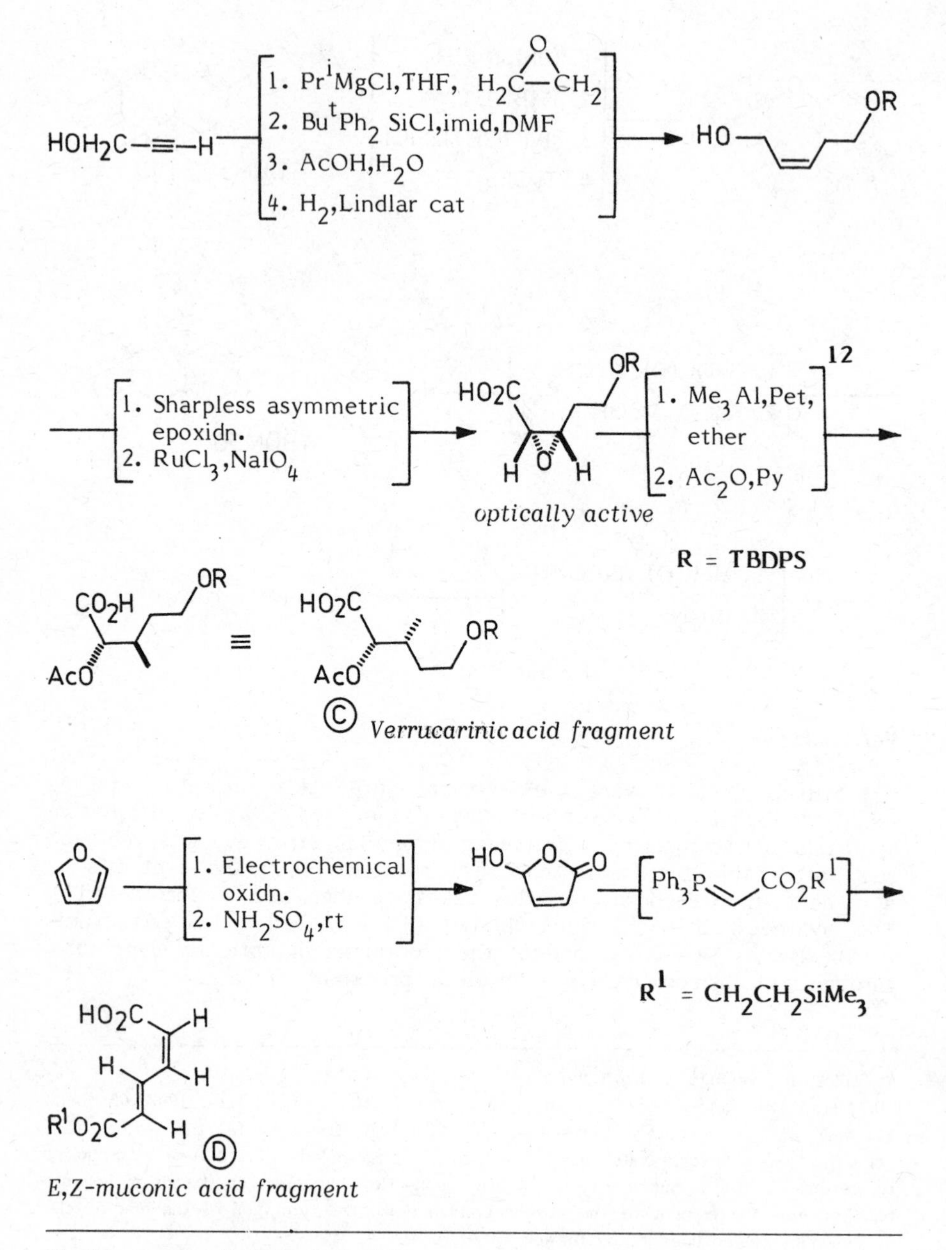

12. This reagent affected only β-addition.

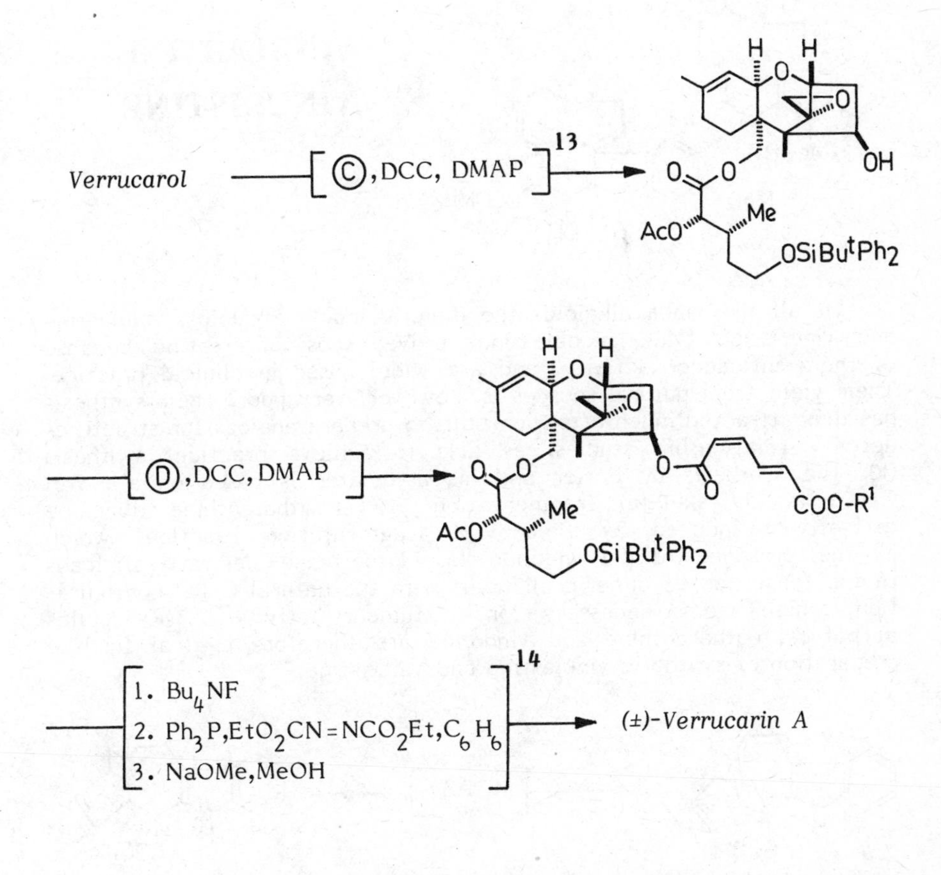

---

13. Hassner, A.; Alexanian, V. Tetrahedron Lett., 1978, 4475.
14. For cyclisation: Kurihara, T.; Nakajima, Y.; Mitsunobu, O. Tetrahedron Lett., 1976, 2445.

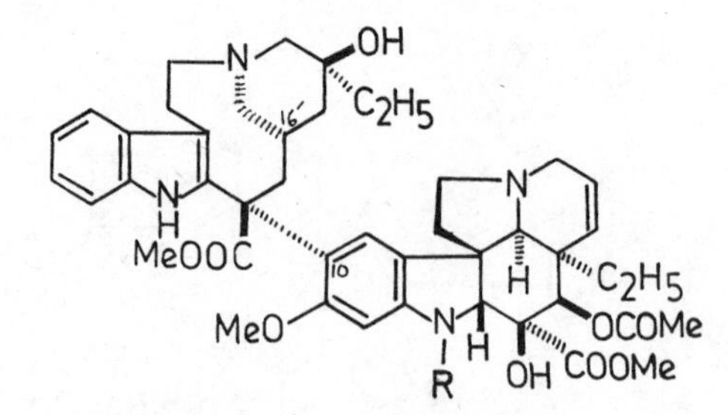

VINBLASTINE
VINCRISTINE

Of all the vinca alkaloids the dimeric indole alkaloids, vinblastine and vincristine (vincaleucoblastine) proved most interesting because of their anticancer activity, and are widely used in clinical practice. Their yield from natural sources is, however, very poor. Their synthesis has thus attracted much attention both to prepare analogs for structure-activity relationship studies, as also to achieve practical synthesis (1). The synthesis by Potier and his associates is based on a novel C-16', -C-21' skeletal fragmentation of (+)-catharanthine (ibogaine derivatives) induced by Polonovski fragmentation reaction, which in the presence of (-)-vindoline (aspidospermane derivatives) leads to the formation of dimeric alkaloids with the natural C-16' configuration, which seems necessary for antitumour activity (2,3). Supplies af natural catharanthine and vindoline are therefore critical for the preparation of synthetic vinblastine and vincristine.

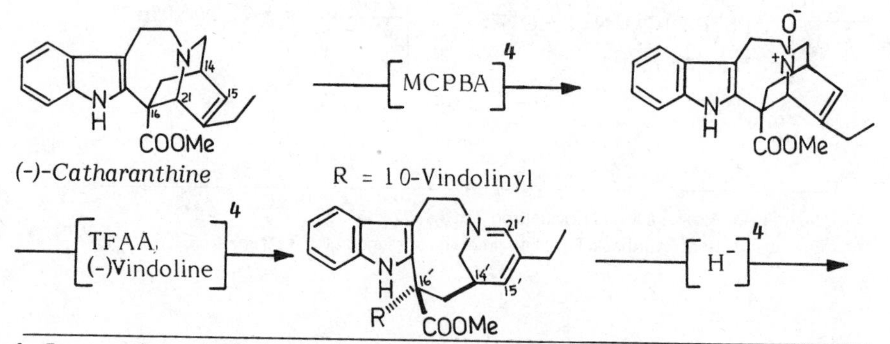

1. For a review of the synthesis of dimeric indole alkaloids see: Kutney,J.P.; Lloydia, 1977, 41, 107; Lect. Heterocyclic Chem., 1978, 4, 59; Potier,P. J. Nat. Prod., 1980, 43, 72; Ann. Symprm., 1984, 2, 65, 67; Chem. Abst., 1984, 101, 130936z, 130937a; Rahman, A.U. Proc. of 4th Asian Symp. Med. Plants Spices, 1980, 1, 222; J. Chem. Soc. Pak., 1979, 1, 81.

2. For earlier work leading to this synthesis see: Langlois,N.; Gueritte,F.; Langlois,Y.; Potier,P. J. Am. Chem. Soc., 1976, 98, 7017; Langlois,N.; Potier,P. J. Chem. Soc. Chem. Comm., 1978, 102.

3. Mangeney,P.; Andriamialisoa,R.Z.; Langlois,N.; Langlois,Y.; Potier,P. J. Am. Chem. Soc., 1979, 101, 2243. For an earlier partial synthesis of vinblastine designed on a similar strategy see: Rehman,A.U.; Bash,A.; Gkazaia,M. Tetrahedran Lett., 1976, 27, 2351.

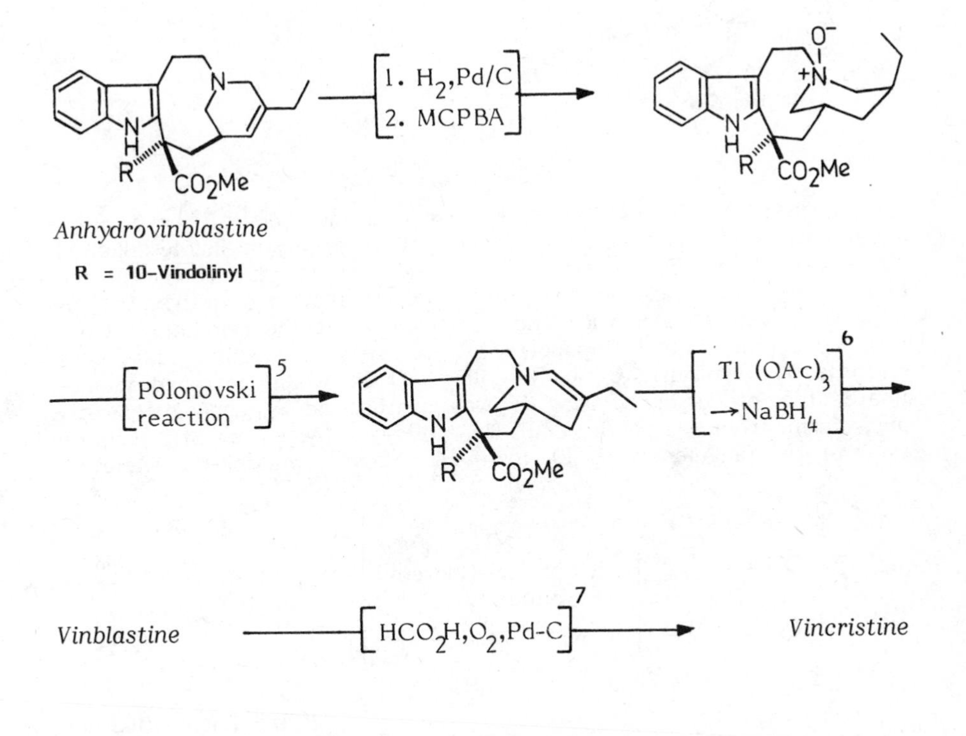

---

4. These three steps can be carried out as one pot reaction (5).

5. Raucher et al. have observed that the coupling of (±)-catharanthine with (-)-vindoline by the modified Polonovski reaction gave (16'S,14'R) diastereomer, (+)-anhydrovinblastine (A), which results from the coupling of (+)-catharanthine and (-)-vindoline in 46% yield based on (+)-catharanthine, along with (16'R,14'S) diastereomer, (-)-anhydrovincovaline (B), which results from the coupling of (-)-catharanthine and (-)-vindoline and was isolated in 54% yield based on catharanthine; (A) and (B) could be easily separated by flash chromatography; (±)-catharanthine can therefore be used for this synthesis: Raucher,S.; Bray,B.L.; Lawrence,R.F. J. Am. Chem. Soc., 1987, 109, 442.

6. Use of OsO_4 as an oxidant followed by $NaBH_4$ gave instead the 21'-epimer leurosidine; this difference in stereoselectivity in oxidation may be due to the size of the oxidant, and bulky OsO_4 would attack from the less hindered -side.

7. Gideon Richter, Belgian Patent, 823560, April 16, 1975; Chem. Abst., 1976, 84, 59835p.

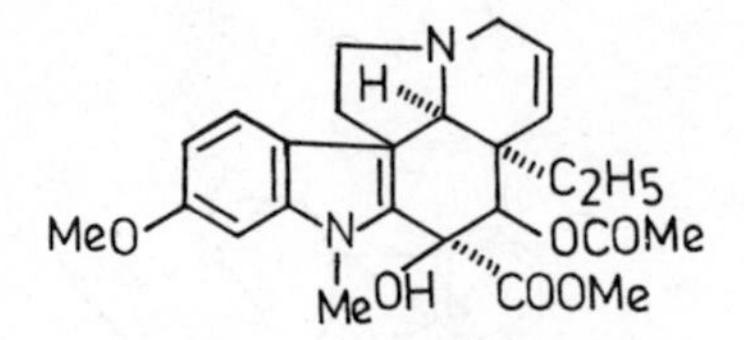

VINDOLINE

Vindoline is one of the most important members of the Aspidosperma alkaloids, and is part of the potent oncolytic indole alkaloid vinblastine. A number of synthesis of (±)-vindoline have been reported all of which are linear in conception and begin with 6-methoxytryptamine or a related structure and construct on it the remaining structure (1). Feldman and Rapoport have recently reported a convergent chirospecific synthesis of (-)-vindoline (2) in which tetrahydroquinolinate, the precursor of ring E was synthesised in enantiomerically pure form from l-aspartic acid (3), followed by an enantioselective skeletal rearrangement of Ⓔ to the required aspidosperma skeleton.

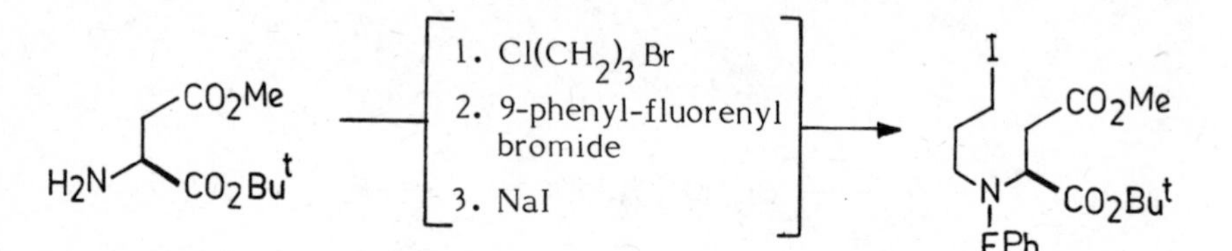

FPh = Phenyl-fluorene

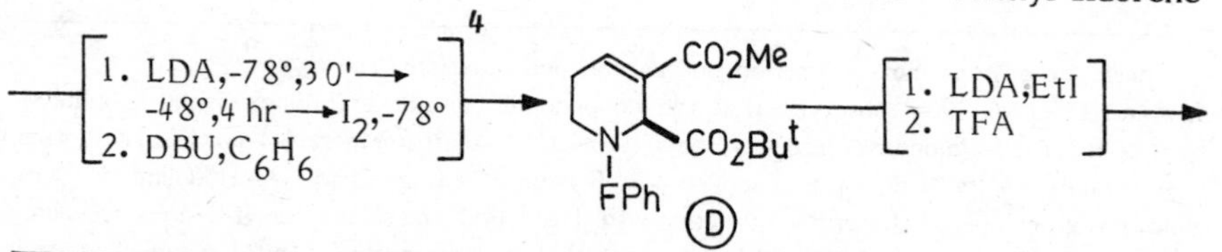

---

1. Total synthesis of (±)-vindoline: (a) Ando,M.; Buchi,G.; Ohnuma,T. J. Am. Chem. Soc., 1975, 97, 6880; (b) Kutney,J.P.; Bunzli-Treppo,U.; Chan,K.K.; deSouza,J.P.; Fujise,Y.; Honda,T.; Katsube,J.; Klein,F.K.; Leutwiler,A.; Morehead,S.; Rohr,M.; Worth,B.R. J. Am. Chem. Soc., 1978, 100, 4220; (c) Andriamialisoa,R.Z.; Langlois,N.; Langlois,Y. J. Org. Chem., 1985, 50, 961.
2. Feldman,P.L.; Rapoport,H. J. Am. Chem. Soc., 1987, 109, 1603.
3. Feldman,P.L.; Rapoport,H. J. Org. Chem., 1986, 51, 3382.
4. LDA/I_2 treatment gave a mixture of A,B&C in a ratio of 3:12:85. DBU treatment converted only isomer Ⓒ to form Ⓓ as stereoelectronic for E_2 reaction of Ⓒ was more favourable; the mixture could be separated to provide Ⓓ, while mixture of Ⓐ & Ⓑ was recycled.

(A)

(B); R'=CO_2Me; R=I (C) R' = I; R = CO_2Me

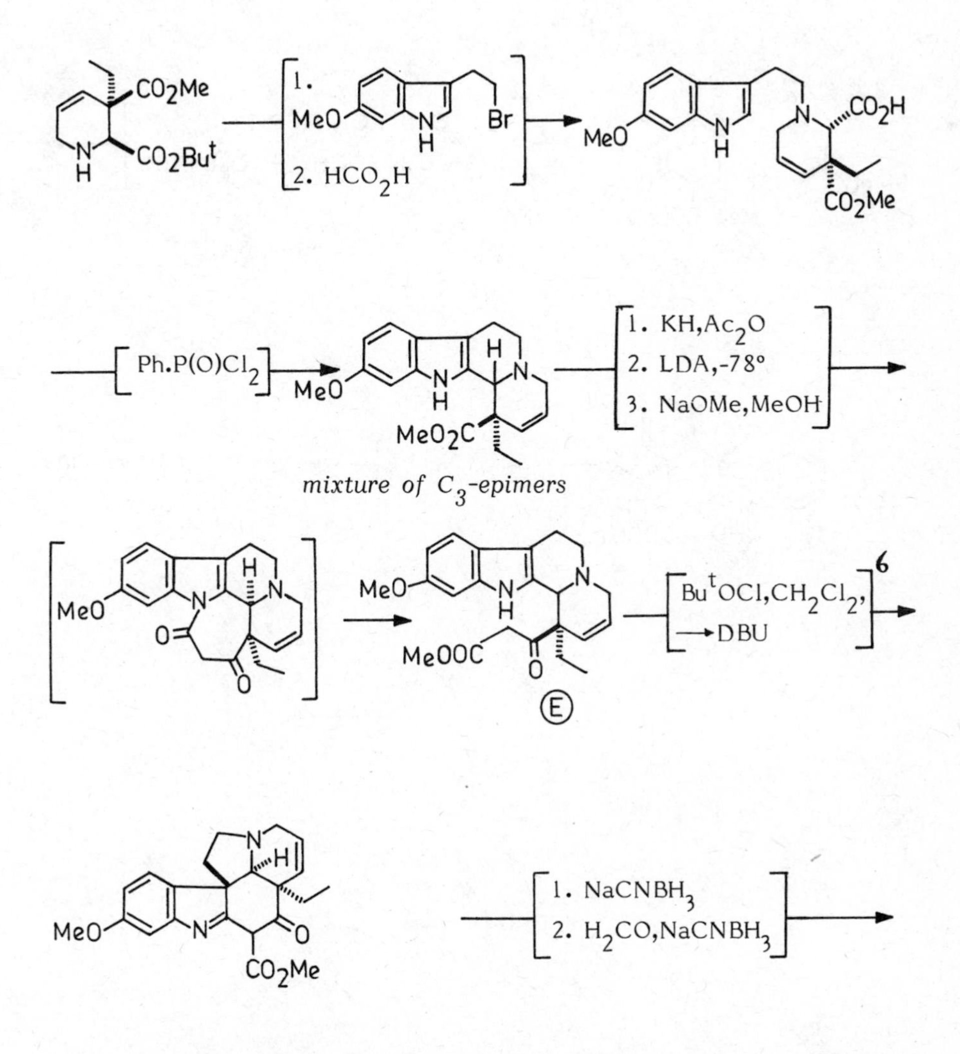

5. Feldman,P.L.; Rapoport,H. Synthesis, 1986, 735.

6. The experimental conditions for this skeletal rearrangement are critical for preserving optimal purity, as a reversible Mannich reaction is very facile in hexahydroindoloquinolizine systems with highly electrophilic carbon C-15.

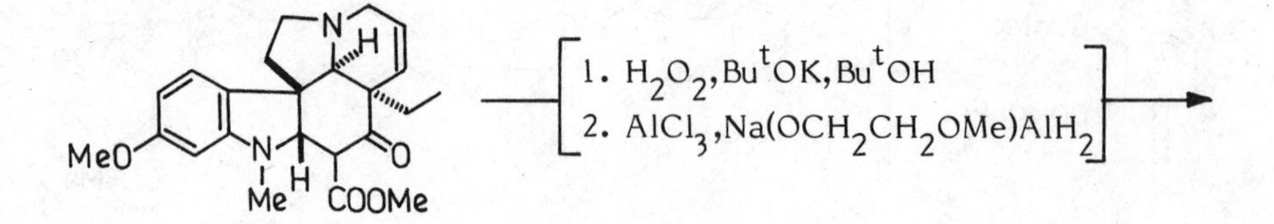
N
H
MeO
N
O
Me
H
COOMe
1. H2O2, ButOK, ButOH
2. AlCl3, Na(OCH2CH2OMe)AlH2

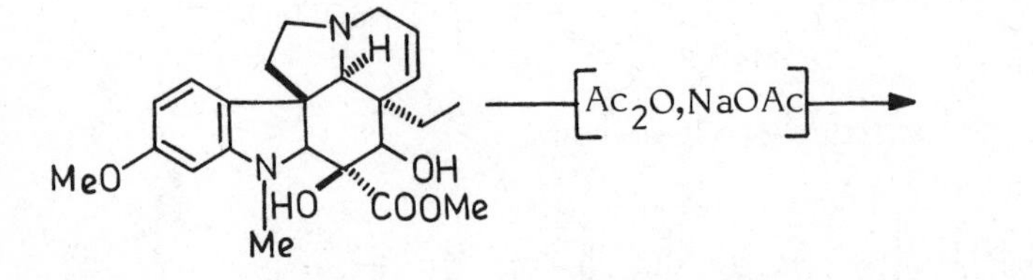
N
H
MeO
N
OH
HO
COOMe
Me
Ac2O, NaOAc

(-)-Vindoline

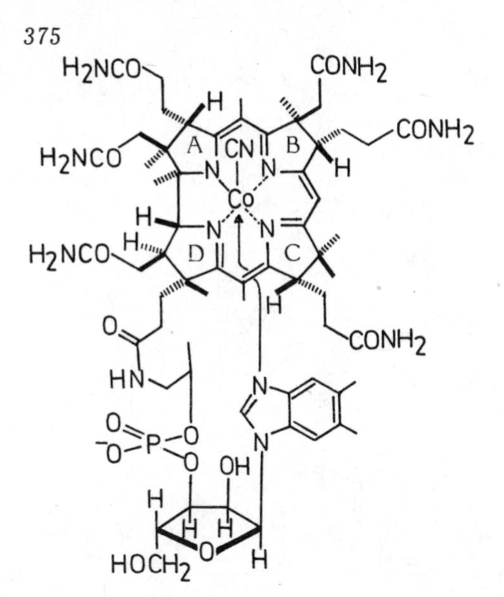

VITAMIN B_{12}

The elucidation of the structure of this complex molecule with 64 atoms and 9 asymmetric centres would stand out as one of the most significant achievements of X-ray crystallography(1) and its synthesis the most outstanding achievement of organic synthesis(2, 3). The completion of this mammoth task, which spanned over a decade, was the outcome of an intercontinental collaboration between the groups led by Woodward at Harvard University, Cambridge, and Eschenmoser at ETH, Zürich. The synthesis culminated in the preparation of Cobyric acid, the simplest of B_{12} derivatives, which had previously been converted to the vitamin by Bernhauer & associates(4), and thus constituted a formal total synthesis.

---

1. Hodgkin, D.C., Pickworth, J., Robertson, J.H., Trueblood, K., Prosen, R.J., White, J.G. Nature, 1956, 176, 325.

2. Woodward, R.B. Pure Appl. Chem., 1968, 17, 519; 1971, 25, 283; 1973, 33, 145; Eschenmoser, A. Proc. of the Robert A. Welch Foundation Conferences on Chemical Research XII, 1968, 9; Quart. Revs., 1970, 24, 366; XXIIIrd Int. Congress Pure Appl. Chem., 1971, 2, 69.

3. The synthesis of vitamin B_{12} and the related work on corrins provides a very instructive example of how a project directed towards the synthesis of a complex natural product has an impact much beyond the immediate structural boundaries of the specific synthetic problem. Innumerable synthetic methods of general applicability emerged from this monumental work, and the principle of orbital symmetry conservation (Woodward-Hoffmann rules), which is of immense consequence to chemistry as a whole, arose directly from the early studies on vitamin B_{12} synthesis. Chem. Soc. Publ. 1967, 21, 217.

4. Freidrich, W., Gross, G., Bernhauer, K., Zeller, P. Helv., 1960, 43, 704.

The overall plan of synthesis was to construct the unit structures of the molecule, designated A,B,C&D, in optically active form of the specified absolute configuration, and bring together the units to generate the needed stereochemical relations of the whole molecule.

A B | + | D C → A B | | D — C → A — B | | D — C

Preparation of A-D component

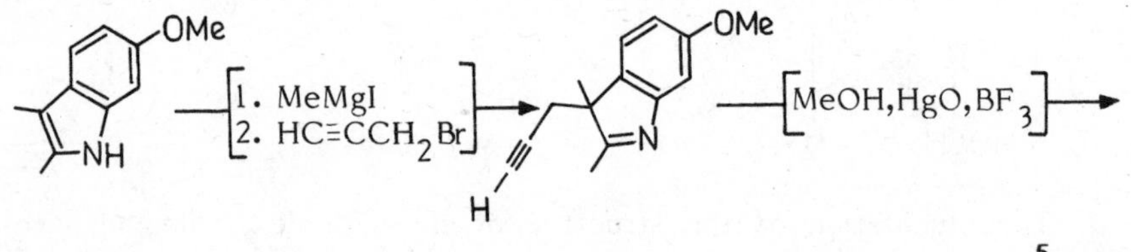

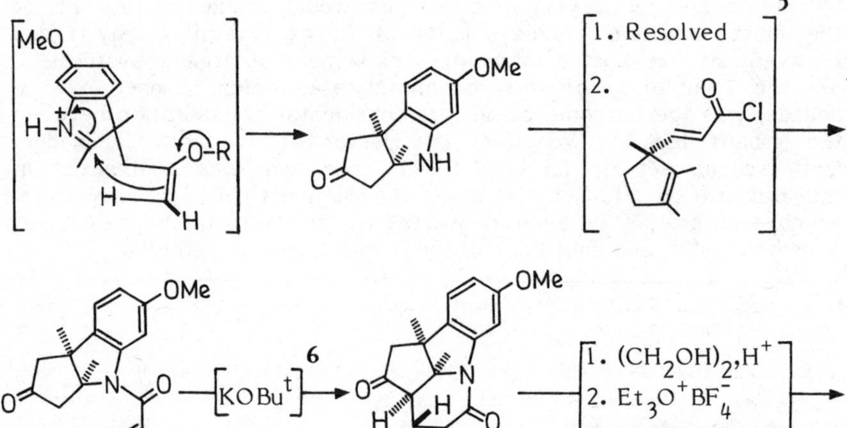

5. The resolution was carried out via the phenylethyl urea derivative. The diastereomers could easily be separated and then converted to the optically active amines by pyrolysis.

6. The alternative mode of cyclization, which would result in a β-bridgehead hydrogen is precluded on steric grounds.

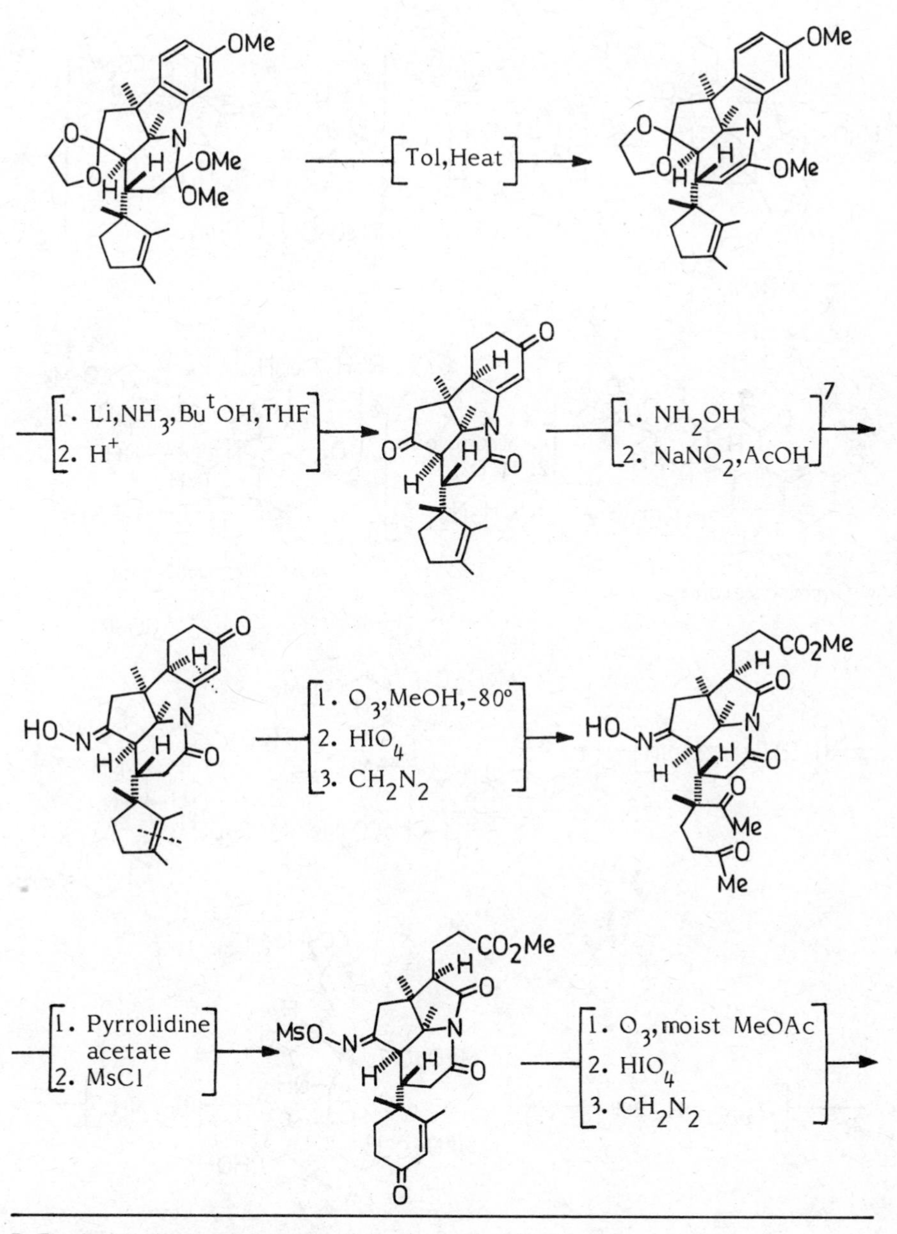

7. The less hindered oxime is selectively cleaved by nitrous acid.

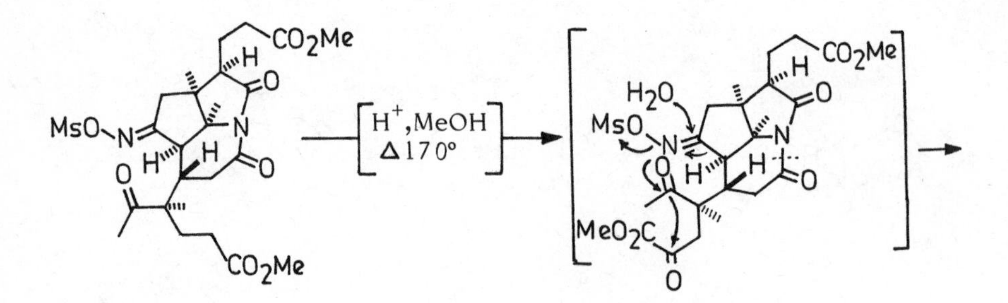

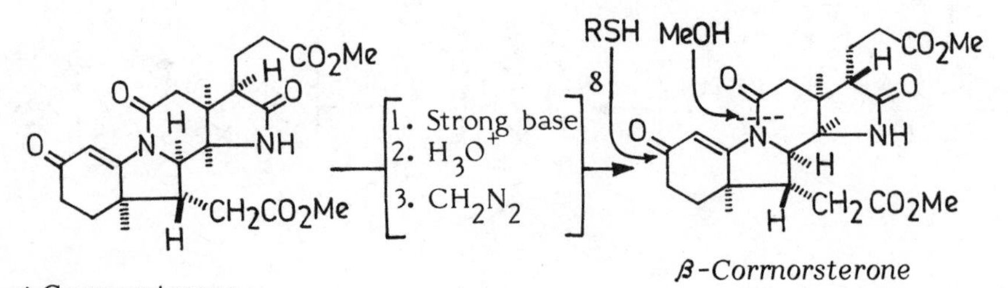

α-Cornorsterone

β-Cornorsterone

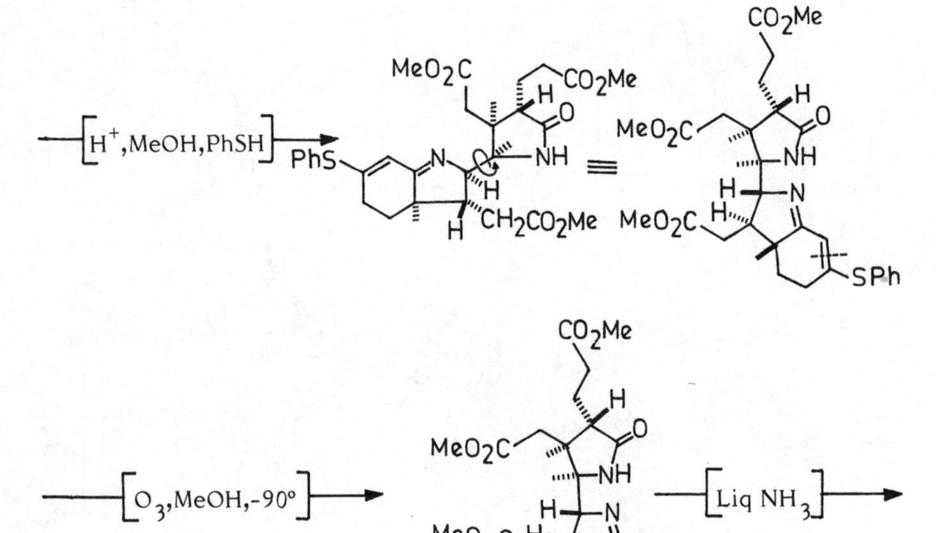

---

8. The β-isomers in contrast to α-isomers undergo facile ring opening.

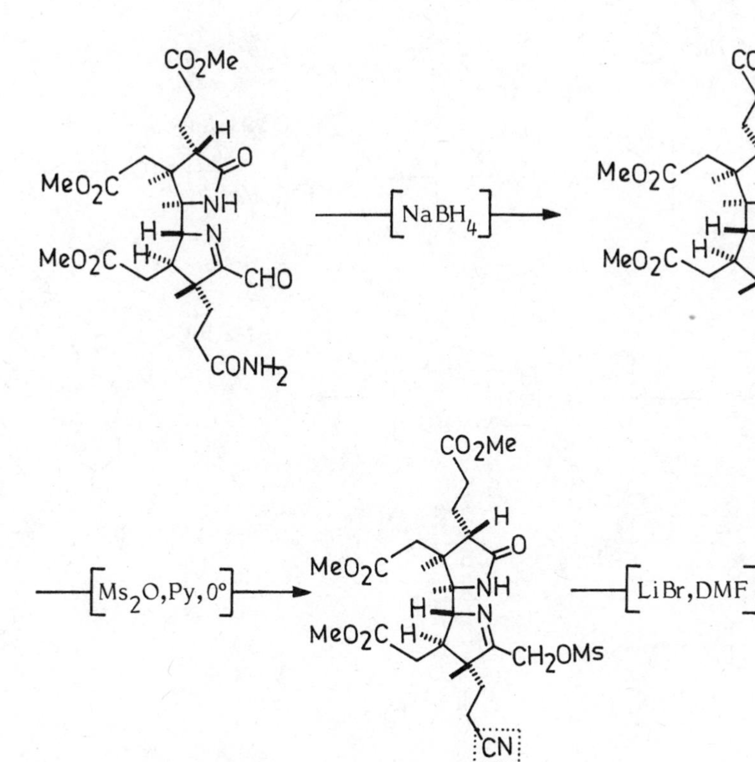

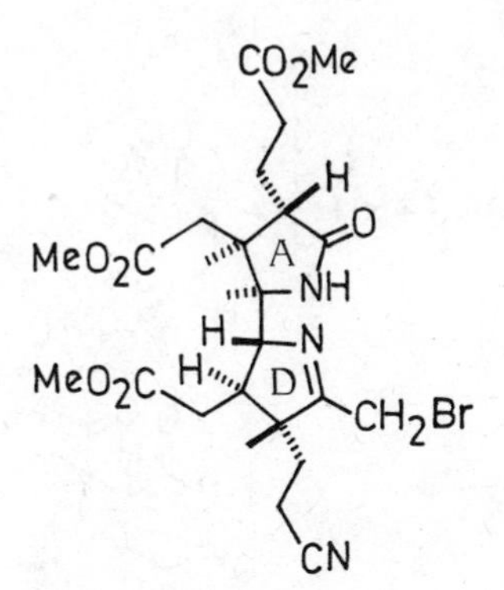

Cyanobromide

Preparation of Unit Ⓑ

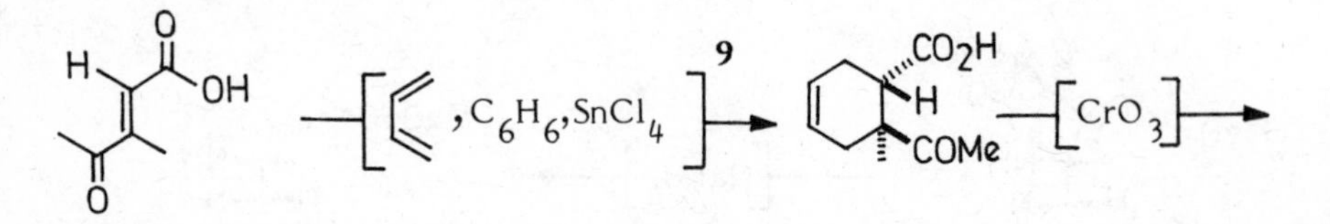

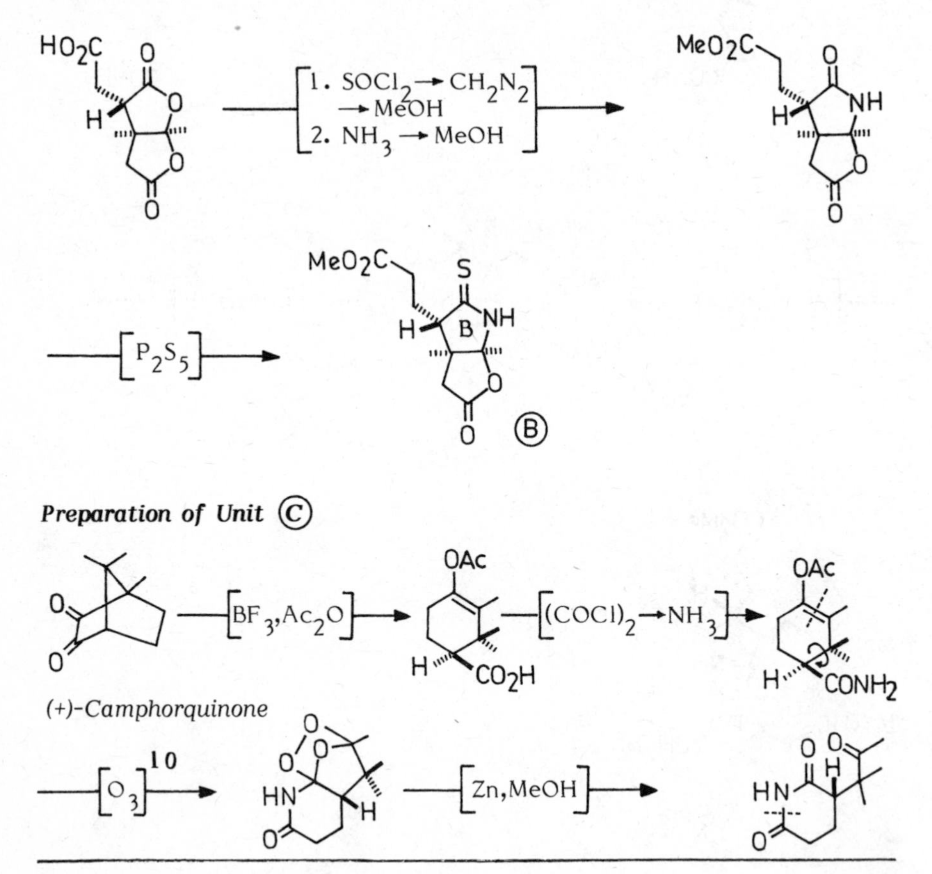

9. The adduct is resolved via α-phenylethylamine salt.

10. Ozone transforms the double bond to a mixed anhydride at one end and a ketone oxide at the other. Subsequent interaction of the amide group with the anhydride, results in a succinimide wherein the ketone oxide undergoes cycloaddition with one of the lactam carbonyls, to yield the 'false' ozonide.

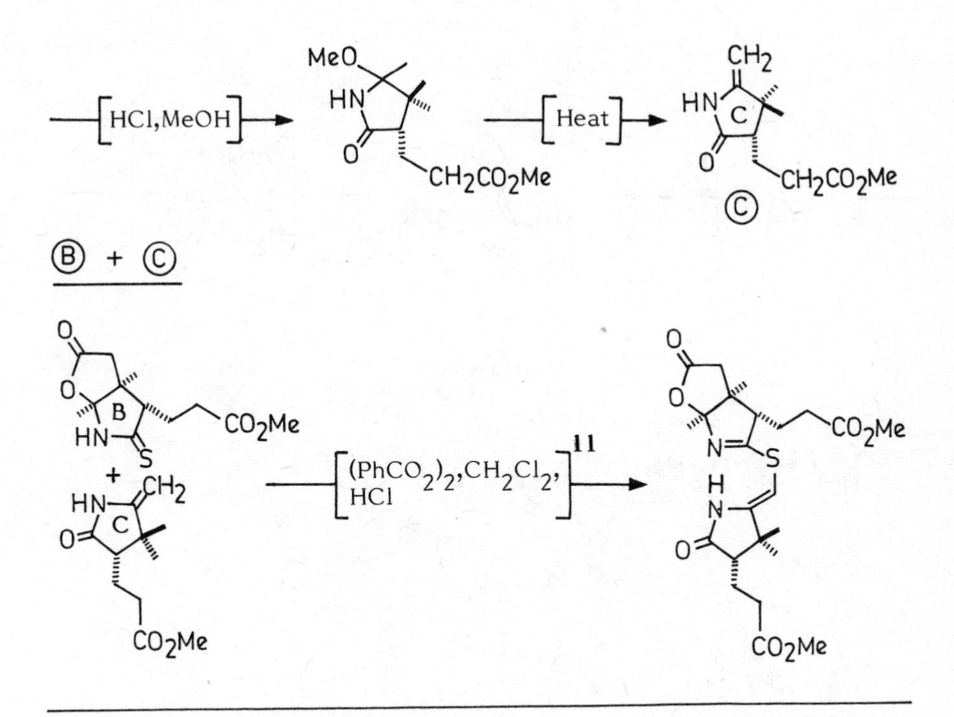

11. The central problem in construction of this component, and in corrin synthesis in general, was the production of the vinylogous amidine system, which was solved using the method of sulfide contraction via oxidative and alkylative coupling of the units B&C. The reaction involves oxidation of the thiolactam to a disulfide (i), followed by nucleophilic attack by the methylidene carbon of the enamide to form the thioiminoenamide (ii), which is suitably disposed for intramolecular enamide-imine attack to construct the critical C-C bond resulting in an episulfide formation (iii); the latter undergoes S-extrusion under the influence of a suitable sulfur acceptor e.g. phosphine or phosphite, to form the desired vinylogous amidine system (iv).

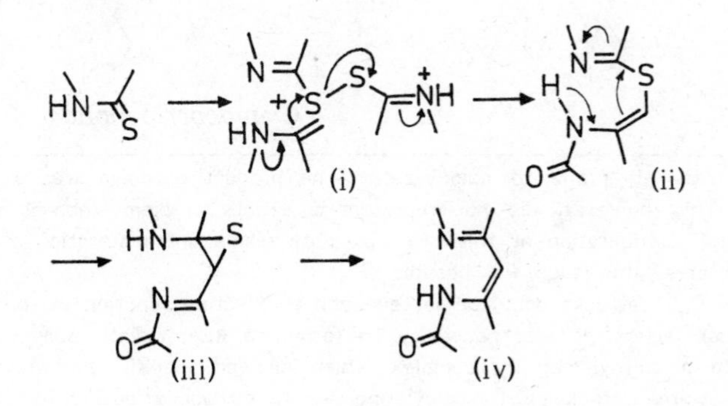

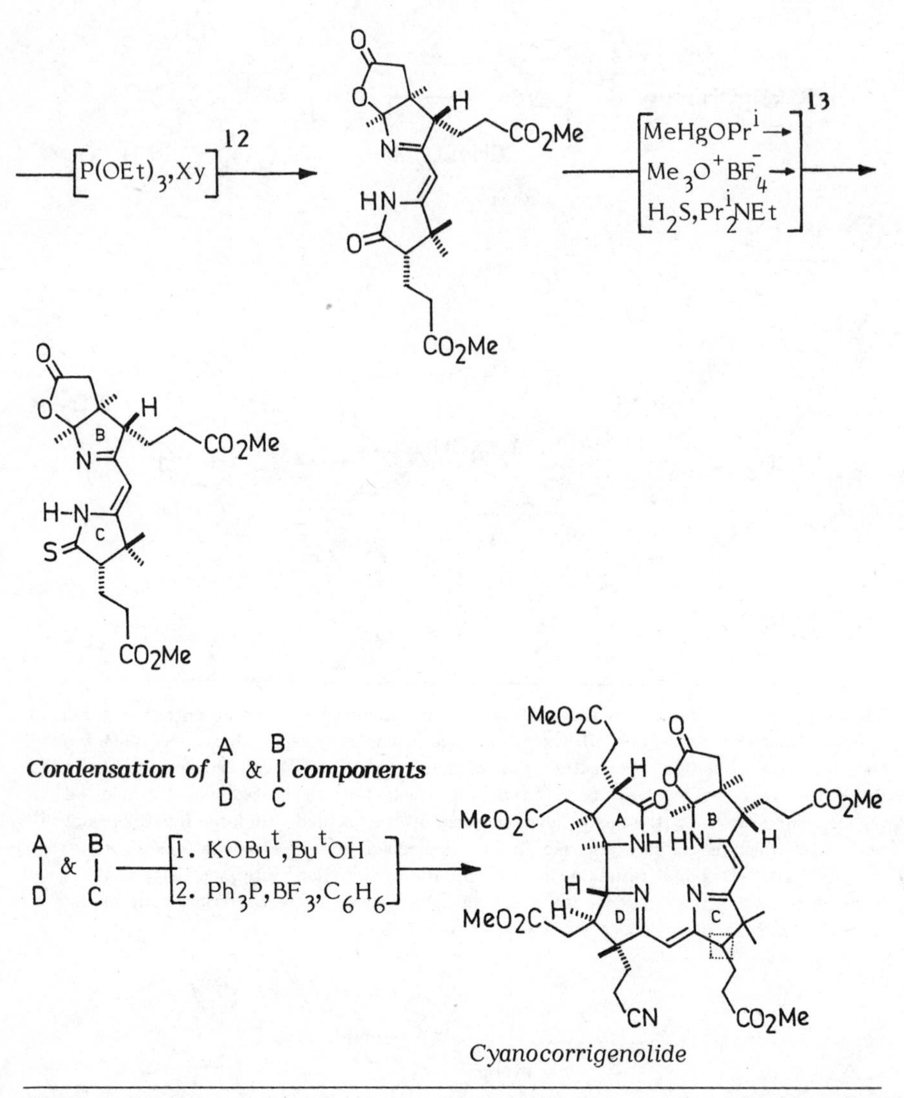

Cyanocorrigenolide

---

12. Since C-8 is prone to epimerization the product obtained was a mixture of C-8 epimers. This, however, did not represent a serious problem since it was known that the natural configuration at this site was more stable and restoration of correct stereochemistry at a later stage was possible.

13. Since P_2S_5 attacks both the lactone and the lactam function in the B/C component with equal facility, it was necessary to resort to intermediate conversion of the free lactam to a methyl-mercury complex which allowed specific activation of the lactam oxygen towards attack by the alkylating agent, without affecting the lactone carbonyl.

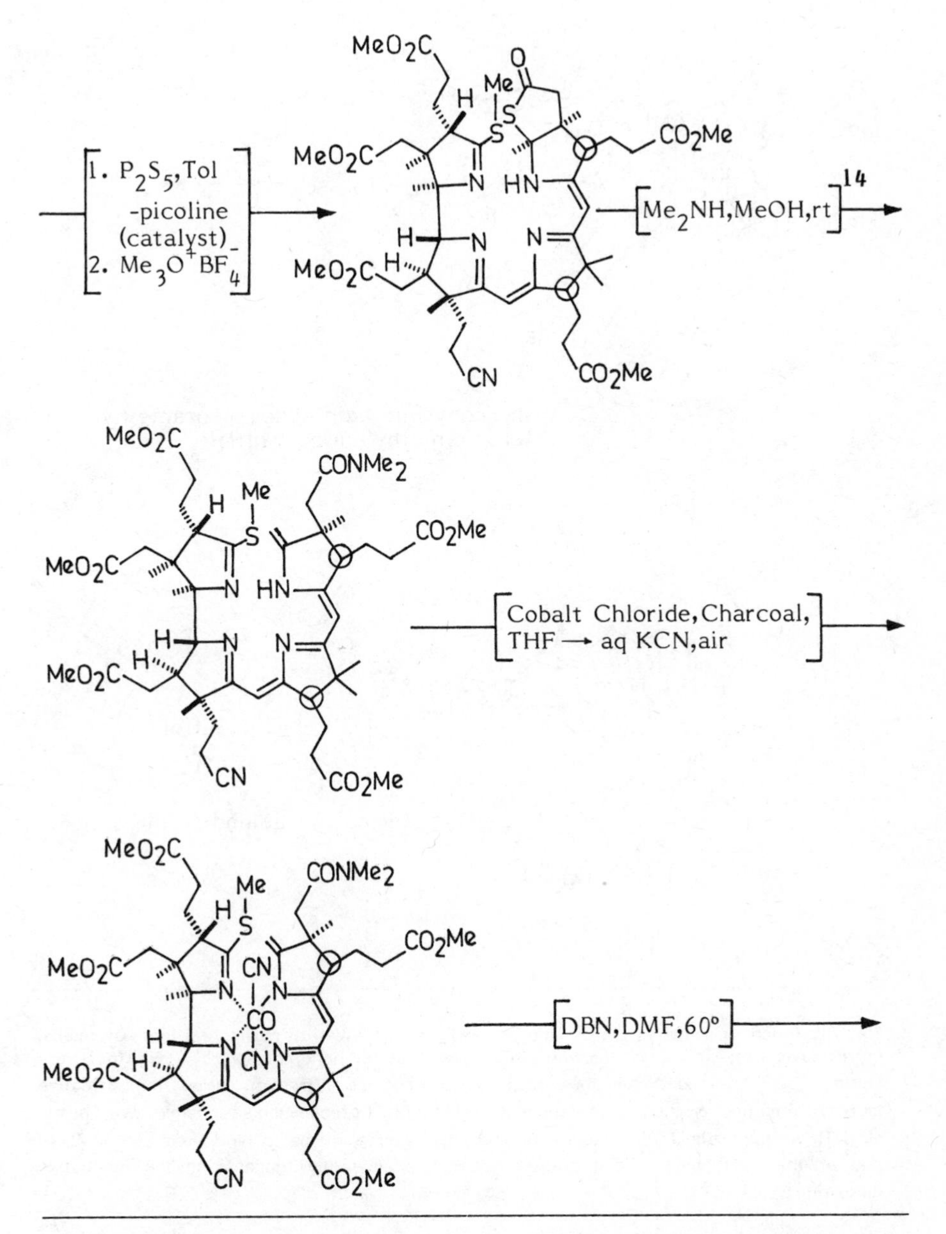

14. The exocyclic ethylidene is not very stable, and under equilibrium conditions results in the undesired endocyclic compound.

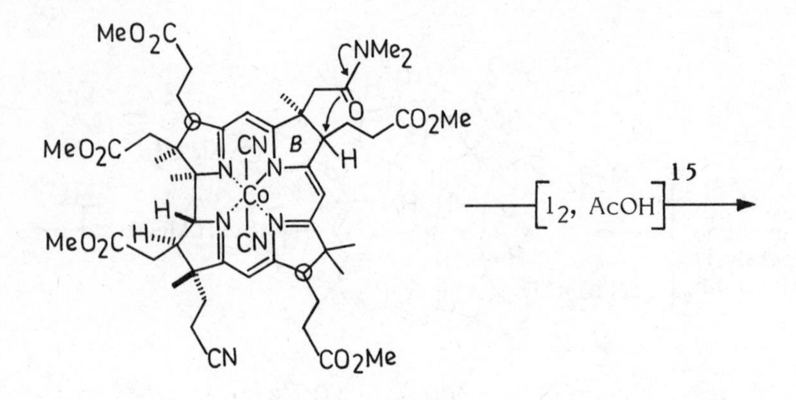

Bisnorcobyrinic acid abdeg pentamethyl ester c dimethylamide f nitrile

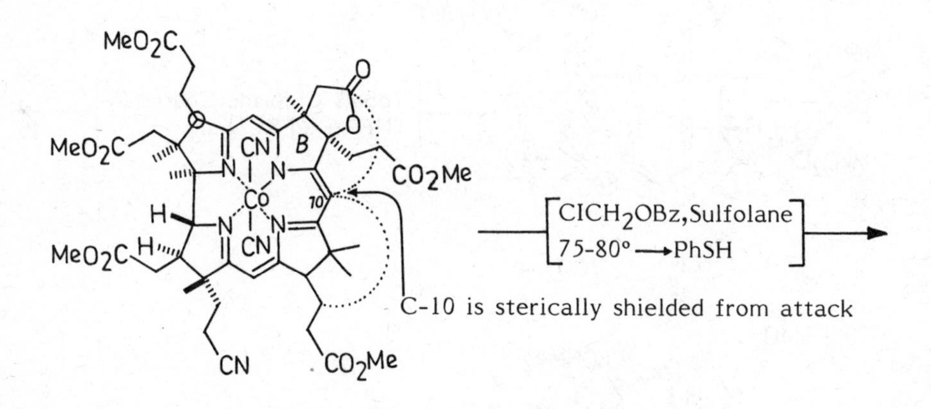

---

15. Alkylation procedures that worked well at unhindered meso positions of model corrin complexes failed in the more complex case in hand, presumably due to severe steric crowding around all the three meso positions. However, using the oxidative lactone formation on ring B, a procedure which finds precedence in vitamin B_{12} chemistry, it was possible to introduce an additional degree of steric hindrance around C-10 and at the same time afford greater access by alkylating reagents to the other two meso positions.

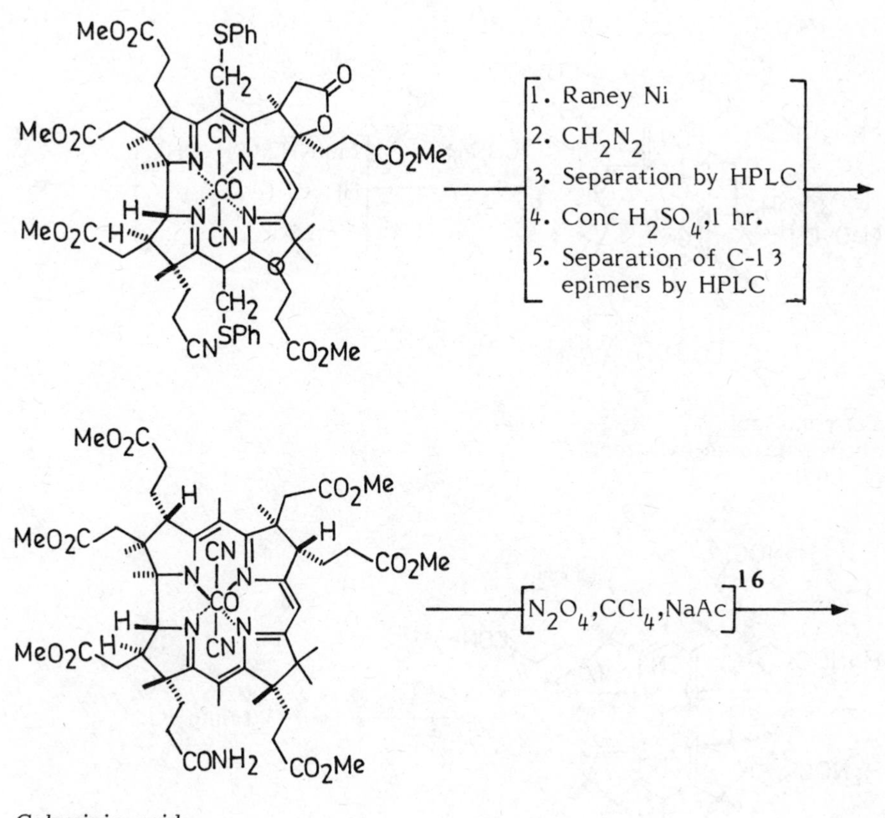

Cobyrinic acid
abcdeg hexamethyl ester f Amide

---

16. Transformation of the f amide to the corresponding f acid was effected conveniently by treatment with nitrogen tetroxide in carbon tetrachloride in the presence of sodium acetate. Conventional deamination of the amide using nitrous acid resulted in nitrosation at C-10.

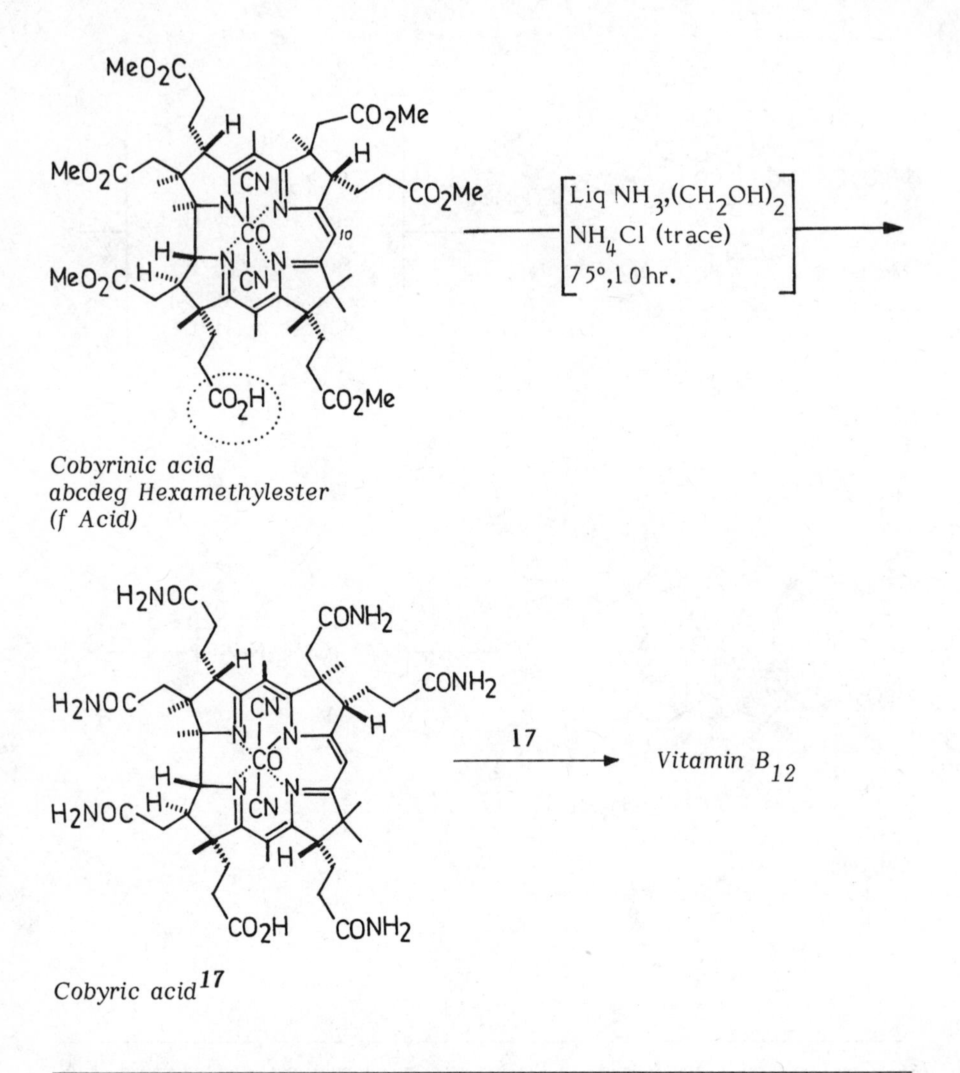

Cobyrinic acid
abcdeg Hexamethylester
(f Acid)

Cobyric acid[17]

---

17. The final stages of the synthesis leading from cobyric acid to vitamin B12 have already been accomplished by Friedrich et al (4).

Acetylfawcettine 223
Adamantane 148
Adrenosterone 1
Aflatoxins 8
Aflatoxin B1 10
Aflatoxin G1 12
Aflatoxin M1 8
Ajmaline 13
Aldosterone 18
(+)-Ambreinolide 28
β-Amyrene 33
δ-Amyrin 27
Androsterone 35
Anhydrovinblastine 370
Annofoline 223
Annulenes 37
Annulenes bridged 39
[14]Annulene 37
[18]Annulene 38
Antheridiogen An 44
Arteannuin 287
Artemesinin 287
Aromatic anions 48
Aspidospermidine 50
Aspidospermine 50
Asteranes 57
Atisine 58
Avermectin B1 64

Bastardane 148
Benzene dimer 70
Z,Z-Benzene dioxide 71
Benzene oxides 71
Z,Z,Z-Benzene trioxide 72
Benzocyclopropene 76
Benzvalene 269
Betweenanenes 74
[10.10]Betweenanene 74
Bicyclo[10.10.0]docos-1(12)-ene 74
Bicyclo[4.1.0.]heptane 77
Bongkrekic acid 78
10,9-Borazaronaphtalin 81
(4R)-4[(E)-2-Butenyl]-4,N-dimethyl-L-threonine 137
Bullvalene 82

Cantharidine 83
Capped porphyrin S 84
Carpanone 86
(-)Carpetimycin A 87
Catenanes 90
Catharanthine 93
Cavitands 95
Cephalosporin C 97
Chelidonine 100
(+)-Chlorin-5 108
(+)-Chlorin-e6, trimethyl ester 112
Chlorophyll-α 103
Cholestanol 113
Cholesterol 109
Clavolonine 223
Clavulones I & II 114
Cobyric acid 386
Coenzyme A 116
iso-Coenzyme A 117
Congressane 149
Connessine 121
(+)-Coriamyrtin 262
Coriolic acid 123
Corrin template 126
α-Corrnorsterone 378
β-Corrnorsterone 378
d-Cortisone 127
Croconic acid 48
Cubane 134
Cyanocorrigenolide 382
Cyclobutadiene 134
Cyclooctatetraene dimer 39
Cyclosporine 136
Cycl[3.2.2]azine 42
Cycl[3.3.3]azine 43
Cyclobutadiene, $Fe(CO)_3$ complex 134
Cytochalasin B 140
Cytochalasin F 145

Deoxyajmalal A&B 17
Desethylaspidospermidine, N-benzyl 56
6-Demethyl-6-deoxytetracycline 341

21-Deoxyajmaline 17
Deoxyarteannuin 290
Deoxyribonucleic acids 179
Dewar benzene 146
E,E-1,4-Diacetoxy buta-1,3-
diene 147
Diamantane 149
Diamond structures 148
Diethyl dreiecksaure 49
15,16-Dihydropyrene,trans-
15,16-dimethyl 42
Dimorphecolic acid 123
Diosgenin 357
1,4-Dioxocin 71
10,22-Dioxokopsane 201
Dodecahedrane 150

Endiandric acids A,B,C,D,E,
F,G 154
Ergotamine 230
Erythromycin A 158
Erythronolide A 162
d-Estrone 165

Gene synthesis 179
Gene, Human Leucocyte
α-Interferon 188
Gene, Yeast Alanine t-RNA 187

Helicenes 193
[13]Helicene 193
Hexahelicene 194
Histidine 196
Homonomarianolic acid 167
D-Homoestrone methyl ether 166
D-Homotestosterone 271

Iceane 197
1,6-Imino[10]annulene 40
Isostrychnine 1

Kekulene 199
Kopsanone 201

Longicamphenilone 207
Longifolene 204

Luciduline 210
Lycodine 220
Lycopodine 214
L-Lysergic acid 223

1,6-Methano[10]annulene 40
Methyl etiochola-$\Delta^{4,9(11),16}$
trienate, 3-ketone 112
Methyl pheophorbide 109
O-Methylorantine 231
Monensin 233

Nonahelicine 194
Norchelidonine 101

Octalene 244
Out-out, out-in bicyclic
systems 245
Ovalicin 246
Oxepin 71
1,6-Oxido[10]annulene 40

Pagodane 249
Pentalenene 252
3-epi Pentalenene 253
Pentaprismane 254
D-Pentatheine 4'-phosphate-116
Perannulenes 256
Peristylanes 258
[3]Peristylanes 258
[4]Peristylane 260
[5]Peristylane 266
(-)-Picrotoxinin 262
Picrotin 264
Polydeoxyribonucleotides 179
Porphyrin 107
Prismane 269
Progesterone 270
[1.1.1]Propellane 277
Prostaglandins 278
Prostaglandins E1,E2,F2α 278
Purpurin 107

Qinghaosu 287
Quassin 291
Quinidine 298

Quinidinone 299
L-Quinine 293
Quininone 298
d-Quinotoxine 293

L-Reserpine 303
Resistomycin 312
Rhodizonic acid 49
(+)-Rifamycin S 314

Sexipyridine 325
Sigma directed π systems 327
Sporidesmin A 330
Squaric acid 48
Strychnine 333
Superphane 337

Terramycin 344
Tetraasterane 57
Tetracyclines 338
Tetrahedrane tetra-tert-butyl 345
Tetrodotoxin 346
(+)-Thienamycin 351
Tigogenin 354
trans,cis,cis-[10.4.4.]Triannulene-16,18-dione 257
Triasterane 57
Triaxane 258
Tropinone 361

Verrucarol 362
Verrucarin A 362
Vincadiffermine 55
Vinblastine 370
Vincaleucoblastine 370
Vincristine 370
Vindoline 373
Vitamin B12 375

The compounds synthesised having asymmetric centres were racemic products unless indicated with the appropriate sign or symbol.

AUTHOR INDEX

Abraham,E.P. 97
Abraham,R.J. 330
Achini,R.S. 32
Ackerman,J. 21
Action,N. 269
Adam,W. 246
Agarwal,K.L. 184,185
Agtarap,A. 233
Akagi,M. 53
Akasaka,K. 233,315
Albertson,N.F. 196
Alder,K. 83
Aldrich,P.E. 14
Alexanian,V. 369
Ali,S.A. 164
Allen,D.S. 167
Almog,J. 84
Altenbach,H.J. 72
Amats,J.S. 351
Amiard,G. 4,97
Amini,N. 308
Ananchenko,S.N. 168,169
Anand,N. 293
Anderson,R.J. 32
Ando,M. 372
Andre,C. 64
Andrews,G.C. 207
Andrews,S.L. 97
Andriamialisoa,R.Z. 372
Ang,S.K. 13
Anner,G. 165
Annis,G.D. 252
Aoki,T. 121
Apgar,J. 185
ApSimon,J. 210
Aratani,M. 346,347,350
Archer,S. 196
Arens,J.F. 129
Arigoni,D. 18,34,118
Armstrong,V.W. 227
Arth,G.E. 127,128,129,130
Asveld,E.W.S. 288
Atkinson,T.C. 187
Au-Yeung,B.W. 158

Ayer,W.A. 214,223

Baasner,B. 244
Back,T.G. 159
Bader,F.E. 303
Baer,H. 204
Bahl,C.P. 180
Bailey,W.F. 327
Baillarge,M. 226
Baker,E.E. 64
Balaram,P. 158
Baldwin,J.E. 84
Bales,G.S. 159
Balls,D.M. 210
Balogh,D.W. 150
Ban,Y. 51,53
Bandaranayake,W.M. 153
Banerjee,D.K. 165
Banfield,J.E. 153
Banner,B.L. 173
Bannister,B. 21
Barborak,J.C. 134
Barkley,L.B. 1,127,131, 132
Barltrop,J.A. 27
Barnes,R.A. 42
Barnette,W. 278
Barrish,J.C. 320
Bartlett,M.I. 17
Bartlett,P.A. 171
Bartlett,W.R. 273
Bartmann,W. 21
Barton,D.H.R. 24,25,29, 30,32,118,338
Basha,A. 370
Basu,N.K. 25
Bates,G.S. 143
Battig,K. 227
Bauer,V.J. 121,272
Baylia,C. 51
Beancage,S.I. 183
Beaton,J.M. 24,30
Behforouz,M. 323
Belagaje,R. 186
Belgaje,R. 179

Berk,H.C. 150
Bernhauer,K. 375
Berson,J.A. 70,210
Bertin,D. 1,168
Besmer,P. 184
Beyler,R.E. 127,128,129
Bhattacharyya,B.K. 21
Bhattacharyya,D. 196
Bianco,E.J. 338
Bickel,H. 303
Biethan,U. 57
Binder,M. 140
Bindra,J.S. 278,281,293
Bindra,R. 278,293
Binkley,E.S. 218
Biskup,M. 39
Black,A.Y. 175
Black,D.St.C. 153
Blankenship,R.M.B. 150
Blondin,G.A. 123
Bloom,B.M. 21
Blumenkopf,T.A. 210
Boekelheide,V. 41,42,337
Boessenkool,I. 322
Boliver,F. 186
Boll,W.A. 39
Booth,A.H. 338
Borel,J.F. 246
Bornmann,W.G. 93
Bosshardt,H. 231
Bouffard,F.A. 88,351
Bowman,W.R. 214
Branca,J.S. 254
Brauman,J.I. 171
Bray,B.L. 93,371
Brener,J.S. 254
Breslow,R. 7,35
Brizzolara,A. 50
Brocksom,T.J. 273
Broka,C.A. 221
Brooks,D.W. 88,352
Brousseau,R. 180
Brown,A.G. 351
Brown,C.A. 123
Brown,D.J. 210
Brown,E.L. 179,186
Brown,L.M. 223
Brown,M. 337
Brown,P. 201
Browne,A.R. 260
Browne,L.J. 158
Brownlie,G. 32
Brunelle,D.J. 162
Brunger,H. 35
Brunner,R.K. 231
Brutcher,F.V. 110
Brutschy,F.J. 204
Bryson,T.A. 272
Buchi,G. 8,93,372
Buchi,H. 184,185
Buchschache,P. 173
Bucourt,R. 1
Buncel,E. 73
Bunzli-Treppo,U. 372
Burg,R.W. 64
Burgmaier,G.J. 327
Burgstahler,A.W. 14,34,
83
Burrell,J.W.K. 108
Butler,K. 338
Buynak,J.D. 231

Cain,P. 174
Calder,I.C. 37
Card,P.J. 158
Cardwell,H.M.E. 109
Carlson,G.R. 261
Carlson,R.M. 147
Carter,C. 231
Caruthers,M.H. 179,183,
184,185
Casey,M. 284
Cashion,P.J. 184
Cassal,Jean-Marie 173
Catlin,J.C. 180
Caton,M.P.L. 278
Cava,M.P. 333
Cavanaugh,R. 175
Cerede,J. 4
Cetenko,W. 308
Chabbale,J.C. 64
Chaiet,L. 64

Chamberlain,J.W. 233
Chan,K.K. 372
Chan,W.K. 143
Chandrasekaran,S. 189
Chapman,O.L. 86
Chauncy,B. 221
Chauvin,M.C.R. 32
Chen,C.H. 158
Chen,Y.Q. 272
Chenevert,R.B. 158
Cheung,H.T. 97
Chinn,L.J. 165
Choong,Tung-chung 102
Chou,T.C. 261
Chou,Wei-Shan 287
Choy,W. 164
Christensen,B.G. 88,351
Christianesen,R.G. 165
Chung,Kyoo-Hyun 75
Clardy,J.C. 86,328
Clark Still,W. 253
Clark,D.A. 321
Clark,V.M. 116
Clement,R.A. 21
Cohen,N. 173
Cole,D.J. 64
Cole,J.E. 165
Cole,T.W.Jr. 135
Collins,J.C. 21
Collum,D.B. 239
Colrin,E.W. 362
Conover,L.H. 338
Constanin,J.M. 127
Coombs,R.V. 271
Cope,A.C. 252
Copper,G.F. 261
Corcoran,R.J. 35
Corcoran,W.J. 159
Corey,E.J. 27,44,65,78, 114,118,128,162,189, 204,246,262,264,278, 281,321
Cornforth,J.W. 109,293
Cornforth,R. 293
Cosyn,J.P. 193
Coulton,S. 227
Cozzareli,N.R. 184
Crabbe,P. 278
Cram,D.J. 9,95,337
Cram,J.M. 337
Cramer,F. 180
Crea,R. 186
Cremer,D. 72
Crimmins,M.T. 252
Crowshaw,K. 278
Cullison,D.A. 261
Cupas,C. 148
Cupas,C.A. 197
Cushman,M. 102

D'Ambra,T.E. 364
Da-Zhong,Huang 289
Daeniker,H.U. 333
Dailey,W.P. 327
Danheiser,R.L. 189
Danieli,N. 354
Danishefsky,S. 174,175, 271
Darling,S.D. 118,129
Darmory,F.P. 53
Das,B. 93
Das,J. 173
Datwyler,P. 231
Dauben,W.G. 83
Davies,J.E. 179
Davis,F.A. 81
Davis,R. 285
Davis,R.F. 70,254
Day,M.J. 25
DeClercg,P. 6
Dehmlow,E.V. 49
DeLoach,J.A. 252
Desai,S.R. 53
Deslongchamps,P. 164
Desouza,J.P. 372
Dev,S. 204
Dewar,M.J.S. 81
Diassi,P.A. 304,309
Diedrich,F. 199
Dietrich,P. 28
DiMaio,G. 37

Distler,J.J. 116
Dittami,J.P. 246
Djerassi,C. 113,127
Doecke,C.W. 151,260
Doering,W.E. 293
Dolak,L.A. 272
Dolby,L.J. 14,333
Dolfini,J.E. 50
Doll,R.J. 35
Donaldson,M.M. 148
Douglas,G.H. 168,176
Drew,M.G.B. 148
Dube,D. 64
Duff,S.R. 109
Durnar,G. 176
Dyer,R.L. 84
Dykstra,S.J. 348

Earley,W.G. 93
Easton,P.E. 150
Eaton,P.E. 135,254,261
Echavarren,A. 323
Eder,U. 174
Edge,M.D. 187
Eggelte,H.J. 246
Egli,C. 13
Eiband,J. 181
Eichel,W.E. 173
Elix,J.A. 37
Emerson,G.F. 134
Endo,J. 308
Endo,M. 44
Engel,M.R. 86
Engl,H. 97
Ermer,O. 150
Ernest,I. 282
Eschenmoser,A. 34,375
Evans,D.A. 210
Everett,G.A. 185

Falck,J.R. 162
Fallon,G.D. 153
Farooqui,F. 196
Farquhar,D. 42
Farrar,M.W. 1,127
Faulkner,D.J. 273
Fayez,M.B.E. 32
Feldman,P.L. 372,373
Fenselau,A.H. 15
Ferrier,B.M. 71,82
Ferrino,S. 291
Fessner,W.D. 249
Feurer,M. 129
Fields,T.L. 338
Fieser,L.F. 18,127
Fieser,M. 18,127
Fishcer,H. 108
Fitzpatrick,J.D. 134
Fitzpatrick,J.M. 52
Flammang-Barbieux,M. 193
Fliri,A. 158
Fores,W.S. 66
Fornefeld,E.J. 224
Forsen,L. 108
Fossel,E.T. 71,82
Foulkes,D.M. 8
Fourrey,J.L. 93
Franck,R.W. 121,272
Francotte,E. 227
Fraser-Reid,B. 68,322
Freedman,P.K. 210
Freidrich,W. 375
Frey,A.J. 230,303
Friary,R.J. 282
Fridkin,M. 184
Fried,J. 97
Fried,J.H. 285
Friedman,L.J. 83
Friedman,M.D. 302
Friedrich,W. 386
Frisch,H.D. 90
Frobel,K. 158
Frye,L.L. 177
Fuchs,P.L. 65
Fujimoto,G.I. 111
Fujise,Y. 372
Fukumoto,K. 170
Fukuyama,T. 233,315,330,
332,346,347
Fullman,E. 73

Funk,R.L. 178
Furst,A. 173

Gais,H.J. 158
Gait,M.J. 181,186,188
Galantay,E. 97
Galbraith,A. 42
Gallagher,T. 54,55,201
Galluci,J.B. 151
Galynker,I. 253
Gaoni,Y. 37
Garratt,D.G. 158
Garratt,P.J. 258
Garratte,P.J. 37
Garvey,D.S. 164,318
Gasic,G.P. 278
Gassen,H.G. 179
Gatehouse,B.M. 153
Gates,M. 301
Gatica,J. 113
Gelbcke,M. 193
Geller,L.E. 24,30
George,G.I. 351
Ghera,E. 27
Ghoshal,M. 312
Gill,H.S. 351
Gizycki,U. 57
Gkazaia,M. 370
Gleicher,G.J. 81,148
Gleiter,R. 328
Godel,T. 208
Goebel,S. 108
Goeddel,D.V. 187
Goldberg,M.W. 35
Goldman,N. 271
Gopalan,B. 189
Gosteli,J. 97,282
Goto,T. 346,347
Gramain,J.C. 56
Gras,Jean-Louis 189
Graves,J.M.H. 168,176
Gravestock,M.B. 272
Greene,A.R. 187
Greenlee,W.J. 140
Greshoff,M. 201
Grethe,G. 293,295,300
Grieco,P.A. 171,291
Grimme,W. 73
Grob,C.A. 61
Grodski,A. 281
Grohmann,K. 37
Gross,G. 375
Gross,H. 337
Grossen,P. 367
Grzejszczak,S. 309
Gueritte,F. 370
Guggisberg,S. 231
Gupta,N.K. 185
Gupta,Y.P. 357
Gurevich,A.I. 338
Guthikonda,N. 308
Guthrie,R.W. 58
Gutsche,C.D. 165
Gutzwiller,J. 295,308

Haas,G. 341
Hadler,H.I. 21
Hagaoka,H. 314
Haginiwa,J. 308
Hajos,Z.G. 173,174
Haley,G.J. 328
Halsall,T.G. 27
Hamon,D.P.G. 198
Hanessian,S. 64,65,68,322
Hanoaka,M. 297
Harada,N. 123
Harayama,T. 210
Hardtmann,G. 341
Harley-Mason,J. 53
Harold,P. 367
Harrison,I.T. 118
Hartenstein,J.H. 71,82
Hartley,D. 168,176
Hartman,G.R. 246
Hasan,I. 315
Hase,T. 321,338
Haslanger,M.F. 162
Hassner,A. 369
Hauel,H. 181
Havel,M. 330
Hayakawa,K. 158

Hayase,Y. 58,143
Haynes,L.J. 154
Heathcliffe,G.R. 187
Heathcock,C.H. 210,218
Hegenberg,P. 48
Heggie,W. 158
Heilbron,I. 154
Hellemann,H. 146
Henderson,T. 295
Hendrickson,J.B. 333
Hertler,W.R. 118
Hess,H.J. 27
Hesse,M. 231
Hesse,R.H. 25
Hesson,D.P. 158
Heusler,K. 18,97,109,127
Heymes,R. 97
Hida,T. 262
Higgins,N.A. 184
Hikkino,H. 231
Hill,R.K. 147
Hirama,M. 164
Hiraoka,T. 351
Hirata,Y. 48,346
Hirose,T. 186
Hobson,J.D. 17
Hodakowski,L. 197
Hodges Paul,J. 64
Hodgkin,D.C. 375
Hoffman,A. 224,230
Hoffmann,R. 147
Hofheinz,W. 287
Hoft,E. 337
Hofteizer,G. 303
Hogrefe,F. 244
Holley,R.W. 185
Holm,R.H. 126
Holtermann,H. 109
Holton,R.A. 32
Homolka,B. 49,49
Honda,T. 372
Hopkins,P.B. 162
Hopla,R.E. 32
Hoppe,D. 158
Hoppe,I. 158
Hsiung,H.M. 180
Huber,W. 278
Huchner,C.F. 196
Hudrlik,A.M. 288
Hudrlik,P.F. 288
Huebner,C.F. 309
Huffman,J.C. 54,55,201
Hughes,G.A. 168,176
Huguenin,R. 136
Hunger,A. 333
Huntenberg,W. 293
Husson,H.P. 56
Hyatt,J.A. 158

Igolen,J. 226
Iguchi,K. 114
Ikeda,D. 158
Ikeyama,Y. 322
Ikota,N. 351
Imanishi,T. 297
Imperiali,B. 164,318
Inone,I. 53
Inone,M. 297
Inoue,I. 51
Inoue,S. 346,347,350
Inubishi,Y. 262
Inubushi,Y. 210
Ireland,R.E. 58,78,165
Isele,G. 90
Ishikawa,H. 170
Isobe,M. 286
Isser,S.J. 205
Itakura,K. 179,180,187
Itoh,K. 286
Izawa,T. 87,351

Jackman,L.M. 108
Jackson,B.G. 97
Jacobi,P.A. 158
Jacobs,W.A. 62
James,D.R. 254
Jansen,A.B.A. 225
Jardine,F.H. 207
Jason,M.E. 327
Jaw,J.Y. 291
Jay,E. 184

Jeger,O. 18,34,118
Jie,Zhu 289
John,S. 231
Johne,S. 231
Johns,W.F. 21,127,128, 129,130
Johnson,A.W. 108
Johnson,I.F. 44
Johnson,J.M. 225
Johnson,M.R. 314,315
Johnson,W.S. 21,109,121, 165,167,171,207,272, 273,274,275
Johnston,D.B.R. 217
Johnston,J.D. 338
Jones Jr.,M. 71,82,210
Jones,E.R.H. 154
Jones,R. 81
Jones,R.G. 224
Joseph,T.C. 214
Joshi,G.S. 357
Julia,M. 226
Jung,S.L. 315
Just,G. 278

Kabayashi,M. 114
Kakoi,H. 346,347
Kaleya,R. 35
Kalojanoff,A. 108
Kametani,T. 170,351
Kanaoka,Y. 51
Kanngiesser,W. 108
Kant,J. 351
Kao,J. 77
Kapla,M. 52
Kaplan,M. 53
Karanewsky,D.S. 233
Karapetyan,M.G. 338
Karbach,S. 95
Karle,J. 148
Karunaratne,V. 252
Katagiri,N. 180
Kathawala,F. 341
Kato,T. 123
Katsube,J. 372
Katz,T.J. 269
Kawamoto,I. 286
Kawashima,K. 289
Keana,J.F.W. 109,121,272
Keay,B.A. 313
Keck,G.E. 189
Keller,K. 100
Kellogg,R.M. 288
Kelly,T.R. 312,323
Kelogg,M.S. 272
Kemp,A.D. 21
Kende,A.S. 338
Keniti,H. 293
Kerady,S. 351
Kerdesky,F.A.J. 164
Kesavan,K. 196
Kessar,S.V. 357
Kessel,C.R. 83
Keuhne,M.E. 93
Khanna,P.L. 35
Khorana,H.G. 116,179,184, 185,186
Kierstead,R.W. 303
Kigoshi,H. 262
Kikuchi,H. 114
Kim,K.S. 158
Kim,S. 162
Kindler,K. 293
Kinoshita,M. 322
Kirby,G.W. 116
Kishi,Y. 233,314,315,322, 330,332,346,347,350
Kleid,D.G. 186
Klein,F.K. 372
Klein,S. 181
Kleinman,E. 218
Kleppe,K. 185
Kleppe,R. 184
Kline,G.B. 224
Klingenberg,M. 78
Klioze,S.S. 53
Klumpp,G. 71,82
Knowles,W.S. 1,127,131, 132
Kobayashi,S. 87
Kobuke,Y. 158
Koelliker,V. 278

Koga,K. 351
Kogan,T.P. 177
Kojima,K. 158
Koleck,M.P. 102
Koller,H. 108
Kolosov,M.N. 338
Konno,C. 231
Konz,W.E. 32
Kornfeld,E.C. 224
Korobko,V.G. 338
Korste,J.J. 338
Korte,F. 11
Korte,S. 73
Korzun,B. 309
Krans,G. 286
Krebs,E.P. 261
Kretchmer,R.A. 214
Kropp,P.J. 21
Krowicki,K. 158
Kubela,R. 173
Kuehne,M.E. 51,309
Kuhn,M. 136
Kukharji,P.C. 132
Kulsa,P. 93
Kumar,A. 184,185
Kuo,C.H. 169
Kurihara,T. 369
Kurono,M. 8
Kutney,J.P. 370,372

Labia,R. 351
Lambert,B.F. 17
Landesman,H. 50
Lang,A. 179
Lange,C. 11
Langlois,N. 370,372
Langlois,Y. 370,372
Laonov,V.N. 168
Larsin,S.D. 189
Lavagnino,E.R. 97
Lawrence,R.F. 93,371
Lawson,A.J. 11
Lawton,R.G. 14,333
Lazary,S. 246
Le Goffic,F. 226
Leaver,D. 42
Lebioda,L. 256
LeBlanc,M. 84
Lederer,E. 28
Lednicer,D. 193
Lee,H.L. 295,300
Lee,H.W. 325
Lee,V.J. 158
LeGoff,M.T. 93
Leonov,V.N. 168
Lett,R. 162
Leutert,T. 158
Leutwiler,A. 372
Lewellyn,M.E. 74
Lex,J. 244
Li,T. 273
Lichti,H. 136
Lier,E.F. 29
Lijima,I. 53
Limanov,V. 168
Limori,T. 87
Lindner,H.J. 337
Lio,H. 314,322
Lis,R. 291
Littlehailes,J.D. 27
Liw,Jing-Ming 287
Loewen,P.C. 184
Loewenthal,H.J.E. 120,
131,132
Logemann,E. 90
Logusch,E. 5,158
Loosli,H.R. 136
Lu,D.L.L. 352
Lu,L.D.L. 88
Lukes,L.M. 127
Lukes,R.M. 127,128,129,
130
Luttringhaus,A. 90
Lyall,J. 165

Maahs,G. 48
MacAlpine,G.A. 173
Maclean,D.B. 213
Madison,J.T. 185
Magdinski,L. 322

Magnus,P. 54,55,201
Magnus,P.D. 338
Mahajan,R.K. 357
Maier,G. 345
Maksimov,V.I. 16
Malchenka,S. 360
Malchenko,S. 158
Mallamo,J.P. 175,177
Mancera,O. 113
Mancrisso,A.J. 315
Mann,M.J. 224
Manske,R.H.F. 304,333
Mansuri,M.M. 159,314
Manzur,Y. 354
Marazano,C. 93
Mark,I. 93
Marker,R.E. 35
Markezich,R.L. 274
Markham,A.F. 187
Marquisee,M. 185
Marshall,J.A. 74, 75,121,
256,272
Martens,J. 158
Martin,D.G. 121,272
Martin,M.M. 252
Martin,R.H. 193,195
Martin,S.F. 53
301,309
Maruyama,H. 351
Masamune,S. 13,58,88,159,
164,318,352
Mascarella,S.W. 252
Massamune,S. 143
Mathieu,J. 1,4,168
Matsumoto,H. 170
Matsumoto,T. 252
Matteucci,M.D. 183
Matthes,H.W.D. 181
Matthews,R.S. 158
Matturro,M.G. 327
Matusch,R. 345
Matz,J.R. 328
Mayer,J. 37
McCarry,B.E. 272,274
McCarthy,P.A. 159
McCluskey,J.G. 17
McDonald,J.H. 239
McDongal,P.G. 360,366
McGhie,J.F. 29
McKennis,J.S. 254
Mckerrey,M.A.Q. 252
Mckervey,M.A. 148
McLamore,W.M. 109,127
McLoughlin,B.J. 168
McLoughlin,J. 176
McMurry,J.E. 205,270,328
Meacock,P.A. 187
Mehrotra,A.K. 210,280
Mehrotra,M.M. 114
Mehrotra,S. 196
Mehta,G. 252
Meier,W. 173
Melntosh,C.L. 86
Melvin Jr.,L.S. 162
Merril,S.H. 185
Meyer,J. 35
Micheli,R.A. 173
Michniewicz,J. 180
Miescher,K. 165
Mijngheer,R. 6
Miles,D.H. 272
Miller,A.C. 53
Miller,B.M. 64
Miller,R.C. 184
Miller,T.W. 64
Milne,G.M. 32
Minamoto,K. 184
Minter,D.E. 302
Misra,R.N. 288
Misumi,S. 252,337
Mitchell,G.F. 8
Mitchell,J. 328
Mith,T. 300
Mitra,R.B. 204
Mitsunobu,O. 369
Mitt,T. 295
Miyahra,Y. 151
Miyano,M. 26
Moerch,R.E. 150
Moffatt,J.G. 15,116
Mohr,P. 367

Molino,B. 322
Mooberry,J.B. 341
Moran,J.R. 95
Morand,P. 165
Morehead,S. 372
Mori,S. 164
Morin,C. 351
Morin,R.B. 97
Morokoshi,I. 360
Morokoshi,N. 360
Morren,G. 195
Morrison,D.E. 224
Morrison,G.C. 308
Mrozik,H. 64
Mueller,R.A. 97
Mueller,R.H. 78,261
Mukharji,P.C. 120
Mukherji,P.C. 131
Mukund,M. 114
Muller,Hans-Jurgen 220
Murata,I. 360
Musso,H. 57
Muxfeldt,H. 338,341
Myers,A.G. 44
Myers,R.F. 272

Naegeti,P. 97
Naf,U. 44
Nagai,M. 51
Nagaoka,H. 315,322
Nagase,H. 262
Nagata,W. 58,61,121
Nagel,A. 174,175
Nagoka,H. 314
Nakahara,Yashiaki 140
Nakahara,Yuko 140
Nakajima,Y. 369
Nakamura,E. 143
Nakanishi,K. 44
Nakashima,T.T. 223
Nakata,J. 315
Nakata,M. 322
Nakatani,Y. 289
Nakatsubo,F. 346
Nakatsuka,N. 13
Nakatsuka,S. 330,332

Namai,T. 123
Nambiar,K.P. 158,162
Narang,S.A. 179,180,181,186
Narisada,M. 58
Naumann,A. 220
Nemoto,H. 170
Nestler,G. 282
Neuberg,R. 38
Neumann,P. 337
Neustaedter,P.J. 275
Newkome,G.R. 325
Newman,M.S. 193
Newton,C.R. 187
Ni,Mu-Yun 287
Nickon,A. 76
Nicolaou,K.C. 153,159,
162,278
Niiyama,K. 262
Nishiyama,H. 286
Nishizawa,M. 286
Niwa,H. 262
Nomine,G. 1,4,97,111,128,
168
Norris,K.E. 186
Noyori,R. 285,286
Nugent,R.A. 362

Ogasawara,K. 93
Ogata,M. 231
Ohashi,M. 271
Ohfune,Y. 252
Ohloff,G. 289
Ohmizu,H. 367
Ohno,M. 87,204,351
Ohnuma,T. 372
Ohtsuka,E. 185
Ohtsuka,T. 252
Oishi,T. 51,53
Okano,K. 351
Okarma,P.J. 327
Oliver,L.K. 16
Ollis,W.D. 333
Ong,B.S. 158
Onoprienko,V.V. 338
Oppolzer,W. 97,100,208,
213,227

Or,Y.S. 254
Osawa,E. 148
Osborn,J.A. 207
Oth,J.F.M. 38,82
Otsubo,T. 337
Ott,H. 230

Panet,A. 184
Pappas,S.P. 146
Pappo,R. 21
Paquette,L.A. 150,151,252,
254,260,261
Park,C.H. 245
Parrish,D.R. 173,174
Parry,R.J. 272
Paterson,I. 159
Pattenden,G. 252
Patterson,I. 314
Patterson,J.W.Jr. 285
Paukstelis,J.V. 77
Pearce,H.L. 262,264
Pearlman,B.A. 305
Pechet,M.M. 24,25,30
Pelletier,S.W. 58,62,210
Penswick,J.R. 185
Peruzzotti,G. 284
Petasis,N.A. 153
Petcher,T.J. 136
Peters,M. 84
Petersen,M.R. 273
Peterson,D. 175
Peterson,J.C. 256
Petit,R. 254
Petrzilka,M. 213
Pettit,R. 134
Pfriems,S. 345
Phillips,G.W. 53
Phillips,J.B. 41
Pickworth,J. 375
Pierdet,A. 1
Piers,E. 252
Pike,J.E. 21
Pinkerton,M. 233
Pizzolato,G. 308
Platonora,A.V. 168
Polonsky,J. 291
Poos,G.I. 127,128,129,130
Popravko,S.A. 338
Portland,L.A. 173
Posner,G.H. 175,177
Potier,P. 370
Pougny,J.R. 322
Powell,D.L. 48
Prashad,M. 68
Prelog,V. 314
Press,J.B. 158
Pressman,B.C. 233
Pretzer,W. 39
Price,P. 284
Prinzbach,H. 249
Prosen,R.J. 375
Proskow,S. 27

Quinkort,G. 176
Quiquerez,C. 136

Rabe,P. 293
Raffelson,H. 1,127,131,
132
Rahman,A.U. 370
Rajan Babu,T.V. 158
Rajbhandary,U.L. 184,185
Rama Rao,A.V. 123
Ramage,R. 97,227
Ramamoorthy,B. 184
Raman,H. 282
Rampal,A.L. 357
Ranganathan,D. 196,210,280
Ranganathan,S. 97,210,280,
293
Rao,K.S. 252
Raphael,R.A. 129,362
Rapoport,H. 372,373
Ratcliffe,R.W. 88,351
Rathi,R. 196
Raucher,S. 93,371
Rebek Jr.,J. 229
Reddy,E.R. 123
Reemer,R.A. 351
Reese,C. 295
Reese,C.B. 179

Reese,F. 295
Reinhart,K.L.Jr. 314
Reynolds,G.D. 221
Richardson,C.C. 184
Rideout,D.C. 7
Rieche,I. 337
Riess,J.G. 84
Rihs,G. 249
Robbiani,C. 100
Roberts,J.S. 362
Roberts,S.M. 278,351
Robins,P.A. 165
Robinson,B.P. 81
Robinson,F.M. 129
Robinson,J.M. 127
Robinson,R. 109,231,293,
333,361
Robortson,J.H. 375
Rodewald,C.B. 148
Rodrigo,R. 313
Rogalski,W. 341
Rogers,N.A.J. 27
Rogier,E.R. 21
Rohr,M. 372
Rona,R.J. 288
Ronaldson,J.W. 330
Rosati,R. 93
Roseman,S. 116
Rosen,P. 271
Rosenkranz,G. 113
Roth,H.D. 39
Rousch,W.R. 364
Rousseau,G. 158
Rubin,M.B. 21
Rubin,R.M. 71,82
Rueger,H. 309
Ruegger,A. 136
Ruest,L. 164
Runzheimer,H.V. 244
Rushton,J.D. 27
Rutsch,W. 314
Ruzicka,L. 34,35
Ryan,M.J. 179
Rzheznikov,V.N. 168
Safe,S. 330
Sakai,S. 308
Sakai,T. 322
Sakamoto,K. 262
Sakan,K. 158
Sakuta,K. 286
Salomon,R.G. 284
Salzmann,T.N. 88,351
Sarett,L.H. 127,128,129,
130
Sargent,M.V. 37
Sarkar,S.K. 13
Sato,Y. 51
Saucy,G. 173
Sauer,G. 174
Saunders,M. 71,82
Sauter,H.M. 158
Saxton,J.E. 333
Scanlon,D.B. 187
Scanlon,W.B. 97
Schaaf,T.K. 278,281
Schafer,W. 146,328
Schaffner,K. 18,118
Schalner,J.L. 150
Schaus,J.M. 307
Scheinmann,F. 278
Schenck,G.O. 83
Schenker,K. 333
Schiess,P. 61
Schill,G. 90
Schilling,W. 143
Schillinger,W.J. 171
Schirch,P.F.T. 337
Schlessinger,R.H. 214,362
Schleyer,P.von R. 148
Schlittler,E. 303,304,309
Schmid,G. 233,287,314,322
Schmid,H. 231
Schmiegel,K.K. 275
Schmuj,N.R. 360
Schneider,D.F. 150
Schneider,J.A. 97,303
Schneider,R.S. 8
Schneider,W.P. 165
Scholz,C.R. 196

Schreiber,W.L. 301
Schroder,G. 38,82
Schuch,W. 187
Schulte-Elte,K.H. 289
Schultze,A. 293
Schumacher,M. 83
Schumann,D. 220
Schurter,J.J. 195
Schwartz,D.A. 164
Schwartz,U. 176
Sciammanna,W. 173
Scott,J.W. 173
Scott,M.A. 173
Scott,W.L. 210
Secrist,J.A. 162
Seebach,D. 64
Seemann-Preising,B. 181
Seiler,M.P. 32
Sekine,V. 337
Sekiya,T. 184
Seliger,H. 181
Semmelhack,M.F. 272
Sensi,P. 314
Sgaramella,V. 185
Shafer,U. 345
Shamma,M. 14
Shani,A. 39
Sharma,G.V.M. 123
Shavel,J.Jr. 308
Sheehan,J.C. 97
Shelberg,W.E 165
Sheldrake,P.W. 162
Shemyakin,M.M. 338
Shidd,L.S. 314
Shiner,C.S. 5
Shirahama,H. 252
Shiroyama,K. 170
Shunk,C.H. 110
Siddal,J. 176
Siddall,J. 168
Siddique,R.H. 13
Siddiqui,S. 13
Siegmann,C.M. 18
Sigg,H.P. 246
Sih,C.J. 284
Simmons,H.E. 245
Simonovitch,C. 278
Singer,P.P. 223
Singh,M. 181
Siret,P. 189
Sitrin,R. 282
Sivanandaiah,K.M. 165
Small,T. 42
Smith,B.H. 337
Smith,G.F. 333
Smith,H. 168,176
Smith,P.J. 214
Sneen,R.A. 128
Snider,B.B. 35,246
Sondheimer,F. 37,39,109,
127,154,354
Sondheimer,P. 27
Sood,R. 284
Spring,F.S. 32
Springer,J.P. 86
Sproat,B.S. 181
Stabb,H.A. 199
Staehelin,H. 246
Stalmann,L. 21
Starratt,A.N. 25,118
Steinrauf,L. 233
Stern,A. 108
Stevens,C.L. 348
Stevens,R.V. 52
Stevenson,R. 32
Stier,E. 108
Still,W.C. 239,320,367
Stojanac,Z. 173
Stoll,A. 108,224
Stork,G. 5,7,34,50,83,97,
118,120,129,131,132,
140,143,214,270,271,
286,308
Stotter,P.L. 302
Strachan,W.S. 32
Strell,M. 108
Suave,G. 164
Suffness,M.I. 32
Suga,K. 32
Sugasawa,T. 58
Sugavanam,B. 301
Sugihara,Y. 360

Sugita,K. 252
Sugiura,S. 346,347
Sultanbawa,M.U.S. 272
Surtees,J.R. 225
Suter,Ch. 282
Suzuki,M. 158,262,285,286
Suzuki,T. 286
Swern,D. 315
Switzer,C. 177
Szabo,A. 97
Szmuszkovicz,J. 21,50
Szpilfogel,S.A. 18
Szychowski,J. 213

Tai,D.F. 229
Takahashi,S. 346
Takahashi,T. 286
Takahashi,Y. 87
Takao,H. 322
Takemura,K.H. 83
Takeya,T. 184
Takigawa,T. 171
Tamm,Ch. 140,367
Tamm,R. 14
Tanaka,K. 262
Tang,S.C. 126
Tanino,H. 347
Tanino,S. 350
Tardella,P.A. 286
Tatsuta,K. 158,322
Taub,D. 109
127,169
Taylor,A. 330
Taylor,E.C. 129,301
Taylor,G.F. 198
Taylor,W.I. 17
Teague,S.J. 252
Terao,T. 185
Terasawa,T. 121
Terashima,M. 51
Ternansky,R.J. 150
Terrell,R. 50
Tessier,J. 1,168
Therien,M.J. 65
Thomas,D.B. 27
Thompson,Q.E. 1,131,132
Titmas,R.C. 181
Todd,A.R. 116
Tolbert,L.M. 158
Tolman,R.L. 64
Tomino,I. 286
Torelli,V. 4
Torgov,I.V. 165,168,169
Tori,M. 367
Toromanoff,E. 1,168
Trecker,D.J. 148
Troin,Y. 56
Tromantano,A. 78
Trost,B.M. 366
Trueblood,K. 95,375
Truesdale,E.A. 158
Truesdale,L.K. 210
Trybulski,E.J. 162
Tsuda,K. 346
Tsuji,J. 271
Tsukitani,Y. 114
Tu,You-You 287
Turner,R.A. 196
Turner,R.B. 111,293

Uchida,I. 158
Uchiyama,F. 262
Ueda,Y. 158
Uenishi,J. 153
Ugolini,A. 64,65
Unich,K.G. 285
Uskokovic,M. 308
Uskokovic,M.R. 293,295,
300
Uyehara,T. 158

Vaga,J. 323
Valenta,Z. 58,173
Valko,J.T. 102
Valls,J. 4,111,128
Van de Sande,J.H. 184,185
Van Der Burg,W.J. 18
Van Dorp,D.A. 18
Van Duyne,G. 328
Van Royen,L.A. 6

Van Tamelen,E.E. 14,16,
32,83,146,333
Vandergrift,J.M. 127
Vasella,A.T. 158
Vatakencherry,P.A. 204
Vedejs,E. 341
Veith,H.J. 231
Velluz,L. 1,4,111,128,168
Vetter,W. 90
Vidari,G. 291
Vignau,M. 168
Vladuchick,W.C. 158
Vogel,E. 39,72,73,244
Volger,G. 293
Volkmann,R.A. 171,207
Vollhardt,K.P.C. 178
Von Doering,W. 71,82
Von Wartburg,A. 136
Vorburggen,H. 97

Wada,Y. 297
Wade,P.A. 158
Wakabayashi,T. 58
Wakamatsu,K. 262
Walsh,E.J. 210
Wamhoff,H. 11
Wang,C.L.J. 233
Wang,E.J. 269
Wang,Tein-Fu. 172
Wang,Y.F. 87
Ward,D.E. 158
Ward,J.S. 254
Warner,P. 327
Wasmuth,D. 64
Wasserman,E. 90
Wasserman,H.H. 231
Watts,L. 134,134
Webber,H.P. 136
Weber,H. 185
Weber,H.P. 246
Weber,W.D. 176
Weedon,B.C.L. 108
Wehrli,P.A. 173
Wei-Shan,Zhou 289
Weinreb,S.M. 8
Weinshenker,N.M. 278
Weinstein,G.N. 126
Weirstock,L.M. 351
Weisenborn,F.L. 304,309
Wender,P.A. 307
Wendler,N.L. 169
Wenger,R.M. 136
Wenkert,E. 221,222,308
Wentland,S.H. 221
Werblood,H.M. 17
Werthemann,L. 273
West,R. 48
Westley,J.W. 233
Wettstein,A. 18
Wharton,P.S. 118
White,A.W. 307
White,D.H. 328
White,E.P. 330
White,J.F. 258
White,J.G. 375
Whitesell,J.K. 282
Wiberg,K.B. 277,327
Wickberg,B. 308
Wiechert,R. 174
Wieland,P. 18
Wierenga,W. 32
Wiesner,K. 58,210,214
Wightman,R.H. 180
Wilds,A.L. 110
Wilkinson,G. 207
Wilkinson,R.G. 338
Willart,A.K. 78
Williams,Jr.von Z. 148
Williams,R.M. 158
Williams,R.V. 260
Williamson,S.A. 309
Willstater,R. 108
Willy,W.E. 274
Wilson,D.R. 9
Windholz,M. 165
Windholz,T.B. 165,217
Winkler,J.D. 5
Wirtz,R. 83
Wirz,H. 35
Withers,G.P. 288

Wolinsky,J. 14
Wolovsky,R. 37
Wong,H.N.C. 158
Woodson,R.E. 303
Woodward,R.B. 97,103,109,
110,127,147,158,204,
224,282,293,303,333,
338,346,375
Worley,S.D. 81
Worth,B.R. 372
Wu,Zhao-Hua 287
Wynberg,H. 21,129
Wyvratt,M.J. 150

Xing-Xiang,Xu 289

Yadav,J.S. 123
Yadgiri,P. 123
Yamada,K. 48,262
Yamada,M. 262,286
Yamada,T. 185
Yamada,Y. 114
Yamaguchi,T. 123
Yamanaka,E. 308
Yamanaka,S. 123
Yamashita,A. 123
Yanagisawa,A. 285
Yasunari,Y. 13
Yokoyama,T. 123
Yonemitsu,O. 51
Yoo,S. 162
Yoo,Y. 162
Yoshino,O. 351
Yoshioka,M. 58
Young,J.F. 207
Younken,H.W. 303
Yu-Fen 287
Yu-Lin-Wu 287
Yyehara,T. 123

Zamir,A. 185
Zeller,P. 375
Zhou,Wei-Shan 289
Ziegler,F.E. 172
Zimmerman,R.L. 52
Zimmerman,W.A. 246
Zipkin,R.E. 153
Zurcher,C. 90
Zurer,P.St.J. 76

Acetaldehyde 95,352
Acetamide 349
Acetic acid-Potassium-acetate 348
Acetic anhydride 4,21,22,23,36,51,62,82,87,113,119,140,175,225,226,229,294,373,380
Acetic anhydride-Dimethyl sulfoxide 162,163
Acetic anhydride-Pyridine 10,19,20,26,36,61,75,87,95,110,112,115,121,143,163,183,281,304,317,319,331,334,335,347,348,365,368
Acetic anhydride-Sodium acetate 4,11,91,120,374
Acetone 97,110,241
Acetonecyanhydrin 309
Acetone dicarboxylic acid 221,361
1-Acetoxy-3-methyl-1,3-butadiene 364
2-Acetoxypropene 354
Acetyl bromide 174
Acetyl chloride 160,208,307,323,358
Acrylamide 215
Acrylonitrile 52,111,120
Allyl bromide 202,252,323
Allyl iodide 316
Allyllithium 74
Allylmagnesium bromide 9,66
Alumina activated 19,159,360
Aluminum amalgam 98,229,266
Aluminum bromide 148,149
Aluminum chloride 9,29,149,224,229,322,374
Aluminum hydride 100,310
Aluminum isopropoxide 145,166,303,347
Amberlite IR-120 363
Ammonia 80,81,281,283,349,378,386
Ammonium chloride 52,80,176,351,386
Ammonium hydroxide 350
Amyl bromide 125
Amylmagnesium bromide 124
Arylseleno cyanide 364
Ascorbic acid 324

Barium hydroxide 2,232,289,299
Benzaldehyde 65,137
Benzenesulfonic acid 43,181
Benzophenone 219,257
1,4-Benzoquinone 166,254
Benzoyl bromide 65,66
Benzoyl chloride 224,297
1H-1,2,3-Benztriazol-1-yloxy-tris(dimethyl-amino)phosphonium hexa-fluorophosphonate 139
Benzyl alcohol-Calcium-carbonate 9
Benzyl bromide 8,65,137,233,234
Benzyl chloride 2,5,14,93,124,125,163,172,237,262,263,289,294,300
Benzyl methylacetoacetate 319
3-Benzyloxybutyl iodide 133
2-Benzyloxycarbonylamino-ethylmercaptan 353
Benzyloxycarbonyl chloride 100,208
(+)-γ-Benzyloxymethoxy-isobutyraldehyde 239

Benzyloxymethyl bromide 240
Benzyloxymethyl chloride 278,280,384
Benzyloxymethyllithium 243
3-Benzyloxypropylamine 218
Bis(2-hydroxyethyl)-sulfide 163
Bis-(2-pentenyl)chromium 315
Bis-trimethylsilyl-acetamide 94
Bis-trimethylsilyl-acetylene 178
Bistrimethylsilyl-trifluoracetamide 171
Bis[hexa-2-ynyl]copper-lithium 208
9-Borabicyclo[3.3.1]-nonane 65,252,317
Borane 65,66,240,251
Boron trichloride 235,323
Boron trifluoride 30,54, 65,66,67,149,172,205, 253,260,263,291,298, 316,322,330,332,346, 376,380,382
Bromine 23,24,30,31,40, 41,70,71,75,79,93, 113,120,122,163,304, 321,327,345,366
1-Bromo-3-Chloropropane 56
1-Bromo-3-methyl-3-butene 248
1-Bromo-3-propanol 221
N-Bromoacetamide 23,281
Bromoacetone 196,234,236
Bromochloromethane 96
Bromoform 206
1-Bromohexane-5-one ketal 221
3-Bromopropargyl alcohol 124
N-Bromosuccinimide 30,34, 41,42,62,71,72,137, 193,198,236,240,243, 244,248,262,265,267, 299,304,326,330,363, 363,365
But-2-enyl iodide 315
1,3-Butadiene 109,213, 346,380
1-(1-Butenyl)pyrrolidine 50
t-Butoxycarbonyl anhydride 231
α-t-Butyl B-methyl l-aspartate 372
t-Butyl chloroformate 97
t-Butyl chromate 276
t-Butyl glyoxalate 339
t-Butyl hydroperoxide 66, 79,135,142,145,248, 255,265,315,363,367, 368
t-Butyl hypochlorite 373
t-Butyl α-lithioacetate 307
Butyl mercaptan 170
t-Butyl propiolate 43
t-Butyl thiopropionate 160
4-t-Butyl-N-isopropyl-2-imidazolyldisulfide 164
t-Butylchlorodimethyl-silane 67,68,69,141, 142,144,162,240,352, 367
t-Butylchlorodiphenyl-silane 65,66,154,316, 317,368
3-t-Butyldimethyl-silyloxy-1-octene-lithium 286
t-Butyldimethylsilyl trifluoromethane-sulfonate 318

n-Butyllithium 67,69,74, 75,78,114,141,144, 172,190,233,248,256, 259,279,284,309,323, 325,331,343
t-Butyllithium 56,160, 163,248,255,277,320, 345
t-Butylmagnesium chloride 351
t-Butyloxycarbonyl chloride 68

Calcium-ammonia 30,120, 121
Camphorsulfonic acid 68, 235,237,238,315,315, 316,324,347
Camphorsulfonyl chloride 267
Carbon tetrabromide-Triphenylphosphine 67, 142,154,155,228,320
Carbon tetrachloride 72
1,1'-Carbonyldiimidazole 88,138,352
4-Carboxybenzenesulfonyl azide 353
4-(3-Carboxyphenyl)-iodobenzene 35
Ceric sulfate 141
Cerium (4+) 134
Cerous chloride 341
Cesium propionate 367
Chloracetic acid 261
Chloracetyl chloride 51
Chlorine 330,339
1-Chloro-2,6-dimethyl-hepta-2,6-diene 172
4-Chloro-2-butenol 366
α-Chloroacrylonitrile 210,278
α-Chloroacryloyl chloride 190
2-Chlorocarbonylmethyl-3-sulfolene 53
m-Chloroperbenzoic acid 55,67,88,156,164,171, 176,191,201,202,206, 211,221,235,255,255, 260,265,266,267,268, 268,279,282,282,288, 310,324,332,347,348, 362,363,364,365,366, 370,371
2-Chloropyridine methiodide 144
N-Chlorosuccinimide 57, 158,161,259,296,317
Chromium (II) 85
Chromium (VI) 261
Chromium trioxide 23,24, 29,53,88,112,119,120, 122,129,130,140,141, 169,174,175,198,199, 205,206,213,217,233, 235,236,237,239,241, 242,244,244,251,271, 273,279,280,281,289, 304,310,319,322,347, 356,362,365,366,380, 380
Cis-2-Butenyl-diethyl-aluminum 239
Cobalt carbonyl 178
Cobalt chloride 383
Collidine 113
Copper 82,198
Copper (II) 269,342
Copper chromite 42
Copper sulfate 110,240, 241
Copper tetrafluoroborate 278
Crown ethers 155,236
Cupric acetate 37,38,156, 241,241
Cupric chloride 70,219
Cupric iodide 286
Cupric nitrate 91

Cuprous chloride 124,154, 214,219,229
Cuprous iodide 137,170, 242,256,264
Cuprous oxide 251
Cyanogen bromide 100
Cyclohexanone 65
Cyclopentadienyl sodium 150
Cyclopentadienylcobalt-carbonyl 178
Cyclopentadienyllithium 284
Cyclopentenylmagnesium bromide 197

Di(dimethylphenyl)-silazanelithium 318
Di-n-hexylborane 263
Di-t-butyl malonate 43
Di-t-butylaliminum hydride 291
Dialkyl phthallate 337
Dialkylborane 74,75,198, 257
Diallyltin 316
1,3-Diaminopropane 126
1,5-Diazabicyclo[4.3.0]-non-5-ene 383
1,5-Diazabicyclo[5.4.0]-undec-5-ene 265,266, 366,372,373
Diazomethane 28,57,74,82, 90,97,101,102,104, 107,112,124,125,155, 163,198,234,239,255, 257,277,290,296,304, 307,310,318,319,334, 356,366,377,378,380, 385
Dibenzoyl peroxide 381
Dibenzyl malonate 22,23
Dibenzyl phosphoro-chloridate 116,117
Diborane 62,81,99,163, (Contd.) 226,234,287,289,291, 320,321
Dibromocarbene 258
1,3-Dibromopropane 325
1,1-Dichloro-2,2-difluoroethylene 39
1,3-Dichloro-2-butene 2
2,3-Dichloro-5,6-dicyano-1,4-benzoquinone 85, 126,195,200
o-Dichlorobenzene 170, 324,348
Dichlorocarbene 40
Dichloromethyllithium 300
1,25-Dichloropentacosan-13-one 90
Dichlorophenylphosphine oxide 373
N,N-Dichloro p-toluene-sulfonamide 172
Dicyclohexylborane 162
Dicyclohexylcarbodiimide 69,99,117,232,304, 305,369
Dicyclohexylcarbodiimide-Dimethyl sulfoxide 138, 203,222,308
N,N′-Dicyclohexyl-4-morpholinecarboxamidine 117
Diethoxyphospinyl-hydrazide 328
Diethyl acetamidomalonate 196
Diethyl acetylenedicarb-oxylate 147
Diethyl diazodicarb-oxylate 35,369
Diethyl N-acetyliminodi-thiocarbonimidate 349
Diethyl oxalate 24
Diethyl phosphorochlori-date 256
D-Diethyl tartrate 66
Diethylaluminum chloride 58

Diethyl cinnamylphosphonate 154,155
Diethylzincmethylene iodide 257
Di[ethyoxycarbonyl]-methyl-3-cyclopentene 13
1,1-Difluoroethane 276
Dihydrofuran 265
Dihydropyran 19,123,142, 189,279,282,285
Diisoamylborane 80
Diisobutylaluminum hydride 66,75,79,137, 143,152,154,155,172, 222,280,291,298,299, 300,315,317,318,321, 330,331,367
Diisopropyl D-tartrate 79
Diisopropyl ethoxycarbonylmethylphosphonate 315
Diisopropylcarbodiimide-Dimethyl sulfoxide 79
1,1-Dimethoxyethane 364
2,2-Dimethoxypropane 137, 159,240,316,324
3,3-Dimethoxypropyl bromide 262
4,4′-Dimethoxytriphenylmethyl chloride 180
N,N-Dimethylacetamide 31
Dimethyl acetylenedicarboxylate 42,73,150
Dimethylamine 75,120,122, 228,333,340
Dimethylaminophosphorodichloridate 75
4-Dimethylaminopyridine 66,67,69,115,141,141, 144,145,160,160,161, 162,163,183,191,263, 264,265,307,310,353
N,N-Dimethylaniline 234
Dimethyl azodicarboxylate 97
Dimethylcadmium 113
Dimethyl carbonate 118, 296
Dimethylcopperlithium 176,206
Dimethyl α-cyanoethylphosphonate 317
Dimethyldilithiocopper cyanide 66
Dimethyl ethylphosphonate 154
N,N-Dimethylhydrazine 262
Dimethyllithiumcopper 79, 315,320
Dimethyl malonate 275
Dimethyl oxalate 19,20, 130,141,141,339
Dimethyl succinate 1
Dimethyl sulfate 8,10, 106,120,239,344
Dimethyl sulfide 65,210, 220,240
Dimethylsulfonium methylide 74
Dimethyl sulfoxide-Acetic anhydride 15,162,163
Dimethyl tetrahydropyranyloxymethylenephosphonate 143
Dinitrogen tetroxide 385
1,3-Dioxalane-2-one 276
Dipentylzirconium chloride 318
Diphenyl phosphochlororidate 353
Diphenyl disulfide 67
Dipotasssium azodicarboxylate 247
Dipyridyl 251
Dipyridyl disulfide 163
Dowex-50 65

Electrochemical oxidation 368
Electrolysis 327
Esterase 87,285
Ethane-1,2-dithiol 205, 261,275,291
(S)-3-[α-Ethoxy-ethoxy]-1-lithio-trans-oct-1-ene 285
3-Ethoxy-1,3-pentadiene 127
Ethoxyacetylenemagnesium bromide 129
2-Ethoxycarbonyl-3-bromo-2-cyclopentene-1-one 10
Ethoxycarbonylmethylene-triphenylphosphorane 317
Ethyl acetoacetate 214
Ethyl acrylate 214
Ethyl α-bromomethyl-acrylate 229
Ethyl 2-bromopropionate 355
Ethyl 7-bromopentanoate 284
Ethyl N-t-butylmalona-mate-magnesium 340
Ethyl chloroformate 144, 218,282,320
Ethyl diazoacetate 57,82, 244
Ethyl formate 1,110,111, 170,261,320
Ethyl glyoxalate 333
Ethyl iodide 14
Ethyl nitrite 294
Ethyl phosphorodi-chloridate 74
Ethyl vinyl ketone 110
2-Ethyl-2-lithio-1,3-dithiane 222
Ethyl 3-(1-pyrrolidinyl)-crotonate 270
Ethyl2-triphenylphos-
(Contd.)
phoranylidenepropionate 79
Ethylaluminum dichloride 93,291
Ethylene 52
Ethylene bromohydrin 84
Ethylene chlorhydrin 62
Ethylene glycol 119,122, 122,219,261,261,350, 376
Ethylene glycol-Boron trifluoride 347
Ethylene glycol-p-Toluenesulfonic acid 3, 51,61,119,122,128, 132,158,175,204,205, 211,218,219,260,261, 266,270,271,274,290, 318,328,356,376
Ethylene glycol-potassium hydroxide 90
Ethylene oxide 309,368
4-Ethylenedioxypentyl-magnesium bromide 5
Ethylidenetriphenyl-phosphorane 204
Ethylmagnesium bromide 243

Ferric chloride 41 85
Fluoboric acid 318
Formaldehyde 75,228,293, 333,363,364,373
Formaldehyde-Formic acid 101,121,122,227,281, 282
Formic acid 141,216,288, 288,340,371,373
1-Formyl-1-ethoxycar-bonyl-cyclopropane 52
Furfuraldehyde 22

Hexachlorobicyclo[2.2.1]-hepta-2,5-diene 249

Hexamethyldisilazane 330
Hexamethyldisilazane-lithium 252,307
Hexamethyldisilazane-potassium 239
Hexamethyldisilazane-sodium 202
High pressure 83
Hydrazine 9,57,90,231, 329,359
Hydrazine-Hydrogen peroxide 152
Hydrogen bromide 90,197, 197,220,229,245,341,358
Hydrogen bromide-Acetic acid 119,336
Hydrogen chloride 15
Hydrogen cyanide 58,60, 107,122,131,225
Hydrogen fluoride 78,239, 286,318,339
Hydrogen iodide 241,334
Hydrogen-Lindlar catalyst 37,80,114,124,125, 154,156,309,324,336, 368
Hydrogen-Palladium 2,4, 20,22,31,38,90,137, 151,163,230,261,327
Hydrogen-Palladium on barium sulfate 7,67
Hydrogen-Palladium on calcium carbonate 19,20
Hydrogen-Palladium on carbon 9,11,15,16,22, 31,33,35,52,53,87,88, 101,119,141,144,164, 174,175,205,209,219, 221,222,223,234,238, 239,241,271,279,307, 308,310,319,320,322, 335,339,340,353,371
Hydrogen-Palladium on strontium carbonate 21, 110,120
Hydrogen peroxide 19,65, 75,145,151,155,159, 160,160,163,163,167, 198,217,234,251,257, 263,283,285,287,289, 291,320,321,374
Hydrogen-Platinum 15,75, 105,112,113,122,149, 160,209,219,245,294, 296
Hydrogen-Platinum on alumina 56
Hydrogen-Raney Nickel 159,160,168,224,271, 294
Hydrogen-Rhodium on alumina 144,241,296
Hydrogen-Rhodium on carbon 119,190
Hydrogen-Ruthenium 119
Hydrogen sulfide 104,290, 290,328,329,382
Hydrogen-tris(triphenylphosphine)iodorhodium 276
Hydroxylamine 1,14,119, 124,160,164,213,220, 248,322,326,346,377
N-Hydroxylbenzotriazole 139,232
Hypobromous acid 3,85

Imidazole 65,66,69,240, 317,318,368
Iodine 24,100,105,106, 131,150,154,162,166, 183,193,194,195,200, 241,248,251,259,320, 372,384
Iodine azide 232
1-Iodo-3-t-butyldiphenylsilyloxypropane 155
2-Iodoacetamide 56
1-Iodooct-2-yne 114
Iodosobenzne diacetate 332

Iron 48
Iron carbonyl 134
Isobutenyl iodide 129
L-Isoleucine 66
Isopropenylmagnesium bromide 211
Isopropyl chloroformate 340,341
Isopropylmagnesium chloride 368

Kinase,DNA 183

Lead acetate basic 343
Lead carbonate 167
Lead perchlorate 161
Lead tetraacetate 21,30, 57,98,146,190,241, 251,263,266,268,322, 328,329,342
Lead trifluoracetate 178
Ligase,T4 DNA 183
2-Lithio-2-trimethylsilyl-1,3-dithiane 352
Lithium 250
Lithium acetylide 262
Lithium aluminum hydride 6,13,20,24,26,28,31, 32,51,52,55,56,59,75, 78,82,90,102,110,119, 122,124,128,130,137, 140,158,160,189,195, 197,203,205,208,213, 214,216,217,218,219, 220,223,233,234,235, 236,241,244,258,259, 260,266,270,273,274, 301,307,316,318,320, 333,336,337,356,363, 365
Lithium aluminum hydride-Aluminum chloride 41, 235,358
Lithium amide-Ammonia 123,125
Lithium-Ammonia 2,22,29, 59,62,67,74,75,151, 152,203,216,237,240, 244,254,271,288,315, 377
Lithium azide 160
Lithium boro hydride 321
Lithium boro tetrafluoride 151
Lithium bromide 23,28, 122,236,379
Lithium diisopropyl amide 64,69,79,87,115,115, 145,145,154,155,158, 163,191,208,228,232, 232,239,241,243,262, 275,287,289,292,292, 296,296,319,352,362, 363,364,372
Lithium-Ethoxide 195
Lithium-Ethylamine 34,36, 260
Lithium2-ethylhexanoate 310
Lithium-Ethylenediamine 364
Lithium hydroxide 155, 160,163,239
Lithium methylacetylide 162
Lithium perchlorate 204
Lithium thiobenzyloxide 160
Lithium triethoxyaluminum hydride 15
Lithium trimethoxylaluminum hydride 364
Lutidine 183,231

Magnesium 28,242
Magnesium oxychloride 108
Magnesium bromide 237, 238,239,243
Magnesium-Carbon dioxide 245,258

Magnesium iodide 324
Magnesium methoxide 339
Malic acid 66,240
Mandelic acid 78
Manganese dioxide 124, 125,317,359
Mercuric acetate 65,147, 162,311
Mercuric chloride 222, 290,317,352
Mercuric oxide 222,263, 277,327,376
Mercuric trifluoracetate 264
Mesitaldehyde dimethyl acetal 161
Mesitylenesulfonyl chloride 180
1-Mesitylenesulfonyl-3-nitro-1,2,4-traizole 182,188
Mesityllithium 159
Methanesulfonic acid 151, 198,229,312
Methanesulfonic anhydride 379
Methanesulfonyl chloride 59,61,62,68,98,158, 159,160,189,207,236, 243,254,265,272,283, 317,352,377
Methanesulfonyl fluoride 199,213
2-Methoxy-1-propene 163, 164
4-Methoxy-3-trimethylsilyloxy-1,3-butadiene 363
Methoxyamine 13,228
Methoxybenzyl bromide 323
3-Methoxymethoxy-1-propyne 309
Methoxymethyl chloride 66,189,287,330,343
Methoxymethyl iodide 159
Methoxymethylenetriphenylphosphorane 253
Methoxyphenylmagnesium bromide 235
Methoxypropylmagnesium bromide 245
Methoxysulfonyl fluoride 229
Methyl acetate 338
Methyl -acetoxyacrylate 307
Methyl acrylate 27,50, 328,339
Methyl acrylonitrile 22
2-Methylallyl bromide 18
Methylamine 212,361
Methylaminoacetone ketal 224
Methylaniline 111
(-)-α-Methylbenzylamine 191
(-)-α-Methylbenzylisocyanate 234
Methyl bromoacetate 173, 338,353
3-Methylbut-3-enylmagnesium bromide 240
Methyl -chloroacrylate 93
Methyl 3-chlorocarbonylpropionate 104
Methyl chloroformate 93, 161
2-Methylcyclopent-2-ene-1-one 176
2-Methylcyclopentane-1,3-dione 4,168,175,264
Methyl dimethylaminochlorophosphite 183
Methylene blue 363
Methylene iodide 209,333
Methylenetriphenylphosphorane 62,65,75, 190,198,265,267,301, 363,367

Methylenezinc iodide 77
(-)-N-Methylephedrine 144
Methyl formate 158,228, 251
Methyl iodide 1,3,20.59, 64,65,68,75,79,100, 128,132,163,164,167, 168,176,205,226,228, 229,233,253,271,276, 294,306,310,322,344
Methyl cis-7-iodo-5-heptenoate 286
Methyllithium 29,61,65, 68,121,137,175,205, 207,208,211,223,237, 266,269,274,363
Methylmagnesium bromide 111,132,237,240,272
Methylmagnesium iodide 120,206,214,290,342, 376
o-Methyl-L-mandelyl chloride 162
Methyl mercaptan 316
Methylmercury isopropoxide 382
Methyl -methylvinyl ketone 119
N-Methylmorpholine 139
N-Methylmorpholine-N-oxide 144,162,190
Methyl 3-oxoglutaramate 343
Methyl 5-oxo-6-heptenoate 131
3-Methyl-4-pentene-magnesium bromide 356
Methylpentylmagnesium bromide 113
4-Methyl-1,2,4-triazoline-3,5-dione 70
2-Methyl-5-trimethylsilyl-1,3-cyclopentadiene 364
Methyl -trimethylsilyl-vinyl ketone 289
Methyl triphenylphoranylideneacetate 66
Methyl 11-(triphenylphoranylidene)undecynoate 90
Methyl vinyl ketone 50, 128,295,364
Molecular sieve 56,66, 201,213,252
Molybdenum carbonyl 142, 145,367
Molybdenum pentoxide 266, 292
Mono p-nitrobenzyl malonatemagnesium 88, 352

Nickel 261
Nickel acetate 126
Ninhydrin 66
2-Nitro-1,3,5-triisopropylbenzene 255
Nitroethane 171,240
Nitroethylene 280
Nitrogen tetroxide 296
Nitromethane 68,105,228
Nitrophenoxycarbonyl chloride 160
o-Nitrophenylselenyl cyanide 159,160
Nitrosyl chloride 25,30, 40
Nitrous acid 31

Osmium 235
Osmium tetroxide 9,14,19, 62,110,130,159,190, 204,237,272,290,304, 316,322,342,350
Oxalic acid 2,289
Oxalyl chloride 9,22,57, 198,218,230,331,380
Oxalyl chloride-Dimethyl

(Contd.)

sulfoxide 68,69,309,
315,316,320
Oxygen 71,114,154,189,
194,263,263,290,298,
344
Oxygen-Palladium 252
Oxygen-Palladium on
carbon 371
Ozone 19,22,54,66,92,119,
121,140,159,160,176,
217,218,220,234,237,
240,241,243,274,290,
315,316,334,342,356,
362,364,377,340

Palladium chloride 86,323
Palladium on carbon 42,
81,250
Paraformaldehyde 212
Penta-2,4-dienyl alcohol
189
Pentachlorophenol 181,232
Pentenylmagnesium bromide
322
Peracetic acid 163,198,
236
Perbenzoic acid 22,40,62,
354,358
Perchloric acid 122,174,
253,290,311,355
Performic acid 306
Periodic acid 14,19,112,
130,144,208,235,260,
304,377
Pertrifluoracetic acid
23,307
Phenoxyacetic anhydride
160
Phenoxymethyl chloride
151
4-Phenyl-1,2,4-triazo-
line-3,5-dione 36,269
2-Phenyl-thiazoline-5-one
343
L-Phenylalanine 174
4-Phenylbenzoyl chloride
161,279,282
9-Phenylfluorenyl bromide
372
Phenyliodonium dichloride
36
Phenylisocyanate 68
Phenyllithium 276
Phenylmagnesium bromide
277
Phenylselenium chloride
145
Phosgene 191,242
Phosphoric acid 168
Phosphorus red 334
Phosphorus oxychloride
28,119,121,131,164,
270,304
Phosphorus pentachloride
28,346
Phosphorus pentasulfide
380,383
Phosphorus pentoxide
151,229
Phosphorus tribromide
90,245
Photolysis 25,30,36,39,
49,56,57,76,82,94,
106,134,146,149,152,
193,194,195,200,208,
250,251,254,255,257,
259,260,290,296,306,
337,345,360
3-Phthallimidopropyl
bromide 231
Piperidine 293,342
Pivalic acid 305
Pivaloyl chloride 67,316
Platinum on alumina 152
Polyphosphoric acid 229,
295
Potassium 190
Potassium acetate 100,
105,131
Potassium amide 100

Potassium-Ammonia 21,130, 130,169,258
Potassium t-amyloxide 266
Potassium t-butoxide 17, 18,19,23,27,28,33,37, 38,41,51,59,62,86, 119,128,129,132,167, 177,199,206,207,217, 219,261,263,267,268, 272,297,298,315,317, 321,330,359,376,382
Potassium carbonate-Methanol 24
Potassium chlorate 342
Potassium cyanide 90,137, 138,156,383
Potassium ferricyanide 232,324
Potassium fluoride 155, 172
Potassium hydride 56,65, 66,160,234,253,287, 316,373
Potassium hydrogen sulfate 172,355
Potassium iodide 91,279
Potassium isocyanate 294
Potassium oxide 236
Potassium periodate 316
Potassium periodide 242
Potassium permanganate 48,72,131,366
Potassium selectride 352
Potassium thioacetate 330
Potassium tri-n-butylboro hydride 223
Potassium tri-t-butoxy-aluminum hydride 77
L-Proline 159
Propane-1,3-diol 328
Propane-1,3-dithiol 223, 321,325
Propargyl bromide 376
Propenylmagnesium bromide 264
Propionic acid 273
Propylene oxide 190
Propylmagnesium bromide 13
Pyridine hydrochloride 167,313,322
Pyridine tosylate 365
Pyridine-2-thiol 242
Pyridinium bromide perbromide 224,261
Pyridinium chlorochromate 65,66,78,144,152,234, 235,252,263,266,288, 318
Pyridinium dichromate 114,124,125,164,310, 317
Pyridyl chlorothioformate 161
Pyrrole 85
Pyrrolidine 215,295
Pyrrolidine acetate 377
4-Pyrrolidinopyridine 369

Raney Nickel 52,55,56,68, 83,127,211,223,226, 281,335,356,385
Rhodium acetate 88
Rose Bengal 114,290
Ruthenium tetroxide 208, 254,261,277,313
Ruthenium trichloride 268,368

Salcomine 189
Selenium dioxide 11,53, 196,217,260,264,322, 323,325,336,347,363
Silica gel 343
Silicon dioxide 161
Silver fluoride 161
Silver hydroxide 75
Silver nitrate 156
Silver nitrite 100
Silver oxide 65,100,124,
(Contd.)

100,124,125,300,306,
310,346
Silver perchlorate 206
Silver trifluoroacetate
79,242
Silver trifluoromethane-
sulfonate 69,161
Sodium 57,245
Sodium acetate 86,98,105,
111,132
Sodium acetylide 336
Sodium amalgam 67,69,161
Sodium amide-1,3-Diamino-
propane 123
Sodium amide-Ammonia 194
Sodium ammonia 40,72,116,
167,174,206,226,290
Sodium azide 98,198,230,
295
Sodium bisulfite 190
Sodium borohydride 2,4,
21,33,47,54,61,65,68,
87,93,101,105,112,
113,120,122,130,159,
160,175,189,196,207,
225,226,259,271,282,
283,289,291,299,304,
308,309,316,317,319,
334,337,340,347,352,
357,359,366
Sodium cyanide 23,34,74,
75,78,79,154,155,225,
244,317,333
Sodium cyanoborohydride
94,151,208,232,373,
Sodium di(B-methoxy-
ethoxy)aluminum hydride
276
Sodium dialkoxyaluminum
hydride 171,275,302
Sodium dialkoxyboro
hydride 141
Sodium dichromate 112,113
Sodium diethylphosphonate
141
Sodium ethoxide 1,48,225
Sodium-Ethylamine 363
Sodium ethylanilide 90
Sodium hexafluoro-
phosphate 126
Sodium hydride 65,68,79,
100,118,154,159,170,
176,193,222,229,231,
251,266,270,277,323,
329,330,338,339,341
Sodium hydride-Dimethyl
sulfoxide 242
Sodium hydrogen arsenate
225
Sodium hydrosulfite 346
Sodium hypochlorite 263
Sodium iodate 268
Sodium iodide 89,113,142,
243,244,254,275,309,
352,372
Sodium methoxide 6,31,52,
80,101,108,110,130,
222,237,261,271,304,
320,335,373,376
Sodium 2-methoxyethoxy-
aluminum hydride 374
Sodium methylsulfinyl-
methylide 206,228,236
Sodium nitrite 25,164,
270,281,284,358,377
Sodium perchlorate 263
Sodium periodate 9,247,
257,318,322,350,362,
368,377
Sodium-Potassium alloy
74,198,251,257
Sodium ruthenate 191
Sodium sulfide 211
Sodium t-amyloxide 2,3,
363
Sodium thiobenyloxide 335
Sodium thiomethoxide 319
Sodium thiophenoxide 211,
236
Sodium thiopropoxide 191

Sodium 4-triphenylphosphoranyldenebutyrate 280
Stannic chloride 34,171, 213,346,380
Stannous chloride 40,316
Succinaldehyde 361
Sulfur trioxide-pyridine 155
Suloflane 384
Sulphur dioxide liquid 225

D-Tartaric acid 144,315, 368
Tetra-n-butylammonium fluoride 65,66,67,68, 69,74,115,143,145, 155,164,256,316,317, 318,320,364,366,367, 369
Tetra-n-butylammonium hydroxide 80
Tetra-n-butylammonium iodide 68
Tetrachlorothiophene-1,1-dioxide 250
1,1,3,3-Tetraethoxypropane 81
8-Tetrahydropyranyloxyoctylmagnesium bromide 125
Tetramethylammonium hydroxide 172
Tetramethylammonium mesitoate 208
Tetramethyleneglycol 359
N,N,N′,N′-Tetramethylethylenediamine 160,248
Thallium salt 324
Thallium triacetate 371
Thermolysis 82,258,337, 366
Thexylborane 320,321
4,4′-Thiobis-(t-butyl-3-methylphenol) 347
Thionyl chloride 69,82, 101,102,104,113,130, 135,196,205,207,208, 224,225,229,259,270, 272,337,380
Thiophenol 183,378,384
Thiophenoxyacetyl chloride 55,210
Thiophenoxyethylamine 55
Thiourea 158,199
Titanium 75,152
Titanium chloride 176
Titanium tetrachloride 76,363
Titanium tetraisopropoxide 66,79,315,368
Titanium trichloride 101, 176,190,228,248,328, 329
p-Toluenesulfonic acid 24,65,119,123,124, 125,131,137,140,142, 152,158,159,162,169, 170,174,189,208,240, 241,243,248,261,275, 279,285,290,310,328, 340,363,364
p-Toluenesulfonyl azide 88,251,257
p-Toluenesulfonyl chloride 23,28,69,74, 75,77,119,130,191, 204,204,212,235,254, 255,260,275,309,319, 334,335
p-Toluenesulfonylhydrazide 202,211,244, 259
p-(Tolylsulfinyl)-2-cyclopentenone 176
Tri(triphenylphosphine)-iodocopper 285
Tri-t-butylaluminum 98
Tri-n-butylamine 230

Tri-n-butyltin chloride 248
Tri-n-butyltin hydride 65,72,163,198,232, 264,277,279
Tribenzylammonium trifluoracetate 190
Tri-n-butylphosphine 159, 160,228,364
Trichloracetic anhydride 190
Trichloroacetyl chloride 237
Trichlorobenzene 228
2,2,2-Trichloroethanol 98,99
2,2,2-Trichloroethoxy-carbonylaminoadipic acid 99
2,2,2-Trichloroethoxy-carbonylmethylenemalon-dialdehyde 98
Trichloroethyl chloro-foramte 54,201,217,231
Trichlorosilane 191
1,1,1-Triethoxyethane 137,235,273,363
Triethylaluminum 60,122
Triethylammonium acetate 342
Triethyl orthformate 348
Triethyloxonium fluoro-borate 94,229,231,349, 376
Triethyl phosphite 328, 329,382
Triethylsilyloxy chlorate 239
Triethyltin hydride 176
Trifluoroacetic acid 55, 66,71,76,81,99,101, 138,159,161,173,178, 202,203,208,227,262, 263,288,296,307,312, 331,347,366,372,372
Trifluoroacetic anhydride 55,191,265,319,352, 370
Trifluoroacetic anhyd-ride-Dimethyl sulfoxide 159
Trifluoroethanol 243
Trifluoromethanesulfonic acid 312
Trifluoromethanesulfonic anhydride 79
1-Triisopropylphenyl-sulfonyl)-1,2,3,4-tetrazole 181
2,4,6-Tri-(p-methoxy-phenyl)-1,3,5-trithiane 330
Trimethyl orthoformate 53,158,237,243,290
Trimethylaluminum 368
Trimethylamine-N-oxide 162
Trimethylbromosilane 67
Trimethylchlorosilane 67, 69,74,77,87,144,256, 257,318,323,362,366
Trimethyliodosilane 94, 308
Trimethyloxonium fluoro-borate 382,383
Trimethylsilyl cyanide 172
Trimethylsilyl hydride 260
Trimethylsilylethoxy-carbonylmethylene-triphenylphosphorane 368
Trimethylsilylmethoxy-methyllithium 287
4-Trimethylsilyloxy-2-methyl-1,3-butadiene 366
3-Trimethylsilyloxy-propane-2-one 218

Triphenylethylidenephosphorane 175
Triphenyl-p-methoxyphenoxymethylene-phosphorane 58
Triphenylmethyl chloride 68
Triphenylmethylsodium 14, 205
Triphenylphosphine 35,67, 142,154,155,163,164, 193,198,214,228,236, 240,241,275,286,301, 369,382
Triphenylphosphine-Rhodium chloride 207
Triphenyltin chloride 286
Triton B 111,120,128,131, 164,339

Vanadyl acetylacetonate 248,265
Vinylmagnesium bromide 78,165,170

Zinc 11,24,39,55,62,70, 99,107,110,140,142, 154,166,176,191,201, 201,217,217,226,232, 259,261,263,268,274, 304,304,310,311,340, 355,380
Zinc amalgam 176
Zinc borohydride 163,279
Zinc bromide 191,208,289
Zinc carbonate 10,11
Zinc-Copper couple 75, 151,243,266,328,329
Zinc iodide 172
Zinc-Silver couple 176, 209,218,365
Zirconium tetrapropoxide 5

The unspecified alkyl groups possess unbranched (normal,n) chains.

I. CARBON-CARBON BOND FORMING REACTIONS

A. Carbon-Carbon Single bonds

1. Alkylation of Aldehydes, Ketones, and Their Derivatives 1,6,7,18,23,47,56,59,62,128,129,130,133,205,252,253,270,286,287

2. Alkylations of Nitriles, Acids, and Acid Derivatives 14,44,64, 100,172,239,277,330,372

3. Alkylation of -Dicarbonyl and -Cyanocarbonyl Systems and Other Active Methylene Compounds 13,114,323,366.

4. Alkylation ofN-, S-, and Se-Stabilized Carbanions 33,56,104,105, 202,221,226,228,247,262,325,331,352

5. Alkylation of Organometallic Reagents 1,125,129,137,163,256, 368,376

6. Other Alkylation 1,278,283,296,309

7. Nucleophilic addition to Electron-Deficient Carbon

a. 1,2-Additions
i. Aldol-Type Condensations
- Intermolecular 1,22,29,61,79,87,88,90,105,112,115,118,119, 120,133,151,160,162,214,228,229,233,238,239,241,243,247, 287,302,318,319,335,352,361,363,364,
- Intramolecular 28,50,52,120,159,168,175,190,221,222,225,253, 262,263,271,338,339,340,342

ii. Addition of N-, S-, or Se-Stabilized Carbanions 28,67,68
iii. Grignard-Type Additions 5,9,28,61,65,66,74,78,111,113,120,121, 124,125,133,165,173,205,206,211,235,237,239,240,242,243,245, 261,262,264,266,272,273,277,300,315,316,342,355,356,363

b. Conjugate Additions

1. Enolate-Type Carbanions 1,5,6,17,21,27,33,50,52,110,111,119, 128,132,168,169,174,176,177,205,264,275,289,295,339,359,376
2. Organometallic Reagents 68,79,170,208,214,219,261,285,286,315, 320,322,
3. Other Conjugate Additions 58,98,122,215

8. Other Carbon-Carbon Single Bond Forming Reactions 17,26,29, 34,37,38,41,44,54,55,60,74,78,86,94,100,124,150,152,154,173, 245,253,259,270,274,276,289,304,308,311,312,334,337,343,376

B. Carbon-Carbon Double Bonds

1. Wittig Type of Olefination Reactions 58,62,65,66,67,69,75,90, 154,155,171,190,194,198,204,214,227,228,233,236,242,265,267, 276,279,280,284,301,315,317,365,367,368

2. Eliminations
a. Alcohols and Derivatives 10,19,21,26,28,33,236,255,272,275, 381
b. Halides 45,71,243,255,297,372
c. Other Eliminations 45,75,294

3. Other Carbon-Carbon Double Bond Forming Reactions 47,61,74, 75,76,81,86,88,100,176,190,200,211,241,251,259,260,371

C. Cyclopropanations

1. Carbene or Carbenoic Addition to Multiple Bonds 45,77,82,206, 209,244,257,258,269
2. Other Cyclopropanations 281

D. Carbon-Carbon Triple Bonds 67,320

E. Cyclocondensations

1. Thermol Cycloadditions 7,39,42,43,46,48,52,53,54,55,57,68,70, 73,82,83,86,93,100,101,109,127,134,147,150,155,166,170,178, 189,190,190,201,202,208,210,213,227,228,249,250,254,269, 278,280,291,303,306,307,310,323,342,346,363,364,366,380

2. Photolytic 49,56,57,146,193,307,345,360

3. Others 145,211,327,337

F. Aromatic Substitutions Forming a New Carbon-Carbon Bond
1. Friedel-Crafts-Type Reactions 95,102,104,171,229,322,331,339, 346,370
2. Other Aromatic Substitutions 10,11,12,216,232,322,373

G. Synthesis via Organometallics 75,178,315,316,386
- Misc. 148,149

II. OXIDATIONS

A. C-O Oxidations

1. Alcohol Ketone, Aldehyde 1,7,15,20,23,24,26,46,47,65,66,69, 78,114,119,122,124,125,129,138,174,206,234,235,239,242,251, 252,255,291,304,308,316,318,319,320,346,356,359

2. Alcohol, Aldehyde Acid, Acid Derivative 46,88,112,124,125, 191,261,292,309,

B. C-H Oxidations
1. C-H C-O 11,29,53,120,132,196,263,265,266,276,292,298,323,325, 336,344,374,
2. Other C-H Oxidations 25,30,158

C. C-N Oxidations 221,225
D. Amine Oxidations 370,371

E. Sulfur Oxidations 55,67,89

F. Oxidative Additions to C-C Multiple Bonds

1. Epoxidations 19,34,40,45,62,66,71,74,79,163,164,206,211,235,260, 263,265,266,268,306,310,315,346,348,354,358,362,363,364,367, 368
2. Hydroxylation 72,110,147,159,162,204,237,242,323,350,371,377
3. Other 190,285

G. Phenol Quinone Oxidation 91,189,346

H. Oxidative Cleavages 9,14,16,19,22,36,54,66,92,112,167,208,217,234, 237,239,240,241,243,277,290,316,334,342,350,356,362,364,366, 377,378,380

I. Photosensitized Oxygenations 114,288,290

J. Dehydrogenation 225,226

K. Other Oxidations 98,151,231,236,252,255,260,279,330,332,368

III REDUCTIONS

A. C=O Reductions 1,11,20,24,28,44,47,51,52,54,55,56,59,65,66,75,78, 79,82,90,100,101,102,110,112,119,127,151,152,163,166,189,195, 197,216,222,223,225,226,233,240,253,259,260,270,315,316,321, 331,340,346,352,355,374

B. Nitrile Reductions 15,52,79,137,154,155,172,333

C. N-O Reductions 91,105,119,229

D. C-C Multiple Bond Reactions

1. C=C Reductions 1,7,11,20,21,28,29,30,31,35,52,90,110,112,119, 127,130,166,167,168,169,175,176,226,241,246,271,276,327,355
2. C=C Reductions 37,67,114,124,125,154,156,320
3. Reduction of Aromatic Rings 1,22,40,44,119,217,221,224,244, 294,377

E. Hydrogenolysis of Hetero Bonds

1. C-O C-H 9,11,15,19,22,30,65,67,74,75,110,120,190,219,222,230, 238,239,288,339,340,
2. C-Hal C-H 33,39,40,62,70,72,151,154,163,197,206,218,226,250, 254,258,263,264,265,268,277,279,281,304,327,363,365
3. C-S C-H 34,160,211,223,385

F. Hydroboration 62,65,74,163,198,234,251,252,256,263,287,289,291, 320,321

IV. PROTECTING GROUPS

A. Hydroxyl 7,44,114,123,137,151,154,159,162,171,189,230,234,235,239, 240,265,343,366,367,

B. Amine 161,224,225,232,308,331,349

C. Carboxyl 158,369

D. Ketone, Aldehyde 1,23,104,128,158,204,205,218,222,223,247

V. USEFUL SYNTHETIC PREPARATIONS

A. Functional Group Preparations

1. Acids, Acid Halides 245,255,277,294,318,385
2. Alcohols, Phenols 198,282
3. Alkyl, Aryl Halides 18,26,28,36,62,137,154,155,172,196,224,228, 240,241,243,245,251,259,262,265,267,275,277,279,296,320,339, 352,366
4. Amides 32,295
5. Amines 13,16,17,32,98,100,160,227,232
6. Amino Acids and Derivatives 138
7. Carbenes 88
8. Ethers 161
9. Ketones and Aldehydes 31,58,101,160,161,167,228,253,278,300
10. Nitriles 14,225,333
11. Nitro 91,100,358
12. Olefins, Acetylenes 33,67,79,100,105,123,236,241,243,320
13. Vinyl Halides, Vinyl Ethers, Vinyl Esters 162,218,247
14. Sulfur Compounds 67,94

B. Ring Enlargement and Contraction

1. Enlargement 77,151,204,244,378
2. Contraction 48,77,133,272,284,327

VI. SYNTHESIS OF HETEROCYCLES

1. Azabicyclo[2.2.2]octane 307
2. 1-Aza-5-thia-8-oxo[4.2.0]oct-2-ene 99
3. 1-Aza-5-thia-8-oxo[4.2.0]oct-3-ene 99
4. Azepino[4,5-b]indole 94
5. Aziridine 40
6. Benzophenanthridine 101,102
7. Borazaronaphthalene 81
8. Carbazole 56
9. Chroman 86
10. Cyclazin[3.3.2] 42
11. Cyclozine[3.3.3] 43
12. Furano[3,4-b]coumarine 10
13. Pyrano[4,3-b]coumarin 12
14. 2,2'-Difuran 236

15. Dioxocin 72
16. 1,7-Dioxaspiro[5,5]undecane 67,164
17. 1,6-Dioxaspiro[4,5]decane 238,243
18. Pyran 19
19. Endoperoxide 288,290
20. Ergoline 225,226,227,229
21. Furan, 2-Furanone 9,22,25,26,46,150,191,192,235,236,263,267,279, 281,284,304,306
22. Furano[2,3-b]benzofuran 11
23. 2-Furano-2-pyrone-2 242
24. Isoxazole 1,270
25. Isoquinoline 307,310
26. Julolidone 219,222
27. -Lactams 88,98
28. Lulolidine 51,52,53
29. Macrolactam 139,232
30. Macrolactone 69,143,145,161,164
31. Oxazinone 281
32. Oxazolone 138
33. Oxazolidinone 230
34. Oxepinone 163
35. Oxirane (including asymmetric epoxidation) 19,22,40,45,66,71,72, 74,137,142,145,163,164,172,206,247,248,263,265,268,299,315, 347,354,365,367
36. Pyridines 51,59,219,222
37. Porphyrin 105
38. Pyran 66
39. 2-Pyrone 5,12,28,79,87,111,132,135,163,190,239,241,282,284
40. Pyrrole, Pyrroline, Pyrrolidine 42,51,52,88,119,122,141,143,225
41. Pyrido[3,2-a]carbazole 54,201
42. Pyrrolopyrimidine 196
43. Pyrrolo[3,2-b]carbazole 56
44. Pyrazole 259,269
45. Quinoline 51,101,102,215
46. Quinuclidine 295,298,299,301
47. 2,2',2"-Trifuran 236
48. 2,2',2"-Trifurano-2-pyrone 237
49. Thiazolidine 97
50. 1,4,7-Trioxacyclontrienone 72
51. 1,5,10-Triazacyclotridecane-2-one
52. Xanthene 86
53. Yohimbane 305,308